Immunology Guidebook

Immunology Guidebook

Julius M. Cruse
Robert E. Lewis
Huan Wang

With Contributions From

Geziena M. Th. Schreuder
Steven G.E. Marsh
Lorna J. Kennedy

ELSEVIER
ACADEMIC
PRESS

AMSTERDAM • BOSTON • HEIDELBERG • LONDON
NEW YORK • OXFORD • PARIS • SAN DIEGO
SAN FRANCISCO • SINGAPORE • SYDNEY • TOKYO
Academic Press is an imprint of Elsevier

Permissions may be sought directly from Elsevier's Science & Technology Rights
Department in Oxford, UK: phone (+44) 1865 843830, fax: (+44) 1865 853333,
e-mail: permissions@elsevier.co.uk. You may also complete your request on-line via
the Elsevier homepage (http://www.elsevier.com), by selecting 'Customer Support'
and then 'Obtaining Permissions'

Elsevier Academic Press
525 B Street, Suite 1900, San Diego, California 92101-4495, USA
http://www.elsevier.com

Elsevier Academic Press
84 Theobald's Road, London WC1X 8RR, UK
http://www.elsevier.com

Library of Congress Catalog Number: 2004105675

British Library Cataloguing in Publication Data
A catalogue record for this book is available from the British Library

ISBN 0-12-198382-X

Printed and bound in Great Britain
03 04 05 06 07 08 9 8 7 6 5 4 3 2 1

Contents

Contributors

Julius M Cruse, BA, BS, D.Med.Sc., MD, PhD, Dr. *h.c.*,
FAAM, FRSH, FRSM
Guyton Distinguished Professor
Professor of Pathology
Director of Immunopathology and Transplantation
Immunology
Director of Graduate Studies in Pathology
Department of Pathology
Distinguished Professor of the History of Medicine
Associate Professor of Medicine and Associate Professor
of Microbiology
University of Mississippi Medical Center
2500 North State Street, Jackson, USA
Investigator of the Wilson Research Foundation
Mississippi Methodist Rehabilitation Center
Jackson, Mississippi

Robert E. Lewis, B.S., M.S., PhD, FRSH, FRSM
Professor of Pathology
Director of Immunopathology and Transplantation
Immunology
Department of Pathology
University of Mississippi Medical Center
2500 North State Street, Jackson
Investigator of the Wilson Research Foundation
Mississippi Methodist Rehabilitation Center
Jackson, USA

Huan Wang, MD, PhD
Associate Professor, Department of Hand Surgery
Hua Shan Hospital, Fudan University, 12 Wulumuqi
Zhong Road, P.R., China
Formerly Postdoctoral Fellow and Research Associate in
Immunology
University of Mississippi Medical Center
2500 North State Street, Jackson
USA

Dr. Geziena M. Th. Schreuder & Associates (Chapter 4)
Department of Immunohematology and Blood
Transfusion
E3-Q Leiden University Medical Center, Leiden
The Netherlands

Dr. Steven G.E. Marsh & Associates (Chapter 5)
Anthony Nolan Research Institute, Royal Free Hospital,
Pond Street
London, UK

Dr. Lorna J. Kennedy & Associates (Chapter 6)
ARC Epidemiology Unit, University of Manchester,
Stopford Building
Oxford Road, UK

Preface

Immunology Guidebook is designed to provide the busy investigator or practitioner immediate access in one volume to immunological information that is difficult to find. The subject matter is divided into 18 chapters that correspond to the contents of a standard immunology textbook. Following a brief synopsis of the current knowledge on each topic, tables of essential data are designed for quick and easy reference to information that might otherwise not be readily available.

The book begins with a consideration of molecules, cells and tissues of immunity in Chapter 1, followed by antigens, immunogens, vaccines and immunization in Chapter 2. Chapter 3 is devoted to both murine and human clusters of differentiation (CD) antigens and Chapter 4 contains the HLA dictionary by Dr. G.M.Th. Schreuder and associates. Dr. Steven G.E. Marsh has kindly provided the "nomenclature for factors of the HLA system, 2002" for inclusion in Chapter 5. Chapter 6 on "nomenclature for factors of the dog histocompatibility system (DLA)" is provided by Dr. Lorna Kennedy. Antigen presentation (Chapter 7) is followed by Chapter 8 on B cells, immunoglobulin genes and immunoglobulin structure. T lymphocytes and the thymus comprise Chapter 9, followed by the most recent data on cytokines and chemokines in Chapter 10. Chapters 11, 12 and 13 are devoted to complement, hypersensitivity and microbial immunity, respectively. Immunoregulation, tolerance and therapeutic immunology comprise Chapter 14, followed by immunohematology and transfusion medicine in Chapter 15. The final three chapters, 16, 17 and 18, are devoted to immunological diseases and immunopathology, congenital and acquired immunodeficiencies and transplantation, respectively.

The authors are grateful to our editor, Mrs. Margaret Macdonald and our publisher, Academic Press, for their unstinting support during preparation of the manuscript. If the contents of this volume save users valuable time in seeking obscure immunological facts, the effort in preparing this book will have been worthwhile.

Acknowledgments

The following individuals, publications, institutions, and corporations are thanked for their kind permission to reproduce the materials listed below. Every effort has been made to trace and contact copyright owners but if any has been overlooked the authors and publishers will be happy to make the appropriate arrangements at the earliest opportunity.

The material in **Tables 1.2 –1.6** and **Table 1.8** is posted on the following website: http://www.neuro.wustl.edu/neuromuscular/lab/adhesion.htm (tables of adhesion molecules – molecules, ligands, distribution) and is used courtesy of Alan Pestronk, MD, Professor of Neurology, Department of Neurology, Washington University in St Louis, USA.

The material in **Table 1.7** is posted on the following website: http://yakko.bme.virginia.edu/biom892/adhesionii.pdf (endothelial-leukocyte adhesion molecules) and is used courtesy of Klaus F Ley, MD, Department MD-Biom Biomedical Engineering, University of Virginia, USA.

The material in **Figure 1.1** is posted on the following website: http://www.med.virginia.edu/medicine/basic-sci/biomed/ley/main.html (Inflammation: The Leukocyte Adhesion Cascade) and is used courtesy of Klaus F Ley, MD, Department MD-Biom Biomedical Engineering, University of Virginia, USA.

Table 1.13 and **Figure 1.2** are adapted from Toll-like Receptor Family, Death Receptor Guide, Costimulation and B7 Family, and Mouse Cell Surface Antigens and are reproduced by permission of eBioscience, San Diego, USA.

Figure 1.3 is reproduced from *Pathway Archives of The Biotech Journal: AKT Signaling, MAPK Signaling, NF-κB Signaling, TNF Signaling, Toll-like Receptors*, by permission of the publisher and editor, Samik Singh, The Biotech Journal, San Diego, USA.

Table 2.5 is adapted from Alternate Names of Mouse Leukocyte Antigens, pp. 160–162; Mouse Leukocyte Antigen Distribution Chart, pp. 163–163; Mouse Leukocyte Alloantigens Chart, pp. 168–169; in *BD Biosciences Product Catalogue*, 2003, and is reproduced by permission of BD Biosciences Pharmingen, San Diego, CA, USA.

Table 2.8 and **Table 2.9** are prepared from the *Immunization Handbook*, 2002 published by the Ministry of Health, New Zealand and posted on the following websites: http://childrensvaccine.org/files/mmwr_vaccines.pdf and http://childrensvaccines.org/files/New-Zealand-Imm-Handbook-2000.pdf. The material is used by permission of the Ministry of Health, Wellington, New Zealand. © 2002 Ministry of Health (New Zealand).

Tables 2.10–2.17 are based on materials developed by ACIP, AAP, and AAFP for the Centers for Disease Control and Prevention, Atlanta, GA, USA and available from http://www.cdc.gov.nip/, and are reproduced from the website of the Immunization Action Coalition at http://www.immunize.org/

Table 3.1 and **Table 3.5** are adapted from Pharmingen Mouse CD Chart and the Official Poster of the 7th International Workshop on Human Leukocyte Differentiation Antigens, and are reproduced by permission of BD Bioscience Pharmingen, San Diego, CA, USA.

Table 3.2 and **Table 3.4** are adapted from Alternate Names of Mouse Leukocyte Antigens, pp. 160–162; Mouse Leukocyte Antigen Distribution Chart, pp. 163–167; Mouse Leukocyte Alloantigens Chart, pp. 168–169; in *BD Biosciences Product Catalog*, 2003, and are reproduced by permission of BD Biosciences Pharmingen, San Diego, CA, USA.

Table 3.3 and **Figures 3.1–3.11** are adapted from Toll-like Receptor Family, Death Receptor Guide, Costimulation and B7 Family, and Mouse Cell Surface Antigens, and are reproduced by permission of eBioscience, San Diego, USA.

Tables 4.1–4.8 are reproduced from *Human Immunology* 2001; 62: 826–849 courtesy of the American Society for Histocompatibility and Immunogenetics (ASHI) and Dr GMTh Schreuder.

Tables 5.1–5.13 are modified from those originally published in Marsh SGE, Bodmer JG, Albert ED, Bodmer AT, Bontrop RE, Dupont B, Erlich HA, Hansen JA, Mach B, Mayr WK Parham P, Petersdorf EW, Sasazuki T, Schreuder GM, Strominger JL, Svejgaard A, Terasaki PI (2001) Nomenclature for factors of the HLA system, 2000. *Tissue Antigens* 57: 236–283.

Tables 6.1–6.5 and **Figures 6.1–6.8** were originally published in Kennedy LJ, Angles JM, Barnes A, Carter SD, Francis O, Gerlach JA, Happ, GM, Ollier WE, Thomson W, Wagner JL (2001) Nomenclature for factors of the dog major histocompatibility system (DLA), 2000: Second Report of the ISAG DLA Nomenclature Committee. *Tissue Antigens* 58: 55–70.

Figure 7.1 and **Tables 7.2–7.4** are adapted from Toll-like Receptor Family, Death Receptor Guide, Costimulation and B7 Family, and Mouse Cell Surface Antigen, and are reproduced by permission of eBioscience, San Diego, USA.

Tables 7.5–7.7 are adapted from Periodic Table of Human Cytokines and Chemokines, Fibroblast Growth Factor Mini-Guide, Human Chemokine Receptor Mini-Guide, Dendritic Cell Maturation, and The Matrix Metalloproteinases and Cytokines, and are reproduced by permission of R&D Systems Inc., Minneapolis, MN, USA.

Figure 8.1 and **Figure 8.2** are reproduced from Flow Chart of Common Cellular Antigens, *in: CD Antigens and Antibodies* (poster), by permission of DAKO Corporation, Carpinteria, CA, USA.

Figures 10.1–10.6, and **10.8–10.16** are adapted from material in Cytokine Secretion and Signaling, Chemokine Activation of Leukocytes, Phosphoregulation of Signal Transduction Pathways III, and are reproduced by permission of Biosource International Inc., Camarillo, CA, USA.

Figure 10.7 and **Figure 10.17** are reproduced from Chemokine Signaling; Pro-Inflammatory Pathways; IL-12, IL-18 Cellular Activation; Th1, Th2 Differentiation Pathway, *in: Biosource Research Products Catalog*, 2003, p. 132–135 by permission of authors/designers Judith Billingsley and Valerie Bressler-Hall, and Biosource International, Inc., Camarillo, CA, USA.

Tables 10.1, 10.6, 10.7, 10.11, and **10.12** are adapted from Periodic Table of Human Cytokines and Chemokines, Fibroblast Growth Factor Mini-Guide, Human Chemokine Receptor Mini-Guide, Dendritic Cell Maturation, and The Matrix Metalloproteinases and Cytokines, and are reproduced by permission of R&D Systems Inc., Minneapolis, MN, USA.

Tables 10.2, 10.3, and **10.5** are adapted from Expression of Chemokine Receptors on Different Cell Types, Chemokines and Their Receptors, Pro-inflammatory Cytokines and Their Receptors, *in: Chemokines, Inflammatory Mediators and Their Receptors*, and are reproduced by permission of BD Biosciences Pharmingen, San Diego, CA, USA.

Tables 10.4, 10.8, 10.9, and **10.10** are reproduced from Table 1 – Cytokines in Wound Healing, p. 3, p. 8; Cells in Chemokine Research: Ligands, pp. 35–39; Cells in Chemokine Research: Receptors, pp. 40–47; Cell Lines in Chemokine Receptor Research: Cont, p. 48; *in: R&D Systems Catalog*, 2002, by permission of R&D Systems Inc., Minneapolis, MN, USA.

Table 11.7 is reproduced from GD Ross (1998) Complement receptors, Table 1, p. 630; **Table 11.8** and **Table 11.9** are reproduced from C Rittner and PM Schneider (1998) Complement genetics, Table 1, p. 620 and Table 2, p. 621 respectively; **Table 11.10** is reproduced from KO Rother and UA Rother (1998) Complement deficiencies, Table 1, p. 614–615; *in:* PJ Delves and IM Roitt (eds) (1998) *Encyclopedia of Immunology*, 2nd edn, vol. 1. © 1998 Academic Press, with permission from Elsevier.

Table 12.3 is adapted from Table 1: Some allergens with known amino acid sequences, *Bulletin of the World Health Organization* 1994; 72 (5): 797–806, and is reproduced by permission of WHO, Geneva, Switzerland.

Table 12.4 is reproduced from Martin P Chapman (1998) Molecular properties of common allergens, Table 3, p. 67, *in:* PJ Delves and IM Roitt (eds) (1998) *Encyclopedia of Immunology*, 2nd edn, vol. 1, Academic Press. © 1998, with permission from Elsevier.

Table 15.1 is reproduced from Table 15-1, p. 316; **Table 15.5** is reproduced from Table 14-1, p. 297; **Table 15.10** is reproduced from Table 16-1, p. 343; **Table 15.11** is reproduced from Table 25-4, p. 555; **Table 15.12** is reproduced from Table 27-1, pp. 586–589, *in: AABB Technical Manual*, 14th edn, by permission of the American Association of Blood Banks.

Tables 17.2, 17.3, and **17.4** are reproduced from Table 20-2 p. 458, Table 20-3D p. 459, and Table 20-4 p. 461, respectively, *in:* Abul K Abbas and Andrew H Lichtman (2003) *Cellular and Molecular Immunology*, 5th edn, WB Saunders. © 2003, with permission from Elsevier.

Tables 18.3, 18.4, and **18.5** are reproduced from LA Baxter Lowe, Histocompatibility testing, Tables 19-3, 19-7, and 19-8, respectively, *in:* TG Parslow, DP Stites, AI Terr, and JB Imboden (2001) *Medical Immunology*, 10th edn, Lange Medical Books/McGraw–Hill, by permission of McGraw–Hill, New York.

MOLECULES, CELLS, AND TISSUES OF IMMUNITY

- ● ADHESION MOLECULES

- ● NATURAL AND ADAPTIVE IMMUNITY

- ● ORGANS AND TISSUES OF THE IMMUNE RESPONSE

ADHESION MOLECULES

Adhesion molecules mediate cell adhesion to their surroundings and to neighboring cells. In the immune system, adhesion molecules are critical to most aspects of leukocyte function, including lymphocyte recirculation through lymphoid organs, leukocyte recruitment into inflammatory sites, antigen-specific recognition and wound healing. There are five principal structural families of adhesion molecules:

- ● Selectins
- ● Integrins
- ● Immunoglobulin superfamily (IgSF) proteins
- ● Cadherins
- ● Mucins

Classification of these major adhesion molecules and their structures and functions are summarized in **Table 1.1**.

Selectins

Selectins are a group of cell adhesion molecules that are glycoproteins and play an important role in the relationship of circulating cells to the endothelium. The members of this surface molecule family have three separate structural motifs. They have a single N-terminal (extracellular) lectin motif preceding a single epidermal growth factor repeat and various short consensus repeat homology units. They are involved in lymphocyte migration. These carbohydrate-binding proteins facilitate adhesion of leukocytes to endothelial cells. There is a single chain transmembrane glycoprotein in each of the selectin molecules with a similar modular structure, that includes an extracellular calcium-dependent lectin domain. There are three separate groups of selectins:

- ● L-selectin (CD62L), expressed on leukocytes
- ● P-selectin (CD62P), expressed on platelets and activated endothelium

- ● E-selectin (CD62E), expressed on activated endothelium

Under shear forces their characteristic structural motif is comprised of an N-terminal lectin domain, a domain with homology to epidermal growth factor (EGF) and various complement regulatory protein repeat sequences.

Characteristics, receptors/ligands, cellular affinities, distribution, function, and other related data such as the expression and regulation of selectins are summarized in **Table 1.2**.

Integrins

Integrins are a family of cell membrane glycoproteins that are heterodimers comprised of α and β chain subunits. They serve as extracellular matrix glycoprotein receptors. They identify the RGD sequence of the β subunit, which consists of the arginine-glycine-aspartic acid tripeptide that occasionally also includes serine. The RGD sequence serves as a receptor recognition signal. Extracellular matrix glycoproteins, for which integrins serve as receptors, include fibronectin, C3, and lymphocyte function-associated antigen 1 (LFA-1), among other proteins. Differences in the β chain serve as the basis for division of integrins into three categories. Each category has distinctive α chains. The β chain provides specificity. The same 95-kD β chain is found in one category of integrins that includes lymphocyte function-associated antigen 1 (LFA-1), p150, 95, and complement receptor 3 (CR3). The same 130-kD β chain is shared among VLA-1, VLA-2, VLA-3, VLA-4, VLA-5, VLA-6, and integrins found in chickens. A 110-kD β chain is shared in common by another category that includes the vitronectin receptor and platelet glycoprotein IIb/IIIa. There are four repeats of 40 amino acid residues in the β chain extracellular domains. There are 45 amino acid residues in the β chain intracellular domains. The principal function of integrins is to link the cytoskeleton to extracel-

Table 1.1   Major classes of adhesion molecules

	Structure	Function
Selectin	Single transmembrane polypeptide composed of an extracellular lectin-like domain, an EGF motif, 62 amino acid repeats, a transmembrane region and a cytoplasmic tail	Slow intravascular leukocytes before transendothelial migration: initiators of leukocyte adhesion to endothelium; serve as signal transducing receptors
Integrin	Noncovalent $\alpha\beta$-heterodimers with 1 α chain and 1 β chain which are both transmembrane; 16 α chains and 8 β chains identified, resulting in a minimum of 22 different combinations	Mediate cell adhesion, mediate interactions with extracellular matrix components and with other cells
Immunoglobulin superfamily	Cell surface protein of a variable number of related 70-110 amino acid Ig-like domains, transmembrane segment and cytoplasmic tail	Engage in homotypic interaction, neurite outgrowth, and myelination; serve as ligands for β_1 and β_2 integrins to form firm adhesion of leukocytes
Cadherin	Single-pass transmembrane glycoprotein composed of about 700-750 residues, with extracellular domain containing 5 tandem repeats and calcium binding sites	Maintaining tissue integrity, cell sorting in development, epithelial integrity
Mucin	High molecular weight glycoprotein characterized by extensive and dense array of carbohydrates. The carbohydrates linkages are primarily O-linked with sulfated core groups, termed sialyl-Lewis x (sLe^x)	Serve as counter-receptors for selectins

Table 1.2   Selectins

Molecule	Ligands	Distribution
L-selectin (CD62L)	Sulfated: GlyCAM 1; CD34; MAdCAM-1	Leukocytes (homing receptor)
E-selectin (CD62E)	Tetrasaccharides: sialyl-Lewis^x; sialyl-Lewis^a cutaneous lymphocyte-associated antigen	Endothelial cells
P-selectin (CD26P)	Tetrasaccharides: sialyl-Lewis^x P-selectin glycoprotein ligand-1	Endothelial cells Platelets

General characteristics:

Expression: Only in vertebrates; in circulatory cells (endothelium and blood cells)

Structure
- Single transmembrane polypeptide
- N-terminal: Homologous to Ca^{++}- dependent lectins
- EGF motif
- 62 amino acid repeats: Homology to complement regulatory proteins
- Transmembrane region
- Cytoplasmic tail

Activation: Induced, then rapidly downregulated

Adhesion
- Transient

- Architecture
- Binding site: Amino-terminal domain
- Connecting arm: Contains EGF-like domain and peptide repeats
- Ca^{++}-dependent
- Ligands: Sialated glycans (similar pattern of binding of sialoadhesins)

Functions
- Slow intravascular leukocytes before transendothelial migration
- E-selectin: Mediates initial PMN adhesion to endothelial cells
- Adhesion is rolling, not firm
- Firm adhesion via LFA/ICAM-1 and VLA-4/VCAM-1

lular ligands. They also participate in wound healing, cell migration, killing of target cells, and in phagocytosis. Leukocyte adhesion deficiency syndrome occurs when the β subunit of LFA-1 and Mac-1 is missing. VLA proteins facilitate binding of cells to collagen (VLA-1, -2, and -3), laminin (VLA-1, -2, and -6), and fibronectin (VLA-3, -4, and -5). The cell to cell contacts formed by integrins are critical for many aspects of the immune response, such as antigen presentation, leukocyte-mediated cytotoxicity and myeloid cell phagocytosis. Integrins comprise an essential part of an adhesion receptor cascade that guides leukocytes from the bloodstream across endothelium and into injured tissue in response to chemotactic signals.

Characteristics, receptors/ligands, cellular affinities, distribution, function, and other related data such as the expression and regulation of integrins are summarized in **Table 1.3**.

Immunoglobulin superfamily

The immunoglobulin superfamily is a group of cell surface proteins characterized by the presence of a variable number of related 70–110 amino acid Ig-like domains originally described in the Ig variable and constant regions. Included are CD2, CD3, CD4, CD7, CD8, CD28, T cell receptor (TCR), MHC class I and MHC class II molecules, leukocyte function-associated antigen 3 (LFA-3), the IgG receptor, and a dozen other proteins. These molecules share in common with each other an immunoglobulin-like domain, with a length of approximately 100 amino acid residues and a central disulfide bond that anchors and stabilizes antiparallel β strands into a folded structure resembling immunoglobulin. Immunoglobulin superfamily members may share homology with constant or variable immunoglobulin domain regions. Various molecules of the cell surface with polypeptide chains whose folded structures are involved in cell to cell interactions belong in this category. Single gene and multigene members are included.

Characteristics, receptors/ligands, cellular affinities, distribution, function, and other related data such as the expression and regulation of the immunoglobulin superfamily are summarized in **Table 1.4**.

Cadherins

Cadherins belong to a family of cell adhesion molecules that enable cells to interact with their environment. Cadherins help cells to communicate with other cells in immune surveillance, extravasation, trafficking, tumor metastasis, wound healing, and tissue localization. Cadherins are calcium-dependent. The five different cadherins include N-cadherin, P-cadherin, T-cadherin, V-cadherin, and E-cadherin. Cytoplasmic domains of cadherins may interact with proteins of the cytoskeleton. They may bind to other recep-

tors based on homophilic specificity, but they still depend on intracellular interactions linked to the cytoskeleton.

Characteristics, receptors/ligands, cellular affinities, distribution, function, and other related data such as the expression and regulation of cadherins are summarized in **Table 1.5**.

Mucins

Mucins are heavily glycosated serine and threonine-rich proteins that serve as ligands for selectins. They contribute to another major group of adhesion molecules.

Other adhesion molecules

Other adhesion molecules that do not fall into these major classes are summarized in **Table 1.6**.

Endothelial–leukocyte interactions

Table 1.7 summarizes adhesion molecules involved in endothelial–leukocyte interactions. Their ligands, expression, regulation, and functions are listed.

Inflammation

Adhesion molecules play an important role in inflammation. The leukocyte adhesion cascade is a sequence of adhesion and activation events that is mediated by different adhesion molecules in different steps of capture: rolling, slow rolling, firm adhesion, and transmigration. The leukocyte adhesion cascade in inflammation is demonstrated in **Figure 1.1**. The adhesion molecules involved in different steps of the cascade are summarized in **Table 1.8**.

NATURAL AND ADAPTIVE IMMUNITY
Natural immunity

Natural or innate immunity comprises the inborn immune mechanisms that do not depend upon previous exposure to an antigen. It is present from birth and is designed to protect the host from injury or infection without previous contact with the infectious agent. It includes the skin, mucous membranes, and other barriers to infection; lysozyme in tears, stomach acid, other antibacterial molecules, and numerous other factors belong to innate immunity. Phagocytes, natural killer cells, complement and cytokines represent key participants in natural innate immunity. **Table 1.9** lists the effector mechanisms of natural immunity, including their components and functions. The cells of natural immunity are summarized as **Table 1.10**. The functions, structure, and membrane markers of these cells are also compared in the table.

Table 1.3   Integrins

Molecule	Ligands	Distribution
α1β1 (VLA-1, CD49a/CD29)	Laminin; collagen; tenascin, common form	NK, B, and activated T cells; fibroblasts; glial perineurium; Schwann cells; endothelial cells
α2β1 (VLA-2, CD49b/CD29)	Laminin; collagen	NK, B, and activated T cells; platelets; endothelial cells; fibroblasts; epithelium astrocytes; Schwann cells; ependymal
α3β1 (VLA-3, CD29c/CD29)	Laminin; collagen; fibronectin	Activated T cells; thymocytes; endothelium; fibroblasts; epithelium; astrocytes
α4β1 (VLA-4, CD49d/CD29)	α4β1;α4β7; fibronectin; VCAM-1; MAdCAM-1; TSP-1	NK, B, and T cells; eosinophils; endothelial cells; muscle; fibroblasts; neural-crest derived **Function:** T cell transendothelial migration
α5β1 (VLA-5, CD49e/CD29)	Fibronectin; murine L1	Activated B and T cells; memory T-cells; thymocytes; fibroblasts; epithelium; platelets; endothelial cells; astrocytes **α-5 disease:** Myopathy in chimeric mouse
α6β1 (VLA-6, CD49f/CD29)	Laminin	Leukocytes; thymocytes; epithelial; T cells (memory and activated); glial; fibroblasts; endothelial cells **α-6 disease:** Junctional epidermolysis bullosa
α7β1	Laminin	Skeletal and cardiac muscle; melanoma **α-7 disease:** Congenital MD Knockout→myopathy
α8β1 (CD-/CD29; VLA-8)	Fibronectin; vitronectin; tenascin, common form	Epithelium; neurons; oligodendroglia
α9β1	Tenascin	Epithelium (airway); muscle
αvβ1	Vitronectin; fibronectin; collagen; von Willebrand factor; fibrinogen	Oligodendroglia
αLβ2(LFA-1α; CD11a/CD18)	ICAM-1; ICAM-2; ICAM-3	Leukocytes; thymocytes; macrophages; T cells; microglia **β-2 disease:** Leukocyte adhesion deficiency
αMβ2 (Mac-1, CD11b/CD18)	ICAM-1; Factor X; iC3b; fibrinogen	Myeloid; B cells (activated); NK cells; macrophages; microglia; B leukemic cells
αXβ2 (P150,95, CD11c/CD18)	iC3b; fibrinogen	Myeloid; dendritic cells; B cells (activated); macrophages; microglia; B leukemic cells
αIIbβ3(gpIIb/IIIa, CD41/CD61)	Fibronectin; vitronectin; von Willebrand factor; thrombospondin	Platelets **β-3 disease:** Glanzmann thrombasthenia, Type B
αVβ3(CD51/CD61,Vitronectin-R)	Fibronectin; osteoponin; von Willebrand factor; PE-CAM-1 vitronectin; fibrinogen; human L1 thrombospondin; collagen	B cells (activated); T cells (activated and γδ); endothelium; monocytes; tumors; glia; Schwann cells; endothelium
α6β4	Laminin	Schwann cells; perineum; endothelium; epithelium; fibroblasts Not in immune cells **β-4 disease:** Junctional epidermolysis bullosa

Table 1.3 Integrins (*continued*)

Molecule	Ligands	Distribution
αvβ5 (CD51/CD-)	Vitronectin; fibronectin; fibrinogen	Fibroblasts; monocytes; macrophages; epithelium; oligodendroglia; tumors
αvβ6	Fibronectin	
αvβ8	Fibronectin	Oligodendroglia; Schwann cells; brain synapses
α4β7 (CD49d-)	Fibronectin; VCAM-1; MAdCAM-1	NK, B, and T cells Not in neural cells
αIELβ7 (CD103)	E-cadherin	Intraepithelial T cells (IEL) (intestinal)
α11		Uterus; heart; skeletal muscle; smooth muscle containing tissue

General characteristics:

Structure
- Heterodimers with 1 α chain and 1 β chain, 16 α chains and 8 β chains, 22 different heterodimers identified
- Transmembrane adhesion molecules, both subunits transmembrane
- Non-covalently bound
- Usually in low affinity conformation

Activation
- Exist in variable activation states

Activated by
- ○ Conformational changes: induced by ligand binding or intracellular processes
- ○ ? Expression at transcriptional level

Adhesion
- Binding site: On β subunit

- Modified by metal binding to α subunit
  - ○ α submit may mediate specificity of ligand binding
- Binding also influenced by divalent cations
- Often occurs after selectin binding
- Extracellular ligands: Binding is low affinity
  - ○ Often single specific IgCAM
  - ○ Subset of extracellular matrix molecules: Fibronectin; laminins
- Intracellular ligands: Talin; α-actinin
- Intracellular ligands then linked to
  - ○ Structural proteins; vinculin; actin microfilaments
  - ○ Signaling pathways
    - ▪ Partly via pp125^{FAK}, a focal adhesion-associated kinase
- Effects of extracellular ligand binding
  - ○ Receptor clustering
  - ○ Autophosphorylation of tyrosine residues
  - ○ Loss of integrin interaction may induce apoptosis

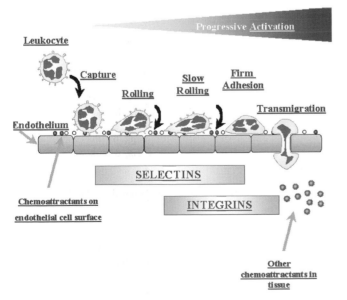

Figure 1.1 Adhesion molecules in leukocyte adhesion cascade of inflammation

Adaptive immunity

Adaptive or acquired immunity is the protection mechanism from an infectious disease agent as a consequence of clinical or subclinical infection with that agent or by deliberate immunization against that agent with products from it. This type of immunity is mediated by B and T cells following exposure to a specific antigen. It is characterized by specificity, immunological memory, and self/nonself recognition. The response involves clonal selection of lymphocytes that respond to a specific antigen. T cells and B cells are the two major components of adaptive immunity. Comparison of these two cell types is presented in **Table 1.11**.

Adaptive immunity has features in contrast to innate immunity. **Table 1.12** compares the characteristics, cellular receptors, functions, markers, and other features of these two limbs of the immune response.

Table 1.4 Immunoglobulin (Ig) superfamily

NEURAL-SPECIFIC IgCAMS		
Molecule	**Ligands**	**Distribution**
Adhesion molecule on glia (AMOG)		Glial neural migration
L1CAM	Axonin	Neural
Myelin-associated glycoprotein (MAG)	MAG	Myelin
Myelin-oligodendrocyte glycoprotein (MOG)		Myelin; oligodendrocytes
NCAM-1 (CD56)	NCAM-1 via polysialic acid; modulated by sialyltransferase X; polysialyltransferase	Neural cells
NrCAM	Ig superfamily	Neural
OBCAM	Opioids (μ); acidic lipids	Brain
P_0 protein	P_0	Myelin
PMP-22 protein	PMP-22	Myelin
Also neurofascin and NgCAM	Tenascin-R, axonin-1, F11	Neural
SYSTEMIC IgCAMS		
Molecule	**Ligands**	**Distribution**
ALCAM (CD166)	CD6; CD166; NgCAM; 35 kD protein	Neural; leukocytes
Basigin (CD147)		Leukocytes; RBCs; platelets; endothelial cells
BL-CAM (CD22)	Sialylated glycoproteins LCA (CD45)	B cells
CD44	Hyaluronin; ankyrin; fibronectin; MIP1β osteopontin	Lymphocytes; epithelial; WM perivascular astrocytes; glial tumors (malignant); metastases (CD44v splice variant)
ICAM-1 (CD54)	αLβ2; LFA-1	Leukocytes; endothelial cells; dendritic cells; fibroblasts; epithelium; synovial cells **Disease:** Lys29Met mutation é Susceptibility to cerebral malaria
ICAM-2 (CD102)	αLβ2 (LFA-1)	Endothelial cells; lymphocytes; monocytes
ICAM-3 (CD50)	αLβ2	Leukocytes
Lymphocyte function antigen-2 (LFA-2) (CD2)	LFA-3	Lymphocytes; thymocytes
LFA-3 (CD58)	LFA-2	Leukocytes; stroma endothelial cells; astrocytoma
Major histocompatibility complex (MHC) molecules		

Table 1.4 Immunoglobulin (Ig) superfamily (*continued*)

NEURAL-SPECIFIC IgCAMS		
Molecule	**Ligands**	**Distribution**
MAdCAM-1	α4β7; L-selectin	Mucosal endothelial cells
PECAM (CD31)	CD31; αvβ3	Leukocytes; synovial cells; endothelial cells
T cell receptor (C-region)		
VCAM-1	α4β1; α4β7	Satellite cells; monocytes; synovial cells; activated endothelial cells

General characteristics:
Expression: Evolutionarily ancient; widely expressed
Structure
- 1 or more repeats of Ig fold of 60–100 amino acids: form sites of adhesion
- Ig domain: No somatic hypermutations
  - Sandwiches of 2 β sheets held together by hydrophobic interactions
- Constitutive or long-term upregulated
- Anchor: Transmembrane segment and cytoplasmic tail

Interactions
- Homophilic: Neural specific Ig cell adhesion molecules (IgCAMs)
- Heterophilic: Systemic IgCAMs

Adhesion
- Sites: Ig fold(s) domains (distal); fibronectin type III (Fn3) domains
- Inhibited by sialylation
- Ca^{++}-independent

Functions
- Neurite outgrowth
- Myelination
- Firm adhesion of leukocytes
  - Via LFA/ICAM-1 and VLA-4/VCAM-1

Toll-like receptors (TLRs)

The innate immune system recognizes a wide spectrum of pathogens without a need for prior exposure. The main cells responsible for innate immunity, neutrophils, monocytes, and macrophages, phagocytose pathogens and trigger the cytokine and chemokine network resulting in inflammation and specific immune responses. Receptors of innate immunity have broad specificity. They recognize many related molecular structures called pathogen-associated molecular patterns (PAMPs). To recognize different types of PAMPs, macrophages have a set of transmembrane receptors called toll-like receptors (TLRs). The TLR family was first discovered in *Drosophila* and has significant homology in its cytoplasmic domain to IL-1 receptor Type I. In macrophages the pathogen is exposed to the TLRs when it is inside the phagosome. The TLR(s) to which it binds determine what the response will be. In this way, the TLRs identify the nature of the pathogen and turn on an appropriate effector response. These signaling cascades lead to the expression of various cytokine genes. For example, TLR2 and TLR4 activate the NF-κB pathway, which regulates cytokine expression, through the adaptor molecule MyD88. Activation of the NF-κB pathway leads to initiation of the adaptive immune response by production of inflammatory cytokines such as IL-1, IL-2, IL-8, TNF-α, IL-12, chemokines and induction of costimulatory molecules. In addition to induction of cytokines, MyD88 binds

FADD and triggers apoptosis through the Caspase cascade. Activation of the apoptosis pathway via TLRs contributes to the defense mechanisms used by innate immunity. To date, ten toll-like receptors have been reported in the human and the mouse. **Table 1.13** shows the chromosomal location, distribution, and function of these TLRs. **Figures 1.2** and **1.3** illustrate the role of the TLR family in immunity against the pathogens.

ORGANS AND TISSUES OF THE IMMUNE RESPONSE

The immune system is found throughout the body and is made up of many different cells, organs, and tissues. The organs and tissues of the system can be classified into two main groups: (1) primary lymphoid organs, in which lymphocytes are generated and undergo development and maturation; and (2) secondary lymphoid organs and tissues, where mature lymphocytes interact with antigen. The vessels of the blood and lymphatic systems connect lymphoid organs and tissues and unite them into a functional whole. Leukocytes, or white blood cells, are found within the blood, lymph, and lymphoid tissues and organs. The vertebrate immune system contains many types of leukocytes, but only the lymphocytes have the attributes of receptor diversity, antigen specificity, and self/nonself recognition that are the hallmarks of adaptive immunity.

Table 1.5   Cadherins

Molecules	Ligands	Distribution
Cadherin E (1)	H	Epithelial
Cadherin N (2)	O	Neural
Cadherin BR (12)	M	Brain
Cadherin P (3)	O	Placental
Cadherin R (4)	P	Retinal
Cadherin M (15)	H	Muscle
Cadherin VE(5) (CD144)	I	Epithelial
Cadherin T and H (13)	L	Heart
Cadherin OB (11)	I	Osteoblast
Cadherin K (6)	C	Brain; kidney
Cadherin 7		
Cadherin 8		Brain
Cadherin KSP (16)		Kidney
Cadherin LI (17)		GI tract; pancreas
Cadherin18		CNS; small cell lung cancer
Cadherin, fibroblasts 1 (19)		Fibroblasts
Cadherin, fibroblasts 2 (20)		Fibroblasts
Cadherin, fibroblasts 3 (21)		Fibroblasts
Cadherin23		Ear
Desmocollin 1		Skin
Desmocollin 2		Epithelium; mucosa; myocardium; lymph nodes
Desmoglein 1		Epidermis, tongue
Desmoglein 2		All
Desmoglein 3		Epidermis; tongue; antibody target in pemphigus
Protocadherin 1, 2, 3, 7, 8, 9		

General characteristics:

Expression: Evolutionarily ancient; widely expressed

Structure
- Extracellular domain: 5 tandem repeats; each comprising sandwich of β sheets
- Often present as dimers
- Anchor: Transmembranse segment; cytoplasmic carboxy terminal domain

Adhesion
- Homophilic: Via most distal cadherin repeats
- Requires: Ca^+; specific intracytoplasmic binding
- Intracellular
  - Cytoplasmic domain binds catenins
  - Catenins then bind to actin cytoskeleton
- Types

○ Interactive with actin cytoskeleton: Cadherins N, P, R, B, E
○ Desmosome-associated: Desmogleins and desmocollins
  ▪ Interact with intermediate filaments
  ▪ Location: In tight junctions
○ Protocadherins
  ▪ Homology to cadherins: Extracellular, but not intracellular, domains

Functions and diseases
- Cadherin E (1): Reduction correlates with tumor malignancy
  ○ Gynecologic malignancies
    ▪ Point mutations in tumor cells
    ▪ Somatic loss of heterozygosity common
  ○ Gastric malignancies
  ○ Susceptibility to Listeria monocytogenes infection
- Cadherin N: Role in establishment of left–right asymmetry
- Cadherin P (3): Congenital hypotrichosis with juvenile muscular dystrophy
- Cadherin23: Deafness
- Catenin β1 (cadherin-associated protein): Mutations in malignancies
  ○ Colon, hepatoblastoma, pilomatricoma, ovarian (endometrioid)
- Desmoglein 3: Antibody target in pemphigus

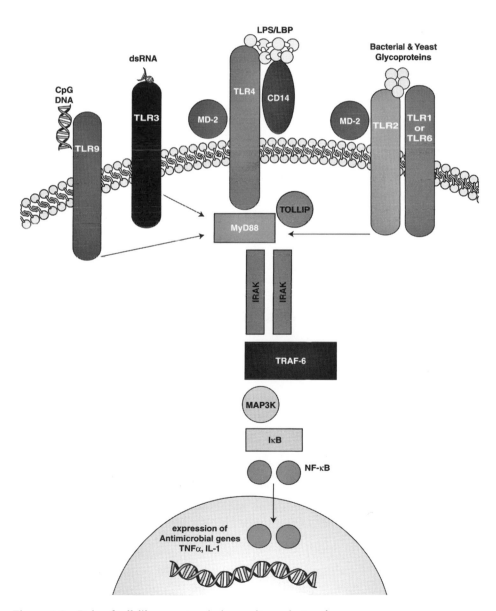

Figure 1.2   Role of toll-like receptors in immunity against pathogens

Table 1.6   Other adhesion molecules

Molecules	Ligands	Distribution
Agrin (neural)	MuSK; NCAM; laminin; heparin-binding proteins; heparan sulfate proteoglycan	Nerve
CD34	MAdCAM-1; L-selectin	Immature lymph/ myeloid
GlyCAM-1	L-selectin	Lymph nodes
Oligodendrocyte-myelin glycoprotein (OMGP)		Myelin; oligodendrocytes

Table 1.7   Endothelial-leukocyte adhesion molecules

INTEGRINS				
Molecule(s)	Ligand(s)	Distribution	Regulation	Function
$\alpha_L\beta_2$ LFA-1 CD11a/CD18	ICAM-1, ICAM-2, ICAM-3	Lymphocytes > neutrophils; monocytes	Conformationally activated by chemoattractants	Lymphocyte, granulocyte adhesion to resting endothelium
$\alpha_M\beta_2$ Mac-1 CD11c/CD18	ICAM-1, C3bi unidentified endothelial ligand	Macrophages > monocytes; neutrophils	Upregulated by degranulation; activated by TNF, chemoattractants	Granulocyte, monocyte adhesion to resting and activated endothelium transmigration; C3bi binding; leukocyte activation; phagocytosis
$\alpha_X\beta_2$ p150,95 CD11c/CD18	C3bi, others?	Macrophages; monocytes	Upregulated by degranulation	Complement fixation; phagocytosis?
$\alpha_4\beta_1$ CD49d/CD29	VCAM-1	Lymphocytes; monocytes; eosinophils	Constitutively active (?) affinity regulation; higher expression on memory T cells	Lymphocyte, monocyte adhesion to endothelium and fibronectin
$\alpha_4\beta_7$	MAdCAM-1	Lymphocytes	Constitutive	Mucosal lymphocyte homing
SELECTINS				
Molecule(s)	Ligand(s)	Distribution	Regulation	Function
E-selectin CD62E	Sialyl-Lewis^x glycoproteins, glycolipids (?) on PMN	Activated endothelium	Upregulated by TNF, IL-1 Max. expression at 4–6 hr	Adhesion and rolling of neutrophils; adhesion of T cells; monocytes; slow (3 μm/s) leukocyte rolling *in vivo*
L-selectin CD62L	GlyCAM-1, CD34 on lymph node HEV; evidence for extralymphatic ligand(s)	Resting neutrophils; most lymphocytes	Constitutive, shed from surface by chemoattractant	Lymphocyte homing to lymph nodes; sustained leukocyte rolling *in vivo*; critical for PMN recruitment; mediates leukocyte capture

Table 1.7 Endothelial-leukocyte adhesion molecules (*continued*)

P-selectin CD62P	PSGL-1 on myeloid cells	Activated platelets; thrombin-, histamine-, cytokine-activated endothelium	Surface expressed by degranulation, transcriptional regulation by TNF, IL-1	Platelet-leukocyte adhesion; leukocyte rolling *in vivo*; critical for PMN recruitment

IMMUNOGLOBULINS				
Molecule(s)	**Ligand(s)**	**Distribution**	**Regulation**	**Function**
ICAM-1 CD54	LFA-1, Mac-1	Endothelium; smooth muscle fibroblasts; T cells	Constitutive, increased expression by IL-1, IFN-γ	Cytotoxic T-cell conjugates, leukocyte-endothelial adhesion, transmigration?
ICAM-2 CD102	LFA-1	Endothelial cells	Constitutive	Leukocyte adhesion to resting endothelium
ICAM-3 CD50	LFA-1	Resting T cells	Constitutive	Homotypic lymphocyte aggregation
VCAM-1	VLA-4 ($\alpha_4\beta_1$ integrin), $\alpha_4\beta_7$ integrin	Cytokine-stimulated endothelium	IL-1, IL-4, TNF upregulate expression, max. at 4–24 hr	Adhesion of monocytes, eosinophils, lymphocytes to activated endothelium; expressed in atherosclerotic plaques
PECAM-1 CD-31	PECAM-1 (homotypic) other ligand(s)?	High on endothelium, moderate on monocytes, PMN, platelets	Constitutive, redistribution to cell borders in confluent EC	Critically involved in neutrophil transmigration; endothelial monolayer integrity; neutrophil adhesion (?)
MAdCAM-1	$\alpha_4\beta_7$ integrin, L-selectin (glycosylation-dependent)	Mucosal venules; Peyer's patch HEV; mesenteric lymph node	Constitutive, different glycosylation in PP and MLN	Lymphocyte homing to mucosal sites; lymphocyte rolling (contains mucin domain)

SELECTIN LIGANDS				
Molecule(s)	**Ligand(s)**	**Distribution**	**Regulation**	**Function**
PSGL-1 CD162	P-selectin	All leukocytes	Constitutive; dimer	Adhesion of myeloid cells to P-, L-selectin
GlyCAM-1	L-selectin	Lymph node HEV; serum	Constitutive; secreted; expression depends on lymph flow	Homing of lymphocytes to lymph node HEV (?); modulation of L-selectin-dependent adhesion?
CD34	L-selectin	Lymph node HEV; other endothelia	Constitutive; expression depends on lymph flow	Homing of lymphocyte to lymph node HEV
ESL-1	E-selectin	Murine myeloid cells	Constitutive	Function unknown, homologous to FGF-receptor
Unidentified	L-selectin	Activated endothelial cells	Induced by TNF *in vitro*, by tissue trauma *in vivo*	Mediates neutrophil adhesion; L-selectin-dependent rolling

Central lymphoid organs

Central lymphoid organs are requisite for the development of the lymphoid and, therefore, the immune system. These include the thymus, bone marrow, and bursa of Fabricus. Central lymphoid organs are also termed primary lymphoid organs. They are sites where lymphocytes are generated. Both T and B cells originate in the bone marrow but only B cells mature there. Human T cells mature in the thymus.

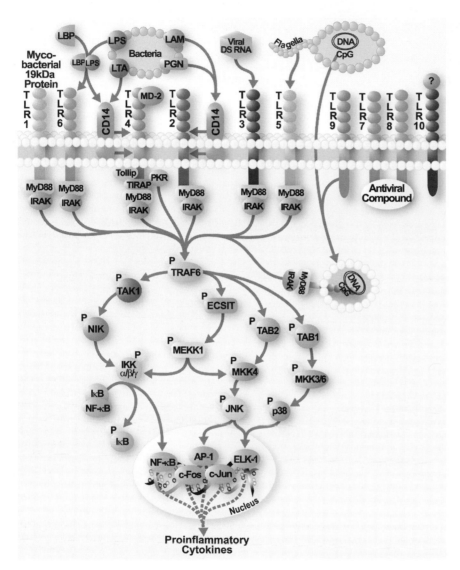

Figure 1.3 Pathways of toll-like receptors

Table 1.8 Adhesion molecules involved in leukocyte emigration from vessels during inflammation

Rolling	Stopping	Aggregation and shape change	Migration through vessel wall
• Sialyl-Lewis^x	• β_2 integrins	• CD11b/CD18 (Mac-1)	• $\beta2$ integrins
• L-selectin	• VLA-4 ($\alpha_4\beta_1$)	• CD11a/CD18 (LFA-1)	• ICAM-1
• P-selectin	• ICAM-1	• P-selectin	• VCAM-1
• E-selectin	• VCAM-1		• PECAM-1
	• $\alpha_4\beta_7$ integrins		
	• MadCAM-1		

Table 1.9 Effector mechanisms of natural immunity

Site	Component	Function
Skin	Squamous cells, sweat	Desquamation; antimicrobial secretions, e.g. fatty acids; flushing
Eye	Tears	Flushing; lysozyme
Nasopharynx	Mucus, saliva	Flushing; lysozyme
Lung	Tracheal cilia	Mucociliary elevator; surfactants
GI tract	Columnar cells	Stomach acidity, bile salts, fatty acids, peristalsis
Blood and lymphoid organs	K cells, LAK cells, NK cells	Direct and antibody-dependent cytolysis
Other serous fluids	Lactoferrin, transferrin; interferons, TNF-α; fibronectin, complement; lysozyme	Iron deprivation; phagocyte activation; opsonization; enhanced phagocytosis; peptidoglycan hydrolysis

Table 1.10 Cells of natural immunity

	Function	Structure characteristics	Membrane marker
Neutrophils	Phagocytosis, intracellular killing, inflammation and tissue damage	Characteristic nucleus and cytoplasm	CD67
Macrophages	Phagocytosis, intracellular and extracellular killing, tissue repair, antigen presentation for specific immune response	Characteristic nucleus	CD14
Natural killer (NK) cells	Kill infected cells and malignant cells; LAK cells kill transformed cells and malignant cells	Also known as large granular lymphocytes (LGL); activated by IL-2 and IFN to become LAK cells	CD56, CD16
K cells	Recognize antibody-coated targets	Morphologically undefined	Could be NK cells (IgG), macrophages (IgG), eosinophils (IgE) or other cells (IgG)

Table 1.11 Cells of adaptive immunity

	B cells	T cells
Origin	Bone marrow	Bone marrow
Site of maturation	Bone marrow	Thymus
Antigen receptor	B cell receptor (BCR)	T cell receptor (TCR)
Target of binding	Soluble antigens	Biomolecular complex displayed at the surface of APC
Branch of immune response	Antibody-mediated immune response	Cell-mediated and antibody-mediated immune response
MHC and antigen presentation	Class II MHC molecules	Class I MHC molecules (CD8$^+$ T cells) and class II MHC molecules (CD4$^+$ T cells)

Table 1.12 Characteristics of innate and adaptive immunity

	Innate immunity	Adaptive immunity
Components	Physical barriers (skin, gut villi, lung cilia, etc.); protein and non-protein secretions; phagocytes, NK cells, eosinophils, K cells	Immunoglobulins (antibodies), T cells, B cells
Antigen dependent	No: system in place prior to exposure to antigen	Yes: induced by antigens
Antigen specific	No: lacks discrimination among antigens	Yes: shows fine discrimination
Time lag	No: immediate response	Yes: delayed (3–5 days) response
Immunologic memory	No	Yes
Pathogen	Essentially polysaccharides and polynucleotides	Most derived from polypeptides (proteins)
Pathogen recognition	Pathogen recognized by receptors encoded in the germline	Pathogens recognized by receptors generated randomly
Receptor specificity	Broad specificity, i.e. recognize many related molecular structures called pathogen-associated molecular patterns	Very narrow specificity, i.e. recognize a particular epitope
Receptors	Pattern recognition receptors	B cell receptors and T cell receptors
Enhancement	Can be enhanced after exposure to antigen through effects of cytokines	Enhanced by antigens
Existence	Occurs in all metazoans	Occurs in vertebrates only

Table 1.13 Toll-like receptors

Name	Chromosome	Distribution (mRNA/protein)	Function/comments
TLR1 (TIL, Rsc786)	Human: 4 Mouse: 5	Leukocytes; upregulated on macrophages	Associates with and regulates TLR2 response
TLR2 (TIL4)	Human: 4 Mouse: 3 or 8	Monocytes, granulocytes; upregulated on macrophages	Interacts with microbial lipoproteins and peptidoglycans: CD14-dependent and -independent response to LPS; NF-κB pathway
TLR3	Human: 4 Mouse: 3 or 8	Dendritic cells; upregulated on endothelium and epithelium	Interacts with dsRNA, activates NF-κB pathway; induces production of type I interferons, MyD88-dependent and -independent response to poly (I:C)
TLR4 (h Toll, Ly87, Ras12-8, RAN/ M1)	Human: 9 Mouse: 4	Monocytes	Interacts with microbial lipoproteins; CD14-dependent response to LPS; NF-κB pathway
TLR5 (TIL3)	Human: 1 Mouse: 1	Leukocytes, prostate, ovary, liver, lung	Interacts with microbial lipoproteins; NF-κB, response to Salmonella

Table 1.13 Toll-like receptors (*continued*)

Name	Chromosome	Distribution (mRNA/protein)	Function/comments
TLR6	Human: 4 Mouse: 5	Leukocytes, ovary, lung	Interacts with microbial lipoproteins; protein sequence most similar to hTLR1; associates with and regulates TLR2 response
TLR7	Human: X Mouse: X?	Spleen, placenta, lung; upregulated on macrophages	Low similarity to other TLR family members
TLR8	Human: X Mouse: X?	Leukocytes, lung	
TLR9	Human: 3 Mouse: 6	Leukocytes	Receptor for CpG bacterial DNA, weakly similar to TLR3; may mediate protein–protein interaction
TLR10	?	Lymphoid tissues	Most closely related to TLR1 and TLR6
RP105 (CD180, Ly78)	Human: 5 Mouse: 13	Mature B cells	B cell activation; LPS recognition
MD-1 (Ly64, Irrp, 14/A10)	Human: 6	Mature B cells	Associates and regulates surface expression of RP105
MD-2 (Ly96)		Macrophages	Associates and regulates surface expression of TLR4; signals LPS presence

Peripheral lymphoid organs

Peripheral lymphoid organs are not required for ontogeny of the immune response. They are sites where adaptive immune responses are initiated and where lymphocytes are maintained. Peripheral lymphoid organs are also termed secondary lymphoid organs. They include the lymph nodes, spleen, tonsils, and mucosal-associated lymphoid tissues in which immune responses are induced.

Table 1.14 depicts the constituents of the primary and secondary lymphoid organs.

Table 1.14 Organs and tissues of the immune response

	Primary lymphoid organs	Secondary lymphoid organs
Component	Bone marrow, fetal liver, thymus	Spleen, lymph nodes, and mucosa-associated lymphoid tissue (MALT) including tonsils, adenoids, respiratory, genitourinary, and gastrointestinal tracts
Proliferation and differentiation	Antigen-independent	Antigen-dependent
Product	Immunocompetent cells (B cells and T cells)	Effector cells (antibody-secreting plasma cells for humoral immune response and T helper and T cytotoxic cells for cell-mediated immune response)
Event	Development and maturation of B and T cells	Induction of immune response: encounter of antigens and antigen-presenting cells (APC) with mature B and T cells, generation of effector cells, and memory cells

Antigens, Immunogens, Vaccines, and Immunization

2

- ● ANTIGENS
- ● VACCINE
- ● IMMUNIZATION

ANTIGENS

Immunogen

The 'traditional' definition of antigen is a substance that may stimulate B and/or T cell limbs of the immune response and react with the products of that response, including immunoglobulin antibodies, and/or specific receptors on T cells. This 'traditional' definition of antigen more correctly refers to an immunogen. A complete antigen is one that both induces an immune response and reacts with the products of it, whereas an incomplete antigen or hapten is unable to induce an immune response alone, but is able to react with the products of it, e.g., antibodies. **Table 2.1** compares the characteristics of immunogen and hapten.

Presently, antigen is considered to be one of many kinds of substances with which an antibody molecule or T cell receptor may bind. These include sugars, lipids, intermediary metabolites, autocoids, hormones, complex carbohydrates, phospholipids, nucleic acids, and proteins.

Antigenic determinant

An antigenic determinant is the site of an antigen molecule that is termed an epitope and interacts with the specific antigen-binding site in the variable region of an antibody molecule known as a paratope. The excellent fit between epitope and paratope is based on their three-dimensional interaction and noncovalent union. An antigenic determinant or epitope may also react with a T cell receptor for which it is specific. A lone antigen molecule may have several different epitopes available for reaction with antibody or T cell receptors. There are two types of antigenic determinants: conformational determinants and linear (sequential) determinants. The characteristics of these two types of epitopes, including their location, composition, antigen-antibody reaction and availability, are listed in **Table 2.2**.

Antigenic determinants are also categorized in **Table 2.3** according to B and T cell recognition.

Thymus-dependent antigen

A thymus-dependent antigen is an immunogen that requires T cell cooperation with B cells to synthesize specific antibodies. Presentation of thymus-dependent antigen to T cells must be in the context of MHC class II molecules. Thymus-dependent antigens include proteins, polypeptides, hapten-carrier complexes, erythrocytes, and many other antigens that have diverse epitopes. T-dependent antigens contain some epitopes that T cells recognize and others that B cells identify. T cells produce cytokines and cell surface molecules that induce B cell growth and differentiation into antibody-secreting cells. Humoral immune responses to T-dependent antigens are associated with isotype switching, affinity maturation, and memory. The response to thymus-dependent antigens shows only minor heavy chain isotype switching or affinity maturation, both of which require helper T cell signals.

Thymus-independent antigen

Thymus-independent antigen is an immunogen that can stimulate B cells to synthesize antibodies without participation by T cells. These antigens are less complex than are thymus-dependent antigens. They are often polysaccharides that contain repeating epitopes or lipopolysaccharides derived from Gram-negative microorganisms. Thymus-independent antigens induce IgM synthesis by B cells without cooperation by T cells. They also do not stimulate immunological memory. Murine thymus-independent antigens are classified as either TI-1 or TI-2 antigens. Lipopolysaccharide (LPS), which activates murine B cells without participation by T or other cells, is a typical TI-1 antigen. Low concentrations of LPS stimulate synthesis of specific antibody, whereas high concentrations activate

Table 2.1 Types of antigens

	Immunogen	Hapten
	Complete antigen	Incomplete antigen
Immunogenicity	Induce immune response, react with antibodies	Unable to induce immune response, but react with antibodies
Molecular weight	At least 10 000	Smaller molecules
Chemistry	Proteins, or polysaccharides	Simple chemicals, highly reactive chemical groupings

Table 2.2 Antigen determinants (epitopes)

	Conformational determinant	Linear determinant
Location	Most globular proteins and native nucleic acids	Most polysaccharides, fibrilar proteins, and single-stranded nucleic acids
Composition	Amino acid residues brought into proximity to one another by folding	Adjacent amino acid residues in the covalent sequence
Antigen-antibody reaction	Dependent on 3-dimensional structure	Dependent on linear structure of 6 amino acids
Availability for antibody interaction	Usually associated with native proteins	Become available upon denaturation of proteins

Table 2.3 Antigenic determinants

	Recognized by B cells and Ab	Recognized by T cells
Composition	Proteins, polysaccharides, nucleic acids	Proteins
Configuration	Linear/conformational determinants	Linear determinants
Size	4–8 residues	8–15 residues
Number	Limited, located on the external surface of the antigen	Limited to those that can bind to MHC

essentially all B cells to grow and differentiate. TI-2 antigens include polysaccharides, glycolipids, and nucleic acids. When T cells and macrophages are depleted, no antibody response develops against them. A comparison of thymus-dependent and thymus-independent antigens is shown in **Table 2.4**.

Alloantigen

Alloantigen is an antigen present in some members or strains of a species, but not in others. Alloantigens include blood group substances on erythrocytes and histocompatibility antigens present in grafted tissues that stimulate an alloimmune response in the recipient not possessing them,

as well as various proteins and enzymes. Two animals of a given species are said to be allogeneic with respect to each other. Alloantigens are commonly products of polymorphic genes. **Table 2.5** gives an example of mouse alloantigens.

VACCINE

A vaccine is a live attenuated or killed microorganism or parts or products from them which contain antigens that can stimulate a specific immune response consisting of protective antibodies and T cell immunity. A vaccine should stimulate a sufficient number of memory T and B cells to yield effector T cells and antibody-producing B cells from mem-

Table 2.4 Comparison of T cell-dependent with T cell-independent antigens

	TD antigens	TI antigens
Activation of B cells	Can only activate B cells in the presence of Th cells	Can activate B cells in the absence of Th cells
Structural properties	Complex	Simple
Presence in most pathogenic microbes	Yes	No
Antibody class-induced	IgG, IgM, IgA, IgD, IgE	IgM
Immunological memory response	Yes	No
Examples	Microbial proteins, non-self or alter-self proteins	Pneumococcal polysaccharide, lipopolysaccharide, flagella

ory cells. Viral vaccine should also be able to stimulate high titers of neutralizing antibodies. Vaccines can be prepared from weakened or killed microorganisms; subcellular segments; inactivated toxins; toxoids derived from microorganisms; or immunologically active surface markers extracted from microorganisms. They can also be classified as viral vaccines and bacterial vaccines according to the pathogens to which they are directed. **Table 2.6** shows the major categories of vaccines in use. They can be administered intramuscularly, subcutaneously, intradermally, orally or intranasally; as single agents or in combination. An ideal vaccine should be effective, well tolerated, easy and inexpensive to produce, easy to administer and convenient to store. Diphtheria, tetanus, and pertussis vaccine data are presented in **Table 2.7**. (**Table 2.17** below lists the vaccine products licensed for use in the United States.)

IMMUNIZATION

Vaccination

Vaccination is the induction of active (protective) immunity in man or other animals against infectious disease by the administration of vaccines (inoculation).

Vaccine preventable diseases

Many diseases are preventable by vaccination. **Table 2.8** and **Table 2.9** define the vaccine preventable diseases and list the microbiological and serological tests used in the diagnosis of vaccine-preventable diseases.

Immunization schedules

Rules of childhood immunization are summarized in **Table 2.10**. **Table 2.11** provides childhood and adolescent immunization schedules recommended in the United States. **Table 2.12** and **Table 2.13** provide supplemental information for children and adolescents who start late or who are >1 month behind. A summary of adult immunization is shown in **Table 2.14**. **Table 2.15** provides the adult immunization schedule recommended in the United States. Adults with medical conditions have a dedicated immunization regimen, which is depicted in **Table 2.16**.

Table 2.5   Mouse leukocyte alloantigens chart

Mouse strains	MHC haplotype	H-2K	I-Aα	I-Aβ	I-E	H-2D	H-2L	Qa-2	Qa-1	Igh-C haplotype	Igh-6 (IgM)	Igh-5 (IgD)	Igh-4 (IgG1)	Igh-3 (IgG2b)	Igh-1 (IgG2a)	Igh-7 (IgE)	Igh-2 (IgA)	CD5 (Ly-1)	CD8a (Ly2)	CD8b (Ly3)
101/-	k	k	k	k	k	k	-								b					
129/-	b	b	b	b	-	b	-	a(+)			a	a			a		a	2	2	2
A/J	a	k	k	k	k	d	d	a(lo)	a	e	e	e	a	e	e	a	d	2	2	2
A2G	a	k	k	k	k	d	d											2	2	2
AKR/J	k	k	k	k	k	k	-	b(-)	b	d	n	a	d	d	d	a	d	2	1	1
AL/N	a	k	k	k	k	d	d	a(lo)		o	e	e	a	d	d		d/e			
AU/SsJ	q	q	q	q	-	q	q													
BALB/cAnN	d	d	d	d	d	d	d	b(-)		a	a	a	a	a	a			2	2	2
BALB/cJ	d	d	d	d	d	d	d	a(lo)	b	a	a	a	a	a	a	a	a	2	2	2
BDP/J	p	p	p	p	p	p	-	b(-)		h	a	a	a	a	h			2	1	2
BUB/BnJ	q	q	q	q	-	q	q								a			2	2	
BXSB/Mp	b	b	b	b	-	b	-											2	2	2
CB-17	d	d	d	d	d	d	d	a(lo)	b	b	b	b	b	b	b	b	b	2	2	2
C3H/Bi	k	k	k	k	k	k	-	b(-)			a				j			1	1	
C3H/He	k	k	k	k	k	k	-	b(-)	b	j	a	a	a	a	j	a		1	1	2
C57BL/-	b	b	b	b	-	b	-	a(hi)	b	b	b	b	b	b	b	b	b	2	2	2
C57BR/cd	k	k	k	k	k	k	-	b(-)	a						a			3	2	2
C57L/-	b	b	b	b	-	b	-	a(+)	b						a			3	2	2
C58/-	k	k	k	k	k	k	-	b(-)		a	a	a	a	a	a			2	1	1
CBA/Ca	k	k	k	k	k	k	-	b(-)		j	a	a	a	a	j	a	a			
CBA/J	k	k	k	k	k	k	-	b(-)	b	j	a	a	a	a	j	a	a	1	1	2
CBA/N	k	k	k	k	k	k	-	b(-)		j	a	a	a	a	j	a	a	1	1	2
CE/-	k	k	k	k	k	k	-	b(-)		f	a	a	a	f	f		f	2	1	2
DA/HuSn	qp1	q	q	q	-	s									g			2	2	2
DBA/1	q	q	q	q	-	q	q	a(lo)		c	a	a	a	a	c			1	1	2
DBA/2	d	d	d	d	d	d	d	a(lo)	b	c	a	a	a	a	c		c	1	1	2
FVB/N	q	q	q	q	-	q	q				a	a						2	2	2
GRS/J	dx	d	f	f	-	w3												2	1	2

Table 2.5 Mouse leukocyte alloantigens chart (*continued*)

Mouse strains	MHC haplotype	H-2K	I-Aα	I-Aβ	I-E	H-2D	H-2L	Qa-2	Qa-1	Igh-C haplotype	Igh-6 (IgM)	Igh-5 (IgD)	Igh-4 (IgG1)	Igh-3 (IgG2b)	Igh-1 (IgG2a)	Igh-7 (IgE)	Igh-2 (IgA)	CD5 (Ly-1)	CD8a (Ly2)	CD8b (Ly3)
HRS/J	k	k	k	k	k	k	—													
I/-	j	j	j	j	j	b	—								c			2	1	2
LP/J	b	b	b	b	—	b	—			b	b	b	b	b	b			2	2	2
MA/	k	k	k	k	k	k	—	b(−)							a			1	1	
MRL/Mp	k	k	k	k	k	k	—											2	1	1
NOD	g7	d	d	g7	—	b		a			b	b				b			Not 1		
NZB/-	d	d	d	d	d	d	—	a(+)	a	n	n	a	d	e	e		d	2	2	2
NZW/-	z	u	u	u	u	z	—	b(−)		n		a			e					
P/-	p	p	p	p	p	p	—	b(−)		h	a				h			2	1	2
PL/J	u	u	u	u	u	d	d			j	a	a	a	a	j			2	1	1
RF/J	k	k	k	k	k	k	—	b(−)		c	a				c			2	1	1
RIII/-	r	r	r	r	r	r		b(−)	c	g	a	a	a	g	g		g	2	2	2
SEC/-	d	d	d	d	d	d	d								h				2	
SJL/J	s	s	s	s	—	s		a(+)		b	b	b	b	b	b			2	2	2
SM/J	v	v	v	v	v	v				b	b	b	b	b	b				1	2
ST/6J	k	k	k	k	k	k	—			a	a				a			2	2	2
SWR/J	q	q	q	q	—	q	q	a(+)	a	p		a		f	c		b	3	2	2

Table 2.5   Mouse leukocyte alloantigens chart (*continued*)

Mouse strains	CD22 (Lyb-8)	CD44 (Ly-24)	CD45 (Ly-5)	CD90 (Thy-1)	CD157 (BP-3)	CD178 (Fas Ligand)	CD244 (2B4)	ART2.2 Rt6-2)	B2 microglobulin	Ly-6	Ly-9 (CD229)	Lyb-2 (CD72)	Mtv Proviruses	NK-1.1	Vα TCR	Vβ TCR
101/-																
129/-		2 (Low)	2	2			1	Low	a	2 (High)	1	2	3, 8, 9, 11, 13, 17	–		
A/J	2	2 (Low)	2	2	–		1		a	1 (Low)	1	2	6, 8, 13, 23, 50	–	a	b
A2G			2	2												
AKR/J	Not 2	2 (Low)	2	1	+		1	Med	a	2 (High)	1	3	7, 8, 9, 17, 23, 30	–	a	b
AL/N	2									2 (High)		Not 1				
AU/SsJ														–		a
BALB/cAnN	2	1 (High)		2		Not 1				1 (Low)	1	2	6, 8, 9	–	a	b
BALB/cJ	2	1 (High)	2	2	+		1	Med	a	1 (Low)	1	2	6, 8, 9	–	a	b
BDP/J			2	1						2 (High)	1	Not 2				
BUB/BnJ				1												
BXSB/Mp				2									3, 6, 8, 9, 17			
CB-17	2	1 (High)	2	2	+				a	1 (Low)	1	2	6, 8, 9	–	a	b
C3H/Bi				2					a			1				
C3H/He	2	2 (Low)	2	2	+	1	1	Low	a	1 (Low)	1	2	1, 6, 8, 11, 14	–	a	b
C57BL/-	2	2 (Low)	2	2	+	1	2	High	b	2 (High)	2	2	8, 9, 17	+	b	b
C57BR/cd		2 (Low)	2	2					b	2 (High)	2	1	8, 9, 11, 17, 29	–	b	a
C57L/-		2 (Low)	2	2					b	2 (High)	2	1	8, 9, 11, 17, 29		b	a
C58/-	2	2 (Low)	2	2			2		a	2 (High)	2	1	3, 7, 17	?		

Table 2.5 Mouse leukocyte alloantigens chart (*continued*)

Mouse strains	CD22 (Lyb-8)	CD44 (Ly-24)	CD45 (Ly-5)	CD90 (Thy-1)	CD157 (BP-3)	CD178 (Fas Ligand)	CD244 (2B4)	ART2.2 Rt6-2	B2 microglobulin	Ly-6	Ly-9 (CD229)	Lyb-2 (CD72)	Mtv Proviruses	NK-1.1	Vα TCR	Vβ TCR
CBA/Ca			2	2								2	8, 9, 14		a	
CBA/J	2	1 (High)	2	2				Med	a	1 (Low)	1	1	6, 7, 8, 9, 14, 17	−	a	b
CBA/N	2	2 (Low)	2	2	+		1			1 (Low)	1	2				
CE/-	2	2 (Low)	2	2					a	1 (Low)	1	3	7, ?	+		
DA/HuSn			1	2												
DBA/1	Not 2	1 (High)	2	2		Not 1	1	Med	a	1 (Low)	1	1		−	d	b
DBA/2	Not 2	1 (High)	2	2	+	Not 1	1	Low	a	2 (High)	1	1	1, 6, 7, 8, 9. 11, 13, 14, 17		d	b
FVB/N				1										+		
GRS/J			2	2								3				
HRS/J												2	?			
I/-				2					a			1				
LP/J			2	2					a	1 (Low)	1		6, 8, 9, 11, 13, 17			
MA/				1					a	2 (High)		2	8, 9, 17, 29, 43	+		
MRL/Mp				2		1				2 (High)			8, 9, 11, 14, 17, 23	−		
NOD	2		1	2		1		Med			1	3	3, 17, 31, 42, 45	−	c	b
NZB/-	Not 2	2 (Low)	2	2	+	1	1	Med		1 (Low)	1	2	3, 7, 9, 14, 17, 27, 28	+	c	
NZW/-						1		−				2	3, 6, 8, 17, 31, 44	+	d	
P/-				1						2 (High)				+		
PL/J	Not 2		2	1						2 (High)	1	2	?			
RF/J	2	2 (Low)	2	1					a	2 (High)	1	3	7, ?			
RIII/-			1		2					1 (Low)	1	2				c

Table 2.5   Mouse leukocyte alloantigens chart (*continued*)

Mouse strains	CD22 (Lyb-8)	CD44 (Ly-24)	CD45 (Ly-5)	CD90 (Thy-1)	CD157 (BP-3)	CD178 (Fas Ligand)	CD244 (2B4)	ART2.2 Rt6-2)	B2 microglobulin	Ly-6	Ly-9 (CD229)	Lyb-2 (CD72)	Mtv Proviruses	NK-1.1	Vα TCR	Vβ TCR
SEC/-				2								2	?	+		
SJL/J	2	2 (Low)	1	2	+	1	1		a	2 (High)	1	3	8, 29, 51	+	c	a
SM/J				2								Not 2	6, 7, 8, 14, 17	−		
ST/bJ			2	2						1 (Low)		Not 2	7	+		
SWR/J	2	2 (Low)	2	2					a	2 (High)	1	1	7, 8, 14, 17		c	a

Table 2.6   Bacterial and viral vaccines

	Viral vaccines	Bacterial vaccines
Live attenuated	Vaccinia, measles, mumps, rubella, yellow fever, polio (OPV), adeno, varicella zoster, cytomegalo, hepatitis A, influenza, dengue, rota, respiratory syncytial, parainfluenza, Japanese encephalitis	Bacille–Calmette–Guerin (BCG), *S. typhi* (Ty21a), *V. cholerae*, *S. typhi* (Aro A)
Inactivated	Polio (IPV), influenza, rabies, Japanese encephalitis, hepatitis A	*V. cholerae*, *B. pertussis* (P), *S. typhi*, *V. cholerae* plus B subunit of CT[a], *M. leprae* with/without BCG

Table 2.7   Summary of diphtheria, tetanus, and pertussis vaccine

	Diphtheria, tetanus toxoids, and whole cell pertussis vaccine	Diphtheria, tetanus toxoids, and acellular pertussis vaccine	Diphtheria and tetanus toxoid (pediatric)	Tetanus and diphtheria toxoids (adult)	Diphtheria and tetanus toxoids and Hib conjugate, and whole cell pertussis vaccines
Synonyms	DTP, DTWP	DTP, DTaP	DT	Td	DTwP-Hib
Concentration (per 0.5 ml) Diphtheria Tetanus Pertussis Hib	6.5–12.5 Lfu 5–5.5 Lfu 4 u None	6.7–7.5 Lfu 5 Lfu either 46.8 μg or 300 Hau None	6.6–12.5 Lfu 5–7.6 Lfu None None	2 Lfu 2–5 Lfu None None	6.7 or 12.5 Lfu 5 Lfu 4 u 10 μg
Appropriate age range	2 months to <7 years	18 months to <7 years	2 months to <7 years	7 years to adult	Typically 2–15 months
Standard schedule	Five 0.5 ml doses: at 2, 4, 6, and 18 months and 4–6 years of age	For doses 4 and 5: at 18 months and at 4–6 years of age	Three 0.5 ml doses: at 2, 4, and 10–16 months of age	Three 0.5 ml doses: the second 4–8 weeks after the first and the third 6–12 months after the second	Four 0.5 ml doses: at 2, 4, 6, and 15 months of age

Table 2.7   Summary of diphtheria, tetanus, and pertussis vaccine (*continued*)

	Diphtheria, tetanus toxoids, and whole cell pertussis vaccine	Diphtheria, tetanus toxoids, and acellular pertussis vaccine	Diphtheria and tetanus toxoid (pediatric)	Tetanus and diphtheria toxoids (adult)	Diphtheria and tetanus toxoids and Hib conjugate, and whole cell pertussis vaccines
Routine additional doses	None	None	None	Every 10 years	None
Route	IM	IM	IM	IM, jet	IM

Table 2.8   Case definition for vaccine-preventable diseases

Disease	Clinical description	Probable case	Confirmed case
Diphtheria	An upper respiratory tract illness characterized by pharyngitis or laryngitis, low grade fever, with or without an adherent membrane of the tonsils, pharynx and/or nose, and/or toxic (cardiac or neurological) symptoms. Cutaneous diphtheria is not notifiable, but should be discussed with the Medical Officer of Health	A clinically compatible illness that is not laboratory confirmed	A clinically compatible illness that is laboratory confirmed
Haemophilus influenzae type b (Hib) invasive disease	Invasive disease due to Hib may cause septicemia, meningitis, epiglottitis, cellulitis, septic arthritis, pneumonia or osteomyelitis	A clinically compatible illness with positive laboratory test, or a confident diagnosis of epiglottitis by direct vision, laryngoscope or X-ray	A clinically compatible illness with isolation of Hib from a normally sterile site
Hepatitis B	An illness with variable symptoms including fever, malaise, anorexia, jaundice and/or elevated serum aminotransferase levels. The acute illness but not the carrier state is to be notified	A clinically compatible illness with a positive HBsAg test	A clinically compatible illness that is laboratory confirmed with a positive anti-HBc IgM test.
Measles	Cases must meet all the following criteria: • fever 38 °C or higher • generalized maculopapular rash lasting 3 or more days • cough or coryza or conjunctivitis or Koplik spots	A clinically compatible illness	A clinically compatible illness that is epidemiologically linked to a confirmed case, or is laboratory confirmed
Mumps	An illness with acute onset of fever and unilateral or bilateral tender, self limited swelling of the parotid or other salivary glands, lasting more than 2 days, and without other apparent cause	A clinically compatible illness	A case with laboratory confirmation or a clinically compatible illness that is epidemiologically linked to another case

Table 2.8 Case definition for vaccine-preventable diseases (*continued*)

Disease	Clinical description	Probable case	Confirmed case
Pertussis	A disease characterized by a cough lasting longer than 2 weeks, and one or more of the following: • paroxysms of cough • cough ending in vomiting or apnoea • inspiratory whoop	Cough lasting longer than 2 weeks and one or more of the following: • paroxysmal cough, cough ending in vomiting or apnoea • inspiratory whoop for which there is no other known cause.	A clinically compatible illness that is laboratory confirmed or that is epidemiologically linked to a confirmed case
Rubella	An illness with a generalized maculopapular rash and fever and one or more of the following: • arthralgia/arthritis • lymphadenopathy • conjunctivitis Rubella often presents atypically and is difficult to diagnose clinically with certainty. If accurate diagnosis is important it must be laboratory confirmed	A case that meets the clinical case definition	A clinically compatible illness that is laboratory confirmed or has a close epidemiological link to a laboratory confirmed case
Rubella (congenital)	A live or stillborn infant with clinically compatible defects (cataracts, congenital heart disease, hearing defects, microcephaly, mental retardation, purpura, hepatosplenomegaly)	A clinically compatible illness	A clinically compatible illness that is laboratory confirmed
Poliomyelitis	A disease with no other apparent cause, characterized by: • acute flaccid paralysis of one or more limbs with decreased or absent deep tendon reflexes in affected limbs • no sensory or cognitive loss • may affect bulbar muscles. Vaccine associated paralytic poliomyelitis (VAPP): a vaccine associated case is defined as one occurring, in a vaccine recipient 7–30 days after receiving oral polio vaccine, or occurring in a contact of a vaccinee 7–60 days after the vaccinee received oral polio vaccine	A clinically compatible illness.	A clinically compatible illness in which the neurological deficit persists 60 days after the onset of symptoms or the individual has died, with no other cause
Tetanus	Acute onset of hypertonia and/or painful muscular contractions, most commonly of the jaw and neck, which may proceed to generalized muscle spasms. The clinical presentation of tetanus may be subtle	Nil	A clinically compatible case

Table 2.9 Microbiological and serological tests used in the diagnosis of vaccine-preventable disease

Disease	Laboratory basis for diagnosis	Specimen	When to take specimens
Diphtheria	Isolation of toxigenic *Corynebacterium diphtheriae* from a clinical specimen	Swab from area of the lesion (eg, throat swab or skin in case of ulcer)	At presentation of illness: must state 'query diphtheria' to ensure appropriate laboratory testing
Haemophilus influenzae type b (Hib)	Isolation of Hib from a normally sterile site OR detection of a positive antigen test in CSF	CSF and/or blood culture or aspirate from normally sterile site	At presentation of illness
Hepatitis B (acute)	Serology (HBsAg positive and anti-HBc IgM positive) and abnormal LFTs	Blood	At presentation of illness, but may need a second specimen one week after presentation
Measles	Demonstration of measles-specific IgM antibody	Blood	(The preferred test) Single specimen taken 3–4 days after onset of rash if negative a repeat test may be required
	OR a significant rise in measles antibody titer (IgG), OR isolation of measles virus from a clinical specimen	Blood Urine; nasopharyngeal swab/ saliva swab for virus	One specimen taken at onset of illness and a second taken at least 14 days later At initial presentation of illness (note: culture of virus takes up to 35 days and viral transport medium is required). Serology is preferred
Mumps	A positive serologic test for mumps IgM antibody except following vaccine, OR	Blood	At initial presentation of illness
	a significant rise in mumps antibody level by any standard serological assay, except following vaccination, OR	Blood	One specimen taken at onset of illness and a second taken at least 14 days later
	isolation of mumps virus from a clinical specimen	Saliva or viral swab taken from mouth or throat	At presentation. Note: viral transport medium is required
Pertussis	Isolation of *Bordetella pertussis* from a pernasal swab*, OR PCR	Isolation: pernasal swab PCR: nasopharangeal swab; for PCR ensure correct swab is used, i.e., not with wooden handle and not cotton tipped	At initial presentation of clinically compatible illness
Poliomyelitis	Two fecal specimens collected at least 24 hours apart 0–14 days after the onset of paralysis are to be sent to ESR (Acute poliomyelitis titers may assist diagnosis, but viral isolation and identification is required to confirm a case of poliomyelitis)	Feces Blood	At initial presentation of illness and a second specimen collected at least 24 hours later At initial presentation and 14 days later

Table 2.9 Microbiological and serological tests used in the diagnosis of vaccine-preventable disease (*continued*)

Disease	Laboratory basis for diagnosis	Specimen	When to take specimens
Rubella	Demonstration of rubella specific IgM antibody, except following immunization, OR	Blood	Four days after onset of illness
	a four fold rise in rubella antibody titer between acute and convalescent sera, OR	Blood	One specimen at onset of illness and second specimen 14 days later
	isolation of rubella virus from a clinical specimen	Nasopharyngeal swab	Taken within 3 days of initial presentation of illness. (Note: rubella virus isolation rate is poor and takes 4 weeks. Viral transport medium is required. Serology is preferred test)
Rubella (congenital)	Isolation of rubella virus from a clinical specimen from the infant, OR	Throat swab	At birth. (Note: rubella virus isolation rate is poor and takes 4 weeks. Viral transport medium is required. Serology is preferred test)
	demonstration of rubella specific antibody (IgM) in the infant's serum, OR	Blood	Cord blood specimen
	persistence of rubella specific IgG antibody of titer higher than expected from passive transfer of maternal antibody, OR	Blood	One specimen at birth and second specimen 14–21 days later
	laboratory confirmed maternal rubella infection in the first trimester of pregnancy	Blood	Two maternal blood tests in first trimester of pregnancy (see rubella diagnosis)
Tetanus	None	None	

*When testing for pertussis, alternative serological tests may be available. Serology is not accepted as a confirmatory test for surveillance in the *Communicable Disease Control Manual 1998* (Ministry of Health [UK]). A case diagnosed from clinical findings and positive serology would be classified as 'probable' and not 'confirmed'. Blood should be taken at the initial clinical presentation and a second specimen taken at least 4 days later. A positive serological test for pertussis IgA and/or IgM or rising titers would be indicative of recent infection.

Vaccine	Ages usually given and other guidelines	If child falls behind	Contraindications
DTaP (Diphtheria, tetanus, acellular pertussis) *Give IM*	• Give at 2 mth, 4 mth, 6 mth, 15–18 mth, 4–6 yr of age • May give dose 1 as early as 6 wk of age • May give dose 4 as early as 12 mth of age if 6 mth have elapsed since dose 3 and the child is unlikely to return at age 15–18 mth • Do not give DTaP to children ≥7 yr of age (give Td) • May give with all other vaccines • It is preferable but not mandatory to use the same DTaP product for all doses	• Dose 2 and dose 3 may be given 4 wk after previous dose • Dose 4 may be given 6 mth after dose 3 • If dose 4 is given before 4th birthday, wait at least 6 mth for dose 5 (4–6 yr of age) • If dose 4 is given after 4th birthday, dose 5 is not needed • Do not restart series, no matter how long since previous dose	**Contraindication for DTaP only:** Previous encephalopathy within 7 days after DTP/DTaP **Precautions for DTaP:** The following are precautions, not contraindications. When these conditions are present, the individual child's disease risk should be carefully assessed. In situations when the benefit outweighs the risk (e.g., community pertussis outbreak), vaccination should be considered • T≥105°F (40.5°C) within 48 hr after previous dose • Continuous crying lasting ≥3 hr within 48 hr after previous dose • Previous convulsion within 3 days after immunization • Pale or limp episode or collapse within 48 hr after previous dose • Unstable progressive neurologic problem (defer until stable)
DT *Give IM*	• Give to children <7 yr of age if child had a serious reaction to 'P' in DTaP/DTP or if parents refuse the pertussis component • May give with all other vaccines		
Td *Give IM*	• Use Td, not TT, for persons ≥7 yrs of age for all indications • A booster dose is recommended for children 11–12 yr of age if 5 yr have elapsed since last dose. Then boost every 10 yr • May give with all other vaccines	• For those never vaccinated or with unknown vaccination history: give dose 1 now, give 2nd dose 4 wk later, give 3rd dose 6 mth after dose 2, then give booster every 10 yr • Do not restart series, no matter how long since prior dose	
MMR (Measles, mumps, rubella) *Give SC*	• Give dose 1 at 12–15 mth of age. Give dose 2 at 4–6 yr of age • Make sure that all children (and teens) over 4–6 yr of age have received both doses of MMR • If a dose was given before 12 mth of age, it doesn't count as the first dose, so give dose 1 at 12–15 mth of age with a minimum interval of 4 wk between these doses • May give with all other vaccines • If MMr and Var (and/or yellow fever vaccine) are not given on the same day, space them ≥28 days apart	• 2 doses of MMR are recommended for all children ≤18 yr of age • Dose should be given whenever it is noted that a child is behind. Exception: If MMR and Var (and/or yellow fever vaccine) are not given on the same day, space them ≥28 days apart • Dose 2 can be given at any time if at least 28 days have elapsed since dose 1 and both doses are administered after 1 yr of age • Do not restart the series, no matter how long since previous dose	• Pregnancy or possibility of pregnancy within 4 weeks (use contraception) • If blood, plasma, and/or immune globulin were given in past 11 mth, see ACIP statement *General Recommendations on Immunization* regarding time to wait before vaccinating • HIV is NOT a contraindication unless severely immunocompromised • Immunocompromised persons (e.g., because of cancer, leukemia, lymphoma) **Note:** For patients on high-dose immunosuppressive therapy, consult ACIP recommendations regarding delay time **Note:** MMR is not contraindicated if a PPD test was recently applied. If PPD and MMR were not given on same day, delay PPD for 4–6 wk after MMR

Do not give if patient (1) has had an anaphylactic reaction to a prior dose or to any vaccine component or (2) has a moderate or severe acute illness. (Minor illness is not a reason to postpone vaccination)

Table 2.10   Summary of rules for childhood immunization (*continued*)

Vaccine	Ages usually given and other guidelines	If child falls behind	Contraindications
Varicella (Var) (Chickenpox) *Give SC*	• Routinely give at 12–18 mth of age • Vaccinate all children ≥12 mth of age, including all adolescents who have not had chickenpox • May use as post-exposure prophylaxis if given within 3–5 days • May give with all other vaccines • If Var and MMR (and/or yellow fever vaccine) are not given on the same day, space them ≥28 days apart • Do not withhold vaccine from children or pregnant women	• Do not give to children <12 mth of age • Susceptible children <13 yr of age should receive 1 dose • Susceptible persons ≥13 yr of age should receive 2 doses 4–8 wk apart • Do not restart series, no matter how long since previous dose	• Pregnancy or possibility of pregnancy within 4 wk • If blood, plasma, and/or immune globulin (IG or VZIG) were given in past 11 mth, see ACIP statement *General Recommendations on Immunization* regarding time to wait before vaccinating • Persons immunocompromised due to high doses of systemic steroids, cancer, leukemia, lymphoma, or immunodeficiency. **Note:** For patients with humoral immunodeficiency, HIV infection, or leukemia, or for patients on high doses of systemic steroids, see ACIP recommendations • For children taking salicylates, see ACIP recommendations
Influenza *Give IM*	Vaccinate children ≥6 mth of age with risk factors and encourage vaccination of all children aged 6–23 mth when feasible. Consult the current year's ACIP statement *Prevention and Control of Influenza* for details		
Meningococcal *Give SC*	Vaccinate children ≥2 yr of age with risk factors. Discuss disease risk and vaccine availability with college students. Consult ACIP statement on meningococcal disease (30 June 2000) for details		

Table 2.10 Summary of rules for childhood immunization (*continued*)

Vaccine	Ages usually given and other guidelines	If child falls behind	Contraindications
			Do not give if patient (1) has had an anaphylactic reaction to a prior dose or to any vaccine component or (2) has a moderate or severe acute illness. (Minor illness is not a reason to postpone vaccination)
Polio (IPV) *Give SC or IM*	• Give at 2 mth, 6–18 mth, and 4–6 yr of age • May give dose 1 as early as 6 wk of age • Not routinely recommended for those ≥18 yr of age (except certain travelers) • May give with all other vaccines	• All doses should be separated by at least 4 wk • If dose 3 of an all-IPV or all-OPV series is given at ≥4 yr of age, dose 4 is not needed • Those who receive a combination of IPV and OPV doses must receive all 4 doses • Do not restart series, no matter how long since previous dose	
Hib *Give IM*	• HibTITER (HbOC) and ActHib or OmniHib (PRP-T): give at 2 mth, 4 mth, 6 mth, 12–15 mth (booster dose) • PedvaxHIB or Comvax (containing PRP-OMP); give at 2 mth, 4 mth, 12–15 mth • Dose 1 of Hib vaccine may be given as early as 6 wk of age but no earlier • The last dose (booster dose) is given no earlier than 12 mth of age and a minimum of 8 wk after the previous dose • May give with all other vaccines • Hib vaccines are interchangeable; however, if different brands of Hib conjugate vaccines are administered, a total of three doses are necessary to complete the primary series in infants • Any Hib vaccine may be used for the booster dose • Hib is not routinely given to children ≥5 yr of age	**Rules for all Hib vaccines:** • If dose 1 was given at 12–14 mth, give a booster dose in 8 wk • Give only 1 dose for unvaccinated children ≥15 mth and <5 yr of age • Do not restart series, no matter how long since previous dose **Rules for HibTITER, ActHib, and OmniHib:** • Dose 2 and dose 3 may be given 4 wk after previous dose • If dose 1 was given at 7–11 mth, only 3 doses are needed; dose 2 is given 4–8 wk after dose 1, then boost at 12–15 mth **Rules for PedvaxHib:** • Dose 2 may be given 4 wk after dose 1	
Hepatitis B *Give IM*	• Vaccinate all newborns prior to hospital discharge. Give dose 2 at 1–4 mth, and dose 3 at 6–18 mth. After the first dose, the series may be completed with single-antigen vaccine or up to 3 doses of Comvax, e.g., 2 mth, 4 mth, 12 mth of age. Dose 1 can be given as late as 2 mth of age if the mother is assured to be HBsAg negative, but this is not the preferred schedule • Vaccinate all children 0 through 18 yr of age • For older children/teens, schedules include: 0–, 1–, 6–mth; 0–, 2–, 4–mth; 0–, 1–, 4–mth • Children born (or whose parents were born) in countries of high HBV endemicity or who have other risk factors should be vaccinated asap • **If mother is HBsAg-positive:** give HBIG + dose 1 within 12 hr of birth, dose 2 at 1–2 mth, and dose 3 at 6 mth of age • **If mother's HBsAg status is unknown:** give dose 1 within 12 hr of birth, dose 2 at 1–2 mth, and dose 3 at 6 mth of age. If mother is later found to be HBsAg positive, give infant HBIG within 7 days of birth • **Note:** For premature infants, hepatitis B vaccination recommendation may be different. Consult the *2000 Red Book* (p.54) • May give with all other vaccines	• Do not restart series, no matter how long since previous dose • 3-dose series can be started at any age • Minimum spacing for children and teens: 4 wk between dose 1 and dose 2, and 8 wk between dose 2 and dose 3. Overall there must be ≥16 wk between dose 1 and dose 3 • The last dose in infant hepatitis B series should not be given earlier than 6 mth of age **Dosing of hepatitis B vaccines:** Vaccine brands are interchangeable for 3-dose schedules For Engerix-B, use 10 µg for 0 through 19 yr of age For Recombivax HB, use 5 µg for 0 through 19 yr of age **Alternative dosing schedule for unvaccinated adolescents aged 11 through 15 yr:** Give Recombivax HB two 10 µg doses (adult dosage) spaced 4–6 mth apart (Engerix-B is not licensed for a 2-dose schedule)	

Table 2.10 Summary of rules for childhood immunization (*continued*)

Vaccine	Ages usually given and other guidelines	If child falls behind	Contraindications
Hepatitis A *Give IM*	• Vaccinate children ≥2 yr old who live in areas with consistently elevated rates of hepatitis A, as well as children who have specific risk factors. (See ACIP statement and column 3 of this table for details) • Children who travel outside of the US (except to Western Europe, New Zealand, Australia, Canada, or Japan) • Dose 2 is given a minimum of 6 mth after dose 1 • Dose 1 may not be given earlier than 2 yr of age • May give with all other vaccines	• Do not restart series, no matter how long since previous dose • Hepatitis A vaccine brands are interchangeable • Consult your local/state public health authority for information regarding your city, county, or state hepatitis A rates. States with consistently elevated rates (average ≥10 cases per 100 000 population from 1987–1997) include the following: AL, AZ, AK, CA, CO, ID, MO, MT, NV, NM, OK, OR, SD, TX, UT, WA, and WY	
PCV *Give IM* **Pneumococcal**	• Give at 2 mth, 4 mth, 6 mth, and 12–15 mth of age • Dose 1 may be given as early as 6 wk of age For unvaccinated high-risk children [1]24–59 mth of age, give 2 doses. If PPV not previously given, administer ≥8 wk after final dose of PCV • For unvaccinated moderate-risk children [2]24–59 mth of age, consider giving 1 dose • May give 1 dose to unvaccinated healthy children 24–59 mth • PCV is not routinely given to children ≥5 yr of age • May give with all other vaccines	• Minimum interval for infants <12 mth of age is 4 wk, for ≥12 mth of age is 8 wk • For infants 7–11 mth of age: if unvaccinated, give dose 1 now, give dose 2 4–8 wk later, and boost at 12–15 mth. If infant has had 1 or 2 previous doses, give next dose now, and boost at 12–15 mth • For infants 12–23 mth: if not previously vaccinated or only one previous dose before 12 mth, give 2 doses ≥8 wk apart. If infant previously had 2 doses, give booster dose ≥8 wk after previous dose • Do not restart series, no matter how long since previous dose	
PPV *IM or SC*	Give PPV to high-risk children ≥2 yr of age as recommended in the ACIP statement *Prevention of Pneumococcal Disease* (4 April 1997)		

Notes:
1 **High-risk children:** Those with sickle cell disease; anatomic or functional asplenia; chronic cardiac, pulmonary, or renal disease; diabetes mellitus; CSF leak; HIV infection; or immunosuppression.
2 **Moderate-risk children:** Children aged 24–35 mth; children aged 24–59 mth who attend group daycare centers or are of Alaska Native, American Indian, or African American descent.
Adapted from ACIP, AAP, and AAFP by the Immunization Action Coalition, July 2002.

Table 2.11 Recommended childhood and adolescent immunization schedule – United States, 2003

Vaccine ▼ / Age ▶	Birth	1 mth	2 mth	4 mth	6 mth	12 mth	15 mth	18 mth	24 mth	4-6 yr	11-12 yr	13-18 yr
Hepatitis B[1]	HepB #1	only if mother HBsAg (·)								HepB series		
		HepB #2				HepB #3						
Diphtheria, tetanus, pertussis[2]			DTaP	DTaP	DTaP		DTaP			DTaP	Td	
Haemophilus influenzae Type b[3]			Hib	Hib	Hib	Hib						
Inactivated polio			IPV	IPV		IPV				IPV		
Measles, mumps, rubella[4]						MMR #1				MMR #2	MMR #2	
Varicella[5]						Varicella				Varicella		
Pneumococcal[6]			PCV	PCV	PCV	PCV				PCV	PPV	
Hepatitis A[7]										Hepatitis A series		
Influenza[8]						Influenza (yearly)						

Vaccines below this line are for selected populations

Key: range of recommended ages   catch-up vaccination   preadolescent assessment

This schedule indicates the recommended ages for routine administration of currently licensed childhood vaccines, as of 1 December 2002, for children up to age 18 years. Any dose not given at the recommended age should be given at any subsequent visit when indicated and feasible. ▨▨ Indicates age groups that warrant special effort to administer those vaccines not previously given. Additional vaccines may be licensed and recommended during the year. Licensed combination vaccines may be used whenever any components of the combination are indicated and the vaccine's other components are not contraindicated. Providers should consult the manufacturers' package inserts for detailed recommendations.

Notes:
1. **Hepatitis B vaccine (HepB)**. All infants should receive the first dose of hepatitis B vaccine soon after birth and before hospital discharge; the first dose may also be given by age 2 months if the infant's mother is HBsAg-negative. Only monovalent HepB can be used for the birth dose. Monovalent or combination vaccine containing HepB may be used to complete the series. Four doses of vaccine may be administered when a birth dose is given. The second dose should be given at least 4 weeks after the first dose, except for combination vaccines which cannot be administered before age 6 weeks. The third dose should be given at least 16 weeks after the first dose and at least 8 weeks after the second dose. The last dose in the vaccination series (third or fourth dose) should not be administered before age 6 months.

     Infants born to HBsAg-positive mothers should receive HepB and 0.5 ml Heptatitis B Immune Globulin (HBIG) within 12 hours of birth at separate sites. The second dose is recommended at age 1–2 months. The last dose in the vaccination series should not be administered before age 6 months. These infants should be tested for HBsAg and anti-HBs at 9–15 months of age.

     Infants born to mothers whose HBsAg status is unknown should receive the first dose of the HepB series within 12 hours of birth. Maternal blood should be drawn as soon as possible to determine the mother's HBsAg status; if the HBsAg test is positive, the infant should receive HBIG as soon as possible (no later than age 1 week). The second dose is recommended at age 1–2 months. The last dose in the vaccination series should not be administered before age 6 months.
2. **Diphtheria and tetanus toxoids and acellular pertussis vaccine (DTaP)**. The fourth dose of DTaP may be administered as early as age 12 months, provided 6 months have elapsed since the third dose and the child is unlikely to return at age 15–18 months. Tetanus and diphtheria toxoids (Td) is recommended at age 11–12 years if at least 5 years have elapsed since the last dose of tetanus and diphtheria toxoid-containing vaccine. Subsequent routine Td boosters are recommended every 10 years.
3. *Haemophilus influenzae* **type b (Hib) conjugate vaccine**. Three Hib conjugate vaccines are licensed for infant use. If PRP-OMP (PedvaxHiB® or ComVax® [Merck]) is administered at ages 2 and 4 months, a dose at age 6 months is not required. DTaP/Hib combination products should not be used for primary immunization in infants at ages 2, 4 or 6 months, but can be used as boosters following any Hib vaccine.
4. **Measles, mumps, and rubella vaccine (MMR)**. The second dose of MMR is recommended routinely at age 4–6 years but may be administered during any visit, provided at least 4 weeks have elapsed since the first dose and that both doses are administered

Table 2.11 Recommended childhood and adolescent immunization schedule – United States, 2003 Notes: (*continued*)

beginning at or after age 12 months. Those who have not previously received the second dose should complete the schedule by the 11–12-year-old visit.

5. **Varicella vaccine**. Varicella vaccine is recommended at any visit at or after age 12 months for susceptible children, i.e. those who lack a reliable history of chickenpox. Susceptible persons aged ≥13 years should receive two doses, given at least 4 weeks apart.

6. **Pneumococcal vaccine**. The heptavalent pneumococcal conjugate vaccine (PCV) is recommended for all children age 2–23 months. It is also recommended for certain children age 24–59 months. Pneumococcal polysaccharide vaccine (PPV) is recommended in addition to PCV for certain high-risk groups. See *MMWR* 2000;49(RR-9);1-38.

7. **Hepatitis A vaccine**. Hepatitis A vaccine is recommended for children and adolescents in selected states and regions, and for certain high-risk groups; consult your local public health authority. Children and adolescents in these states, regions, and high risk groups who have not been immunized against hepatitis A can begin the hepatitis A vaccination series during any visit. The two doses in the series should be administered at least 6 months apart. See *MMWR* 1999;48(RR-12);1-37.

8. **Influenza vaccine**. Influenza vaccine is recommended annually for children age ≥6 months with certain risk factors (including but not limited to asthma, cardiac disease, sickle cell disease, HIV, diabetes, and household members of persons in groups at high risk; see *MMWR* 2002;51(RR-3);1-31), and can be administered to all others wishing to obtain immunity. In addition, healthy children age 6–23 months are encouraged to receive influenza vaccine if feasible because children in this age group are at substantially increased risk for influenza-related hospitalizations. Children aged ≥12 years should receive vaccine in a dosage appropriate for their age (0.25 ml if age 6–35 months or 0.5 ml if aged ≥3 years). Children aged ≤8 years who are receiving influenza vaccine for the first time should receive two doses separated by at least 4 weeks.

Table 2.12 Catch-up schedule for children age 4 months to 6 years

Dose 1 (Minimum age)	Minimum interval between doses			Dose 4 to Dose 5
	Dose 1 to Dose 2	Dose 2 to Dose 3	Dose 3 to Dose 4	
DTaP (6 wk)	4 weeks	4 weeks	6 months	6 months[1]
IPV (6 wk)	4 weeks	4 weeks	4 weeks[2]	
HepB[3] (birth)	4 weeks	8 weeks (and 16 weeks after first dose)		
MMR (12 mths)	4 weeks[4]			
Varicella (12 mth)				
Hib[5] (6 wk)	4 weeks: if first dose given at age <12 mth 8 weeks (as final dose): if first dose given at 12–14 mth No further doses needed: if first dose given at age ≥15 mth	4 weeks[6]: if current age <12 mth 8 weeks (as final dose)[6]: if current age ≥12 mth and second dose given at age <15 mth No further doses needed: if previous dose given at age ≥15 mth	8 weeks (as final dose): this dose only necessary for children age 12 mth–5 yr who received 3 doses before age 12 mth	
PCV[7]: (6 wk)	4 weeks: if first dose given at age <12 mth and current age <24 mth 8 weeks (as final dose): if first dose given at age ≥12 mth or current age 24–59 mth No further doses needed: for healthy children if first dose given at age >24 mth	4 weeks: if current age <12 mth 8 weeks (as final dose): if current age ≥12 mth No further doses needed: for healthy children if previous dose given at age ≥24 mth	8 weeks (as final dose): this dose only necessary for children age 12 mth–5 yr who received 3 doses before age 12 mth	

Notes:

For children and adolescents who start late or who are >1 mth behind

1 **DTaP:** The fifth dose is not necessary if the fourth dose was given after the 4th birthday.

2 **IPV:** For children who received an all-IPV or all-OPV series, a fourth dose is not necessary if third dose was given at age ≥4 years. If both OPV and IPV were given as part of a series, a total of four doses should be given, regardless of the child's current age.

3 **HepB:** All children and adolescents who have not been immunized against hepatitis B should begin the hepatitis B vaccination series during any visit. Providers should make special efforts to immunize children who were born in, or whose parents were born in, areas of the world where hepatitis B virus infection is moderately or highly endemic.

4 **MMR:** The second dose of MMR is recommended routinely at age 4–6 years, but may be given earlier if desired.

5 **Hib:** Vaccine is not generally recommended for children age ≥5 years.

6 **Hib:** If current age <12 months and the first 2 doses were PRP-OMP (PedvaxHIB or ComVax), the third (and final) dose should be given at age 12–15 months and at least 8 weeks after the second dose.

7 **PCV:** Vaccine is not generally recommended for children age ≥5 years.

Table 2.13 Catch-up schedule for children age 7 to 18 years

Minimum interval between doses		
Dose 1 to dose 2	**Dose 2 to dose 3**	**Dose 3 to booster dose**
Td: 4 weeks	**Td:** 6 months	**Td[1]: 6 months:** If first dose given at age <12 mth and current age <11 yr **5 years:** if first dose given at age ≥12 mth and third dose given at age <7 yr and current age ≥11 yr **10 years:** if third dose given at age ≥7 yr
IPV[2]: 4 weeks	**IPV:** 4 weeks	**IPV[2]**
HepB: 4 weeks	**HepB:** 8 weeks (and 16 weeks after first dose)	
MMR: 4 weeks		
Varicella[3]: 4 weeks		

Notes:
For children and adolescents who start late or who are >1 month behind
1 **Td:** For children age 7–10 years, the interval between the third and booster dose is determined by the age when the first dose was given. For adolescents age 11–18 years, the interval is determined by the age when the third dose was given.
2 **IPV:** Vaccine is not generally recommended for persons age ≥18 years.
3 **Varicella:** Give 2-dose series to all susceptible adolescents age ≥13 years.

Table 2.14 Summary of recommendations for adult immunization

Vaccine name and route	For whom it is recommended	Schedule for routine and 'catch-up' administration	Contraindication (mild illness is not a contraindication)
Influenza *Give IM*	• Adults who are 50 yr of age or older • People 6 mth to 50 yr of age with medical problems such as heart disease, lung disease, diabetes, renal dysfunction, hemoglobinopathies, immunosuppression, and/or people living in chronic care facilities • People ≥6 mth of age working or living with at-risk people • Pregnant women who have underlying medical conditions should be vaccinated before influenza season, regardless of the stage of pregnancy • Healthy pregnant women who will be in their 2nd or 3rd trimesters during influenza season • All health care workers and those who provide key community services • Travelers who go to areas where influenza activity exists or who may be among people from areas of the world where there is current influenza activity (e.g., on organized tours) • Anyone who wishes to reduce the likelihood of becoming ill with influenza	• Given every year • October to November is the *optimal* time to receive an annual flu shot to maximize protection • Influenza vaccine may be given at any time during the influenza season (typically December to March) or at other times when the risk of influenza exists • May give with all other vaccines	• Previous anaphylactic reaction to this vaccine, to any of its components, or to eggs • Moderate or severe acute illness **Note:** Pregnancy and breast feeding are not contraindications to the use of this vaccine
Pneumococcal polysaccharide (PPV23) *Give IM or SC*	• Adults who are 65 yr of age or older • People 2 to 64 yr of age who have chronic illness or other risk factors, including chronic cardiac or pulmonary diseases, chronic liver disease, alcoholism, diabetes mellitus, CSF leaks, as well as people living in special environments or social settings (including Alaska Natives and certain American Indian populations). Those at highest risk of fatal pneumococcal infection are people with anatomic asplenia, functional asplenia, or sickle cell disease; immunocompromised persons including those with HIV infection, leukemia, lymphoma, Hodgkin's disease, multiple myeloma, generalized malignancy, chronic renal failure, or nephrotic syndrome; persons receiving immunosuppressive chemotherapy (including corticosteroids); and those who received an organ or bone marrow transplant. Pregnant women with high-risk conditions should be vaccinated if not done previously	• Routinely given as one-time dose; administer if previous vaccination history is unknown • One-time revaccination is recommended 5 yr later for people at highest risk of fatal pneumococcal infection or rapid antibody loss (e.g. renal disease) and for people ≥65 yr of age if the first dose was given prior to age 65 and ≥5 yr have elapsed since previous dose • May give with other vaccines	• Previous anaphylactic reaction to this vaccine or to any of its components • Moderate or severe acute illness **Note:** Pregnancy and breastfeeding are not contraindications to the use of this vaccine

Table 2.14 Summary of recommendations for adult immunization (*continued*)

Vaccine name and route	For whom it is recommended	Schedule for routine and 'catch-up' administration	Contraindication (mild illness is not a contraindication)
Hepatitis B (Hep-B) *Give IM* Brands may be used interchangeably	• All adolescents • High-risk adults, including household contacts and sex partners of HBsAg-positive persons; users of illicit injectable drugs; heterosexuals with more than one sex partner in 6 months; men who have sex with men; people with recently diagnosed STDs; patients receiving hemodialysis and patients with renal disease that may result in dialysis; recipients of certain blood products; health care workers and public safety workers who are exposed to blood; clients and staff of institutions for the developmentally disabled; inmates of long-term correctional facilities; and certain international travelers **Note:** Prior serologic testing may be recommended depending on the specific level of risk and/or likelihood of previous exposure. **Note:** In 1997, the NIH Consensus Development Conference, a panel of national experts, recommended that hepatitis B vaccination be given to all anti-HCV positive persons **Ed. note:** Provide serologic screening for immigrants from endemic areas. When HBsAg-positive persons are identified, offer appropriate disease management. In addition, screen their sex partners and household members and, if found susceptible, vaccinate	• Three doses are needed on a 0, 1, 6 mth schedule • Alternative timing options for vaccination include 0, 2, 4 mth and 0, 1, 4 mth. • There must be 4 wk between doses 1 and 2, and 8 wk between doses 2 and 3. Overall there must be at least 16 wk between doses 1 and 3. • **Schedule for those who have fallen behind:** If the series is delayed between doses, DO NOT start the series over. Continue from where you left off. • May give with all other vaccines.	• Previous anaphylactic reaction to this vaccine or to any of its components • Moderate or severe acute illness **Note:** Pregnancy and breastfeeding are not contraindications to the use of this vaccine
Hepatitis A (Hep-A) *Give IM* Brands may be used interchangeably	• People who travel outside of the US (except for Western Europe, New Zealand, Australia, Canada, and Japan) • People with chronic liver disease, including people with hepatitis C; people with hepatitis B who have chronic liver disease; illicit drug users; men who have sex with men; people with clotting-factor disorders; people who work with hepatitis A virus in experimental lab settings (not routine medical laboratories); and food handlers when health authorities or private employers determine vaccination to be cost effective **Note:** Prevaccination testing is likely to be cost effective for persons >40 yr of age as well as for younger persons in certain groups with a high prevalence of hepatitis A virus infection	For Twinrix™ (hepatitis A and B combination vaccine [GSK]) three doses are needed on a 0, 1, 6 mth schedule • Two doses are needed • The minimum interval between dose 1 and dose 2 is 6 mth • If dose 2 is delayed, do not repeat dose 1. Just give dose 2 • May give with all other vaccines	• Previous anaphylactic reaction to this vaccine or to any of its components • Moderate or severe acute illness • Safety during pregnancy has not been determined, so benefits must be weighted against potential risk **Note:** Breastfeeding is not a contraindication to the use of this vaccine

Table 2.14   Summary of recommendations for adult immunization (*continued*)

Vaccine name and route	For whom it is recommended	Schedule for routine and 'catch-up' administration	Contraindication (mild illness is not a contraindication)
Td (Tetanus, diphtheria) *Give IM*	• All adolescents and adults • After the primary series has been completed, a booster dose is recommended every 10 yr. Make sure your patients have received a primary series of 3 doses • A booster dose as early as 5 yr later may be needed for the purpose of wound management, so consult ACIP recommendations	• Give booster dose every 10 yr after the primary series has been completed • For those who are unvaccinated or behind, complete the primary series (spaced at 0, 1–2 mth, 6–12 mth intervals). Don't restart the series, no matter how long since the previous dose • May give with all other vaccines	• Previous anaphylactic or neurologic reaction to this vaccine or to any of its components • Moderate or severe acute illness **Note:** Pregnancy and breastfeeding are not contraindications to the use of this vaccine
MMR (Measles, mumps, rubella) *Give SC*	• Adults born in 1957 or later who are ≥18 yr of age (including those born outside the US) should receive at least one dose of MMR if there is no serologic proof of immunity or documentation of a dose given on or after the first birthday • Adults in high-risk groups, such as health care workers, students entering colleges and other posthigh school educational institutions, and international travelers, should receive a total of two doses • Adults born before 1957 are usually considered immune but proof of immunity may be desirable for health care workers • All women of childbearing age (i.e., adolescent girls and premenopausal adult women) who do not have acceptable evidence of rubella immunity or vaccination • Special attention should be given to immunizing women born outside the United States in 1957 or later	• One or two doses are needed • If dose 2 is recommended, give it no sooner than 4 wk after dose 1 • May give with all other vaccines • If varicella vaccine and MMR are both needed and are not administered on the same day, space them at least 4 wk apart • If a pregnant woman is found to be rubella-susceptible, administer MMR postpartum	• Previous anaphylactic reaction to this vaccine, or to any of its components • Pregnancy or possibility of pregnancy within 4 weeks (use contraception) • Persons immunocompromised due to cancer, leukemia, lymphoma, immunosuppressive drug therapy, including high-dose steroids or radiation therapy. **Note:** HIV positivity is NOT a contraindication to MMR except for those who are severely immunocompromised • If blood, plasma, and/or immune globulin were given in past 11 mth, see ACIP statement *General Recommendations on Immunization* regarding time to wait before vaccinating • Moderate or severe acute illness **Note:** Breastfeeding is not a contraindication to the use of this vaccine **Note:** MMR is not contraindicated if a PPD test was recently applied. If PPD and MMR not given on same day, delay PPD for 4–6 wk after MMR

Table 2.14   Summary of recommendations for adult immunization (*continued*)

Vaccine name and route	For whom it is recommended	Schedule for routine and 'catch-up' administration	Contraindication (mild illness is not a contraindication)
Varicella (Var) (Chickenpox) *Give SC*	All susceptible adults and adolescents should be vaccinated. It is especially important to ensure vaccination of the following groups: susceptible persons who have close contact with persons at high risk for serious complications (e.g., health care workers and family contacts of immunocompromised persons) and susceptible persons who are at high risk of exposure (e.g, teachers of young children, day care employees, residents and staff in institutional settings such as colleges and correctional institutions, military personnel, adolescents and adults living with children, non-pregnant women of childbearing age, and international travelers who do not have evidence of immunity) **Note:** People with reliable histories of chickenpox (such as self or parental report of disease) can be assumed to be immune. For adults who have no reliable history, serologic testing may be cost-effective since most adults with a negative or uncertain history of varicella are immune	• Two doses are needed • Dose 2 is given 4–8 wk after dose 1 • May give with all other vaccines • If varicella vaccine and MMR are both needed and are not administered on the same day, space them at least 4 wk apart • If the second dose is delayed, do not repeat dose 1. Just give dose 2	• Previous anaphylactic reaction to this vaccine or to any of its components • Pregnancy or possibility of pregnancy within 4 weeks (use contraception) • Immunocompromised persons due to malignancies and primary or acquired cellular immunodeficiency including HIV/AIDS. (See *MMWR* 1999, Vol. 28, No. RR-6). **Note:** For those on high-dose immunosuppressive therapy, consult ACIP recommendations regarding delay time • If blood, plasma, and/or immune globulin (IG or VZIG) were given in past 11 mth, see ACIP statement *General Recommendations on Immunization* regarding time to wait before vaccinating • Moderate or severe acute illness **Note:** Breastfeeding is not a contraindication to the use of this vaccine **Note:** Manufacturer recommends that salicylates be avoided for 6 wk after receiving varicella vaccine because of a theoretical risk of Rey's syndrome
Polio (IPV) *Give IM or SC*	Not routinely recommended for persons 18 yr of age and older **Note:** Adults living in the US who never received or completed a primary series of polio vaccine need not be vaccinated unless they intend to travel to areas where exposure to wild-type virus is likely. Previously vaccinated adults can receive one booster dose if traveling to polio endemic areas	• Refer to ACIP recommendations regarding unique situations, schedules, and dosing information • May give with all other vaccines	• Previous anaphylactic or neurologic reaction to this vaccine or to any of its components • Moderate or severe acute illness **Note:** Pregnancy and breastfeeding are not contraindications to the use of this vaccine
Meningococcal *Give SC*	Vaccinate people with risk factors. Discuss disease risk and vaccine availability with college students. Consult ACIP statement on meningococcal disease (30 June 2000)		

Adapted from the ACIP recommendations by the Immunizations by the Immunization Action Coalition, June 2002

Table 2.15   Recommended adult immunization schedule, United States, 2002–2003

Age group ▶ Vaccine ▼	19–49 yr	50–64 yr	65 yr and older
Tetanus, diphtheria (Td)*	1 dose booster every 10 years [1]		
Influenza	1 dose annually for persons with medical or occupational indications, or household contacts of persons with indications [2]	1 annual dose	
Pneumococcal (polysaccharide)	1 dose for persons with medical or other indications. (1 dose revaccination for immunosuppressive conditions) [3,4]		1 dose for unvaccinated persons [3] 1 dose revaccination [4]
Hepatitis B*	3 doses (0, 1–2, 4–6 months) for persons with medical, behavioral, occupational, or other indications [5]		
Hepatitis A*	2 doses (0, 6–12 months) for persons with medical, behavioral, occupational, or other indications [6]		
Measles, mumps, rubella (MMR)*	1 dose if measles, mumps, or rubella vaccination history is unreliable; 2 doses for persons with occupational or other indications [7]		
Varicella*	2 doses (0, 4–8 weeks) for persons who are susceptible [8]		
Meningococcal (polysaccharide)	1 dose for persons with medical or other indications [9]		

Key: ▢ For all persons in this group   ▨ Catch-up on childhood vaccinations   ▧ For persons with medical/exposure indications

Notes:

1   **Tetanus and diphtheria (Td):** A primary series for adults is 3 doses: the first 2 doses given at least 4 weeks apart and the third dose 6–12 months after the second. Administer 1 dose if the person had received the primary series and the last vaccination was 10 years ago or longer. *MMWR* 1991;40 (RR-10): 1-21. The ACP Task Force on Adult Immunization supports a second option: a single Td booster at age 50 years for persons who have completed the full pediatric series, including the teenage/young adult booster. *Guide for Adult Immunization*, 3rd edn ACP 1994: 20.

2   **Influenza vaccination:** Medical indications: chronic disorders of the cardiovascular or pulmonary systems including asthma; chronic metabolic diseases including diabetes mellitus, renal dysfunction, hemoglobinopathies, immunosuppression (including immunosuppression caused by medications or by human immunodeficiency virus [HIV]), requiring regular medical follow-up or hospitalization during the preceding year; women who will be in the 2nd or 3rd trimester of pregnancy during the influenza season. Occupational indications: healthcare workers. Other indications: residents of nursing homes and other long-term care facilities; persons likely to transmit influenza to persons at high-risk (in-home caregivers to persons with medical indications, household contacts and out-of-home caregivers of children birth to 23 months of age, or children with asthma or other indicator conditions for influenza vaccination, household members and caregivers of elderly and adults with high-risk conditions); and anyone who wishes to be vaccinated. *MMWR* 2002;51 (RR-3): 1-31.

Table 2.15   Recommended adult immunization schedule, United States, 2002–2003 (*continued*)

3. **Pneumococcal polysaccharide vaccination:** Medical indications: chronic disorders of the pulmonary system (excluding asthma), cardiovascular diseases, diabetes mellitus, chronic liver diseases including liver disease as a result of alcohol abuse (e.g., cirrhosis), chronic renal failure or nephrotic syndrome, functional or anatomic asplenia (e.g., sickle cell disease or splenectomy), immunosuppressive conditions (e.g., congenital immunodeficiency, HIV infection, leukemia, lymphoma, multiple myeloma, Hodgkin's disease, generalized malignancy, organ or bone marrow transplantation), chemotherapy with alkylating agents, anti-metabolites, or long-term systemic corticosteroids. Geographic/other indications: Alaskan Natives and certain American Indian populations. Other indications: residents of nursing homes and other long-term care facilities. *MMWR* 1997; 47 (RR-8): 1–24.

4. **Revaccination with pneumococcal polysaccharide vaccine:** One time revaccination after 5 years for persons with chronic renal failure or nephrotic syndrome, functional or anatomic asplenia (e.g., sickle cell disease or splenectomy), immunosuppressive conditions (e.g., congenital immunodeficiency, HIV infection, leukemia, lymphoma, multiple myeloma, Hodgkin's disease, generalized malignancy, organ or bone marrow transplantation), chemotherapy with alkylating agents, anti-metabolites, or long-term systemic corticosteroids. For persons 65 and older, one-time revaccination if they were vaccinated 5 or more years previously and were aged less than 65 years at the time of primary vaccination. *MMWR* 1997; 47 (RR-8): 1–24.

5   **Hepatitis B vaccination:** Medical indications: hemodialysis patients, patients who receive clotting-factor concentrates. Occupational indications: healthcare workers and public-safety workers who have exposure to blood in the workplace, persons in training in schools of medicine, dentistry, nursing, laboratory technology, and other allied health professions. Behavioral indications: injecting drug users, persons with more than one sex parner in the previous 6 months, persons with a recently acquired sexually transmitted disease (STD), all clients in STD clinics, men who have sex with men. Other indications: household contacts and sex partners of persons with chronic HBV infection, clients and staff of institutions for the developmentally disabled, international travelers who will be in countries with high or intermediate prevalence of chronic HBV infection for more than 6 months; inmates of correctional facilities. *MMWR* 1991; 40 (RR-13): 1-25. (www.cdc.gov/travel/diseases/hbv.htm)

6   **Hepatitis A vaccination:** For the combined HepA–HepB vaccine use 3 doses at 0, 1, 6 months). Medical indications: persons with clotting-factor disorders or chronic liver disease. Behavioral indications: men who have sex with men, users of injecting and noninjecting illegal drugs. Occupational indications: persons working with HAV-infected primates or with HAV in a research laboratory setting. Other indications: persons traveling to or working in countries that have high or intermediate endemicity of hepatitis A. *MMWR* 1999; 48 (RR-12): 1–37. (www.cdc.gov/travel/diseases/hav.htm)

7   **Measles, Mumps, Rubella vaccination (MMR):** Measles component: Adults born before 1957 may be considered immune to measles. Adults born in or after 1957 should receive at least one dose of MMR unless they have a medical contraindication, documentation of at least one dose or other acceptable evidence of immunity. A second dose of MMR is recommended for adults who:

    *   are recently exposed to measles or in an outbreak setting
    *   were previously vaccinated with killed measles vaccine
    *   were vaccinated with an unknown vaccine between 1963 and 1967
    *   are students in post-secondary educational institutions
    *   work in health care facilities
    *   plan to travel internationally

    Mumps component: 1 dose of MMR should be adequate for protection. **Rubella component:** Give 1 dose of MMR to women whose rubella vaccination history is unreliable and counsel women to avoid becoming pregnant for 4 weeks after vaccination. For women of child-bearing age, regardless of birth year, routinely determine rubella immunity and counsel women regarding congenital rubella syndrome. Do not vaccinate pregnant women or those planning to become pregnant in the next 4 weeks. If pregnant and susceptible, vaccinate as early in postpartum period as possible. *MMWR* 1998; 47 (RR-8): 1–57.

8   **Varicella vaccination:** Recommended for all persons who do not have reliable clinical history of varicella infection, or serological evidence of varicella zoster virus (VZV) infection; healthcare workers and family contacts of immunocompromised persons, those who live or work in environments where transmission is likely (e.g., teachers of young children, day care employees, and residents and staff members in institutional settings), persons who live or work in environments where VZV transmission can occur (e.g., college students, inmates and staff members of correctional institutions, and military personnel), adolescents and adults living in households with children, women who are not pregnant but who may become pregnant in the future, international travelers who are not immune to infection. **Note:** Greater than 90% of US born adults are immune to VZV. Do not vaccinate pregnant women or those planning to become pregnant in the next 4 weeks. If pregnant and susceptible, vaccinate as early in postpartum period as possible. *MMWR* 1996; 45 (RR-11): 1–36, *MMWR* 1999; 48 (RR-6): 1–5.

9   **Meningococcal vaccine (quadrivalent polysaccharide for serogroups A, C, Y, and W-135):** Consider vaccination for persons with medical indications: adults with terminal complement component deficiencies, with anatomic or functional asplenia. Other indications: travelers to countries in which disease is hyperendemic or epidemic (meningitis belt of sub-Saharan Africa, Mecca, Saudi Arabia for Hajj). Revaccination at 3–5 years may be indicated for persons at high risk for infection (e.g., persons residing in areas in which disease is epidemic). Counsel college freshmen, especially those who live in dormitories, regarding meningococcal disease and the vaccine so that they can make an educated decision about receiving the vaccination. *MMWR* 2000; 49 (RR-7): 1–20. **Note:** The AAFP recommends that colleges should take the lead on providing education on meningococcal infection and vaccination and offer it to those who are interested. Physicians need not initiate discussion of the meningococcal quadravalent polysaccharide vaccine as part of routine medical care.

Table 2.16 Recommended immunizations for adults with medical conditions, United States, 2002–2003

Medical conditions ▼ / Vaccine ▶	Tetanus-diphtheria (Td)*	Influenza	Pneumococcal (polysacch-aride)	Hepatitis B*	Hepatitis A	Measles, mumps, rubella (MMR)*	Varicella*
Pregnancy		A					
Diabetes, heart disease, chronic pulmonary disease, chronic liver disease, including chronic alcoholism		B	C		D		
Congenital immunodeficiency, leukemia, lymphoma, generalized malignancy, therapy with alkylating agents, antimetabolites, radiation or large amounts of corticosteroids			E				F
Renal failure/end stage renal disease, recipients of hemodialysis or clotting factor concentrates			E	G			
Asplenia including elective splenectomy and terminal complement component deficiencies			E, H, I				
HIV infection			E, J			K	

Key:
- For all persons in this group
- Catch-up on childhood vaccinations
- For persons with medical/exposure indications
- Contraindicated

A   If pregnancy is at second or third trimester during influenza season.

B   Although chronic liver disease and alcoholism are not indicator conditions for influenza vaccination, give 1 dose annually if the patient is ≥50 years, has other indications for influenza vaccine, or if the patient requests vaccination.

C   Asthma is an indicator condition for influenza but not for pneumococcal vaccination.

D   For all persons with chronic liver disease.

E   Revaccinate once after 5 years or more have elapsed since initial vaccination.

F   Persons with impaired humoral but not cellular immunity may be vaccinated. *MMWR* 1999; 48 (RR-06): 1–5.

G   Hemodialysis patients: Use special formulation of vaccine (40 µg/ml) or two 1.0 ml 20 µg doses given at one site. Vaccinate early in the course of renal disease. Assess antibody titers to help B surface antigen (anti-HBs) levels annually. Administer additional doses if anti-HBs levels decline to <10 millinternational units (mIU)/ml.

H   Also administer meningococcal vaccine.

I   Elective splenectomy: vaccinate at least 2 weeks before surgery.

J   Vaccinate as close to diagnosis as possible when CD4 cell counts are highest.

K   Withhold MMR or other measles containing vaccines from HIV-infected persons with evidence of severe immunosuppression. *MMWR* 1996; 45: 603–606. *MMWR* 1992; 41 (RR-17): 1–19.

Table 2.17   Vaccines and related products distributed in the United States, 2003

Vaccine/biologic	Brand name	Manufacturer	Type	How supplied
Diphtheria, tetanus, acellular pertussis	Infanrix	GlaxoSmithKline	Inactivated	Single-dose vial or syringe
Diphtheria, tetanus, acellular pertussis	Tripedia	Aventis Pasteur	Inactivated	Single-dose vial
Diphtheria, tetanus, acellular pertussis	Daptacel	Aventis Pasteur	Inactivated	Single-dose vial
Diphtheria, tetanus, acellular pertussis + Hib	TriHIBit	Aventis Pasteur	Inactivated	Single-dose vial
Diphtheria, tetanus, acellular pertussis + Hep B + IPV	Pediarix	GlaxoSmithKline	Inactivated	Single-dose vial or syringe
Diphtheria, tetanus (DT; pediatric <7 yr)	Generic	Aventis Pasteur	Inactivated	10-dose vial
Tetanus, diphtheria, adsorbed (Td; ≥7 yr)	Generic	Aventis Pasteur	Inactivated	Single-dose syringe and 10-dose vial
Tetanus, diphtheria, adsorbed (Td; ≥7 yr)	Generic	U of Mass Labs	Inactivated	15-dose vial
Tetanus toxoid (TT; ≥7 yr), adsorbed	Generic	Aventis Pasteur	Inactivated	10-dose vial
Tetanus toxoid (TT; adult booster use only)	Generic	Aventis Pasteur	Inactivated	15-dose vial
Tetanus immune globulin (TIG)	BayTet	Bayer	Human immunoglobulin	Single-dose syringe
Measles, mumps, rubella (MMR)	M-M-R II	Merck	Live attenuated	Single-dose vial
Rubella	Meruvax II	Merck	Live attenuated	Single-dose vial
Varicella	Varivax	Merck	Live attenuated	Single-dose vial
Hemophilus b conjugate (PRP-T)	ActHIB	Aventis Pasteur	Inactivated	Single-dose vial
Hemophilus b conjugate (HbOC)	HibTITER	Wyeth	Inactivated	Single-dose vial
Hemophilus b conjugate (PRP-OMP)	PedvaxHIB	Merck	Inactivated	Single-dose vial
Hemophilus b conjugate (PRP-OMP) + Hep B	Comvax	Merck	Inactivated	Single-dose vial
Pneumococcal 7-valent conjugate	Prevnar	Wyeth	Inactivated	Single-dose vial
Polio (E-IPV)	IPOL	Aventis Pasteur	Inactivated	Single-dose syringe and 10-dose vial
Hepatitis B: pediatric/adolescent formulation	Engerix-B	GlaxoSmithKline	Inactivated	Single-dose vial or syringe
Hepatitis B: pediatric/adolescent formulation	Recombivax HB	Merck	Inactivated	Single-dose vial or syringe
Hepatitis B: adult formulation	Engerix-B	GlaxoSmithKline	Inactivated	Single-dose vial or syringe
Hepatitis B: adult/adolescent formulation	Recombivax HB	Merck	Inactivated	Single-dose vial or syringe
Hepatitis B: dialysis formulation	Recombivax HB	Merck	nactivated	Single-dose vial

Table 2.17 Vaccines and related products distributed in the United States, 2003 (*continued*)

Vaccine/biologic	Brand name	Manufacturer	Type	How supplied
Hepatitis B immune globulin (HBIG)	BayHep B	Bayer	Human immunoglobulin	Single-dose vial or syringe
Hepatitis B immune globulin (HBIG): pediatric formulation	BayHep B	Bayer	Human immunoglobulin	Single-dose neonatal syringe
Hepatitis B immune globulin (HBIG)	Nabi-HB	Nabi	Human immunoglobulin	Single-dose vial
Hepatitis A: pediatric/adolescent formulation	Havrix	GlaxoSmithKline	Inactivated	Single-dose vial or syringe
Hepatitis A: pediatric/adolescent formulation	Vaqta	Merck	Inactivated	Single-dose vial or syringe
Hepatitis A: adult formulation	Havrix	GlaxoSmithKline	Inactivated	Single-dose vial or syringe
Hepatitis A: adult formulation	Vaqta	Merck	Inactivated	Single-dose vial or syringe
Hepatitis A immune globulin	BayGam	Bayer	Human immunoglobulin	Single-dose vial
Hepatitis A and B: adult formulation	Twinrix	GlaxoSmithKline	Inactivated	Single-dose vial or syringe
Influenza	Fluvirin	Evans	Inactivated	Single-dose syringe and 10-dose vial
Influenza	Fluzone	Aventis Pasteur	Inactivated	Single-dose syringe and 10-dose vial
Influenza: pediatric use (preservative-free)	Fluzone	Aventis Pasteur	Inactivated	Single-dose syringe (0.25 and 0.5 ml)
Pneumococcal polysaccharide, 23-valent	Pneumovax 23	Merck	Inactivated	Single dose vial or syringe and 5-dose vial
Meningococcal vaccine	Menomune	Aventis Pasteur	Inactivated	Single- and 10-dose vial
Rabies	Imovax	Aventis Pasteur	Inactivated	Single-dose vial
Rabies	RabAvert	Chiron	Inactivated	Single-dose vial
Rabies immune globulin (RIG)	Imogam Rabies-HT	Aventis Pasteur	Human immunoglobulin	2 ml and 10 ml vials
Rabies immune globulin (RIG)	BayRab	Bayer	Human immunoglobulin	Single-dose vial
Japanese encephalitis	JE-VAX	Aventis Pasteur	Inactivated	Single- and 10-dose vial
Typhoid vaccine	Typhim Vi	Aventis Pasteur	Inactivated	Single-dose syringe and 20-dose vial
Typhoid vaccine live oral Ty21	Vivotif Berna	Berna	Live attenuated	4-capsule package
Varicella-zoster immune globulin (VZIG)	Generic	U of Mass Labs	Human immunoglobulin	125 unit and 625 unit vials
Yellow fever vaccine	YF-VAX	Aventis Pasteur	Live attenuated	Single- and 5-dose vial
Anthrax vaccine, adsorbed	BioThrax	BioPort	Inactivated	Multi-dose vial

CLUSTER OF DIFFERENTIATION (CD) ANTIGENS

3

- ● MOUSE CD ANTIGENS
- ● HUMAN CD ANTIGENS

Leukocytes express distinct assortments of molecules on their cell surfaces, many of which reflect either different stages of their lineage-specific differentiation or different states of activation or inactivation. These cell surface molecules of leukocytes are routinely detected with anti-leukocyte monoclonal antibodies. Clusters of antigens on the surface of leukocytes can be designated by their reactions with monoclonal antibodies. This designation of the antigens is called clusters of differentiation (CDs). Using different combinations of mAbs, it is possible to chart the cell surface immunophenotypes of different leukocyte subpopulations, including the functionally distinct mature cell subpopulations of B cells, helper T cells (TH), cytotoxic T cells (TC), and natural killer (NK) cells. Some CD antigens have a well-known function, but other CD antigens have no known function.

MOUSE CD ANTIGENS

Mouse CD antigens are listed in **Table 3.1**. Their gene, molecular weight, ligands, distribution, and functions are shown in the table. For reference, alternate names of mouse leukocyte antigens are listed in **Table 3.2**. Non-CD antigens are listed in alphanumeric order in **Table 3.3**.

Table 3.4 is a detailed summary of mouse leukocyte antigen distribution depicting the presence of surface antigens on different subsets. Antigen distribution on hematopoietic stem cells, erythrocytes, epithelial cells, endothelial cells, NK cells, monocytes/macrophages, T cells, B cells, granulocytes, megakaryocytes/platelets, and dendritic cells is illustrated graphically in **Figures 3.1–3.11**.

HUMAN CD ANTIGENS

CD antigens established in the 7th International Workshop of Human Leukocyte Differentiation Antigens are listed in **Table 3.5**. This table provides information regarding their molecular weight, gene locus, ligands/receptors, functions, and distribution. An addendum describing HLDA family and main antigen expression is provided as **Table 3.6**. A list of abbreviations can be found inside the back cover of this book.

Table 3.1 Mouse CD antigen chart

Key

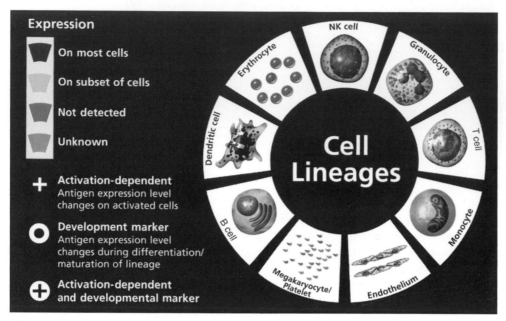

The distribution of activation-dependent and developmental cluster of differentiation (CD) markers on various cell types is presented in the following pages

Table 3.1 Mouse CD antigen chart (*continued*)

Antigen — clones offered by BD Biosciences, alternate names of antigen	Gene	Component of / Molecular Weight / Family/Superfamily	Ligands/Substrates	Antigen Distribution	Functions of Antigens
CD1d — 1B1, 3C11; CD1.1, CD1.2, Ly-38	Cd1d1	43–49 kDa / CD1/MHC	lipid/glycolipid Ag		Antigen presentation
CD2 — RM2-5; LFA-2, Ly-37	Cd2	45–58 kDa / CD2 Ig	CD48		Activation/costimulation; adhesion
CD3 — 17A2; T3		T-cell receptor			Signal transduction
CD3δ — CD3d, CD3 δ chain	Cd3d	T-cell receptor / 20 kDa / Ig			Signal transduction
CD3ε — 145-2C11, 500A2; CD3e, CD3 ε chain	Cd3e	T-cell receptor / 20 kDa / Ig			Signal transduction
CD3γ — CD3g, CD3 γ chain	Cd3g	T-cell receptor / 25 kDa / Ig			Signal transduction *For CD3ζ see CD247
CD4 — GK1.5, H129.19, RM4-5, RM4-4; L3T4	Cd4	T-cell receptor complex / 55 kDa / Ig	MHC class II		Signal transduction; receptor/coreceptor
CD5 — 53-7.3; Ly-1, Lyt-1, Ly-12	Cd5	67 kDa / SRCR	CD72		Adhesion; immunoregulation
CD5.1 — H11-86.1; Ly-1.1	Cd5a	67 kDa / SRCR	CD72		Immunoregulation
CD6	Cd6	100- and 128–130 kDa / SRCR	CD166		Activation/costimulation; adhesion; differentiation/development
CD7	Cd7	40 kDa / Ig			Immunoregulation
CD8a — 53-6.7, 5H10-1; Ly-2, Lyt-2	Cd8a	T-cell receptor complex / 38 kDa / Ig	MHC class I		Signal transduction; receptor/coreceptor
CD8b — H35-17.2; Ly-3, Lyt-3	Cd8b	T-cell receptor complex / 30 kDa / Ig	MHC class I		Signal transduction; receptor/coreceptor
CD8b.2 — 53-5.8; Ly-3.2	Cd8bb	T-cell receptor complex / 30 kDa / Ig	MHC class I		Signal transduction; receptor/coreceptor
CD9 — KMC8; p24	Cd9	21, 24 kDa / TM4			Activation/costimulation

Table 3.1 Mouse CD antigen chart (*continued*)

Antigen clones offered by BD Biosciences alternate names of antigen	Gene	Component of / Molecular Weight / Family/Superfamily			Ligands/ Substrates	Antigen Distribution	Functions of Antigens
CD10 R103 CALLA, MME, NEP	Mme		100 kDa Metalloproteinase		Peptides		Enzymatic activity; differentiation/ development
CD11a 2D7, M17/4 Ly-15, Ly-21, Integrin α_L chain	Itgal	LFA-1 180 kDa Integrin			CD54; CD102		Adhesion; differentiation/ development
CD11b M1/70 Integrin α_M chain, Ly-40	Itgam	Mac-1 (aka CR3) 170 kDa Integrin			CD54; iC3b; fibronectin		Adhesion
CD11c HL3 Integrin α_X chain	Itgax	p150, 95 (aka CR4) 150 kDa Integrin			iC3b; fibronectin		Adhesion
CD13 R3-242 Aminopeptidase N, gp150	Anpep	140-150 kDa Metalloproteinase			L-leucyl-β-naphthylamine		Enzymatic activity
CD14 rmC5-3 Mo2, LPS Receptor	Cd14	53-55 kDa Leucine-rich repeat			LPS/LPB complex		Receptor/coreceptor
CD15 SSEA-1, FAL, Lewis X	Fut4				CD62E?		Adhesion
CD16 2.4G2 FcγRIII, FcγRIIIα, Ly-17	Fcgr3	40-60 kDa Ig			mouse IgG		Ig Fc Receptor
CD18 GAME-46, C71/16, M18/2 Integrin β_2 chain	Itgb2	LFA-1, Mac-1, & p150,95 95 kDa Integrin			varies, see CD11a, b, c		Signal transduction; adhesion
CD19 1D3, MB19-1 B4	Cd19	CD19/CD21/CD81 complex 95 kDa Ig					Signal transduction; receptor/coreceptor
CD20 Ly-44, B1	Ms4a2	33-37 kDa CD20/FcεRIβ					Activation/costimulation; differentiation/ development
CD21 7G6 CR2	Cr2	CD19/CD21/CD81 complex 150 kDa RCA			C3d		C' regulation
CD22.2 Cy34.1 Lyb-8.2, Siglec-2	Cd22b	140-160 kDa Siglec			N-glycolyl neuraminic acid		Adhesion; immunoregulation; receptor/coreceptor
CD23 B3B4 FcεRII, Ly-42	Fcer2a	45-49 kDa C-type lectin			IgE		Ig Fc Receptor
CD24 30-F1, J11d, M1/69 Heat Stable Antigen, Ly-52, Nectadrin	Cd24a	35-52 kDa			CD62P		Activation/costimulation; adhesion

Table 3.1 Mouse CD antigen chart (*continued*)

Antigen — clones offered by BD Biosciences — alternate names of antigen	Gene	Component of — Molecular Weight — Family/Superfamily	Ligands/ Substrates	Antigen Distribution	Functions of Antigens
CD25 PC61, 7D4, 3C7 — Ly-43, IL-2 Receptor α chain, p55	*Il2ra*	IL-2 receptor / 50-60 kDa / CCP-like	IL-2		Activation/costimulation; receptor/coreceptor
CD26 H194-112 — Dipeptidyl peptidase, DPP IV, THAM	*Dpp4*	/ 220 kDa / Dipeptidyl-peptidase	Polypeptides		Activation/costimulation; adhesion; enzymatic activity
CD27 LG.3A10	*Tnfrsf7*	Homodimer / 45 kDa / TNFR	CD70		Activation/costimulation; receptor/coreceptor
CD28 37.51	*Cd28*	Homodimer / 65-80 kDa / Ig	CD80; CD86		Signal transduction; activation/costimulation; receptor/coreceptor
CD29 9EG7, KMI6, Ha2/5, HMβ1-1 — Integrin β₁ chain, VLAβ, gpIIa	*Itgb1*	VLA-1-VLA-6; α_vβ_1, α_xβ_1, α_8β_1, α_9β_1 integrins / 130 kDa / Integrin	varies, see CD49a-f and CD51		Signal transduction; adhesion;differentiation/ development
CD30 mCD30.1 (*aka* 2SH12-5F-2D) — Ki-1	*Tnfrsf8*	/ 105-120 kDa / TNFR	CD153		Immunoregulation; receptor/coreceptor; cytotoxicity?
CD31 MEC 13.3, 390 — PECAM-1, gpIIa, endoCAM	*Pecam*	/ 130-140 kDa / Ig	CD38; vitronectin receptor		Adhesion; angiogenesis
CD32 2.4G2 — FcγRII, Ly-17, Ly-m20	*Fcgr2b*	/ 40-60 kDa / Ig	mouse IgG		Ig Fc Receptor; phagocytosis
CD33 Siglec-3	*Cd33*	/ 67 kDa / Siglec	Sialylated glyco-proteins?		Adhesion
CD34 RAM34 (*aka* 49E8) — Mucosialin	*Cd34*	/ 90, 105-120 kDa / Sialomucin	CD62L		Adhesion
CD35 8C12, 7G6 — CR1, C3b receptor	*Cr2*	/ 190 kDa / RCA	C3b		C' regulation; phagocytosis
CD36 Scavenger receptor	*Cd36*	/ 88 kDa / Class B scavenger receptor	oxidized LDL		Adhesion; receptor/coreceptor; phagocytosis
CD37	*Cd37*	/ / TM4			
CD38 90 — T10	*Cd38*	/ 42 kDa /	CD31		Activation/costimulation; enzymatic activity
CD39 NTPDase-1	*Entpd1*	/ /	ATP; ADP		Enzymatic activity

Table 3.1   Mouse CD antigen chart (*continued*)

Antigen clones offered by BD Biosciences alternate names of antigen	Gene	Component of Molecular Weight Family/Superfamily	Ligands/ Substrates	Antigen Distribution	Functions of Antigens
CD40 HM40-3, 3/23 gp39 receptor	Tnfrsf5	 45–50 kDa TNFR	CD154		Activation/costimulation; immunoregulation
CD41 MWReg30 Integrin α$_{IIb}$ chain	Itga2b	α$_{IIb}$β$_3$ integrin (GPIIb-IIIa) 105 kDa Integrin	Fibronectin; fibrinogen; von Willebrand factor; thrombospondin		Adhesion; hemostasis
CD42a GPIX	Gp9	GPIb/IX/V Complex 20 kDa Leucine-rich repeat			Adhesion; hemostasis
CD42b GPIbα	Gp1ba	GPIb/IX/V Complex 145 kDa Leucine-rich repeat	von Willebrand factor		Adhesion; hemostasis
CD42c GPIbβ	Gp1bb	GPIb/IX/V Complex 24 kDa Leucine-rich repeat			Adhesion; hemostasis
CD42d GPV	Gp5	GPIb/IX/V Complex 88 kDa Leucine-rich repeat			Adhesion; hemostasis
CD43 S7, 1B11 Leukosialin, Ly-48, sialophorin	Spn	 115 and 130 kDa Sialomucin	CD54		Signal transduction; adhesion
CD44 IM7, KM114, TM-1 Pgp-1, Ly-24	Cd44	 85–95 kDa Core/link proteoglycan	hyaluronate; collagen; fibronectin; laminin; osteopontin		Activation/costimulation; adhesion
CD45 30-F11, 69 Ly-5, T200, LCA	Ptprc	 180–240 kDa RPTP			Signal transduction
CD45.1 A20 Ly-5.1	Ptprc^a	 180–240 kDa RPTP			Signal transduction
CD45.2 104 Ly-5.2	Ptprc^b	 180–240 kDa RPTP			Signal transduction
CD45R RA3-6B2 B220	Ptprc	 220 kDa RPTP			Signal transduction
CD45RA 14.8	Ptprc	 220, 235 kDa RPTP			Signal transduction
CD45RB 16A (*aka* C363.16A)	Ptprc	 200–240 kDa RPTP			Signal transduction
CD45RC DNL-1.9	Ptprc	 200–240 kDa RPTP			Signal transduction

Table 3.1 Mouse CD antigen chart (*continued*)

Antigen — clones offered by BD Biosciences, alternate names of antigen	Gene	Component of / Molecular Weight / Family/Superfamily	Ligands/ Substrates	Antigen Distribution	Functions of Antigens
CD45RO	Ptprc	— / 180 kDa / RPTP			Signal transduction
CD46	Mcp	— / 41 kDa / RCA	C3b		C' regulation
CD47 miap301, Integrin-associated protein (IAP)	Itgp	β₃ integrins / 50 kDa / Ig	CD172a		Signal transduction?; activation/costimulation; adhesion
CD48 HM48-1, BCM1, sgp-60	Cd48	— / 45 kDa / CD2 Ig	CD2; CD244; enteric bacteria		Activation/costimulation; adhesion
CD49a Ha31/8, Integrin α₁ chain	Itga1	VLA-1 / 200 kDa / Integrin	laminin; collagen		Adhesion; differentiation/ development
CD49b HMα2, Ha1/29, DX5, Integrin α₂ chain	Itga2	VLA-2 / 165 kDa / Integrin	laminin; collagen; fibronectin		Adhesion; differentiation/ development
CD49c 42, Integrin α₃ chain	Itga3	VLA-3 / — / Integrin	fibronectin; laminin; collagen		Adhesion; differentiation/ development
CD49d R1-2, 9C10(MFR4.B), DATK32, SG31, Integrin α₄ chain	Itga4	VLA-4 (LPAM-2), LPAM-1 / 150-155 kDa / Integrin	VCAM-1; fibronectin; MAdCAM-1; invasin		Adhesion; differentiation/ development
CD49e 5H10-27(MFR5), HMα5-1, Integrin α₅ chain	Itga5	VLA-5 / 135 kDa / Integrin	fibronectin		Adhesion; differentiation/ development
CD49f GoH3, Integrin α₆ chain	Itga6	VLA-6, α₆β₄ integrin (TSP-180) / 120 kDa / Integrin	laminin		Adhesion; differentiation/ development
CD50 ICAM-3	Icam3	— / — / Ig	Unknown in mouse		
CD51 H9.2B8, RMV-7, 21, Integrin α_V chain	Itgav	Vitronectin receptor; α_Vβ₁, α_Vβ₅, α_Vβ₆, and α_Vβ₈ integrins / 125 kDa / Integrin	vitronectin; fibronectin; fibrinogen; thrombospondin; von Willebrand factor; CD31		Activation/costimulation; adhesion; differentiation/ development
CD52 CAMPATH-1, B7	Cd52	— / 12 kDa / —			
CD53 OX-79	Cd53	— / 35-45 kDa / TM4			Signal transduction?; differentiation/ development?

Table 3.1 Mouse CD antigen chart (*continued*)

Antigen clones offered by BD Biosciences alternate names of antigen	Gene	Component of Molecular Weight Family/Superfamily	Ligands/ Substrates	Antigen Distribution	Functions of Antigens
CD54 3E2 ICAM-1, Ly-47, MALA-2	Icam1	85-110 kDa Ig	LFA-1; Mac-1; CD43		Adhesion
CD55 Decay accelerating factor (DAF)	Daf1	 RCA	C3b; CD97		C' regulation
CD56 N-CAM13, 12F8, 12F11 N-CAM	Ncam	120, 140, 180 kDa Ig			Adhesion; differentiation/ development
CD59 Complement inhibitor	Cd59a	20-26 kDa Ly6	C5b-8		C' regulation
CD61 2C9.G2 Integrin β₃ chain	Itgb3	Vitronectin receptor, α_{IIb}β₃ integrin (GPIIb-IIIa) 95 kDa Integrin	varies, see CD51 and CD41		Signal transduction; adhesion
CD62E 10E9.6 E-selectin, ELAM	Sele	110 kDa C-type lectin	E-selectin Ligand-1 (ESL-1)		Adhesion
CD62L MEL-14 L-selectin, LECAM-1, Ly-22	Sell	74 kDa (lymphocytes); 95 kDa (neutrophils) C-type lectin	PNAd (CD34, GlyCAM-1, MAdCAM-1)		Adhesion
CD62P Polyclonal, RB40.34 P-selectin, GMP-140, PADGEM	Selp	140 kDa C type lectin	CD162; CD24		Adhesion
CD63 ME491	Cd63	53 kDa? TM4			Differentiation/ development?
CD64 Fcγl Receptor	Fcgr1	70 kDa Ig	mouse IgG		Ig Fc Receptor
CD66a BGP, CEA-1	Ceacam1	 Ig	CD62E?		Signal transduction; adhesion; angiogenesis
CD66b CGM6, CEA-3	Psg18	 Ig			
CD68 Macrosialin, lysosomal glycoprotein	Cd68	87-115 kDa Sialomucin, Lamp			Phagocytosis
CD69 H1.2F3 Very Early Activation Antigen	Cd69	85 kDa C-type lectin			Activation/costimulation; differentiation/ development

Table 3.1 Mouse CD antigen chart (*continued*)

Antigen clones offered by BD Biosciences alternate names of antigen	Gene	Component of / Molecular Weight / Family/Superfamily	Ligands/ Substrates	Antigen Distribution	Functions of Antigens
CD70 FR70 CD27 Ligand	Tnfsf7	30-33 kDa / TNF	CD27		Activation/costimulation
CD71 C2 (C2F2) Transferrin receptor	Trfr	180-190 kDa	transferrin		Activation/costimulation; metabolism
CD72 10-1.D.2, K10.6, JY/93 Lyb-2, Ly-m19	Cd72	90 kDa / C-type lectin	CD5; CD100		Activation/costimulation; differentiation/ development
CD73 TY/23 NT, ecto-5'-nucleotidase	Nt5	69 kDa	NMP		Enzymatic activity
CD74 In-1 Ia-associated invariant chain (Ii)	Ii	Ia-associated chondroitin sulfate proteoglycan / 31, 41 kDa	CD44; MHC class II		Antigen presentation; differentiation/ development
CD79a HM47 Igα, mb-1, Ly-54	Iga	B-cell receptor complex / 30-35 kDa / Ig			Signal transduction
CD79b HM79b Igβ, B29	Igb	B-cell receptor complex / 35-40 kDa / Ig			Signal transduction; differentiation/ development
CD80 16-10A1, 1G10 B7/BB1, B7-1, Ly-53	Cd80	55 kDa / Ig	CD28; CD152		Activation/costimulation; immunoregulation
CD81 2F7, Eat1, Eat2 TAPA-1	Cd81	CD19/CD21/CD81 complex / 26 kDa / TM4			Activation/costimulation; adhesion; differentiation/ development
CD82 C33 Ag, KAI1	Kai	TM4			Activation/costimulation
CD83	Cd83	Ig			Activation/costimulation
CD84	Cd84	CD2 Ig			
CD86 GL1, PO3 B7-2, B70, Ly-58	Cd86	80 kDa / Ig	CD28; CD152		Activation/costimulation; immunoregulation
CD87 uPA Receptor	Plaur		uPA		Adhesion; receptor/coreceptor

Table 3.1　Mouse CD antigen chart (*continued*)

Antigen clones offered by BD Biosciences alternate names of antigen	Gene	Component of Molecular Weight Family/Superfamily	Ligands/ Substrates	Antigen Distribution	Functions of Antigens
CD88 C5a Ligand, C5aR	C5r1	 G-protein coupled	C5a		Activation/costimulation; C' regulation
CD90 G7 Thy-1, θ, T25	Thy1	 25-30 kDa Ig			Signal transduction; activation/costimulation; adhesion; differentiation/ development
CD90.1 HIS51, OX-7 Thy-1.1, θ-AKR	Thy1[a]	 25-30 kDa Ig			Signal transduction; activation/costimulation; adhesion; differentiation/ development
CD90.2 53-2.1, 30-H12 Thy-1.2, θ-C3H	Thy1[b]	 25-30 kDa Ig			Signal transduction; activation/costimulation; adhesion; differentiation/ development
CD91 LRP, A2MR	Lrp1	 600 kDa LDLR	LDL; LRPAP1; α_2M; apo E; Gp96		Antigen presentation; hemostasis; metabolism
CD94 18d3	Klrd1	CD94/NKG2 heterodimers C-type lectin	Qa-1/Qdm		Antigen recognition; immunoregulation
CD95 Jo2, 13 Fas, APO-1	Tnfrsf6	 45 kDa TNFR	CD178		Apoptosis
CD97	Cd97	 EGF-TM7	CD55		
CD98 H202-141 4F2, Ly-10, RL-388	Cd98	 120 kDa 			Activation/costimulation; immunoregulation?
CD100 30 Semaphorin H, coll-4	Sema4d	 150 kDa Semaphorin	CD72; Plexin-B1		Immunoregulation
CD102 3C4(mIC2/4) ICAM-2, Ly-60	Icam2	 55-68 kDa Ig	LFA-1		Activation/costimulation; adhesion
CD103 2E7, M290 Integrin α_{IEL} chain	Itgae	$\alpha_{IEL}\beta_7$ integrin 150 kDa (and 20 kDa?) Integrin	E-cadherin		Activation/costimulation; adhesion; differentiation/ development
CD104 346-11A Integrin β_4 chain	Itgb4	$\alpha_6\beta_4$ integrin (TSP-180) 205 kDa 	Laminin		Adhesion
CD105 MJ7/18 Endoglin	Eng	 180 kDa TGFR	TGF-β		Adhesion; receptor/coreceptor

Table 3.1   Mouse CD antigen chart (*continued*)

Antigen clones offered by BD Biosciences alternate names of antigen	Gene	Component of Molecular Weight Family/Superfamily	Ligands/ Substrates	Antigen Distribution	Functions of Antigens
CD106 429 (MVCAM.A) VCAM-1	*Vcam1*	100-110 kDa; 47 kDa GPI-linked Ig	VLA-4		Adhesion; differentiation/ development
CD107a 1D4B LAMP-1	*Lamp1*	110-140 kDa Lamp	collagen?; laminin?; fibronectin?		Adhesion?
CD107b ABL-93 LAMP-2	*Lamp2*	100-110 kDa Lamp			Adhesion?
CD110 Thrombopoietin Receptor, c-mpl	*Mpl*	CKR			Differentiation/ development
CD111 PRR1, nectin-1	*Pvrl1*	Ig	α-herpesviruses		Adhesion
CD112 PRR2, nectin-2	*Pvs*	Ig			Adhesion
CD114 G-CSF Receptor	*Csf3r*	95-125 kDa CKR	G-CSF		Signal transduction; differentiation/ development; receptor/coreceptor
CD115 M-CSF Receptor, CSF-1R, c-fms, Fim-2	*Csf1r*	165 kDa RTK	M-CSF		Signal transduction; differentiation/ development; receptor/coreceptor
CD116 GM-CSF Receptor α chain	*Csf2ra*	GM-CSF Receptor CKR	GM-CSF		Signal transduction; differentiation/ development; receptor/coreceptor
CD117 2B8, ACK45 c-kit, Steel factor receptor, Dominant white spotting	*Kit*	145-150 kDa Ig, RTK	c-Kit Ligand (*aka* Steel, stem cell, or mast cell growth factor)		Signal transduction; adhesion; differentiation/ development; receptor/coreceptor
CD118 IFN-α/β Receptor, Type 1 IFN-R, IFN-a Receptor	*Ifnar*	CKR	IFN-α; IFN-β		Immunoregulation?; receptor/coreceptor
CD119 GR20, 2E2 IFN-γ Receptor α chain	*Ifngr*	IFN-γ Receptor 85-95 kDa CKR	IFN-γ		Immunoregulation; receptor/coreceptor
CD120a 55R-170, 55R-593, 55R-286 TNFR1, TNF-R55	*Tnfrsf1a*	55-60 kDa TNFR	TNF; LT-α3 (*aka* TNF-β)		Signal transduction; apoptosis; receptor/coreceptor
CD120b TR75-32, TR75-54, TR75-89 TNFR2, TNF-R75	*Tnfrsf1b*	75-80 kDa TNFR	TNF; LT-α3 (*aka* TNF-β)		Signal transduction; apoptosis; necrosis; receptor/coreceptor

Table 3.1 Mouse CD antigen chart (*continued*)

Antigen (clones offered by BD Biosciences / alternate names of antigen)	Gene	Component of / Molecular Weight / Family/Superfamily	Ligands/Substrates	Antigen Distribution	Functions of Antigens
CD121a 35F5, 12A6, JAMA-147 IL-1 Receptor, Type I	Il1r1	/ 80 kDa / Ig	IL-1α; IL-1β		Signal transduction; activation/costimulation; receptor/coreceptor
CD121b 1F6, 4E2 IL-1 Receptor, Type II	Il1r2	/ 60 kDa / Ig	IL-1α; IL-1β		Immunoregulation; receptor/coreceptor
CD122 TM-β1, 5H4 IL-2 and IL-15 Receptor β chain	Il2rb	IL-2 and IL-15 receptors / 85-100 kDa / CKR	IL-2; IL-15		Signal transduction; immunoregulation; receptor/coreceptor
CD123 5B11 IL-3 Receptor α chain	Il3ra	IL-3 receptor / 60-70 kDa / CKR	IL-3		Differentiation/development?; receptor/coreceptor
CD124 mIL4R-M1 IL-4 Receptor α chain	Il4ra	IL-4 and IL-13 receptors / 138-145 kDa / CKR	IL-4; IL-13		Signal transduction; receptor/coreceptor
CD125 IL-5 Receptor α chain	Il5ra	IL-5 receptor / 60 kDa / CKR	IL-5		Activation/costimulation; immunoregulation?; receptor/coreceptor
CD126 D7715A7 IL-6 Receptor α chain	Il6ra	IL-6 receptor / 80 kDa / CKR, Ig	IL-6		Differentiation/development; immunoregulation; receptor/coreceptor
CD127 B12-1, SB/14 IL-7 Receptor α chain	Il7r	IL-7 receptor / 65-75 kDa / CKR	IL-7		Signal transduction; differentiation/development; receptor/coreceptor
CD128 IL-8 Receptor α chain, CXCR2	Cmkar2	/ / α-chemokine receptor	MIP2; KC; (human IL-8)		Activation/costimulation; receptor/coreceptor
CD130 gp130, Common β chain	Il6st	IL-11, OSM, CNTF & LIF receptors / 130 kDa / CKR			Signal transduction
CD131 JORO50 AIC2A & AIC2B, β_{IL-2} and β_c	Csf2rb1 and Csf2rb2	IL-3, IL-5, & GM-CSF receptors / 110-120 kDa(AIC2A); 120-140 kDa (AIC2B) / CKR	IL-3 (for AIC2A)		Signal transduction; receptor/coreceptor
CD132 4G3, 3E12, TUGm2 Common γ chain	Il2rg	IL-2, IL-4, IL-7, IL-9 and IL-15 receptors / / CKR			Signal transduction
CD133 AC133, Prominin	Prom	/ / 5-TM			
CD134 OX-86 Ly-70, OX-40 antigen, ACT35 antigen	Tnfrsf4	/ 50 kDa / TNFR	OX-40 Ligand		Activation/costimulation

Table 3.1 Mouse CD antigen chart (*continued*)

Antigen clones offered by BD Biosciences alternate names of antigen	Gene	Component of Molecular Weight Family/Superfamily	Ligands/ Substrates	Antigen Distribution	Functions of Antigens
CD135 A2F10.1 Flk-2, Flt3, Ly-72	Flt3	135-150 kDa Ig, RTK	flt3 Ligand		Differentiation/ development; receptor/coreceptor
CD137 1AH2 (aka 53A2) 4-1BB, Ly-63	Tnfrsf9	30 kDa (monomer); 55 kDa (dimer); or 110 kDa (tetramer) TNFR	4-1BBL; fibronectin; laminin; vitronectin; collagen IV		Antigen presentation; signal transduction; activation/costimulation; adhesion
CD138 281-2 Syndecan-1	Sdc1	31-kDa core protein Glycosaminoglycan	interstitial matrix proteins		Adhesion
CD140a APA5 PDGF Receptor α chain, PDGFR-a	Pdgfra	PDGF receptor homod- imer and heterodimer 180 kDa CKR, RTK	PDGF A chain; PDGF B chain		Signal transduction; differentiation/ development; receptor/coreceptor
CD140b 28 PDGF Receptor β chain, PDGFR-b	Pdgfrb	PDGF receptor homod- imer and heterodimer 180 kDa CKR, RTK	PDGF B chain		Signal transduction; differentiation/ development; receptor/coreceptor; chemotaxis
CD141 Thrombomodulin	Thbd	C-type lectin	thrombin		Hemostasis
CD142 Tissue Factor, Coagulation Factor III	F3	Serine protease cofactor	Plasma Factor VII/VIIa		Differentiation/ development?; hemostasis; angiogenesis
CD143 Angiotensin converting enzyme, dipeptidyl peptidase	Ace	Peptidylpeptidase	angiotensin I		Enzymatic activity
CD144 11D4.1 VE-Cadherin	Cdh5	125 kDa Cadherin	CD144		Adhesion; angiogenesis
CD146	Mcam	Ig			Adhesion
CD147 Basigin, HT7, neurothelin, gp42	Bsg	Ig			Adhesion
CD148 PTPβ2, ByP	Ptprj	FNIII, PTP			Signal transduction
CD150 IPO-3	Slam	Ig			Signal transduction
CD151 SFA-1, PETA-3	Cd151	TM4			Hemostasis?

Table 3.1   Mouse CD antigen chart (*continued*)

Antigen (clones offered by BD Biosciences / alternate names of antigen)	Gene	Component of / Molecular Weight / Family/Superfamily	Ligands/ Substrates	Antigen Distribution	Functions of Antigens
CD152 9H10, UC10-4F10-11 CTLA-4, Ly-56	Cd152	33-37 kDa / Ig	CD80; CD86		Immunoregulation
CD153 RM153 CD30 Ligand	Tnfsf8	40 kDa / TNF	CD30		Activation/costimulation; immunoregulation
CD154 MR1 gp39, CD40 Ligand, Ly-62	Tnfsf5	39 kDa / TNF	CD40		Activation/costimulation
CD156a MS2, ADAM 8	Adam8	89 kDa / Metalloproteinase			Adhesion; enzymatic activity
CD156q TACE, ADAM17	Adam17	130 kDa / Metalloproteinase	TNF-α; APP; CD62L		Adhesion; enzymatic activity; receptor/coreceptor
CD157 Ly-65, BP-3, BST-1	Bst1	38-48 kDa / ADP-ribosylcyclase			Adhesion?
CD159a 20d5 NKG2A, NKG2B	Klrc1	CD94/NKG2 heterodimers / 38 kDa / C-type lectin	Qa-1/Qdm		Antigen recognition; signal transduction
CD160 BY55	Cd160	Ig			
CD161a NKR-P1A	Ly55a				
CD161b PK136 NKR-P1B	Ly55b	81 kDa / C-type lectin			
CD161c PK136 NKR-P1C, NK-1.1, Ly-55	Ly55c	76-80 kDa / C-type lectin			Activation/costimulation
CD162 2PH1 P-selectin-IgG fusion protein P-selectin glycoprotein Ligand (PSGL-1)	Selpl	160 kDa / Sialomucin	CD62P		Adhesion
CD163	Cd163	SRCR			
CD164 MGC-24, A115, A24	Cd164				Adhesion
CD166 ALCAM, DM-GRASP	Alcam	120 kDa / Ig	CD6		Activation/costimulation; adhesion; differentiation/ development?

Table 3.1 Mouse CD antigen chart (*continued*)

Antigen (clones offered by BD Biosciences, alternate names of antigen)	Gene	Component of / Molecular Weight / Family/Superfamily	Ligands/ Substrates	Antigen Distribution	Functions of Antigens
CD167a Cak, Nep	Ddr1	RTK			Adhesion
CD168 RHAMM	Hmmr				Adhesion
CD169 Sialoadhesin, Siglec-1	Sn	Siglec	CD43; CD162		Adhesion
CD170 Siglec-5	Siglec5	Siglec			Adhesion
CD171 17, L1	L1cam				Adhesion
CD172a P84, SIRPα, SHPS-1, BIT, P84 Antigen	Ptpns1	77, 86 kDa / Ig	CD47		Signal transduction; adhesion
CD178 MFL3, MFL4, 33, CD95L, Fas Ligand	Tnfsf6	TNF	CD95		Signal transduction; activation/costimulation; differentiation/development; apoptosis; cytotoxicity?
CD178.1 Kay-10, mFasL.1	Tnfsf6	TNF	CD95		Signal transduction; activation/costimulation; differentiation/development; apoptosis; cytotoxicity?
CD179a VpreB	Vpreb1	Pre-B cell receptor / 16 kDa / Ig			Differentiation/ development
CD179b LM34, λ5	Vpreb2	Pre-B cell receptor / 22 kDa / Ig			Differentiation/ development
CD180 RP/14, RP105	Ly78	RP105/MD-1 complex / 105 kDa / Leucine-rich repeat			Signal transduction
CD183 CXCR3	Cmkar3	Chemokine receptor	IP-10; 6Ckine; Mig; I-TAC		Receptor/coreceptor; chemotaxis
CD184 2B11/CXCR4, CXCR4	Cmkar4	Chemokine receptor	SDF-1		Receptor/coreceptor; chemotaxis
CD195 C34-3448, CCR5	Cmkbr5	Chemokine receptor	MIP-1α; MIP-1β; RANTES; MCP-1		Receptor/coreceptor; chemotaxis
CD197 CCR7	Cmkbr7	Chemokine receptor	SLC		Receptor/coreceptor; chemotaxis

Table 3.1   Mouse CD antigen chart (*continued*)

Antigen clones offered by BD Biosciences alternate names of antigen	Gene	Component of Molecular Weight Family/Superfamily	Ligands/ Substrates	Antigen Distribution	Functions of Antigens
CD200 OX-90 OX-2 Ag	Mox2	Ig	CD200 receptor		Immunoregulation
CD201 CCD41, EPCR, Protein C Receptor	Procr	CD1/MHC	Protein C		Receptor/coreceptor; hemostasis
CD202 33 Endothelial-specific receptor tyrosine kinase, Tie2	Tek	RTK			Differentiation/ development
CD203c Ly-41, PC-1	Enpp1	220 kDa E-NPP	Extracellular nucleotides		Enzymatic activity
CD204 Macrophage scavenger receptor	Scvr	Class A scavenger receptor	LPS; collagen; LDL?		Adhesion
CD205 DEC-205, Ly-75	Ly75	205 kDa C-type lectin			Antigen presentation
CD206 Macrophage mannose receptor	Mrc1	175 kDa C-type lectin	High-mannose carbohydrates		Antigen presentation
CD210 1B1.3a IL-10 receptor, CRF2-4	Il10ra, Il10rb	CKR	IL-10		Immunoregulation; receptor/coreceptor
CD212 114 IL-12R β chain	Il12rb1	IL-12 receptor CKR	IL-12		Immunoregulation; receptor/coreceptor
CD213a1 NR4, IL-13R α1 chain	Il13ra1	IL-13 and IL-4 receptors CKR	IL-13		Immunoregulation; receptor/coreceptor
CD213a2 IL-13R α2 chain	Il13ra2	IL-13 receptor CKR	IL-13		Immunoregulation; receptor/coreceptor
CD217 IL-17R	Il17r		IL-17; vIL-17		Immunoregulation; receptor/coreceptor
CD220 46, Polyclonal Insulin receptor	Insr	130 kDa, 95 kDa RTK	Insulin		Receptor/coreceptor; metabolism
CD221 IGF-IR	Igf1r	RTK	IGF-I; Insulin		Receptor/coreceptor; metabolism
CD222 IGF-IIR, CI-MP R	Igf2r	220-250 kDa RTK	IGF-II; mannose-6-phosphate residues; retinoic acid; TGF-β LAP		Receptor/coreceptor; metabolism

Table 3.1 Mouse CD antigen chart (*continued*)

Antigen (clones offered by BD Biosciences, alternate names of antigen)	Gene	Component of / Molecular Weight / Family/Superfamily	Ligands/ Substrates	Antigen Distribution	Functions of Antigens
CD223 C9B7W, Ly-66, LAG3	Lag3	Ig	MHC class II		Immunoregulation
CD224 γ-glutamyl transpeptidase	Ggtp		glutathione		Enzymatic activity
CD227 EMA	Muc1				
CD228 MTf, p97	Mfi2				
CD229 Lgp100, T100, Ly-9	Ly9	100 kDa, 150 kDa / CD2 Ig			
CD229.1 30C7, Lgp100, Ly-9.1	Ly9a	100 kDa / CD2 Ig			
CD230 Prion protein	prnp				Differentiation/ development
CD231 TALLA, A15	Tm4sf2	TM4			
CD232 Plexin C1, Vespr	plxnc1	Plexin			
CD233 AE1, band 3	slc4a1	SLC			Metabolism
CD234 Duffy blood group, DARC	Dfy		Chemokines		Receptor/coreceptor; chemotaxis
CD235a Glycophorin A	Gypa				
CD236R Glycophorin C	Gypc				
CD238 Kell blood group, endothelin-3-converting enzyme	Kel	Kell/Kx antigen complex / 110 kDa / Neutral endopeptidase	big ET3		Enzymatic activity
CD239 Lutheran blood group	Lu	Ig	Laminin 10/11		Adhesion

Table 3.1 Mouse CD antigen chart (*continued*)

Antigen clones offered by BD Biosciences alternate names of antigen	Gene	Component of / Molecular Weight / Family/Superfamily	Ligands/ Substrates	Antigen Distribution	Functions of Antigens
CD240 Rh30, Rh antigen	*Rhced*	Rh blood group / / Rh			Metabolism
CD241 Rh50, Rh-associated glycoprotein	*Rhag*	Rh blood group / 50 kDa / Rh			Metabolism
CD243 P- glycoprotein1, Mdr1	*Abcb1*	/ / ABC transporter, MDR/TAP	Drugs, dyes		Enzymatic activity; metabolism
CD244.2 2B4 2B4 Antigen	*Nmrkb*	/ 66 kDa / CD2 Ig	CD48		Signal transduction
CD246 Anaplastic lymphoma kinase	*alk*	/ / RTK	Unknown		Enzymatic activity
CD247 1ζ3A1, 1η4F2, 8d3 CD3z, CD3 ζ chain	*Cd3z*	T-cell receptor / 16, 21, 32, 42 kDa /			Signal transduction

Table 3.2   Alternate names of mouse leukocyte antigens

Common names for mouse leukocyte antigens	Specificity in mouse leukocyte catalog section
2B4 antigen	CD244
4-1BB	CD137
4F2	CD98
5E6	Ly-49C and Ly-49I
6C3 antigen	Ly-51
A1	Ly-49A
ACT35 antigen	CD134
AIC2A and AIC2B	CD131
Aminopeptidase N	CD13
APO-1	CD95
APO-2 ligand	TRAIL
B220	CD45R/B220
B29	CD79b
B7-1, B7/BB1	CD80
B7-2, B70	CD86
BCM1	CD48
β_{IL-2} and β_c	CD131
BIT	CD172a
BP-1 antigen	Ly-51
BP-3	CD157
BST-1	CD157
c-Kit	CD117
ClqRp	Early B lineage
C3b receptor	CD35
CALLA	CD10
CCR5	CD195
CD1.1	CD1d
CD3 ε chain	CD3ε
CD3 ζ chain	CD247
CD21b	CD35
CD27 ligand	CD70

Table 3.2 Alternate names of mouse leukocyte antigens (*continued*)

Common names for mouse leukocyte antigens	Specificity in mouse leukocyte catalog section
CD30 ligand	CD153
CD40 ligand	CD154
CD62L ligand	PNAd
CD95 ligand	CD178
CD161	NK-1.1
coll-4	CD100
Common γ chain	CD132
CR1	CD35
CR2/CR1	CD21/CD35
CRF2-4	CD210
CTLA-4	CD152
CXCR4	CD184
DDP IV	CD26
DX5	CD49b
E-selectin	CD62E
Ecto-5′-nucleotidase	CD73
ELAM-1	CD62E
endoCAM	CD31
Endoglin	CD105
Erythroid cells	TER-119
Fas	CD95
Fas ligand	CD178
Fcε RII	CD23
Fcγ III/II receptor	CD16/CD32
Fibronectin receptor α chain	CD49e
Fibronectin receptor β chain	CD29
Flk-2	CD135
Flt3	CD135
γ_c	CD132
GL7	T and B cell activation antigen
GMP-140	CD62P

Table 3.2 Alternate names of mouse leukocyte antigens (*continued*)

Common names for mouse leukocyte antigens	Specificity in mouse leukocyte catalog section
gp150	CD13
gp39	CD154
gp39 receptor	CD40
gpIIa	CD29
gpIIb	CD41
gpIIIa	CD61
Gr-1	Ly-6G and Ly-6C
H2-DM	HZ-M
H4	ICOS
Heat stable antigen	CD24
HsAg, HSA	CD24
la-associated invariant chain	CD74
IAP	CD47
ICAM-1	CD54
ICAM-2	CD102
IFN γ receptor α chain	CD119
Igα	CD79a
Igβ	CD79b
IgE Fc receptor	CD23
II	CD74
IL-1 receptor type I/p80	CD121a
Il-1 receptor type II/p60	CD121b
IL-2 receptor α chain	CD25
IL-2 and IL-15 receptor β chain	CD122
IL-3 receptor α chain	CD123
IL-4 receptor α chain	CD124
IL-6 receptor α chain	CD126
IL-7 receptor α chain	CD127
IL-10 receptor	CD210
IL-12 receptor β chain	CD212
Insulin receptor	CD220
Integrin α_1 chain	CD49a

Table 3.2 Alternate names of mouse leukocyte antigens (*continued*)

Common names for mouse leukocyte antigens	Specificity in mouse leukocyte catalog section
Integrin α_2 chain	CD49b
Integrin α_3 chain	CD49c
Integrin $\alpha_4 \beta_7$ complex	LPAM-1
Integrin α_4 chain	CD49d
Integrin α_5 chain	CD49e
Integrin α_6 chain	CD49f
Integrin α_{1EL} chain	CD103
Integrin α_{IIb} chain	CD41
Integrin α_L chain	CD11a
Integrin α_M chain	CD11b
Integrin α_V chain	CD51
Integrin α_X chain	CD11c
Integrin β_1 chain	CD29
Integrin β_2 chain	CD18
Integrin β_3 chain	CD61
Integrin β_4 chain	CD104
Integrin-associated protein	CD47
Ki-1	CD30
L-selectin	CD62L
L1	CD171
L3T4	CD4
LAG3	CD223
λ 5	CD179b
Laminin receptor α chain	CD49f
Laminin receptor β chain	CD29
LAMP-1	CD107a
LAMP-2	CD107b
LCA	CD45
LECAM-1	CD62L
Leukocyte common antigen	CD45
Leukosialin	CD43

Table 3.2 Alternate names of mouse leukocyte antigens (*continued*)

Common names for mouse leukocyte antigens	Specificity in mouse leukocyte catalog section
LFA-1 α chain	CD11a
LFA-1 β chain	CD18
LFA-2	CD2
LGL-1	Ly-49G2
Lgp-100	CD299.1
LPAM-1 α chain	CD49d
LPAM-1 β chain	Integrin β$_7$ chain
LPAM-2 α chain	CD49d
LPAM-2 β chain	CD29
Ly-1	CD5
Ly-2	CD8a
Ly-3	CD8b
Ly-4	CD4
Ly-5	CD45
Ly-6E	TSA-1
Ly-9.1	CD229.1
Ly-10	CD98
Ly-12	CD5
Ly-15	CD11a
Ly-17	CD16/CD32
Ly-19	CD72
Ly-21	CD11a
Ly-22	CD62L
Ly-24	CD44
Ly-32	CD72
Ly-35	CD8a
Ly-37	CD2
Ly-38	CD1d
Ly-40	CD11b
Ly-42	CD23
Ly-43	CD25
Ly-44	CD20

Table 3.2 Alternate names of mouse leukocyte antigens (*continued*)

Common names for mouse leukocyte antigens	Specificity in mouse leukocyte catalog section
Ly-47	CD54
Ly-48	CD43
Ly-52	CD24
Ly-53	CD80
Ly-54	CD79a
Ly-55	NK1.1
Ly-56	CD152
Ly-58	CD86
Ly-59	NK1.1
Ly-60	CD102
Ly-61	Ly-6D
Ly-62	CD154
Ly-63	CD137
Ly-63L	4-1BB ligand
Ly-65	CD157
Ly-66	CD223
Ly-67	TSA-1
Ly-68	Early B lineage
Ly-69	Integrin β_7 chain
Ly-70	CD134
Ly-70L	OX-40 ligand
Ly-72	CD135
Ly-73	Flk1
Ly-74	Ep-CAM
Ly-76	TER-119
Ly-77	T and B cell activation antigen
Ly-78	CD180
Ly-79	Dendritic cells
Ly-81	TRAIL
Ly-89	PIR-A/B
Ly-90	CD244

Table 3.2 Alternate names of mouse leukocyte antigens (*continued*)

Common names for mouse leukocyte antigens	Specificity in mouse leukocyte catalog section
Ly-92a	ART2.2
Ly-101	PD-1
Ly-115	ICOS
Ly-A	CD5
Ly-B	CD8a
Ly-C	CD8b
Ly-m10	CD98
Ly-m11	β_2 microglobulin
Ly-m19	CD72
Ly-m22	CD62L
Lyam-1	CD62L
Lyb-2	CD72
Lyb-8.2	CD22.2
LyM-1	CD16/CD32
Lym-20	CD16/CD32
Lyt-1	CD5
Lyt-2	CD8a
Lyt-3	CD8b
Mac-1 α chain	CD11b
Mac-1 β chain	CD18
MAFA	KLRG1
MALA-2	CD54
Mast cell factor receptor	CD117
mb-1	CD79a
MECA-32 antigen	Panendothelial cell antigen
Mo2	CD14
Mucosialin	CD34
N-CAM	CD56
Nectadrin	CD24
NKR-P1B	NK-1.1
NKR-P1C	NK-1.1
NT	CD73

Table 3.2 Alternate names of mouse leukocyte antigens (*continued*)

Common names for mouse leukocyte antigens	Specificity in mouse leukocyte catalog section
OX-2 antigen	CD200
OX-40 antigen/receptor	CD134
P-selectin	CD62P
P-selectin glycoprotein ligand	CD162
p150, 95 α chain	CD11c
p150, 95 β chain	CD18
p24	CD9
p55	CD25
P84 antigen	CD172a
PADGEM	CD62P
Pan-NK cells	CD49b
PDGF receptor α chain	CD140a
PDGF receptor β chain	CD140b
PECAM-1	CD31
pglla	CD31
Pgp-1	CD44
Pre-BCR	Pre-B cell receptor
PSGL-1	CD162
pTα	Pre-T cell receptor α chain
RL-388	CD98
RP105	CD180
Rt6-2	ART2.2
Sca-1	Ly-6A/E
Sca-2	TSA-1
Scavenger receptor	CD36
Semaphorin H	CD100
Siglec-2	CD22
sgp-60	CD48
SHPS-1	CD172a
Sialophorin	CD43
SIRPα	CD172a

Table 3.2   Alternate names of mouse leukocyte antigens (*continued*)

Common names for mouse leukocyte antigens	Specificity in mouse leukocyte catalog section
Steel factor receptor	CD117
Stem cell factor receptor	CD117
Syndecan-1	CD138
θ	CD90
T3	CD3
T10	CD38
T200	CD45
TAP	Ly-6A/E
TAPA-1	CD81
THAM	CD26
ThB	Ly-6D
Thy-1	CD90
Thy-1.1	CD90.1
Thy-1.2	CD90.2
TNFR receptor type I/p55	CD120a
TNF receptor type II/p75	CD120b
Transferin receptor	CD71
TSP-180 α chain	CD49f
TSP-180 β chain	CD104
VCAM-1	CD106
VE-cadherin	CD144
VEGF-R2	Flk1
Very Early Activation antigen	CD69
Vitronectin receptor α chain	CD51
Vitronectin receptor β chain	CD61
VLA-1 α chain	CD49a
VLA-2 α chain	CD49b
VLA-3 α chain	CD49c
VLA-4 α chain	CD49d
VLA-5 α chain	CD49e
VLA-6 α chain	CD49f
VLAβ	CD29

Table 3.3 Mouse cell surface antigens: Non-CD antigens

Non-CD antigens by alphanumeric order						
Antigen	Other names	MW	Structure	Chromo-some #	Expression	Function
4-1BBL	Tnfsf9 TNFSF			17	B^{act}, DC^{act}, peritoneal mac^{act}	DC activation, cytokine production
B7-H1	PD-L1				Broad	Cell costimulation, receptor for PD-1
B7-H2	GL50, ICOS-L, B7h, B7RP-1				B, DC, mono	Cell costimulation, receptor for ICOS
B7-DC	PD-L2				Mono, mac, DC subset	Cell costimulation, receptor for PD-1
BP-1	Ly-51, 6C3, Enpep	120–160 kD	Type II TM	3	Early B progenitors, BM stromal cells, thymic epith	Zinc metalloproteinase, glutamyl aminopeptidase
DX5	VLA-2, Integrin α2, Itga 2	165 kD	IntgF	13	NK, T subset	
Flk-1	Kdr, Ly73, VEGFR2		RTK family	5	Endoth	Receptor for VEGF
Flt-4	VEGFR3	170 kD	RTK family	11	Lymphatic endoth	Endoth growth factor receptor, binds VEGF-C
ICOS	Ly115	26 kD	IgSF	1	Thymic medulla, geminal center T cells, T^{act}	Inducible T cell Costimulator, T costimulation, B7-H2 receptor, cytokine production, B help
IgE high affinity receptor					B, mono	High affinity binding to IgE
IgM					Surface expression by mature B cells	
Jagged-1						Receptor for Notch-1
Ly-6A/E	Sca-1	18 kD	GPI-linked		Gran, mono, B, T subset, endoth	T activation
Ly-6B						
Ly-6C		14–17 kD	GPI-linked		Endoth, T, NK, mono, mac	
Ly-6D	ThB, Ly-61	15 kD	GPI-linked		B, T, thymic epith	
Ly-6F						
Ly-6G	Gr-1	21–25 kD	GPI linked	Unknown	Myeloid cells	

Table 3.3 Mouse cell surface antigens: Non-CD antigens (*continued*)

			**Non-CD antigens by alphanumeric order**			

Antigen	Other names	MW	Structure	Chromo-some #	Expression	Function
Ly-49A	A1, Klra1	85 kD	Type II TM C-type lectin	6	T subset, NK subset	Regulation of cytotoxicity, binds MHC class I
Ly-49B	Klra2			6		
Ly-49C	Klra3, 5E6	110 kD	Type II TM C-type lectin	6	T subset, NK subset	Regulation of cytotoxicity, binds MCH class I
Ly-49D	Klra4		Type II TM C-type lectin	6	NK subset	NK activation
Ly-49E	Klra5			6		
Ly-49F	Klra6			6		
Ly-49G	LGL1, Klra7	85 kD	Type II TM C-type lectin	6	T subset, NK subset	Regulation of cytotoxicity
Ly-49H	Klra8		Type II TM C-type lectin	6		
Ly-49I	Klra9		Type II TM C-type lectin	6		
Mac-3		93–110 kD			Mac (surface and intercellular) related to CD107b	
MAd-CAM-1		50 kD	IgSF, Type I TM	10	Endoth subset	Mucosal vascular addressin cell adhesion molecule, adhesion, cells homing, binds CD49d and CD62L
Notch-1	Lin-12, Tan1			2	Developing embryo, variety of adult tissues	Cell–cell interaction, cell fate determination
OX-40 ligand	Tnfsf4, gp34	35 kD	TNFSF	1	B^{act}, cardiac myocytes	T-B interaction, T costimulation
PD-1	Programmed death-1	55 kD			Thymocyte subset, T^{act}, B^{act}	T-B interaction, T costimulation, peripheral tolerance
Sca-1	Sca-1	18 kD	GPI-linked		Gran, mono B, T subset, endoth	T activation
Ter-119	Ly-76				Early proerythroblast to mature erythrocyte	W/ glycophorin, but not a typical glycophorin

Table 3.3 Mouse cell surface antigens: Non-CD antigens (*continued*)

Non-CD antigens by alphanumeric order						
Antigen	**Other names**	**MW**	**Structure**	**Chromo-some #**	**Expression**	**Function**
Tie2	Tek	140 kD	RTK family	4	Stem cells, endoth from early development	Angiogenesis, Angiopoietin-1 receptor
TLR1				5		Activation of AP-1 not NF-kB
TLR2				3 or 8	Mono, mac, adipocytes, γδT	Response to bacterial lipoproteins
TLR3				3 or 8		Binds double stranded RNA, activation of NK-kB
TLR4	Ly87, Rasl2-8			4	Peritoneal mac	Bacterial lipoproteins response, NF-kB and AP-1 activation
TLR5				1	mRNA: liver, lung, lower level in MOLF/Ei mice	Role in Gram-negative bacterial infection
TLR6				5	mRNA: spleen, thymus, ovary, lung	Activation of NF-kB and JunK
TLR7				X		
TLR8				X		
TLR9				6		CpG DNA receptor, TLR9KO resist lethal effect of CpG
TRAIL	Ly-81, APO-2L, Tnfsf10 TNFSF			Unknown	NK^{act}, liver NK	Apoptosis
TCR	αβ				T subset	Antigen recognition
TCR	γδ				T subset	Antigen recognition
TCR-Hy					Transgenic H-Y T cells	

Abbreviations:

Act	Activated	KO	Knock-out mouse
Ag	Antigen	LRRF	Leucine-rich repeat family
BM	Bone marrow	Mac	Macrophages
CCRSF	Complement component receptor superfamily	MHC	Major histocompatibility complex
CHO	Carbohydrate moiety	Mono	Monocytes
CRSF	Cytokine receptor superfamily	NK	Natural killer
DC	Dendritic cells	RTK	Receptor tyrosine kinase
ECM	Extracellular matrix	SRCRSF	Scavenger receptor cysteine-rich superfamily
Endoth	Endothelial cells	TM	Transmembrane
Epith	Epithelial cells	TM12SF	12-transmembrane spanning protein superfamily
FDC	Follicular dendritic cells	TM4SF	4-transmembrane spanning protein superfamily
GPI	Glycophosphatidylinositol	TM7SF	7-transmembrane spanning protein superfamily
Gran	Granulocytes	TNFRSF	TNF receptor superfamily
(H)	Human CD, not defined in mouse	TNFSF	TNF superfamily
IgSF	Immunoglobulin superfamily	TLRSF	Toll-like receptor superfamily
IntgF	Integrin family	W/	Associates with

Table 3.4   Mouse leukocyte antigen distribution chart

Antigen	T cell	B cell	Dendritic cell	NK cell	Monocyte	Granulocyte	Megakaryocyte	Erythrocyte	Endothelial cell
CD1d (CD1.1, Ly-38)	Most	Most	Most	Most	Most	Most	Unkn	Unkn	Unkn
CD2 (LFA-2)	Most/Dev	Sub/Dev	Unkn	Most	Sub	Most	Unkn	Most	Unkn
CD3 molecular complex	Most/Dev	ND	ND	Sub	ND	ND	ND	ND	ND
CD3e (CD3 ε chain)	Most/Dev	ND	ND	Sub	ND	ND	ND	ND	ND
CD4 (L3T4)	Sub/Dev	ND	Sub	Sub	Most/Dev	ND	Unkn	ND	Unkn
CD5 (Ly-1)	Most/Dev	Sub	Unkn	ND	Unkn	Unkn	Unkn	Unkn	Unkn
CD8a (Ly-2)	Sub/Dev	ND	Sub	ND	ND	ND	ND	ND	ND
CD8b (Ly-3)	Sub/Dev	ND	ND	ND	ND	ND	ND	ND	ND
CD9	Sub/Act	Sub	Most	Unkn	Most	Most/Dev	Most/Dev	ND	Unkn
CD11a (integrin α_L chain, LFA-1 α chain)	Most	Most	Most	Most	Most/Dev	Most	Unkn	ND	Unkn
CD11b (integrin α_M chain, Mac-1 α chain)	Sub/Act	Sub	Sub/Act	Sub	Most/Dev	Most/Dev	Unkn	ND	Unkn
CD11c (integrin α_X chain)	Sub/Act	Sub	Most	Most	Most/Dev	Most/Dev	Unkn	ND	Unkn
CD13	ND	ND	Most	Unkn	Sub/Dev	Unkn	Unkn	Unkn	Most
CD14	ND	ND	Unkn	ND	Sub/Act	Most/Act	Unkn	Unkn	Unkn
CD16/CD32 (Fcγ III/II receptor)	ND	ND	Unkn	Most	Most/Dev	Sub/Dev	Unkn	Unkn	Unkn
CD18 (integrin β_2 chain)	Most	Most	Most	Most	Most	Most	Unkn	ND	Unkn
CD19	ND	Most	Unkn	ND	ND	Sub	ND	ND	Unkn
CD21/CD35 (CR2/CR1, CD21a/CD21b)	ND	Most	Unkn	Unkn	Sub	Most/Act	ND	ND	Unkn
CD22.2 (Lyb-8.2)	ND	Most/Dev	Unkn	Unkn	Unkn	Unkn	Unkn	Unkn	Unkn

Table 3.4 Mouse leukocyte antigen distribution chart (*continued*)

Antigen	T cell	B cell	Dendritic cell	NK cell	Monocyte	Granulocyte	Megakaryocyte	Erythrocyte	Endothelial cell
CD23 (FCεRII)	ND	Sub/Act	Sub/Act	Unkn	Most	Sub	Most	Unkn	Unkn
CD24 (heat stable antigen)	Most/Dev	Most/Dev	Sub	ND	Most/Dev	Most	Unkn	Most	Unkn
CD25 (IL-2 receptor α chain, p55)	Most/Act+Dev	Most/Act+Dev	Sub/Dev	ND	Most/Dev	Unkn	Unkn	Unkn	Unkn
CD26 (THAM, DPP IV)	Most/Act	Most	Unkn	Unkn	Unkn	Unkn	Unkn	Unkn	Unkn
CD27	Most	Sub	Unkn	Sub	ND	ND	Unkn	Unkn	Unkn
CD28	Most/Act+Dev	ND	Unkn	Most	Unkn	Unkn	Unkn	Unkn	Unkn
CD29 (integrin β₁ chain)	Most	Most	Most	Most	Most	Most	Most	ND	Most
CD30	Most/Act	Most/Act	Unkn	Unkn	Unkn	Unkn	Unkn	Unkn	Unkn
CD31 (PECAM-1)	Sub	Sub	Most	Most/Act	Sub	Most	Most	Unkn	Most
CD34	ND	ND	ND	ND	ND	ND	ND	ND	Sub
CD35 (CR1, CD21b)	ND	Most	Sub	Unkn	Most	Most/Act	ND	ND	Unkn
CD36 (scavenger receptor)	Unkn	Most/Dev	Unkn	Unkn	Most/Dev	Unkn	Unkn	Unkn	Sub
CD38	Sub	Sub	Unkn	Most	Sub	Unkn	Unkn	ND	Unkn
CD40	Sub	Most	Sub	Unkn	Most/Act	Unkn	Unkn	Unkn	Unkn
CD41 (integrin α₁ᵦ chain)	ND	ND	Unkn	Unkn	Sub	Sub	Most	ND	Unkn
CD43 (Ly-48, leukosialin)	Most	Most/Dev	Unkn	Most	Most	Most	Most	Unkn	Unkn
CD43 activation-associated glycoform	Most/Act	Most/Dev	Unkn	Unkn	Most	Most	Unkn	Unkn	Unkn
CD44 (Pgp-1, Ly-24)	Most/Act+Dev)	Most/Act	Most	Most	Most	Most	Most	Unkn	Unkn
CD45 (leukocyte common antigen, Ly-5)	Most	Most	Most	Most	Most	Most	Most	Most/Dev	ND
CD45R/B220	Sub/Act	Most/Dev	Unkn	Sub/Act+Dev	Sub	ND	ND	ND	ND

Table 3.4 Mouse leukocyte antigen distribution chart (*continued*)

Antigen	T cell	B cell	Dendritic cell	NK cell	Monocyte	Granulocyte	Megakaryocyte	Erythrocyte	Endothelial cell
CD45RA	Sub	Most	ND	Unkn	Unkn	Unkn	Unkn	Unkn	ND
CD45RB	Sub/Dev	Most	Sub	Unkn	Sub	Unkn	Unkn	Unkn	ND
CD45RC	Sub	Most	Unkn	Unkn	Unkn	Unkn	Unkn	Unkn	ND
CD47 (IAP)	Most	Most	Most	Most	Most	Most	Most	Most	Most
CD48 (BCM1)	Most	Most	Unkn	Most	Most	Most	Unkn	ND	ND
CD49a (integrin α_1 chain)	Most/Act	ND	Unkn	Unkn	ND	Unkn	Unkn	ND	Sub
CD49b (integrin α_2 chain)	Most/Act	ND	Unkn	Sub	Unkn	Unkn	Most/Dev	ND	Sub
CD49c (integrin α_3 chain)	Unkn	Unkn	Unkn	Unkn	Unkn	Unkn	Unkn	ND	Unkn
CD49d (integrin α_4 chain)	Most	Most	Unkn	Unkn	Most	Unkn	Unkn	ND	Unkn
CD49e (integrin α_5 chain)	Sub/Act+Dev	ND	Unkn	Unkn	Sub	Sub	Unkn	ND	Unkn
CD49f (integrin α_6 chain)	Sub	Sub	Unkn	Unkn	Sub	Unkn	Most	ND	Sub
CD51 (integrin α_v chain)	Sub/Act	ND	Unkn	Unkn	Sub	Sub	Most/Dev	ND	Unkn
CD53	Most/Dev	Most	Most	Most	Most	Most	Unkn	ND	Unkn
CD54 (ICAM-1)	Sub/Act	Most/Act	Most/Act	Unkn	Most/Act	Most/Act	Unkn	ND	Most/Act
CD56 (N-CAM)	ND	ND	Unkn	ND	Unkn	Unkn	Unkn	Unkn	Unkn
CD61 (integrin β_3 chain)	ND	ND	Unkn	ND	Unkn	Sub	Most	ND	Most
CD62E (E-selectin, ELAM-1)	ND	ND	Unkn	ND	ND	ND	Unkn	Unkn	Most/Act
CD62L (L-seletin)	Most/Act+Dev	Most/Act+Dev	Unkn	Sub	Most	Most/Act	Unkn	Unkn	Unkn
CD62P (P-selectin)	ND	ND	Unkn	ND	ND	ND	Most/Act	Unkn	Most/Act
CD69 (Very Early Activation antigen)	Most/Act+Dev	Most/Act	Unkn	Most/Act	Unkn	Most/Act	Unkn	Unkn	Unkn
CD70	Most/Act	Most/Act	Unkn	Unkn	Unkn	Unkn	Unkn	Unkn	Unkn
CD71 (transferrin receptor)	Most/Act	Most/Act	Unkn	Most/Act	Most/Dev	Most/Dev	Unkn	Most/Dev	Unkn

Table 3.4 Mouse leukocyte antigen distribution chart (*continued*)

Antigen	T cell	B cell	Dendritic cell	NK cell	Monocyte	Granulocyte	Megakaryocyte	Erythrocyte	Endothelial cell
CD72a alloantigen (Lyb2.1)	ND	Most/Dev	Unkn	Unkn	Unkn	Unkn	Unkn	Unkn	Unkn
CD72b alloantigen (Lyb-2.2)	Sub/Act	Most/Dev	Unkn	Unkn	Unkn	Unkn	Unkn	Unkn	Unkn
CD72c alloantigen (Lyb-2.3)	ND	Most	Unkn	ND	ND	Unkn	Unkn	Unkn	Unkn
CD73 (Ecto-5′-nucleotidase)	Sub/Dev	Sub/Dev	ND	Unkn	Most/Dev	Most/Dev	Unkn	ND	Sub
CD74 (Ii)	ND	Most	Most	Unkn	Most/Act	ND	ND	ND	ND
CD79a (Igα, mb-1)	ND	Most/Dev	Unkn	Unkn	Unkn	Unkn	Unkn	Unkn	Unkn
CD79b (Igβ)	ND	Most/Dev	Unkn	Unkn	Unkn	Unkn	Unkn	Unkn	Unkn
CD80 (B7-1)	ND	Sub/Act	Most	Unkn	Most	Unkn	Unkn	Unkn	Unkn
CD81 (TAPA-1)	Most/Dev	Sub/Dev	Most	Unkn	Unkn	Unkn	Unkn	Unkn	Most
CD86 (B7-2)	Sub/Act	Sub/Act	Most/Act	Unkn	Most/Act	Unkn	Unkn	Unkn	Unkn
CD90 (Thy-1)	Most	ND	Sub	Most	Sub	ND	ND	ND	Unkn
CD94	Sub	ND	Unkn	Most	ND	ND	Unkn	ND	Unkn
CD95 (Fas)	Most/Dev	Sub/Act	Sub/Act	Unkn	Sub	Sub	Sub	Sub	Unkn
CD98 (4F2)	Most/Act	Most/Act	Unkn	Unkn	Most/Dev	Most/Dev	Unkn	Most/Dev	Unk
CD100	Unkn	Unkn	Unkn	Unkn	Unkn	Unkn	Unkn	Unkn	Unkn
CD102 (ICAM-2)	Most	Most	Sub	Unkn	Unkn	Unkn	Most	Unkn	Most
CD103 (integrin α$_{1EL}$ chain)	Sub/Dev	ND	Unkn	Unkn	Unkn	Unkn	Unkn	Unkn	ND
CD104 (integrin β$_4$ chain)	Sub/Dev	ND	Unkn	ND	ND	ND	Unkn	ND	Sub
CD105 (endoglin)	ND	ND	ND	ND	ND	ND	ND	ND	Most
CD106 (VCAM-1)	ND	ND	Sub	ND	Sub	ND	Unkn	ND	Most/Act
CD107a (LAMP-1)	Most/Act	Most/Act	Unkn	Unkn	Most/Dev	Most/Dev	Unkn	ND	Most

Table 3.4 Mouse leukocyte antigen distribution chart (*continued*)

Antigen	T cell	B cell	Dendritic cell	NK cell	Monocyte	Granulocyte	Megakaryocyte	Erythrocyte	Endothelial cell
CD107b (LAMP-2)	ND	ND	Unkn	Unkn	Most/Dev	Most/Dev	Unkn	ND	Unkn
CD117 (c-Kit)	Most/Dev	Most/Dev	Most/Dev	Most/Dev	Most/Dev	Most/Dev	Most/Dev	Most/Dev	Unkn
CD119 (IFN-γ receptor α chain)	Most	Most	Most	Most	Most	Most	Most	ND	Most
CD120a (TNFR receptor type I/p55)	Most	Most	Most	Most	Most	Most	Most	Unkn	Most
CD120b (TNFR receptor type II/p75)	Most	Most	Most	Most	Most	Most	Most	Unkn	Most
CD121a (IL-1 receptor type I/p80)	Most	ND	Most/Dev	Unkn	Unkn	Unkn	Unkn	Unkn	Sub
CD121b (IL-1 receptor, type II/p60)	Sub	Most	Sub	Unkn	Most/Dev	Most/Dev	Unkn	Unkn	Unkn
CD122 (IL-2 and IL-15 receptor β chain)	Sub/Dev	Sub	Unkn	Most	Most/Dev	Unkn	Unkn	Unkn	Unkn
CD123 (IL-3 receptor α chain)	Unkn	Most/Dev	Unkn	Unkn	Most	Most	Unkn	Unkn	Unkn
CD124 (IL-4 receptor α chain)	Most	Most	Sub	Unkn	Most	Most	Unkn	Most	Most
CD126 (IL-6 receptor α chain)	Most	Most	Sub	Unkn	Most	Most	Unkn	Most	Unkn
CD127 (IL-7 receptor α chain)	Most/Dev	Most/Dev	Unkn	Unkn	Sub	Unkn	Unkn	Unkn	Unkn
CD131 (β_{IL-3R}/β_c)	Sub	Sub	Unkn	Unkn	Sub	Sub	Unkn	Unkn	Unkn
CD132 (common γ chain; γ_c)	Sub	Sub	Sub	Sub	Sub	Sub	Unkn	ND	Unkn
CD134 (OX-40 antigen)	Most/Act	Sub/Act	Unkn	Unkn	Unkn	Unkn	Unkn	Unkn	Unkn
CD135 (Flk-2/Flt3, Ly-72)	Most/Dev	Most/Dev	Unkn	Unkn	Most/Dev	Most/Dev	Most/Dev	Most/Dev	Unkn
CD137 (4-1BB, Ly-63)	Most/Act	Unkn	Unkn	Most/Act	Unkn	Unkn	Unkn	Unkn	Unkn
CD138 (Syndecan-1)	ND	Most/Dev	Unkn	Unkn	Unkn	Unkn	Unkn	Unkn	Unkn
CD140a (PDGF receptor α chain)	ND	ND	ND	ND	ND	ND	ND	ND	ND
CD140b (PDGF receptor β chain)	Unkn	Unkn	Unkn	Unkn	Unkn	Unkn	Unkn	Unkn	Unkn
CD144 (VE-cadherin)	ND	ND	ND	ND	ND	ND	ND	ND	Most
CD152 (CTLA-4)	Most/Act	Unkn	Unkn	Unkn	Unkn	Unkn	Unkn	Unkn	Unkn
CD153 (CD30 ligand)	Sub/Act	ND	Unkn	Unkn	Unkn	Unkn	Unkn	Unkn	Unkn

Table 3.4 Mouse leukocyte antigen distribution chart (*continued*)

Antigen	T cell	B cell	Dendritic cell	NK cell	Monocyte	Granulocyte	Megakaryocyte	Erythrocyte	Endothelial cell
CD154 (CD40 ligand, gp39)	Sub/Act	ND	Unkn	Sub/Act	Unkn	Unkn	Unkn	Unkn	Unkn
CD157 (BP-3 alloantigen)	Sub/Dev	Sub/Dev	Unkn	Unkn	Most/Dev	Most/Dev	Unkn	Unkn	Unkn
CD162 (PSGL-1)	Sub	Unkn	Unkn	Unkn	Unkn	Sub	Unkn	Unkn	Unkn
CD171 (L1)	Unkn	Unkn	Unkn	Unkn	Unkn	Unkn	Unkn	Unkn	Unkn
CD172a (SIRPα, SHPS-1)	Unkn	Unkn	Unkn	Unkn	Sub	Unkn	Unkn	Unkn	Unkn
CD178 (Fas ligand, CD95 ligand)	Most/Act	Unkn	Unkn	Unkn	Unkn	Unkn	Unkn	Unkn	Unkn
CD179b (λ_5)	ND	Most/Dev	ND	ND	ND	ND	ND	ND	ND
CD180 (RP105)	ND	Most/Dev	Unkn	Unkn	Unkn	Unkn	Unkn	Unkn	Unkn
CD184 (CXCR4)	Most	Most/Act	Unkn	Unkn	Most	Most	Most	Unkn	Unkn
CD195 (CCR5)	Sub/Act	Unkn	Unkn	Most	Most/Act	Unkn	Unkn	Unkn	Unkn
CD200 (OX-2 antigen)	Sub	Most	Sub	ND	ND	ND	Unkn	ND	Sub
CD210 (IL-10 receptor)	Sub	Most	Unkn	Unkn	Sub	Unkn	Unkn	Unkn	Unkn
CD212 (IL-12 receptor β chain)	Sub	Sub	Sub	Most	Sub	Unkn	Unkn	Unkn	Unkn
CD220 (insulin receptor)	Unkn	Sub/Dev	Unkn	Unkn	Unkn	Unkn	Unkn	Unkn	Unkn
CD223 (LAG3)	Most/Act	Unkn	Unkn	Most/Act	Unkn	Unkn	Unkn	Unkn	Unkn
CD229.1 (Ly-9.1)	Most	Most	Unkn	Unkn	Sub	Sub	Sub	Sub	Unkn
CD244.1 (2B4 BALB alloantigen)	Sub/Act	Sub	ND	Most	ND	ND	ND	ND	ND
CD244.2 (2B4 B6 alloantigen)	Sub/Act	ND	ND	Most	ND	ND	ND	ND	ND
CD247 (CD3 ζ chain)	Most	ND	ND	Sub	ND	ND	ND	ND	ND
3G11 (disaloganglioside antigen)	Sub/Act/Dev	ND	Unkn	Unkn	Unkn	Unkn	Unkn	Unkn	Unkn
4-1BB ligand	ND	Sub/Act	Sub	Unkn	Sub/Act	Unkn	Unkn	Unkn	Unkn
ART2.2 (Rt6-2)	Most/Dev+Act	ND	Unkn	Sub	ND	ND	Unkn	Unkn	Unkn

Table 3.4   Mouse leukocyte antigen distribution chart (*continued*)

Antigen	T cell	B cell	Dendritic cell	NK cell	Monocyte	Granulocyte	Megakaryocyte	Erythrocyte	Endothelial cell
β_2 microglobulin	Most/ Dev	Most	Unkn	Unkn	Most	Most	Unkn	Unkn	Unkn
CC chemokine receptor 3 (CCR3)	Sub	Unkn	Unkn	Unkn	Unkn	Sub	Unkn	Unkn	Unkn
Crry/p65	Most	Most	Unkn	Unkn	Sub/ Act	Unkn	Most	Unkn	Sub
Cytokeratins	ND	ND	ND	ND	ND	ND	ND	ND	ND
Dendritic cells	ND	ND	Sub	ND	ND	ND	ND	ND	Unkn
Early B lineage	ND	Sub/Dev	Unkn	Sub	Sub	ND	Most	Unkn	Most
Ep-CAM	Sub/Dev	Sub/Act	Sub	Unkn	Sub	Sub	Unkn	Unkn	Unkn
Flk1 (VEGF-R2, Ly-73)	ND	ND	Unkn	ND	Unkn	Unkn	Unkn	Unkn	Most/ Dev
Follicular dendritic cell	ND	ND	Sub	ND	ND	ND	ND	ND	ND
Forssman antigen	ND	ND	Unkn	ND	Sub	ND	Unkn	Most/ Dev	ND
gp49 receptor	ND	ND	Unkn	Most/ Act	Most	Most	Unkn	Unkn	Unkn
H-2D	Most	Most	Most	Most	Most	Most	Most	Most	Most
H-2K	Most	Most	Most	Most	Most	Most	Most	Most	Most
H-2L	Most	Most	Most	Most	Most	Most	Most	Most	Most
H2-M (H2-DM)	ND	Most	Most	ND	Most	ND	Unkn	ND	Unkn
H2-M3	ND	Most	Sub	Unkn	Sub	ND	Unkn	ND	Unkn
I-A	ND	Most	Most	ND	Most	Unkn	Unkn	ND	Most/ Act
I-E	ND	Most	Most	ND	Most	Unkn	Unkn	ND	Most/ Act
ICOS	Sub/Act	ND	Unkn	ND	ND	ND	Unkn	Unkn	Unkn
Integrin β_7 chain	Most/ Dev	Most/ Dev	Unkn	Unkn	Most	Unkn	Unkn	Unkn	Unkn
Interferon-γ receptor β chain	Sub	Sub	Unkn	Unkn	Most	Unkn	Unkn	Unkn	Unkn
Interleukin-10 receptor	Sub	Sub	Unkn	Unkn	Sub	Unkn	Unkn	Unkn	Unkn
Ki-67	Most/Act	Most/ Act	Most/ Act	Most/ Act	Most/ Act	Most/ Act	Most/ Act	Most/ Act	Most/ Act

Table 3.4   Mouse leukocyte antigen distribution chart (*continued*)

Antigen	T cell	B cell	Dendritic cell	NK cell	Monocyte	Granulocyte	Megakaryocyte	Erythrocyte	Endothelial cell
KLRG1 (MAFA)	Sub	ND	Unkn	Sub	Unkn	ND	Unkn	Unkn	Unkn
LPAM-1 (integrin $\alpha_4\beta_7$ complex)	Most/Dev	Most/Dev	Unkn	Unkn	Most	Unkn	Unkn	Unkn	Unkn
Ly-6A/E (Sca-1)	Sub/Dev	Sub	Unkn	Unkn	Most	Most	Unkn	Unkn	Sub
Ly-6C	Sub/Act+Dev	Sub/Act	Unkn	Sub	Sub	Unkn	Unkn	Unkn	Most
Ly-6D (ThB)	Most/Dev	Most	Unkn	Unkn	Unkn	Unkn	Unkn	Unkn	Unkn
Ly-6G	ND	ND	Unkn	ND	Most/Dev	Most/Dev	Unkn	ND	Unkn
Ly-49	Sub/Act	ND	Unkn	Sub	Unkn	Unkn	Unkn	Unkn	Unkn
Ly-51 (6C3/BP-1 antigen)	ND	Most/Dev	Unkn	ND	ND	ND	Unkn	ND	Unkn
Mac-3	ND	ND	Unkn	Unkn	Sub/Dev	Unkn	Unkn	Unkn	Unkn
MAdCAM-1	ND	ND	Unkn	Unkn	ND	ND	Unkn	ND	Sub
NKCells/3A4	ND	ND	ND	Most	ND	ND	ND	ND	ND
NK-1:1 (NKR-P1B and NKR-P1C)	Sub	ND	ND	Most	ND	ND	ND	ND	ND
NK-T/NK cell antigen	Sub	ND	Unkn	Sub	Unkn	ND	Unkn	ND	Unkn
NKG2A/C/E	Sub	ND	Unkn	Sub	Unkn	Unkn	Unkn	Unkn	Unkn
Notch 1	Sub/Dev	ND	Unkn	ND	ND	ND	Unkn	ND	Unkn
OX-40 ligand	ND	Most/Act	Sub	Unkn	Unkn	Unkn	Unkn	Unkn	Unkn
Panendothelial cell antigen	ND	ND	Unkn	Unkn	ND	ND	Unkn	ND	Most
PD-1	Sub/Dev+Act	Sub/Dev+Act	Unkn	Unkn	Sub/Act	Sub/Act	Unkn	ND	Unkn
PIR-A/B	ND	Most	Sub	ND	Most	Most/Dev	Unkn	Unkn	Unkn
PNAd carbohydrate epitope (CD62L ligand)	ND	ND	Unkn	Unkn	Unkn	Unkn	Unkn	Unkn	Sub/Act
Pre-B cell receptor (Pre-BCR)	ND	Sub/Dev	Unkn	ND	ND	ND	Unkn	ND	Unkn
Pre-T cell receptor α chain (pTα)	ND	Sub/Dev	Unkn	ND	ND	ND	Unkn	ND	Unkn
Qa-1[b]	Most/Act	Most/Act	Unkn	Unkn	Most	Most	Unkn	Unkn	Unkn

Table 3.4 Mouse leukocyte antigen distribution chart (*continued*)

Antigen	T cell	B cell	Dendritic cell	NK cell	Monocyte	Granulocyte	Megakaryocyte	Erythrocyte	Endothelial cell
Qa-2	Sub/Act/Dev	Most	Unkn	Sub/Dev	Unkn	Unkn	Unkn	Unkn	Unkn
Siglec-F	ND	ND	Unkn	ND	Sub/Dev	Sub/Dev	Unkn	Unkn	Unkn
Syndecan-4	Sub	Most/Dev+Act	Unkn	Unkn	Sub/Act	Unkn	Unkn	Unkn	Sub/Act
T and B cell activation antigen (GL7, Ly-77)	Most/Act	Most/Act	Unkn	Unkn	Unkn	Unkn	Unkn	Unkn	Unkn
TCR α chain	Sub	ND	ND	ND	ND	ND	ND	ND	ND
TCR β chain	Sub	ND	ND	ND	ND	ND	ND	ND	ND
TCR γ chain	Sub	ND	ND	ND	ND	ND	ND	ND	ND
TCR δ chain	Sub	ND	ND	ND	ND	ND	ND	ND	ND
TER-119/erythroid cells (Ly-76)	ND	ND	Unkn	ND	ND	ND	ND	Most	ND
Thymic medullary epithelium	ND	ND	ND	ND	ND	ND	ND	ND	ND
TRAIL	ND	ND	ND	Most/Act	ND	ND	Unkn	ND	Unkn
TSA-1 (Sca-2, Ly-6E)	Most/Dev	Most	Unkn	Unkn	Sub	Sub	Unkn	Unkn	Unkn

Abbreviations:

ND	Not detected
Dev	Developmental marker
Most	On most cells
Sub	On subset of cells
Act	Activation-dependent

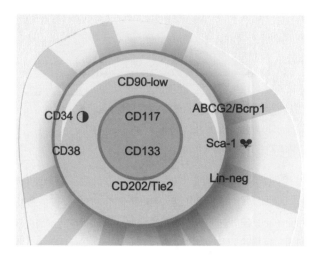

Figure 3.1   Surface antigens of hematopoietic stem cells

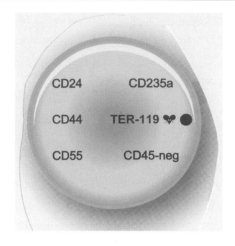

Figure 3.2   Surface antigens of erythrocytes

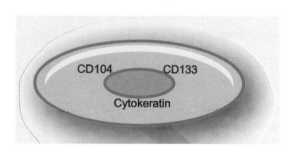

Figure 3.3   Surface antigens of epithelial cells

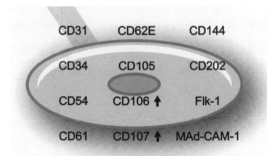

Figure 3.4   Surface antigens of endothelial cells

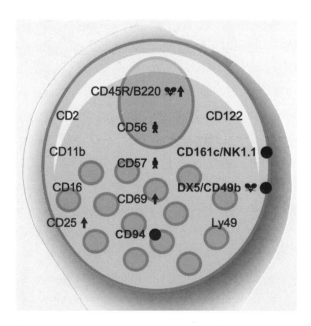

Figure 3.5   Surface antigens of natural killer cells

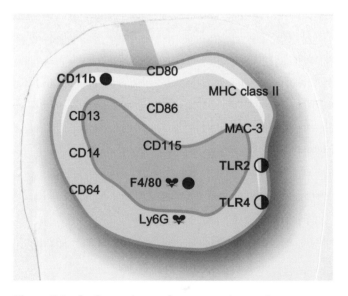

Figure 3.6   Surface antigens of monocytes/macrophages

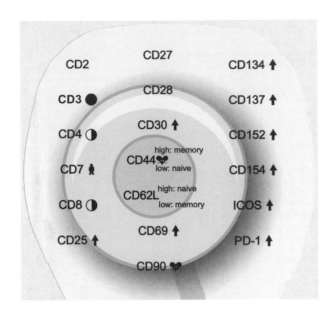

Figure 3.7 Surface antigens of T cells

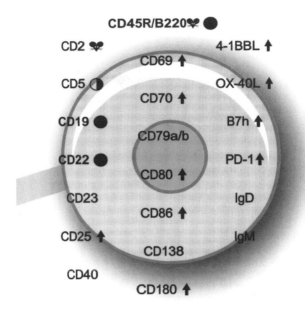

Figure 3.8 Surface antigens of B cells

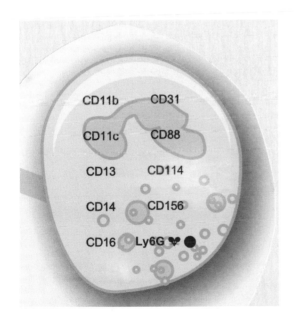

Figure 3.9 Surface antigens of granulocytes

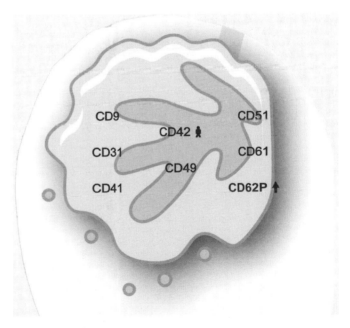

Figure 3.10 Surface antigens of megakaryocytes/platelets

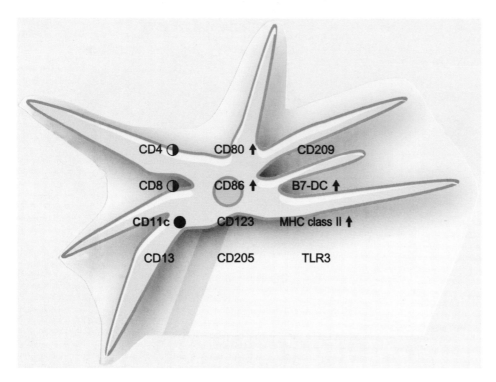

Figure 3.11   Surface antigens of dendritic cells

Table 3.5 Human leukocyte differentiation antigens

CD	Alternative name	HLDA section	Ligand/receptor/substrate/associated molecule	Description and function	MW (kDa)	T cell	B cell	Dendritic cell	NK cell	Stem cell/precursor	Macrophage/monocyte	Granulocyte	Platelet	Erythrocyte	Endothelial cell	Epithelial cell	Gene locus
CD1a	R4	T		Non-peptide antigen presenting molecules; involved in lymphocyte activation; related to thymic T cell development	49/-	⊕	⊕		−		⊕	−	−	−			1q22-q23
CD1b	R1	T		Non-peptide antigen presenting molecules; involved in lymphocyte activation; related to thymic T cell development	45/-	⊕	⊕		−		⊕	−	−	−			1q22-q23
CD1c	M241, R7	T		Non-peptide antigen presenting molecules; involved in lymphocyte activation; related to thymic T cell development. Expressed by a subset of peripheral blood B cells	43/-	⊕	+		−		⊕	−	−	−			1q22-q23
CD1d	R3	T		Non-peptide antigen presenting molecules; involved in lymphocyte activation; related to thymic T cell development		⊕	+		−			−	−	−		+^{c}	1q22-q23
CD1e	R4	T		Non-peptide antigen presenting molecules; involved in lymphocyte activation; related to thymic T cell development		+			−			−	−	−			1q22-q23
CD2	E-rosette R, T11, LFA-2	T		Receptor for CD58, CD48, CD59 and CD15; adhesion and signal-transducing molecule	50/-	+			+	−		−	−	−			1p13
CD3	T3	T		Associated with T cell receptor α/β or γ/δ dimer; signal transduction; assembly and expression of the T cell receptor complex	20-26	+	−		−	−	−	−	−	−	−	−	11q23
CD4	L3T4, W3/25	T	MHC Class II, gp120, IL-16	Co-receptor in antigen-induced T cell activation; thymic differentiation; regulation of T-B cell adhesion; primary receptor for HIV; binds to MHC class II. Also expressed in peripheral blood monocytes, tissue macrophages, granulocytes	55	+	−		−	−	+	−	−	−	−	−	12pter-p12
CD5	T1, Tp67, Leu-1	T	CD72, BCR, gp35-37	Co-stimulatory molecule; receptor for constitutive (CD72) and inducible (gp35-37) B cell-specific molecules	58/67	+	+		−		−	−	−	−			11q13
CD6	T12	T	CD166	Adhesion molecule. In thymocyte resistance to apoptosis and in positive selection; important in T mature cell response to both alloantigen and self-antigen	-/105-130	+	+		−		−	−	−	−			11q13
CD7	gp40	T		Possible co-activation/adhesion modulating molecule	40	+	−		+	+	−	−	−	−		−	17q25.2-q25.3
CD8α	Leu2, T8	T	MHC I, Lck	Co-receptor molecule; binds to MHC class I	68/30-34	+	−		+	−	−	−	−	−		−	2p12

89

Table 3.5 Human leukocyte differentiation antigens (*continued*)

CD	Alternative name	HLDA section	Ligand/ receptor/ substrate/ associated molecule	Description and function	MW (kDa)	T cell	B cell	Dendritic cell	NK cell	Stem cell/precursor	Macrophage/monocyte	Granulocyte	Platelet	Erythrocyte	Endothelial cell	Epithelial cell	Gene locus	CD
CD8β	CD8, Leu2, Lyt3	T		Co-receptor molecule; binds to MHC class I		+	–		–	–	–	–	–	–	–	–	2p12	CD8β
CD9	p24, DRAP-1, MRP-1	Platelet	CD63, CD31, CD82	Modulates cell adhesion and migration; triggers platelet activation; expressed on eosinophils and basophils	–/24,26	⊕	+			–	+	+	+	–	+	+	2p13	CD9
CD10	CALLA, NEP, gp100	B		Zinc metalloprotease; neutral endopeptidase; regulator of B cell growth and proliferation by hydrolysis of peptides with proliferative/anti-proliferative effects	100/–	–	–		–	+	–		–	–		+	3q25.1-q25.2	CD10
CD11a	LFA-1a, Integrin αL	Adhesion structure	ICAM-1,2,3	Intracellular adhesion and co-stimulation; binds to ICAM-1, ICAM-2, ICAM-3; expressed on eosinophils and basophils	170/180	+	+		+	–	+	+	–	–	–		16p11.2	CD11a
CD11b	Integrin αM MAC-1a	Adhesion structure	iC3b, Fibrinogen	Adherence of polymorphonuclear neutrophils and monocytes to fibrinogen, ICAM-1 endothelium; extravasation; chemotaxis; apoptosis	165/170	+	+	+	+	–	+	+	–	–	–		16p11.2	CD11b
CD11c	Integrin αX, p150.95a	Adhesion structure	iC3b	Adherence of polymorphonuclear neutrophils and monocytes to fibrinogen, ICAM-1 endothelium; binds iC3b-coated particles	145/150	+	+	+	+	–	+	+	–	–	–		16p11.2	CD11c
CDw12	p90-120	Myeloid		Function unknown	150–160/120	–	–		+	–	+	+						CDw12
CD13	APN, Gp150	Myeloid		Acts as receptor for coronavirus which causes upper respiratory tract infections; involved in interactions between human CMV and target cells; CD13 auto-Ab associated with GVHD	150/–	–	–		–	–	+	+	–	–	+	+	15q25-q26	CD13
CD14	LPS-R	Myeloid	LPS	Receptor for lipopolysaccharide (endotoxin)	53/55	–	–		–	–	+	+					5q31.1	CD14
CD15	X-hapten, Lewis X	Carbohydrate	CD62 selectin	May be important for direct carbohydrate–carbohydrate interactions		–	–		–	–	+	+	–	–	–			CD15
CD15s	Sialyl Lewis X	Carbohydrate and lectin	E-selectins	Expressed on myelomonocytic leukemia, some lyphocytic leukemia cells, and on adenocarcinomas		+	⊕		+		+	+			+			CD15s
CD15u	3' sulpho Lewis X	Carbohydrate and lectin	P-selectins	CD15 subgroups involved with different carbohydrate to carbohydrate cell adhesion		+	⊕		+		+	+		–	+			CD15u

90

Table 3.5 Human leukocyte differentiation antigens (*continued*)

CD	Alternative name	HLDA section	Ligand/receptor/substrate/associated molecule	Description and function	MW (kDa)	T cell	B cell	Dendritic cell	NK cell	Stem cell/precursor	Macrophage/monocyte	Granulocyte	Platelet	Erythrocyte	Endothelial cell	Epithelial cell	Gene locus	CD
CD15su	6 sulphosialyl Lewis X	Carbohydrate and lectin	L-selectins	CD15 subgroups involved with different carbohydrate to carbohydrate cell adhesion		+	⊕		+		+	+		–	+			CD15su
CD16	FcγRIIIa	NK	Fc	Low affinity receptor for IgG. Major histocompatibility complex	50–65/–	–	–		+	+	+	+	–	–				CD16
CD16b	FcγRIIIb	NK	Fc	Function unknown	N/A	+			+		+	+	–	–	+			CD16b
CDw17	None	Myeloid		Possible role in phagocytosis. Expressed in basophils	150–160/120	+	+	+	–		+	+	+	–	+	+		CDw17
CD18	Integrin β2	Adhesion structure	CD11a, b, c	Leukocyte adhesion	90/95	+	+		+		+	+		–	–		21q22.3	CD18
CD19	B4	B	CD2, CD81, CD225	A critical signal transduction molecule that regulates B cell development, activation and differentiation	90	–	+	+	–	+	–	–	–	–	–	–	16p11.2	CD19
CD20	B1, Bp35	B		Regulation of B cell activation and proliferation by regulating transmembrane Ca2+ conductance and cell cycle progression	37/35	–	+	–	–	–	–	–	–	–	–	–	11q12-q13.1	CD20
CD21	CR2, EBV-R, C3dR	B	C3d, CD23, CD19, CD81	Receptor for EBV and C3d, C3dg, and iC3b; subset of immature thymocytes; CD21 is part of a large signal transduction complex that also involves CD91, CD81, and Leu1	130–145	–	+	–	–	+	–	–	–	–	–	+	1q32	CD21
CD22	BL–CAM, Lyb8	B	p72sky, p53/56lyn, SHP1	Adhesion molecule; signaling molecule; antibody treatment of leukemia and lymphoma	135	–	+	–	–	–	–	–	–	–	–	–	19p13.1	CD22
CD23	FcεRII, B6, BLAST-2	B	IgE, CD21, CD11b, CD11c	Low affinity IgE receptor; regulates IgE synthesis; triggers monokine release; serum soluble CD23 level is a significant prognostic marker in CLL	50–45	–	+	–	–	–	–	–	–	–	–	–	19p13.3	CD23
CD24	BBA-1, HSA	B	P-selectin	Function unknown; homologous to mouse heat stable antigen; P-selectin on human carcinomas is involved in carcinoma binding to platelets	41/38	–	+	–	–	–	–	+	–	–	–	+	6g21	CD24
CD25	Tac antigen, IL-2Rα	CK/CKR	IL-2	IL-2 receptor α chain; associated with CD122 and CD132	55	⊕	+	–	+	–	–	–	–	–	–	–	10p15-p14	CD25

Table 3.5   Human leukocyte differentiation antigens (*continued*)

CD	Alternative name	HLDA section	Ligand/receptor/substrate/associated molecule	Description and function	MW (kDa)	T cell	B cell	Dendritic cell	NK cell	Stem cell/precursor	Macrophage/monocyte	Granulocyte	Platelet	Erythrocyte	Endothelial cell	Epithelial cell	Gene locus	CD
CD26	DPP IV ectoenzyme	T	Adenosine deaminase	Co-stimulatory molecule in T cell activation; associated marker of autoimmune diseases, adenosine deaminase-deficiency and HIV pathogenesis	120	+	+	−	+		−	−	−	−	−	+	2q24.3	CD26
CD27	T14, S152	T	CD70, TRAF5, TRAF2	Mediates a co-stimulatory signal for T cell activation. Involved in murine T cell development	110–120	+	+	−	+		−	−	−	−	−	−	12p13	CD27
CD28	Tp44, T44	T	CD80, CD86	Co-stimulates T cell proliferation and cytokine production with CD3; co-stimulates T cell effector function and T cell-dependent antibody production	90	+	−	−	−		−	−	−	−	−	+	2q33	CD28
CD29	Platelet GPIIa, β-1 integrin	Adhesion structure	VCAM-1 and MAdCAM-1	Critical molecule for embryogenesis and development; essential to the differentiation of hematopoietic stem cells; associated with tumor progression and metastasis/invasion	110–130	+	+		+		+	+		−	+	+	10p11.2	CD29
CD30	Ber-H2, Ki-1	Non-lineage	CD153, TRAF1,2,3,5	Member of TNFR family, involved in negative selection of T cells in thymus and TCR-mediated cell death; expressed on R-5 cells in Hodgkin's lymphomas	120	⊕	⊕	−	⊕		−	−	−	−	−	−	1p36	CD30
CD31	PECAM-1, endocam	Adhesion structure	CD38	Adhesion receptor with signaling function that participates in an adhesion cascade; transendothelial migration cell–cell adhesion	130–140	+	+	−	+		+	+	+	−	+	−	17q23	CD31
CD32	FCγRII	Non-lineage	Phosphatases	Regulates B cell functions; major player in immune complex-induced tissue damage	40	−	+	−	−		+	+	+	−	−		1q23	CD32
CD33	P67	Myeloid	Sugar chains	Diagnosis of acute myelogenous leukemia; negative selection for human self-regenerating hematopoietic stem cells	67	−		−		+	+	+	−	−	−		19q13.3	CD33
CD34	gp 105–120	Adhesion structure	L-selectin	Cell adhesion; CD34 also expressed on embryonic fibroblasts and nervous tissue	105–120	−	−	−	−	+	−	−	−	−	+		1q32	CD34
CD35	CR1, C3b/C4b receptor	Myeloid	C3b, C4b, iC3, iC4	C3b/C4b receptor; promotes phagocytosis (immune adherence); plays a major role in removal of immune complexes; regulates complement activation	160–250	+	+	+			+	+	−	+	−		1q32	CD35
CD36	GpIIIb, GPIV	Platelet	Thrombospondin	Recognition and phagocytosis of apoptotic cells; involved in platelet adhesion and aggregation; cytoadherence of plasmodium falciparum-infected erythrocytes	90	−	−	+		+	+	+	−	+	+	−	7q11.2	CD36
CD37	gp52-40	B	CD53, CD81, CD82, MHC II	Involved in signal transduction	40–52/40–52	+	+	−	−		+	+	−	−			19p13-q13.4	CD37

92

Table 3.5 Human leukocyte differentiation antigens (*continued*)

CD	Alternative name	HLDA section	Ligand/ receptor/ substrate/ associated molecule	Description and function	MW (kDa)	T cell	B cell	Dendritic cell	NK cell	Stem cell/precursor	Macrophage/monocyte	Granulocyte	Platelet	Erythrocyte	Endothelial cell	Epithelial cell	Gene locus	CD
CD38	ADP-ribosyl cyclase, T10	B	CD31	Regulates cell activation and proliferation; involved in lymphocyte and endothelial cell adhesion	45/45	+	+		+	+	+	–					4p15	CD38
CD39	None	B	ADP/ATP	May protect cells from lytic effects of extracellular ATP	80/80	⊕	+	+	+		+	–	–	–	+	+	10q24	CD39
CD40	Bp50	B	CD40L	Involved in B cell growth, differentiation and isotype switching; potent rescue signal from apoptosis; promotes cytokine production	85/48	–	+	+	–	+	+	–		–	+	+	20q12-q13.2	CD40
CD41	GPIIb, αIIb integrin	Platelet	Fg, Fn, vWF	CD41/CD61 complex plays a central role in platelet activation and aggregation	135/120, 23	–	–	–	–	+	+	–	+	–	–	–	17q21.32	CD41
CD42a	GPIX	Platelet	vWF, thrombin	Forms complex with GPIbα GPIbβ, and GPV, which binds to vWF and thrombin	22/17-22	–	–	–	–	+	–	–	+	–	–	–	3q21	CD42a
CD42b	GPIbα	Platelet	vWF, thrombin	Forms complex with GPIX, GPIbβ and GPV, which binds to vWF and thrombin	160/145	–	–	–	–	+	–	–	+	–	–	–	17pter-p12	CD42b
CD42c	GPIbβ	Platelet	vWF, thrombin	Forms complex with GPIX, GPIbα, and GPV, which binds to vWF and thrombin	160/24	–	–	–	–	+	–	–	+	–	–	–	22q11.21	CD42c
CD42d	GPV	Platelet	vWF, thrombin	Forms complex with GPIX GPIbα, and GPIbβ, which binds to vWF and thrombin	82/82	–	–	–	–	+	–	–	+	–	–	–	3	CD42d
CD43	Sialophorin, leukosialin	Non-lineage	Hyaluronan	Anti-adhesion molecules mediates repulsion between leucocytes and other cells; under some circumstances it may act as an adhesion molecule	95-135/ 95-135	+		–		+	+	+	+	–			16p11.2	CD43
CD44	ECMRII, H-CAM, Pgp-1	Adhesion structure	Hyaluronan	An adhesion molecule in lymphocyte-endothelial cell interaction; a differentiation antigen during lymphopoiesis; a potential marker of malignancy and metastasis	85/-	+	+		+		+	+	–	+	+	+	11p13	CD44
CD44R	CD44v, CD44v9	Adhesion structure	Hyaluronan	Involved in adhesion of leukocytes and endothelial cells; leukocyte homing	85/ 200/-	–					⊕		–	–	–	+	11p13	CD44R
CD45	LCA, T200	Non-lineage	p56, p59, Src kinases	Critical requirements for TCR- and BCR-mediated activation; possible requirement for receptor-mediated activation in other leukocytes	180-220/-	–	+	+	+	+	+	+	–	–	–	–	1q31-q32	CD45

Table 3.5 Human leukocyte differentiation antigens (*continued*)

CD	Alternative name	HLDA section	Ligand/receptor/substrate/associated molecule	Description and function	MW (kDa)	T cell	B cell	Dendritic cell	NK cell	Stem cell/precursor	Macrophage/monocyte	Granulocyte	Platelet	Erythrocyte	Endothelial cell	Epithelial cell	Gene locus
CD45RA		Non-lineage	p56, p59, Src kinases	Critical requirement for TCR- and BCR-mediated activation; expressed on resting/naïve T cells; possible requirement for receptor-mediated activation in other leukocytes	220	+	+	+	+	+	+	−	−	−	−	−	1q31-q32
CD45RB		Non-lineage	p45, p59, Src kinases	Critical requirement for TCR- and BCR-mediated activation; possible requirement for receptor-mediated activation in other leukocytes	220	+	+	+	+	+	+	+	−	+	−	−	1q31-q32
CD45RC		Non-lineage	p56, p59, Src kinases	Critical requirement for TCR- and BCR-mediated activation; possible requirement for receptor-mediated activation in other leukocytes	220	+	+	+	+	+	+	−	−	−	−	−	1q31-q32
CD45RO	UCHL-1	Non-lineage	p56, p59, Src kinases	Critical requirement for TCR- and BCR-mediated activation; expressed on activated/memory T cells; possible requirement for receptor-mediated activation in other leukocytes	180	⊕	+	+	+	+	+	+	−	−	−	−	1q31-q32
CD46	MCP	Non-lineage	SCR	Co-factor for factor I proteolytic cleavage of C3b and C4b	52–58/64–68	+	+		+		+	+	+	−	+	+	1q32
CD47	gp42, IAP, OA3	Adhesion structure	SIRP	Adhesion molecule; thrombospondin receptor	45–60/50–55	+	+		+		+	+	+	+	+	+	3q13.1-q13.2
CD47R	MEM-133	Non-lineage		CDw149 mAbs actually recognized with low affinity the CD47 glycoprotein	120/-	+	+		+		+	+	+	−	+	+	3q13.1-q13.2
CD48	Blast-1, Hu lym3	Non-lineage	CD2, lck, fyn	Adhesion molecule; acts as an accessory molecule for γ/δ T-cell recognition; as predicted for α/β T-cell antigen recognition	45/45	+	+		+		+	−	−	−	+	+	1q21.3-q22
CD49a	VLA-1α, α1 integrin	Adhesion structure	Collagen, laminin-1	Adhesion receptor	200/200	⊕	−		⊕		−	−	−	−	−		5
CD49b	VLA-2α, GPIa	Adhesion structure	Collagen, laminin	Adhesion molecule	150/160	⊕	+		+		+	−	+	−	+	+	5q23-31
CD49c	VLA-3α, α3 integrin	Adhesion structure	laminin-5, Fn, collagen	Component of adhesion receptor; associates with TM4 of protein; may be involved in signal transduction	145-150/125,30	−			−		+	−	+	−	+	+	17q21.31
CD49d	VLA-4α, α4 integrin	Adhesion structure	CD106, MAdCAM	Cell adhesion; lymphocyte migration; tethering or rolling and homing of T cells	145/150	+	+	+	+	+	⊕	−	−	−	+		2q31-q32

Table 3.5 Human leukocyte differentiation antigens (*continued*)

CD	Alternative name	HLDA section	Ligand/receptor/substrate associated molecule	Description and function	MW (kDa)	T cell	B cell	Dendritic cell	NK cell	Stem cell/precursor	Macrophage/monocyte	Granulocyte	Platelet	Erythrocyte	Endothelial cell	Epithelial cell	Gene locus
CD49e	VLA-5α, α5 integrin	Adhesion structure	Fibronectin, invasin	Adhesion molecule	160/135, 25	+		+	+	+	+			−	+	+	12q11-q13
CD49f	VLA-6α, α6 integrin, gplc	Adhesion structure	Laminins, invasin	Component of adhesion receptor; CD49f/CD29-mediated T cell binding to laminin receptor provides a co-stimulatory signal to T cells for activation and proliferation	150/125	+	+		+	+	+		+		+	+	2p14-q14.3
CD50	ICAM-3	Adhesion structure	LFA-1, integrin ad/b2	Co-stimulatory molecule; regulates LFA-1-/ICAM-1 and integrin-β1-dependent pathways adhesion; soluble form can be detected in the blood	110/140/-	+	+				+	+	−	−	+	−	19p13.3-p13.2
CD51	Integrin αv, VNR-α	Platelet	Arg-Gly-Asp	Involved in cell adhesion and signal transduction; role in bone metabolism and apoptosis; possible role in infection	150/124, 24						+	−	+		+		2q31-q32
CD52	CAMPATH-1, HE5	Non-lineage		CD52 antibodies are remarkably lytic for target cells, both with human complement and by antibody-dependent cellular cytotoxicity	25-29/25-29	+	+		+		+	+	−	−		+	1p36
CD53		Non-lineage	VLA-4, HLA-DR	Signal transduction; CD53 cross-linking promotes activation of B cells	32-42/	+	+	−	+		+	+	−	−	+	−	1p31-p12
CD54	ICAM-1	Adhesion structure	LFA-1, Mac-1, rhinovirus	Involved in immune reaction and/or inflammation; receptor for Rhinovirus or RBCs infected with malarial parasite; soluble form can be detected in the blood	90/95	+	+		+		+				+		19p13.3-p13.2
CD55	DAF	Non-lineage	SCR, CD97	Complement regulator by decay acceleration; ligand or protective molecule in fertilization; involved in signal transduction; soluble form can be detected in plasma and body fluid	55-70/80	+	+	+	+		+	+	+	+	+	+	1q32
CD56	Leu-19, NKH-1, NCAM	NK	NCAM, Heparin sulfate	Homophilic and heterophilic adhesion	140	+			+								11q23-q24
CD57	HNK1, Leu-7	NK	L-selectin, P-selectin, Laminin	Cell-cell adhesion	110-115				+								
CD58	LFA-3	Adhesion structure	CD2	Mediates adhesion between killer and target cells, antigen-presenting cells and T cells; activation of killer cells; co-stimulatory molecule	55-70	+	+	+	+		+	+		+	+	+	1p13
CD59	IF5Ag, H19	Non-lineage	C8-α, C9, lck, fyn	Associates with C9, inhibiting incorporation into C5b-8 preventing the completion of MAC formation	18-25	+			+		+	+		+			11p13

Table 3.5 Human leukocyte differentiation antigens (*continued*)

CD	Alternative name	HLDA section	Ligand/receptor/substrate/associated molecule	Description and function	MW (kDa)	T cell	B cell	Dendritic cell	NK cell	Stem cell/precursor	Macrophage/monocyte	Granulocyte	Platelet	Erythrocyte	Endothelial cell	Epithelial cell	Gene locus	CD
CD60a	GD3	Carbohydrate and lectin		Induces mitochondrial permeability transition during apoptosis; marker for malignant melanomas		+	+				+	+	+	−				CD60a
CD60b	9-0-acetyl CD3	Carbohydrate and lectin	9-0-acetyl-CD3	mAbs immunoreactive to CD60b have co-mitogenic activity of synovial T cells; also observed on some breast carcinomas and melanomas	90–94/120	+				+	+	+				+		CD60b
CD60c	7-0-acetyl CD3	Carbohydrate and lectin		T cell activation receptor; T cell activation by CD60c does not require co-stimulatory signals		+					+	+						CD60c
CD61	GP IIIa, β3 integrin	Platelet	Fibrinogen	CD41/61 mediates attachment of cells to diverse matrix proteins	90–110	+					+		+		+		17q21.32	CD61
CD62E	E-selectin	Adhesion structure	(CD15s)	Mediates leukocyte rolling on activated endothelium at inflammatory sites; may support cell adhesion during hematogenous metastasis and play a role in angiogenesis	115										+		1q22-q25	CD62E
CD62L	L-selectin	Adhesion structure	CD34, GlyCAM-1, M	Mediates lymphocyte homing to high endothelial venules or peripheral lymphoid tissue and leukocyte rolling on activated endothelium at inflammatory sites	74	+	+		+	−	+	+					1q23-q25	CD62L
CD62P	P-selectin, GMP-140	Platelet	CD162, CD24	Interaction of CD62P and CD162 mediates tethering and rolling of leukocytes on the surface of activated endothelial cells; mediates rolling of platelets on endothelial cells	120								+		+		1q22-q25	CD62P
CD63	LIMP, MLA1, gp55	Platelet	VLA-3, VLA-6, CD81	CD63 gene may play a role in tumor suppression; expression of CD63 in melanoma cells reduces metastasis	40–60			+			+	+	+		+		12-q12-q13	CD63
CD64	FCRI	Myeloid	IgG	Receptor-mediated endocytosis of IgG-antigen complexes; antigen capture for presentation to T cells. ADCC	72	−	−	+	−	+	+		−	−	−	−	1q21.2-q21.3	CD64
CD65	Ceramide, VIM-2	Myeloid	E-selectin	Function unknown							+	+	−	−				CD65
CD65s	Sialylated-CD65, VIM2	Myeloid	Possibly E- or P-selectin	VIM2 antibody has been described to inhibit phagocytosis and to induce phagocyte calcium flux and oxidative burst		−	−		−		+	+		−	−			CD65s

Table 3.5 Human leukocyte differentiation antigens (*continued*)

CD	Alternative name	HLDA section	Ligand/receptor/substrate/associated molecule	Description and function	MW (kDa)	T cell	B cell	Dendritic cell	NK cell	Stem cell/precursor	Macrophage/monocyte	Granulocyte	Platelet	Erythrocyte	Endothelial cell	Epithelial cell	Gene locus
CD66a	NCA-160, BGP	Myeloid		Homophilic and heterophilic adhesion; E-selectin binding; capable of activating granulocytes; functions as a receptor for Neisseria gonorrhea	140–180	−	−		−			+	−	−	−	+	
CD66b	CD67, CGM6, NCA-95	Myeloid		Capable of heterophilic adhesion and transmembrane signaling; capable of activating neutrophils	95–100	−	−		−			+	−	−	−		19q13.2
CD66c	NCA, NCA-50/90	Myeloid		Homophilic and heterophilic adhesion; E-selectin binding; capable of activating granulocytes; functions as a receptor for Neisseria gonorrhea	90	−	−		−			+	−	−	−	+	19q13.2
CD66d	CGM1	Myeloid		Capable of activating granulocytes. Functions as a receptor for Neisseria gonorrhea	35	−	−		−			+	−	−	−		19q13.2
CD66e	CEA	Myeloid		Homphilic and heterophilic adhesion	180–200	−	−		−				−	−	−	+	19q13.2-q13.2
CD66f	SP-1, PSG	Myeloid		Unclear, may be involved in immune regulation and regulation and protection of fetus from maternal immune system; necessary for successful pregnancy	54–72	−	−		−				−		−	+	19q13.2
CD68	gp110, macrosialin	Myeloid	LDL	Lysosomal membrane glycoprotein (LAMP 1 group); possible receptor	110	+c	+c	+c		+c	+c	+c		−	−		17p13
CD69	AIM, EA 1, MLR3, gp34/28	NK		Involved in lymphocyte, monocyte, and platelet activation	60	⊕	⊕		⊕		⊕	⊕	+				12p13-p12
CD70	Ki-24	Non-lineage	CD27	Co-stimulation of T and/or B cells; enhances the proliferation of cytotoxic T cells and cytokine production. Co-stimulates B cell proliferation and Ig production	55–170	⊕	⊕	−	−	−		−		−	−	−	19p13
CD71	T9, Transferrin receptor	Non-lineage	Transferrin	Controls the supply of iron uptake during proliferation	190	−	−	−		+		−	−	−	+	−	3q26.2-qter
CD72	Ly-19.2, Ly-32.2, Lyb-2	B	CD5	Plays a role in downregulation of signaling through the BCR on B cells as a regulator of signaling thresholds	43/39	−	+	+		+		−	−	−	−	−	9p
CD73	Ecto-5-nucleotidase	B	AMP	Hydrolyzes adenosine monophosphate into adenosine; can mediate co-stimulatory signals in T cell activation	69–72	+	+	+				−	−	−	+	+	6q14-q21
CD74	invariant chain	B	HLA-DR, CD44	Intracellular sorting of MHC class II molecules; also known as Class II specific chaperone Ii	41	+	+				+				+	+	5q32

Table 3.5   Human leukocyte differentiation antigens (*continued*)

CD	Alternative name	HLDA section	Ligand/receptor/substrate/associated molecule	Description and function	MW (kDa)	T cell	B cell	Dendritic cell	NK cell	Stem cell/precursor	Macrophage/monocyte	Granulocyte	Platelet	Erythrocyte	Endothelial cell	Epithelial cell	Gene locus	CD
CD75	Sialo-masked lactosamine	Carbohydrate and lectin		CD75 is newly clustered including CDw75 and CDw76, CDw76 has been deleted		−	+		−	−	+	−	−	+	−	−		CD75
CD75s	a2, 6 sialylated lactosamine	Carbohydrate and lectin	CD22 (proposed)	May be involved in regulation of CD95-mediated apoptosis and may be important for infection by a lymphotropic virus		+	+		−	−	+	+	−	+	+	+		CD75s
CD77	Pk antigen/BLA/CTH/Gb3	B	Receptor for Shiga toxin	Cross-linking of CD77 induces apoptosis in Burkitt's lymphoma cells	1	−	+		−	−	−	−	−	−	+	+		CD77
CD79a	Ig α/MB1	B	Ig/CD5/CD19/CD22/CD79b	Transmits signals into cytoplasm upon antigen-binding to surface Igs	40–45	−	+		−	−	−	−	−	−			19q13.2	CD79a
CD79b	Ig β/B29, BCR	B	Ig/CD5/CD19/CD22/CD79a	B cell antigen receptor (BCR) mediates the response of B cells to foreign antigens and determines the fate of B cells during development and differentiation	−/37	−	+		−	−	−	−	−	−	−	+	17q23	CD79b
CD80	B7-1/BB1	B	CD28/CD152 (CTLA-4)	Co-regulation of T-cell activation with CD86	60/−	⊕	⊕	+	−	−	+	−	−	−	−	−	3q13.3-q21	CD80
CD81	TAPA-1	B	Leu-13/CD19/CD21	Member of CD19/CD21/Leu-13 signal transduction complex. #Or ly on eosinophils, not neutrophils	26/−	+	+	+	+	−	−	+#	−	−	+	+	11p15	CD81
CD82	4F9/C33/IA4/KAI1/R2	B		Signal transduction. #Also associates with MHC class I and II, β1 integrins, CD4 and CD8	45–90/−	+	+	+	+	−	+	+	+	−	+	+	11p11.2	CD82
CD83	HB15	B	Unknown	Function unknown	−/43	−	+	+	−	−	−	−	−	−	−	−	6p23	CD83
CD84	None	B	Unknown	Function unkown, some indication that it may be a signaling molecule	68–80	+	+		−	−	+	−	−	−	−	−	1q24	CD84
CD85a*	ILT5/LIR3/HL9	Dendritic cell	HLA class I	Contains ITIM sequences in cytoplasmic tail; involved in the suppression of NK-mediated cytotoxicity		+	−	+	−	−	+	+	−	−	−	−	19q13.4	CD85a*

Table 3.5 Human leukocyte differentiation antigens (*continued*)

CD	Alternative name	HLDA section	Ligand/receptor/substrate/associated molecule	Description and function	MW (kDa)	T cell	B cell	Dendritic cell	NK cell	Stem cell/precursor	Macrophage/monocyte	Granulocyte	Platelet	Erythrocyte	Endothelial cell	Epithelial cell	Gene locus	CD
CD85b*	ILT8	NK	FcRγ	Involved with activation of NK-mediated cytotoxicity													19q13.4	CD85b*
CD85c*	LIR8	NK	FcRγ	Involved with activation of NK-mediated cytotoxicity													19q13.4	CD85c*
CD85d*	ILT4/LIR2/MIR 10	Dendritic cell	HLA class I	Contains ITIM sequences in cytoplasmic tail; involved in the suppression of NK-mediated cytotoxicity	110	–	–	+	–	–	+	+	–	–	–	–	19q13.4	CD85d*
CD85e*	ILT6/LIR4	NK	FcRγ	Involved with activation of NK-mediated cytotoxicity													19q13.4	CD85e*
CD85f*	ILT11	NK	FcRγ	Involved with activation of NK-mediated cytotoxicity. Mainly expressed on PBL													19q13.4	CD85f*
CD85g*	ILT7	NK	FcRγ	Involved with activation of NK-mediated cytotoxicity													19q13.4	CD85g*
CD85h*	ILT1/LIR7	NK	FcRγ	Involved with activation of NK-mediated cytotoxicity. Expressed on myeloid cells and some NK cells													19q13.4	CD85h*
CD85i*	LIR6a	NK	FcRγ	Involved with activation of NK-mediated cytotoxicity													19q13.4	CD85i*
CD85j*	ILT2/LIR1/MIR7	Dendritic cell	HLA class I	Contains ITIM sequences in cytoplasmic tail; involved in the suppression of NK-mediated cytotoxicity	110	+	+	+	+	–	+	+	–	–	–	–	19q13.4	CD85j*
CD85k*	ILT3/LIR5/HM18	Dendritic cell	HLA class I	Ligation of CD85K induces an inhibitory signal via recruitment of SHP-1 phosphatase	60			+		+	+	+			–	–	19q13.4	CD85k*
CD85L*	ILT9	NK	FcRγ	Binds FcRγ													19q13.4	CD85L*
CD85m*	ILT10	NK	FcRγ	Binds FcRγ													19q13.4	CD85m*
CD86	B7-2/B70	B	CD28/CD152 (CTLA-4)	Co-regulator of T cell activation with CD80	–/80	⊕	⊕	+	–		+	–	–	–	+	–	3q21	CD86
CD87	uPAR	Myeloid	uPA/Pro-UPA/vitronectin	CD87 serves as the cellular receptor for pro-uPA and uPA	35–68/32–66	+	–	+	+		+	+	–	–	+	–	19q13	CD87
CD88	C5aR	Myeloid	C5a/C5a(desArg)	C5a-mediated inflammation; activation of granulocytes	43/–	–	–	+	–		+	+	–	–	+	+	19q13.3-q13.4	CD88
CD89	IgA FC receptor	Myeloid	IgA1/IgA2	Induces phagocytosis, degranulation, respiratory burst, and the killing of microorganisms	45–100/45–100	–	–	+	–		+	+	–	–	–	+	19q13.2-q13.4	CD89

Table 3.5 Human leukocyte differentiation antigens (continued)

CD	Gene locus	Epithelial cell	Endothelial cell	Erythrocyte	Platelet	Granulocyte	Macrophage/monocyte	Stem cell/precursor	NK cell	Dendritic cell	B cell	T cell	MW (kDa)	Description and function	Ligand/receptor/substrate/associated molecule	HLDA section	Alternative name	CD
CD90	11q22.3-q23	−	+	−	−	−	−	+	−		−	−	25–35/25–35	May contribute to lymphocyte co-stimulation, inhibition of stem cell proliferation/differentiation and neuron memory formation	CD45/lck/fyn/P100	Endothelial cell	Thy-1	CD90
CD91	12q13-q14	+	−	−		−	+	+#	−		−	−	600/-	Endocytosis-mediating receptor expressed in coated pits. #Expressed on erythroblast/reticulocytes	ALPHA2M/LDLs	Myeloid	ALPHA2M-R/LRP	CD91
CD92		+	+	−		+	+		−		+	+	70/70	Function unkown	Unknown	Myeloid	None	CD92
CDw93		−	+	−		+	+	−	−	−	−	−	110/120	Function unknown	Unknown	Myeloid	None	CDw93
CD94	12q13	−	−	−		−	−	−	+		−	+	70, 30	Assembled with other C-type lectins (NKG2) forms inhibitory or activating receptors for HLA class I	HLA class I	NK	Kp43	CD94
CD95	10q24.1			−		+	+		+		+	+	45, 90, 200/45	Receptor molecule for Fas ligand, which mediates apoptosis-inducing signals	Fas ligand	Cytokine receptor	APO-1, FAS, TNFRF6	CD95
CD96			−	−		−	−	−	⊕		−	⊕	160/-	Adhesion of activated T and NK cells during the late phase of immune response; weakly expressed by peripheral resting NK or T cells, upregulated after activation		NK	TACTILE	CD96
CD97	19p13.2-p13.12			−		+	+		⊕	+	⊕	⊕	/28, 75–85	Member of the EGF-TM7 family; weakly expressed on resting lymphocytes, upregulated by activation	CD55	Non-lineage		CD97
CD98	11q13	+	+	−	+	+	+		+		+	+	125/80, 45	Possible amino acid transporter; broad reactivity on activated and transformed cells, not hematopoietic specific, and found at lower levels on quiescent cells	actin	Non-lineage	4F2, FRP-1, RL-388	CD98
CD99	Xp22.32, Yp11.3		+	+	+	−	+		+		+	+	32/32	Modulates T-cell adhesion; induces apoptosis of double-positive thymocytes; expressed on all hemological cells and present on many other cell types		T	MIC2, E2	CD99
CD99R	9q22-q31			−	−	+	+		+	−	−	+	32/32	Modulates T-cell adhesion; induces apoptosis of double-positive thymocytes		T	CD99 Mab restricted	CD99R
CD100	9q22-q31	−	−			+	+		+	−	+	+	300/150	Co-stimulatory molecule for T-cells; increases PMA, CD3, and CD2 induced T cell proliferation; Soluble form is 120 kD	CD45, serine kinase	Non-lineage	SEMA4D	CD100

Table 3.5 Human leukocyte differentiation antigens (*continued*)

CD	Alternative name	HLDA section	Ligand/receptor/substrate/associated molecule	Description and function	MW (kDa)	T cell	B cell	Dendritic cell	NK cell	Stem cell/precursor	Macrophage/monocyte	Granulocyte	Platelet	Erythrocyte	Endothelial cell	Epithelial cell	Gene locus	CD
CD101	IGSF2, P126, V7	Myeloid		Co-stimulatory molecule; antibodies against CD101 inhibit allogenic T cell responses and co-stimulate T cell proliferation with suboptimal anti-CD3 activation	240/120	⊕	-	+	+		+	+	-	-			1p13	CD101
CD102	ICAM-2	Adhesion structure	LFA-1, CD11b/CD18	Provides co-stimulatory signal in immune response; lymphocyte recirculation; expressed on some resting lymphocytes	55–65/	+			+		+	-	+	-	+		17q23-q25	CD102
CD103	HML-1, integrin αE	Adhesion structure	E-cadherin; integrin β7	Expressed on intestinal intraepithelial lymphocytes, lamina propria T cells in intestine; stimulation of PBL with PHA induce CD103 expression	175/150,25	⊕			-		-	-	-	-	-		17p13	CD103
CD104	β4 integrin chain, TSP-180	Adhesion structure	Laminins (I, II, IV, V), CD49F	Hemidesmosomal CD49f/CD104 (α6β4 integrin) plays an important role in the adhesion of epithelia to basement membranes. #CD4+CD8- pre-T cells	205/220	-			-	+#	+	-	-	-	+	+	17q11-qter	CD104
CD105	Endoglin	Endothelial cell	TGF-β 1, TGF-β 3	Regulatory component of the TGF β receptor complex; modulator of cellular responses to TGF-1 β	180/90	-			-	+		-	-	-	+		9q33-q34.1	CD105
CD106	VCAM-1, INCAM-110	Endothelial cell	Integrin α4β1	Leukocyte adhesion, transmigration and co-stimulation of T cell proliferation; expressed on activated endothelial cells, follicular dendritic cells, and certain tissue macrophages#	110/110	-		+#	-		+#	+#	-	-	⊕		1p32-p31	CD106
CD107a	LAMP-1	Platelet		Possible role in cell adhesion; highly metastatic tumor cells express more LAMP molecules on the cell surface than poorly metastatic cells; expressed on lysosomal membrane	100–120/	⊕			-			⊕	⊕	-	⊕	+	13q34	CD107a
CD107b	LAMP-2	Platelet		Possible role in cell adhesion; highly metastatic tumor cells express more LAMP molecules on the cell surface than poorly metastatic cells; expressed on lysosomal membrane	100–120/	-			-		-	⊕	⊕	-	⊕		Xq24	CD107b
CD108	SEMA7A, JMH	Non-lineage	CD232	Function unknown; carries JMH blood group antigen; expressed at low levels on circulating lymphocytes, at moderately high levels by cells and lymphoblastic cell lines	76/80	+	+				-	-	-	+			15q22.3-q23	CD108
CD109	8A3, 7D1, E123	Endothelial cell		Function unknown	170/170	⊕	-		-	+	-	-	⊕	-	+	+		CD109
CD110	TPO-R, MPL, C-MPL	Platelet	TPO	Receptor for TPO. Receptor binding results in the prevention of apoptosis. stimulation of cell growth and differentiation of megakaryocyte and platelet formation	82–92/	-	-		-	+	-	-	+	-	-		1p34	CD110

Table 3.5 Human leukocyte differentiation antigens (*continued*)

CD	Alternative name	HLDA section	Ligand/receptor/substrate/associated molecule	Description and function	MW (kDa)	T cell	B cell	Dendritic cell	NK cell	Stem cell/precursor	Macrophage/monocyte	Granulocyte	Platelet	Erythrocyte	Endothelial cell	Epithelial cell	Gene locus
CD111	HveC, PRR1, PVRL1, nectin1	Myeloid	gD, nectin3, afadin	Intercellular adhesion molecule; involved in epithelial cell physiology; pan-alphaherpes virus entry receptor	-/75	-	-			+	+	+	-	+	+	+	11q23-q24
CD112	HveB, PRR2, PVRL2, nectin2	Myeloid	PRR3, afadin	Homophilic adhesion receptor that could play a role in the regulation of hematopoietic/endothelial cell functions; involved in cell to cell spreading of viruses	72, 64/ 72, 64	-	+			+	+	+	+	-	+	+	19q13.2-13.4
CD114	CSF3R, G-CSFR, HG-CSFR	Myeloid	G-CSF, Jak1, Jak2	Regulates myeloid proliferation and differentiation	130	-	-		-	+		+		-	+		1p35-p34.3
CD115	c-fms, CSF-1R, M-CSFR	Myeloid	CSF-1	Receptor for CSF-1 (macrophage colony stimulating factor); mediates all of the biological effects of this cytokine	150	-	-		-	+	+		-	-			5q33.2-33.3
CD116	GM-CSFRα	CD/CKR	GM-CSF	Primary binding subunit of GM-CSF with low affinity and binds it with high affinity when it is coexpressed with the common beta subunit CDw131	80	-	-	+	-	+		+	-	-	-	-	Xp22.32 or Yp11.3
CD117	c-KIT SCRF	CK/CKR	SCF, MGF, KL	Growth factor receptor, tyrosine kinase	145	-	-		-	+			-	-			4q11-q12
CDw119	iFN-γR, IFNγRa	CK/CKR	IFNγ	Interferon γ binding	80–95	+	+	+	+			+	+	-	+	+	6q23-q24
CD120a	TNFRI, p55	CK/CKR	TNF, TRADD, TRAF, RiP, LTa	Programmed cell death anti-viral activity; receptor for TNF	55	+	+	+	+			+	+	-	+	+	12p13.2
CD120b	TNFRII, p75, TNFR p80	CK/CKR	TNF, TRADD, TRAF, RiP, LTa	Programmed cell death anti-viral activity; receptor for TNF	75	+	+	+	+			+	-	-	+	+	1p36.3-p36.2
CD121a	IL-1R type 1	CK/CKR	Il-1α and Il-1β	IL-1 signaling	75–85/ 75–85	+	+	-	-			-	-	-	-	+	2q12
CD121b	IL-1R type 2	CK/CKR	IL-1β, IL-1Rα, IL-1α	Negative regulator of IL-1	60–68/ 60–68	+	+				+	+	-	-	+	+	2q12-q22

Table 3.5 Human leukocyte differentiation antigens (*continued*)

CD	Alternative name	HLDA section	Ligand/receptor/substrate/associated molecule	Description and function	MW (kDa)	T cell	B cell	Dendritic cell	NK cell	Stem cell/precursor	Macrophage/monocyte	Granulocyte	Platelet	Erythrocyte	Endothelial cell	Epithelial cell	Gene locus	CD
CD122	IL2Rβ	CK/CKR	IL-2, IL-15, CD25, CD132	Critical component of IL-2 and IL-15-mediated signaling	70–75/-	+	+		+		+	-	-	-	-		22q13.1	CD122
CD123	IL-3Rα subunit	CK/CKR	IL-3	Primary low affinity binding subunit of IL-3 receptor	70	-	-	+	-	+	+	+	-		+		Xp22.3 or Yp11.3	CD123
CD124	IL-4R	CK/CKR	IL-4, IL-13	Receptor subunit for IL-4 and IL-13; #expression on B cells is upregulated by LPs, anti-IgM or IL-4; #on T cells is increased by stimulation with ConA or IL-4	140/-	⊕#	⊕#		-		+	+	-	-	-		16p11.2-12.1	CD124
CD125	IL-5Rα	CK/CKR	IL-5	Low affinity receptor for IL-5; alpha chain of IL-5 receptor; expressed on eosinophils and basophils	60/-		⊕		-		-	+	-	-	-		3p26-p24	CD125
CD126	IL-6R	CK/CKR	IL-6	Required, in association with gp130(CD130), for mediating biological activities of interleukin-6; expressed on hepatocytes and some non-hematopoietic cells	80/80	+	⊕		-		+	-	-	-	-		1q21	CD126
CD127	IL-7Rα	CK/CKR	IL-7, CD132, fyn, yn, Jak1	Specific receptor for IL-7; expression downregulated following T cell activation	65–90/-	+	+		-	+	-	-	-	-	-		5p13	CD127
CD128a	Il-8RA, CXCR1	CK/CKR	IL-8	Critical regulation of IL-8 mediated neutrophil chemotaxis and activation; potential role in angiogenesis	44, 59, 67–70	+			+		+	+	+		+		2q35	CD128a
CD128b	IL-8RB, CXCR2	CK/CKR	IL-8	Critical regulators of IL-8 mediated neutrophil chemotaxis and activation; potential role in angiogenesis	44, 59, 67–70	+	-		+		+	+	+		+		2q35	CD128b
CD130	gp130	CK/CKR	Oncostatin M	Required for transducing biological activities of IL-6, IL-11, LIP, ciliary neurotrophic factor, oncostatin M, and cardiotrophin-1	130–140/130–140	+	-		+	-	+	+	+		+		5q11	CD130
CD131	Common beta subunit	CK/CKR	CD123, CD125, CD116	Key signal transducing molecule of the IL-3, GM-CSF, and IL-5 receptors; expressed on early B cells and early progenitors	120–140/-	-			-	+		+					22q13.1	CD131
CD132	IL-2Rγ	CK/CKR	IL-12	Common subunit of IL-2, IL-4, IL-7, IL-9, IL-15 receptors; mutation causes X-linked severe combined immunodeficiency (XSCID)	65–70/-	+	+		+		+	+	+				Xq13.1	CD132

Table 3.5 Human leukocyte differentiation antigens (*continued*)

CD	Alternative name	HLDA section	Ligand/receptor/substrate/associated molecule	Description and function	MW (kDa)	T cell	B cell	Dendritic cell	NK cell	Stem cell/precursor	Macrophage/monocyte	Granulocyte	Platelet	Erythrocyte	Endothelial cell	Epithelial cell	Gene locus	CD
CD133	AC133	Stem Cell	N/A	Used for positive selection of hematopoietic stem and progenitor cells for transplantation studies	120	-	-	-	-	+	-	-	-	-	+	+	4p16.2	CD133
CD134	OX40	CK/CKR	OX40 ligand	Receptor for OX40 ligand; co-stimulatory signal transducer of T cell receptor-mediated activation, cell adhesion	48–50/-	⊕	-	-	-		-						1p36	CD134
CD135	Flt3, FLK2, STK1	CK/CKR	FL	Receptor tyrosine kinase; co-stimulatory molecule; survival receptor; growth factor receptor for early hematopoietic progenitors	130/155–160	-	-	-	-	+	+	-		-			13q12	CD135
CDw136	MSP-R, RON	CK/CKR	MSP, HGFI	Chemotactic migration, morphological change, cell growth, cytokine induction, phagocytosis, and cell differentiation	180/150, 40						+					+	13p21.3	CDw136
CDw137	4-1BB, ILA	CK/CKR	4-1BB ligand	Receptor for 4-1BB ligand; co-stimulatory molecule	85/39	⊕	+		-		+	-				+	1p36	CDw137
CD138	Syndecan1	B		Extracellular matrix receptor, co-receptor for fibroblast growth factor signaling receptors; #expressed on plasma cells	-/165-150	-	+#	-	-		+	-		-		+	2p24.1	CD138
CD139	None	B		Function unknown	209/228	-	+	+	-	+	+	+		+				CD139
CD140a	PDGF a receptor	Endothelial cell	PDGF	Involved in signal transduction associated with PDGF receptors; expressed on mesenchymal cells	160, 180/-	-	-		-			-	+	-			4q11-q13	CD140a
CD140b	PDGF B receptor	Endothelial cell	PDGF	Involved in signal transduction associated with PDGF receptors; expressed on mesenchymal cells	160, 180/-	-	-		-		+	+	-	-	+		5q31-q32	CD140b
D141	Thrombomodulin	Endothelial cell	Thrombin, protein C	Critical for activation of protein C and initiation of the protein C anticoagulant pathway; co-factor in the thrombin-mediated activation of protein C	75/105	-	-		-	+	+	+	-	-	+		20p12-cen	CD141
CD142	Tissue factor	Endothelial cell	Factor VIIIa, factor Xa/TFPI	Initiator of the blood clotting cascade; cell surface receptor/cofactor for factor VII; can be induced by inflammatory mediators	45–47/45-47	-	-		-		-	-	-	+	+	+	1p22-p21	CD142

Table 3.5 Human leukocyte differentiation antigens (continued)

CD	Gene locus	Epithelial cell	Endothelial cell	Erythrocyte	Platelet	Granulocyte	Macrophage/monocyte	Stem cell/precursor	NK cell	Dendritic cell	B cell	T cell	MW (kDa)	Description and function	Ligand/receptor/substrate/associated molecule	HLDA section	Alternative name	CD
CD143	17q2	+	+	–	–	–	–		–	–	–	+	90, 170/ 90, 170	Angiotensin-converting enzyme, peptidyl dipeptidase, is necessary for spermatozoa to bind to eggs	ANG-1, bradykinin	Endothelial cell	ACE	CD143
CD144		–	+	–	–	–	–		–	–	–	–	135/ 130	Controls endothelial permeability, growth, migration, and contact inhibition of cell growth; expressed only on endothelial cells	β-catenin, p120 CAS	Endothelial cell	VE-cadherin, cadherin-5	CD144
CDw145			+	–	–	–	–		–	–	–	–	25, 90, 110	Highly expressed on endothelial cells; antibodies were originally raised against human urinary bladder carcinoma cells		Endothelial cell	None	CDw145
CD146	11q23.3		+	–	–	–	–		–	–	–	⊕	118/ 130	Potential adhesion molecule; expressed by melanoma, smooth muscle, and intermediate trophoblasts		Endothelial cell	Muc 18 S-endo	CD146
CD147	19p13.3		+	+	+	+	+		+		+	+	50–60/ 55–95	Potential adhesion molecule; involved in regulation of T cell function		Endothelial cell	Neurothelin, OX-47	CD147
CD148	11p11.2		+	–	+	+	+			+	+	+	200–260/ 200–260	HPTP-etc/Dep-1 involved in contact inhibition of cell growth; chromosomal location region frequently detailed in carcinoma		Non-lineage	HPTPn, p260 DEP-1	CD148
CD150	1q22-q23		+	–	–	–		+	–	+	+	+	65–85/ 75–95	An important molecule associated with intracellular adaptor protein SAP. Absence of SAP causes X-linked lymphoproliferative disease	Tyrosine phosphatase CD45	Non-lineage	SLAM-1, IPO-3	CD150
CD151	11p15.5	+	+	–		–	–	+	–	–	–	–	32/–	Integrin-associated protein; transmembrane signaling	β1 integrins	Platelet	PETA-3	CD151
CD152	2q33			–	–	–				–	⊕	⊕	50/33	Receptor for CD80/CD86; negative regulator of T cell activation		T	CTLA-4	CD152
CD153	9q33					+	⊕					+	40	Co-stimulatory signal for peripheral blood T cells	CD30	T	CD30L	CD153
CD154	Xq26		–	–		–				–	–	⊕	33	Essential for germinal center formation and antibody class switching; co-stimulatory molecule; regulator of TH1 generation and function	Ligand for CD40	T	CD40L, gp39, TRAP-1, T-BAH	CD154
CD155	19q13.2				–		+			–			60–90	Possible interaction with CD44	Polio virus receptor	Myeloid	PVR	CD155
CD156a	10q26.3					+	+						–/69	Possible involvement in extravasation of leukocytes	Myeloid	Myeloid	CD156, ADAM8, MS2	CD156a

Table 3.5 Human leukocyte differentiation antigens (*continued*)

CD	Alternative name	HLDA section	Ligand/receptor/substrate/associated molecule	Description and function	MW (kDa)	T cell	B cell	Dendritic cell	NK cell	Stem cell/precursor	Macrophage/monocyte	Granulocyte	Platelet	Erythrocyte	Endothelial cell	Epithelial cell	Gene locus	CD
CD156b	TACE, ADAM17	Adhesion structure	Pro-INF1 pro-TGFα, MAD2	Cleavers the transmembrane form of TNF-α to yield the soluble active form	100–120	+	–	+	–	–	+	+	–	–	+	–	2p25	CD156b
CD157	Mo5, BST-1	Myeloid		A sister molecule of CD38, a type II membrane protein with identical ectoenzyme activity; a distribution complementary to that of CD38	42–45			+		+	+	+			+			CD157
CD158a[†]	KIR2DL1, p58.1	NK	HLA-Cw4, 2,5,6	Contains ITIM sequences in cytoplasmic tail; involved in the suppression of NK-mediated cytotoxicity	58/58	+			+								19q13.4	CD158a[†]
CD158b1[†]	KIR2DL2, p58.2	NK	HLA-3.1, 7, 8	Contains ITIM sequences in cytoplasmic tail; involved in the suppression of NK-mediated cytotoxicity	58/58	+			+								19q13.4	CD158b1[†]
CD158b2[†]	KIR2DL3, p58.3	NK	HLA-Cw3, 1, 7, 8	Contains ITIM sequences in cytoplasmic tail; involved in the suppression of NK-mediated cytotoxicity	58/58	+			+								19q13.4	CD158b2[†]
CD158c1[†]	KIR2DS6, KIRX	NK		Contains ITIM sequences in cytoplasmic tail; involved in the suppression of NK-mediated cytotoxicity		+			+								19q13.4	CD158c1[†]
CD158d[†]	KIR2DL4	NK		Function unknown					+								19q13.4	CD158d[†]
CD158 e1/e2[†]	KIR3DLl/S1, p70	NK	HLA-Bw4	Involved in the suppression of NK-mediated cytotoxicity (KIR3DL1); expressed on subsets of NK and cytotoxic cells	70/70	+			+								19q13.4	CD158 e1/e2[†]
CD158f[†]	KIR2DL5	NK		Contains ITIM sequences in cytoplasmic tail; involved in the suppression of NK-mediated cytotoxicity		+			+								19q13.4	CD158f[†]
CD158g[†]	KIR2DS5	NK	HLA-C	Associated with KARAP/DAP12; involved in the activation of NK-mediated cytotoxicity		+			+								19q13.4	CD158g[†]
CD158h[†]	KIR2DS1, p50.1	NK	HLA-C	Associated with KARAP/DAP12; involved in the activation of NK-mediated cytotoxicity		+			+								19q13.4	CD158h[†]
CD158i[†]	KIR2DS4, p50.3	NK	HLA-C	Associated with KARAP/DAP12; involved in the activation of NK-mediated cytotoxicity	50/50	+			+								19q13.4	CD158i[†]
CD158j[†]	KIR2DS2, p50.2	NK	HLA-C	Associated with KARAP/DAP12; involved in the activation of NK-mediated cytotoxicity		+			+								19q13.4	CD158j[†]
CD158k[†]	KIR3DL2, p140	NK	HLA-A	Contains ITIM sequences in cytoplasmic tail; involved in the suppression of NK-mediated cytotoxicity	140/70	+			+								19q13.4	CD158k[†]

Table 3.5 Human leukocyte differentiation antigens (continued)

CD	Alternative name	HLDA section	Ligand/receptor/substrate/associated molecule	Description and function	MW (kDa)	T cell	B cell	Dendritic cell	NK cell	Stem cell/precursor	Macrophage/monocyte	Granulocyte	Platelet	Erythrocyte	Endothelial cell	Epithelial cell	Gene locus
CD158z†	KIR3DL7, KIRC1	NK		Contains ITIM sequences in cytoplasmic tail; involved in the suppression of NK-mediated cytotoxicity													19q13.4
CD159a	NKG2A	NK	HLA-E	CD94/CD159a heterodimer constitutes a potent negative regulator of NK- T and cell activation programs; expressed on subsets of NK and CD8+ (γδ) cells	70/43	+			+								12p12.3-p13.1
CD160	BY55, NK1, NK28	T	MHC class I	Cross-linking CD160 with certain mAbs triggers co-stimulatory signals in CD8 T cells. CD160 is also expressed on all intestinal intraepithelial lymphocytes	80/27	+	−		+								1q42.3
CD161	NKR-P1A	NK		NK cell cytolytic activity; regulation of thymocyte precursor proliferation	80/40	+	−	−	+				−	−	−	−	
CD162	PSGL-1	Adhesion structure	P-selectin	Binds P- and L-selectins; can mediate leukocyte rolling	160-250/110-120	+	+			+	+	+	−	−	−		
CD162R	PEN5	NK	L-selectin	Post-translational modification of the P-selectin glycoprotein ligand-1 (CD162); developmentally regulated marker of both immune and neural cells	240/140				+								
CD163	M130, GHI/61, RM3/1	Myeloid		Expressed on tissue macrophages and LPS activated monocytes	110	−	−	−			⊕	−	−	−	−		
CD164	MGC-24, MUC-24	Adhesion structure		Facilitating the adhesion of human CD34+ cells to stroma and by negatively regulating CD34+CD38−progenitor cell proliferation	160/80	+	+				+			−	−	+	
CD165	Ad2, gp37	Adhesion structure		Adhesion of thymocytes to thymic epithelial cells; expressed on many T cell acute lymphoblastic leukemia (ALL)	37/42					−	+	−	+	−		+	
CD166	ALCAM, KG-CAM	Adhesion structure	Binds CD6	Adhesion receptor	100-105/100-105	⊕	+				⊕	−			+	+	
CD167a	DDR1	Adhesion structure	Collagen	Adhesion molecule, DDR1 overexpression in several human cancers suggests a function in tumor progression	52-62	+	+	+								+	6p21.3
CD168	RHAMM	Adhesion structure	CD44	Involved in adhesion of early thymocyte progenitors to matrix and its interaction with HA can mediate signals to other cell adhesion molecules	52-125	−	−	+		+	+						5q33.2

Table 3.5   Human leukocyte differentiation antigens (*continued*)

CD	Alternative name	HLDA section	Ligand/receptor/substrate/associated molecule	Description and function	MW (kDa)	T cell	B cell	Dendritic cell	NK cell	Stem cell/precursor	Macrophage/monocyte	Granulocyte	Platelet	Erythrocyte	Endothelial cell	Epithelial cell	Gene locus	CD
CD169	Sialoadhesion/Siglec-1	Adhesion structure	MUC1, CD206	Mediates cell–cell, cell matrix interaction; may facilitate phagocytosis	180/200			+			+						20p13	CD169
CD170	Siglec-5	Adhesion structure	Terminal sialic acid residues	Adhesion molecule; as a pattern or self/non-self recognition receptor and mediates negative signals into the cell	140			+			+	+					19q13.3	CD170
CD171	N-CAM, L1	Adhesion structure	CD56, CD24	Neuronal cell recognition molecule L1 involved in cell adhesion, cell spreading and motility. Also acts as a co-stimulatory molecule on lymphocytes	200–230	+	+	+			+				+		Xq28	CD171
CD172	SIRP-1a	Adhesion structure	CD47	Adhesion molecule; binds to CD47 and may mediate inhibitory signals via the ITIM/SHP-2	65			+		+	+	+		+			20p13	CD172
CD173	Blood group H type 2	Carbohydrate and lectin		Biosynthetic precursor of A and B antigen; carcinoma-associated antigen; may be involved in the homing process of hematopoietic stem cells to the bone marrow					-	+					+			CD173
CD174	Lewis Y	Carbohydrate and lectin		New hematopoietic progenitor cell marker; may be involved in the homing process of hematopoietic stem cells to the bone marrow					-	+						+		CD174
CD175	Tn	Carbohydrate and lectin	TFRA	Tumor-specific antigen expressed on various carcinomas; histo–blood group-related carbohydrate antigen; precursor of the blood groups ABO and TF antigen		-	-		-	+						+		CD175
CD175s	Sialyl-Tn (s-Tn)	Carbohydrate and lectin	TFRA	Tumor-specific antigen expressed on various carcinomas; histo–blood group-related carbohydrate antigen; precursor of the blood groups ABO and TF antigen		-	+		-	+					+	+		CD175s
CD176	TF antigen	Carbohydrate and lectin	TFRA	Pan–carcinoma antigen tumor antigen marker; may be involved in metastasis of tumor cells	120–198		-			+				+	+	+		CD176
CD177	NB1, HNA-2a	Myeloid		NB antigens play a critical role in autoimmune neonatal neutropenia and autoimmune neutropenia; polymorphic; expressed in 89–97% of healthy individuals	49/55/56–64							+						CD177

Table 3.5   Human leukocyte differentiation antigens (continued)

CD	Alternative name	HLDA section	Ligand/receptor/substrate/associated molecule	Description and function	MW (kDa)	T cell	B cell	Dendritic cell	NK cell	Stem cell/precursor	Macrophage/monocyte	Granulocyte	Platelet	Erythrocyte	Endothelial cell	Epithelial cell	Gene locus	CD
CD178	Fas ligand	CKR	CD95(Fas)	Involved in Fas/Fas ligand interaction, apoptosis, regulates immune responses, #Expressed on immature dendritic cells	27–40/27–40	⊕	–	+#	+		–	+	–	–	+	+	1q23	CD178
CD179a	VpreB	B	CD179b, μ heavy chain	Surrogate light chain VpreB is one of the components of the pre-B-cell receptor complex. #Expressed in cytoplasm of pro-B cells and on the surface of pre-B cells	16–18/					+#							22q11.22	CD179a
CD179b	λ5, 14.1	B	CD179a, μ heavy chain	λ5 is one of the components of the pre-B-cell receptor complex. #Expressed in cytoplasm of pro-B cells and on the surface of pre-B cells	–/22					+#							22q11.23	CD179b
CD180	RP105/Bgp95	B	LPS, MD-1	May regulate the LPS signaling in B cells in concert with TLR4; ligation of CD180 induces proliferation of B cells and increases susceptibility to BCR-induced cell death	95–105/95–105		+	+			+						5q12	CD180
CD183	CXCR3	CK/CKR	IP-10, Mig, I-TAC	Involved with inflammation-associated effector T-cell chemotaxis	40–41	+	+	+			+						8p12-p11.2, Xq13	CD183
CD184	CXCR4	CK/CKR	HIV-1	Homing receptor of hematopoietic progenitor cells; co-stimulation of B cells; induces apoptosis; involved with the entry of HIV-1		+	+	+		+	+	+	+	–	+	+	2q21	CD184
CD195	CCR5	CK/CKR	HIV-1	Regulates lymphocyte chemotaxis activation and transendothelial migration during inflammation. Neutralizes HIV infection. #Expressed on immature dendritic cells	37.0/40.6	+	–	+#	–		+	+	–	–	–			CD195
CDw197	CCR7, EBI1, BLR2	CK/CKR	SLC/6Ckine, ELC/MIP-3b	Lymphocytes and dendritic cell homing to lymphoid organs	90	+	+	+	+		+	+	–	–			9p13	CDw197
CD200	OX2	Non-lineage	OX2R	Ig-SF, OX2 shares many biochemical similarities with Thy-1; may regulate myeloid cell activity	40–45		+	+	–	+	–	–	–	–	+			CD200
CD201	EPCR	Endothelial cells	Protein C	Involved in protein C activation	49/25										+		20q11.2	CD201
CD202b	TEK/Tie2	Endothelial cells	Angiopoietin-1,2, and 4	Involved in vascular development	140										+		9p21	CD202b

Table 3.5 Human leukocyte differentiation antigens (*continued*)

CD	Alternative name	HLDA section	Ligand/receptor/substrate/associated molecule	Description and function	MW (kDa)	T cell	B cell	Dendritic cell	NK cell	Stem cell/precursor	Macrophage/monocyte	Granulocyte	Platelet	Erythrocyte	Endothelial cell	Epithelial cell	Gene locus	CD
CD203c	PDNP3, B10, PDIβ, E-NPP3	Myeloid	Oligonucleotides	Multi-functional ectoenzyme involved in the clearance of extracellular nucleotides. #Expressed on basophils, mast cells, and their precursors	270/130, 150					+#		+#					6q22	CD203c
CD204	MSR	Myeloid	LDL	Role in deposition of cholesterol through receptor mediated uptake of LDL; recognition and elimination of pathogenic microorganisms	220						+							CD204
CD205	DEC-205	Dendritic cell	Unknown	Antigen-uptake receptor for mannosylated antigens; present on both CD11c+ blood dendritic cells and in lesser density on surface of T and B cells	198	+	+	+		+	+							CD205
CD206	MMR	Dendritic cell	Sialodinesins and CD45	Mediates endocytosis of glycoconjugates with terminal mannose, fucose N-acetylglucosamine or glucose residues	162–175			+		+	+						10p13	CD206
CD207	Langerin	Dendritic cell		Found on a subset of cultured blood CD11c+ DC and TGF beta differentiated MoDC. Provides new reagent for characterizing Langerhans histiocytosis. Endocytic receptor with functional lectin domain with mannose specificity				+		+	+						2p13	CD207
CD208	DG-LAMP	Dendritic cell		Function unknown. Possible participation in peptide loading onto MHC class II				+		+							3q26.3-q27	CD208
CDw209	DC-SIGN	Dendritic cell		Expressed on MoDC but not on blood DC even after activation; contributes to the initial adhesion interaction between MoDC and naïve T cells, regulation of T cell proliferation	44	–	–	+									19p13	CDw209
CDw210	CK	CK/CKR	IL-10	Receptors involved with cell signaling and immune regulation	90	–	+	–									11q23.3, 21q22.11	CDw210
CD212	CK	CK/CKR	IL-12	Tyrosine kinase membrane receptor for angiopoietin; involved in cell signaling and immune regulation	–/110	⊕	–		+		–	–					19p13.1	CD212
CD213a	CK	CK/CKR	IL-13	Receptors involved in cell signaling and immune regulation. CD213a1, CD213a2		–	–		–		+	–			+		x13	CD213a
CDw217	CK	CK/CKR	IL-17	Involved in inflammation, osteogenesis, and granulopoiesis		+	+				+	+					2p31	CDw217
CD220	Insulin R	Non-lineage	Insulin	Functions in the clearance of ligands rather than intracellular signaling		+	+		–	+	+			+			19p13.3	CD220

Table 3.5 Human leukocyte differentiation antigens (*continued*)

CD	Alternative name	HLDA section	Ligand/receptor/substrate/associated molecule	Description and function	MW (kDa)	T cell	B cell	Dendritic cell	NK cell	Stem cell/precursor	Macrophage/monocyte	Granulocyte	Platelet	Erythrocyte	Endothelial cell	Epithelial cell	Gene locus
CD221	IGF1 R	Non-lineage	Insulin	Functions in the clearance of ligands rather than intracellular signaling		+	+		+	+	+	+					15q25-26
CD222	M6P/IGFII-R	Non-lineage	Plasminogen, M6P and IGFII	Plays role in the transport of newly synthesized acid hydrolases to lysomes	250	+	+		+		+	+		+			
CD223	LAG-3	Non-lineage	MHC class II	Cell activation Gene-3, like CD4, interacts with MHC class II molecules	70	⊕	-	-	⊕	+	-	-					12p13
CD224	GGT	Non-lineage	GSH	G-glutamyl transderase; ectoenzyme; maintains intracellular glutathione (GHS) concentrations and consequently a state of oxidative homeostasis within cellular microenvironments	27–68	+	+		-	+	+	-				+	
CD225	Leu 13	Non-lineage	IFN-γ	Interferon-inducible protein may play role in controlling cell–cell interactions	17	+	+		+	-	-	-			+	-	
CD226	DNAM-1, PTA1	T	LFA-1	Adhesion molecule; cytolytic function mediated by CTL and NK cells; platelet and T cell activation antigen 1	65	+	+		+	+	+	-	+	-	-		18q22.3
CD227	MUC1	Non-lineage	CD54, CD169	Involved in cell surface protection and modulation of adhesion and cell migration	300	⊕	+	+	+	+	⊕					+	
CD228	p97, gp95, MT	Non-lineage		GPI-anchored melanoma-associated protein	97	-	-	-		+	-	-			+		
CD229	Ly9	Non-lineage	SAP protein	In activated T cells, the SAP protein binds to and regulates signal transduction events initiated through the engagement of SLAM, 2B4, CD84, and Ly-9	100	+	+		+	-	-	-		-			1q22
CD230	Prion Protein (PrP)	Non-lineage		Isoform PrPsc (pathological) is present in transmissible spongiform encephalitis (TSE)	33–37	+	+	+	+	+	+	-					
CD231	TALLA-1/A15, TALLA	Non-lineage		Highly expressed on T cell acute lymphoblastic leukemia; can be potentially useful as an anti-tumor agent	30–45										+		
CD232	VESP-R	Non-lineage	CD108	Receptor for CD108 and semaphorin from virus; A39R (protein of semaphorin family) upregulates ICAM-1 and induces cytokine production	200		+		+		+	+		+			
CD233	Band 3/AE1	RBC		Carrier of the Diego blood group system; maintains red cell morphology; Band 3 is essential for terminal erythroid differentiation	95–105	-	-	-	-	-	-	-	-	+	-		17q12-q21

Table 3.5 Human leukocyte differentiation antigens (*continued*)

CD	Alternative name	HLDA section	Ligand/receptor/substrate/associated molecule	Description and function	MW (kDa)	T cell	B cell	Dendritic cell	NK cell	Stem cell/precursor	Macrophage/monocyte	Granulocyte	Platelet	Erythrocyte	Endothelial cell	Epithelial cell	Gene locus	CD
CD234	DARC/Fy-glycoprotein	RBC	IL-8, MGSA RANTES, MCP-1	Carrier of the Duffy blood group system; binds to a number of chemokines to modulate the intensity of inflammatory reactions	34-43	-	-	-	-	+	-	-	-	+	+	+	1q22-23	CD234
CD235a	Glycophorin A	RBC		Major membrane sialoglycoprotein of RBC membrane and carrier of blood group M and N specificites		-	-	-	-	-	-	-	-	+	-		4q28-q31	CD235a
CD235ab	Glycophorin A and B	RBC		Glycophorin B is the carrier of blood group S, s, and N specificities (for Glycophorin A see CD235a)		-	-	-	-	-	-	-	-	+		-	4q28-q31	CD235ab
CD236	Glycophorin C and D	RBC		One of the chored protein of red blood cell skeleton that maintains cell morphology; carrier of Gerbich blood group	30-40					+				+			2q14-q21	CD236
CD236R	Glycophorin C, GYPC	RBC		Plays a role in the invasion and intra-erythrocytic development of *P. falciparum*	40									+			2q14-q21	CD236R
CD238	Kell	RBC	Endothelin-3	Kell is classified as a member of the small neprilysin (M13) family of zinc metalloproteases, which include CD10. Kell antibodies inhibit erythropoiesis	93									+			7q33	CD238
CD239	Lu/B-CAM	RBC	Laminin	Carrier of the Lutheran blood group; receptor for laminin; plays role in terminal erythroid differentiation; facilitates trafficking of more mature RBC	78-85									+	+	+	19q13.2	CD239
CD240		RBC		CD240 includes CD240CE (RhCE), CD240D (RhD), and CD240DCE (RhD/RhCE). Rh system is one of the most polymorphic in the blood group system comprising 45 different antigens Rh antigen may promote export of ammonium	30									+			1p34.3-p36.1	CD240
CD241	RhAG/Rh50	RBC		Promotes export of ammonium that accumulates within erythrocytes; promotes erythrocyte-mediated retention of ammonium from the plasma and its release to detoxifying organs	50	-		-	-	-	-	-	-	+			6p11-p21.1	CD241
CD242	ICAM-4/LW	RBC	LFA-1, Mac-1, VLA-4	Carrier of LW blood group system; involved in red cell senescence; interaction with VLA-4 may stabilize erythroblastic islands in normal BM	37-43	-		-	-	+	-	-	-	+	+		19p13.3	CD242
CD243	MDR-1	Stem/progenitor cells		p-glycoprotein, drug resistance pump	180	-	-		-	+	-	-		-				CD243

Table 3.5 Human leukocyte differentiation antigens (*continued*)

CD	Alternative name	HLDA section	Ligand/receptor/substrate/associated molecule	Description and function	MW (kDa)	T cell	B cell	Dendritic cell	NK cell	Stem cell/precursor	Macrophage/monocyte	Granulocyte	Platelet	Erythrocyte	Endothelial cell	Epithelial cell	Gene locus	CD
CD244	2B4, P38, NAIL	NK	CD48	Engagement of 2B4 with its ligand, CD48, or with specific antibodies enhances NK cell cytokine production and cytolytic function. #Found only on basophils	70/70	+			+		+	+#					1q22	CD244
CD245	p220/240	T	Lymphocyte receptor	Signal transduction and co-stimulation of T and NK cells; function is distinct from CD45 or CD148	220-250	+	+				+	+	+	-				CD245
CD246	ALK	T	Tyrosine kinase R	Expressed in T cell lymphoma subtype; suggested role in cellular proliferation, apoptosis and embryonic neural differentiation	200										+		2p23	CD246
CD247	Zeta chain	T		Essential signal sub-unit of activating receptor on T and NK cells		+			+								2p23	CD247

Key:
+ Positive
⊕ Positive upon activation
+c Positive by cytoplasm staining
− Negative
+# Refer to 'Description and function' column for further details
* A CD nomenclature of detailing LIR/ILT genes (CD85) as well as KIR genes (CD158) has been proposed based on the previous
† CD designation of some members of this family and on the position of the genes on chromosomes 19q13.4 from centromeric to telomeric loci

Abbreviations:
MW   Molecular weight is shown as non-reduced/reduces where available
CK/CKR Cytokine/chemokine receptors

Table 3.6 Human leukocyte differentiation antigens

CD no.	Session	Main antigen expression	Family
CD1a	T	Cortical mature thymocytes, dendritic cell subset, Langerhans cells	IgSF
CD1b	T	Cortical mature thymocytes, dendritic cell subset, Langerhans cells	
CD1c	T	Cortical mature thymocytes, dendritic cell subset, Langerhans cells, B cell subset	
CD1d	T	Cortical thymocytes, dendritic cell subset, Langerhans cells, intestinal epithelium	
CD1e	T	Cortical thymocytes, dendritic cell subset	
CD2	T	Thymocytes, T cells, most NK cells, B cells	IgSF
CD2R	T	Activated T cells, NK cells	
CD3d	T	T cells	T cell receptor complex
CD3e	T	T cells	
CD3g	T	T cells	
CD3z	N	T cells, NK cells, macrophages	ND
CD4	T	Helper/inducer T cells, monocyte subset, thymocyte subset, macrophages	IgSF
CD5	T	Mature T cells, thymocytes, B cell subset	SRCR
CD6	T	Mature T cells, B cell subset, medullary thymocytes	SRCR
CD7	T	Mature T cells, NK cells, immature myeloid cell subset	IgSF
CD8a	T	Cytotoxic/suppressor T cells, NK cell subset, thymocytes	T cell coreceptor
CD8b	T	Cytotoxic/suppressor T cells, NK cell subset, thymocytes	
CD9	P	Platelets, activated T cells, eosinophils, basophils, endothelial cells, pre-B cells	TM4SF
CD10	B	Pre-B cell subset, B cell subset, cortical thymocyte subset, granulocytes, monocyte subset	Zinc metalloprotease
CD11a	Ad	Most of lymphoid and myeloid cells	Integrin a chain
CD11b	Ad	Myeloid cells and NK cells	Integrin a chain
CD11c	Ad	Myeloid cells, NK cells macrophages, activated T cells	Integrin a chain
CD11d	Ad	Leucocytes	Integrin a chain
CDw12	M	Monocytes, granulocytes, platelets, NK cells	–
CD13	M	Monocytes, neutrophils	–
CD14	M	Monocytes, macrophages, Langerhans cells	LRG
CD15	Ad	Neutrophils, eosinophils, monocytes, Reed Sternberg cells	Carbohydrate 2
CD15s	Ad	Neutrophils, basophils, monocytes	Sialylated carbohydrate 2

Table 3.6 Human leukocyte differentiation antigens (*continued*)

CD no.	Session	Main antigen expression	Family
CD15u	Ca		
CD16a	N	NK cells, macrophages, mast cells, monocytes	IgSF
CD16b	N	Granulocytes neutrophil only	IgSF
CDw17	M	Neutrophils, basophils, monocytes, platelets, B cell subset	LacCer
CD18	Ad	Leucocytes	Integrin
CD19	B	Precursor B cells and B cells, follicular dendritic cells	IgSF
CD20	B	Precursor B cell subset, B cells	CD20 family
CD21	B	Mature B cells, follicular dendritic cells, thymocyte subset	RCA
CD22	B	Precursor and mature B cells	IgSF
CD23	B	B cells, monocytes, follicular dendritic cells	C-type lectin
CD24	B	B cells, granulocytes	CD52/CD24/HSA
CD25	Ck	Activated T and B cells, stimulated monocytes/macrophages	–
CD26	X	Mature thymocytes, activated T cells, B cells, macrophages, NK cells	Serine-type exopeptidase
CD27	T	Mature T cells, B cell subset, NK cells	TNF receptor family
CD28	T	Mature thymocytes subset, T cells, plasma cells	IgSF
CD29	Ad	Broad	Integrin
CD30	X	Activated T and B cells, activated NK cells, monocytes, Reed Sternberg cells	TNF receptor family
CD31	Ad	Platelets, endothelial cells, monocytes, NK cells, neutrophils, T cell subset	IgSF
CD32	M	Broad except NK cells	Fc receptor
CD33	M	Pan myeloid, majority of monocytic cells	Sialoadhesin family, IgSF
CD34	M	Hematopoietic precursor cells, endothelial cells	Sialomucin
CD35	M	Neutrophils, eosinophils, monocytes, follicular dendritic cells, B cells, erythrocytes, T cell subset RCA	
CD36	P	Platelets, monocytes, macrophages, early erythroid cells, endothelial cells	–
CD37	B	B cells, weak on T cells, monocytes, granulocytes	Tetraspan
CD38	B	Plasma cells, majority of hemopoietic cells	ADP-ribosyl cyclase
CD39	B	Mantle zone B cells, activated T cells, NK cells, dendritic cells, Langerhans cells, monocytes	Ecto-apyrase
CD40	B	B cells, macrophages, follicular dendritic cells, endothelial cells, platelets	TNF/NGF receptor
CD41	P	Platelets and platelet precursors	Integrin
CD42a	P	Platelets, megakaryocytes	LGR

Table 3.6   Human leukocyte differentiation antigens (*continued*)

CD no.	Session	Main antigen expression	Family
CD42b	P	Platelets, megakaryocytes	LGR
CD42c	P	Platelets, megakaryocytes	LGR
CD42d	P	Platelets, megakaryocytes	LGR
CD43	X	Broad, except resting B cells	Sialomucin
CD44 and CD44S	Ad	Most cell types	Hyaladherin
CD44R	Ad	Epithelial cells, monocytic cells	–
CD45	X	All hematopoietic cells	PTPase
CD45RA	X	Naive resting T cells, medullary thymocytes	
CD45RB	X		
CD45RC	X		
CD45RO	X	Memory-activated T cells, cortical thymocytes	
CD46	X	Broad	RCA
CD47	Ad	Broad	IgSF
CD47R	X	Broad	
CD48	X	Pan leukocyte	IgSF
CD49a	Ad	Activated T cells, monocytes, NK cells, endothelial cells	Integrin
CD49b	Ad	Platelets, megakaryocytes, NK cells, endothelial cells	Integrin
CD49c	Ad	Non-hematopoietic cells	Integrin
CD49d	Ad	Broad	Integrin
CD49e	Ad	Broad	Integrin
CD49f	Ad	Broad (except erythrocytes)	Integrin
CD50	Ad	Leukocytes, endothelial cells, epidermal Langerhans cells	IgSF
CD51	P	Platelets, endothelial cells, activated T cells, B cell subset	Integrin
CD52	X	Thymocytes, lymphocytes, monocytes, macrophages	CD52/CD24/HSA
CD53	X	Pan leukocyte	TM4SF
CD54	Ad	Activated endothelial cells, activated T and B cells, monocytes	IgSF
CD55	X	Broad	RCA
CD56	N	NK cells, T cell subset	IgSF
CD57	N	NK cell subset, T cell subset	–
CD58	Ad	Broad	IgSF

Table 3.6 Human leukocyte differentiation antigens (*continued*)

CD no.	Session	Main antigen expression	Family
CD59	X	Broad	Ly6
CD60a	Ca	T cell subset, platelets	Glycolipid
CD60b	Ca		Glycolipid
CD60c	Ca		Glycolipid
CD61	P	Platelets, megakaryocytes	Integrin
CD62E	Ad	Endothelial cells	Selectin
CD62L	Ad	T and B cells, monocytes, granulocytes, some NK cells	Selectin
CD62P	Ad	Platelets, megakaryocytes, activated endothelial cells	Selectin
CD63	P	Activated platelets, monocytes, degranulated neutrophils, endothelium	Tetraspan, TM4SF
CD64	M	Monocytes, macrophages, dendritic cell subset	IgSF
CD65	M	Granulocytes (monocytes)	poly-N-acetyllactosamine
CD65s	M	Granulocytes, monocytes	poly-N-acetyllactosamine
CD66a	M	Granulocytes and epithelial cells	IgSF, CEA
CD66b	M	Granulocytes	IgSF, CEA
CD66c	M	Granulocytes and epithelial cells	IgSF, CEA
CD66d	M	Granulocytes	IgSF, CEA
CD66e	M	Epithelial cells	IgSF, CEA
CD66f	M	Myeloid cell lines, fetal liver, placental syncytiotrophoblasts	IgSF, CEA
CD67		cancelled: now CD66b	
CD68	M	Monocytes, macrophages, dendritic cells, neutrophils, myeloid progenitor cells	Sialomucin
CD69	N	Activated T and B cells, thymocytes, NK cells, neutrophils, eosinophils	C-type lectin
CD70	X	Activated B and T cells	TNF
CD71	X	Proliferating cells, reticulocytes, erythroid precursors	Transferrin receptor
CD72	B	Pan B, including progenitors	C-type lectin
CD73	B	B and T cell subsets, follicular dendritic cells, epithelial cells, endothelial cells	GPI-anchored
CD74	B	B cells, activated T cells, macrophages, activated epithelial and endothelial cells	–
CD75	Ca	Mature B cells, T cell subset	Lactosamine
CD75s	Ca	Mature B cells, T cell subset	Sialylated lactosamine
CDw76		cancelled: now CD75s	

Table 3.6 Human leukocyte differentiation antigens (*continued*)

CD no.	Session	Main antigen expression	Family
CD77	B	Burkitt's lymphoma cells, germinal center B lymphocytes	Carbohydrate
CDw78		*cancelled*	
CD79a	B	B cells	IgSF
CD79b	B	B cells	IgSF
CD80	B	Activated B and T cells, macrophages	IgSF
CD81	B	Broad hemopoietic, endothelial and epithelial cells	Tetraspan
CD82	B	Broad	Tetraspan
CD83	B	Circulating and interdigitating reticular dendritic cells, Langerhans cells	IgSF, Siglec
CD84	B	Mature B cells, monocytes, macrophages, platelets, thymocytes and T cell subset	–
CD85a	D	Monocytes, macrophages, granulocytes, dendritic cells, T cell subset	IgSF
CD85b	D	Monocytes, macrophages, dendritic cells	IgSF
CD85c	D		IgSF
CD85d	D	Monocytes, macrophages, dendritic cells	IgSF
CD85e	D	Monocytes	IgSF
CD85f	D		IgSF
CD85g	D	Monocytes	IgSF
CD85h	D	Monocytes, dendritic cell subset, macrophages, granulocytes, NK subset	IgSF
CD85i	D	Monocytes	IgSF
CD85j	D	Monocytes, macrophages, dendritic cells, NK subset, T cell subset, B cells	IgSF
CD85k	D	Monocytes, macrophages, dendritic cells	IgSF
CD85l	D		IgSF
CD85m	D		IgSF
CD86	B	Memory B cells, monocytes, dendritic cells, endothelial cells and activated T cells	IgSF
CD87	M	T cells, NK cells, monocytes and neutrophils as well as non-hemopoietic cells	GPI-anchored
CD88	M	Granulocytes, monocytes, dendritic cells	Rhodopsin
CD89	M	Myeloid cells	IgSF, Fc receptor, MIRR
CD90	En	Hemopoietic stem cells, neurons	IgSF, GPI linked
CD91	M	Monocytes, macrophages, neurons, fibroblasts	LDL receptor
CD92	M	Monocytes, granulocytes	–

Table 3.6 Human leukocyte differentiation antigens (*continued*)

CD no.	Session	Main antigen expression	Family
CDw93	M	Monocytes, granulocytes, endothelial cells	
CD94	N	NK cells, g/d and a/b T cell subsets	C-type lectin
CD95	Ck	Broad including activated T and B cells	TNF/NGF receptor
CD96	N	Activated T and NK cells	IgSF
CD97	X	Activated T and B cells, monocytes, granulocytes	EGF-TM7
CD98	X	Broad on activated cells	
CD99	T	Broad, including lymphocytes	
CD99R	T	Restricted hematopoietic expression	
CD100	X	Broad	Semaphorin
CD101	M	Granulocytes, monocytes, dendritic cells, activated T cells	IgSF
CD102	Ad	Resting lymphocytes, monocytes, platelets, vascular endothelial cells	IgSF
CD103	Ad	Mucosa associated T lymphocytes	Integrin
CD104	Ad	Epithelial cells, keratinocytes, Schwann cells, monocytes, endothelial cells	Integrin
CD105	En	Activated monocytes, endothelial cells, stromal cells, pre-B cells	TGF receptor
CD106	En	Follicular dendritic cells, activated endothelium	IgSF
CD107a	P	Degranulated platelets, activated neutrophils, activated T cells	
CD107b	P	Degranulated platelets, activated neutrophils	
CD108	X	Erythrocytes, circulating lymphocytes	
CD109	Ca	Activated T cells and platelets, endothelial cells	
CD110	P	Hematopietic stem and progenitor cells, megakaryocytes, platelets	IgSF
CD111	M	CD34+ hematopoietic progenitors, epithelial and neuronal cells	IgSF
CD112	M	CD34+ hematopoietic progenitors, epithelial and endothelial cells	IgSF
CD113		*NA (reserved)*	
CD114	M	Granulocytes, monocytes, mature platelets, endothelial cells	Class I CK-R
CD115	M	Monocytes, macrophages and their precursors, placenta	IgSF, tyrosine kinase R
CD116	Ck	Macrophages, neutriphils, eosinophils, dendritic cells and their precursors	IgSF, class I CK-R
CD117	Ck	Hematopoietic progenitor cells, tissue mast cells	IgSF, tyrosine kinase R
CD118		*NA (reserved)*	
CDw119	Ck	Broad	IgSF, class II CK-R
CD120a	Ck	Broad	TNF receptor
CD120b	Ck	Broad	TNF receptor

Table 3.6 Human leukocyte differentiation antigens (*continued*)

CD no.	Session	Main antigen expression	Family
CD121a	Ck	Broad	IgSF
CDw121b	Ck	B cells, myeloid cells, some T cells	IgSF
CD122	Ck	NK cells, T cells and B cells, monocytes/macrophages	IgSF, CK-R
CD123	Ck	Myeloid cells including early progenitors endothelial cells	IgSF, class I CK-R
CD124	Ck	Broad	IgSF, CK-R
CDw125	Ck	Eosinophils, activated B cells, basophils	IgSF, CK-R
CD126	Ck	T cells, monocytes, activated B cells	IgSF, class I CK-R
CD127	Ck	T cells, B cell precursors	IgSF, CK-R
CDw128a	Ck	Neutrophils, T cell subset, monocytes, endothelial cells, fibroblasts, platelets	Chemokine receptor
CDw128b	Ck	Neutrophils, T cell subset, monocytes, melanocytes	Chemokine receptor
CD129		NA (reserved IL-9R)	
CD130	Ck	Broad	Class I CK-R
CDw131	Ck	Myeloid cells, early B cells	Class I CK-R
CD132	Ck	T and B cells, NK cells, monocytes/macrophages, neutrophils	Class I CK-R
CD133	S	CD34+ hematopoietic progenitors, neural and endothelial stem cells	5-TM
CD134	Ck	Activated T cells	TNF/NGF receptor
CD135	Ck	Early and lymphoid committed progenitors	Tyrosine kinase receptor
CDw136	Ck	Broad	Tyrosine kinase receptor
CDw137	Ck	T cells	TNF receptor
CD138	Ad	Pre-B cells, plasma cells	Syndecan
CD139	B	B cells, monocytes, granulocytes, follicular dendritic cells	–
CD140a	En	Fibroblasts, smooth muscle cells, platelets	Split-tyrosine kinase
CD140b	En	Fibroblasts, smooth muscle cells, monocytes, neutrophils, endothelial cells	Split-tyrosine kinase
CD141	En	Endothelial cells, monocytes, neutrophils, megakaryocytes, platelets	C-type lectin
CD142	En	Epithelial cells, stromal cells, keratinocytes	Serine protease cofactor
CD143	En	Endothelial and epithelial cells, activated macrophages	Peptidylpeptidase
CD144	En	Endothelial cells	Cadherin
CDw145	En	Endothelial cells, some stromal cells	
CD146	En	Endothelial cells, smooth muscle cells, activated T cells, melanoma cells	IgSF

Table 3.6 Human leukocyte differentiation antigens (*continued*)

CD no.	Session	Main antigen expression	Family
CD147	En	Leukocytes, red blood cells, platelets and endothelial cells	IgSF
CD148	X	Granulocytes, monocytes, resting T cells, dendritic cells, platelets, fibroblasts	RPTPase type III, phosphatase
CDw149		*cancelled: now CD47R*	
CD150	X	Thymocytes, B cells, T cell subset, dendritic cells, endothelial cells	IgSF
CD151	P	Platelets, megakaryocytes, endothelial and epithelial cells	Tetraspan
CD152	T	Activated T and B cells	IgSF
CD153	T	Activated T cells, activated macrophages, neutrophils, B cells	TNF
CD154	T	Activated CD4$^+$T cells	TNF
CD155	M	Monocytes, broad tissue distribution	IgSF
CD156a	M	Monocytes, neutrophils	ADAM
CD156b	M	Broad	ADAM
CD157	M	Granulocytes, monocytes, bone marrow stromal cells	ADP-ribosyl cyclase
CD158a	N	NK cell subset, minor subset of T cells	IgSF
CD158b1	N	NK cell subset, minor subset of T cells	IgSF
CD158b2	N	NK cell subset, minor subset of T cells	IgSF
CD158c	N	NK cell subset, minor subset of T cells	IgSF
CD158d	N	NK cell subset, minor subset of T cells	IgSF
CD158e1	N	NK cell subset, minor subset of T cells	IgSF
CD158e2	N	NK cell subset, minor subset of T cells	IgSF
CD158f	N	NK cell subset, minor subset of T cells	IgSF
CD158g	N	NK cell subset, minor subset of T cells	IgSF
CD158h	N	NK cell subset, minor subset of T cells	IgSF
CD158i	N	NK cell subset, minor subset of T cells	IgSF
CD158j	N	NK cell subset, minor subset of T cells	IgSF
CD158k	N	NK cell subset, minor subset of T cells	IgSF
CD158z	N	NK cell subset, minor subset of T cells	IgSF
CD159a	N	NK cell subset, T cells, thymocytes	C-type lectin
CD160	N	NK cells, cytotoxic T cells	IgSF
CD161	N	NK cells, T cells	C-type lectin
CD162	Ad	Monocytes, granulocytes, T cells, some B cells	Sialomucin
CD162R	N	NK cell subset	

Table 3.6 Human leukocyte differentiation antigens (*continued*)

CD no.	Session	Main antigen expression	Family
CD163	M	Monocytes, macrophages	Scavenger receptor
CD164	Ad	Monocytes, B cells (weak expression), CD34$^+$ progenitor cells, bone marrow stromal cells, epithelial cells	Sialomucin
CD165	Ad	Lymphocyte subset, monocytes, platelets, thymocytes	
CD166	Ad	Activated T cells, activated monocytes, epithelium fibroblasts, neurons	IgSF
CD167a	Ad	Epithelial cells	Kinases
CD168	Ad	Broad	
CD169	Ad	Macrophages	IgSF
CD170	Ad	Neutrophils, macrophages	
CD171	Ad	Neurons, some epithelial cells and some lymphoid and myelomonocytic cells	IgSF
CD172a	Ad	Stem cells, monocytes, T cell subset	
CD173	Ca	Red blood cells, platelets, CD34 stem cell subset	Blood group antigen
CD174	Ca	Epithelial and endothelial cells, granulocytes, CD34 stem cell subset	Blood group antigen
CD175	Ca	Stem cell subset	
CD175s	Ca	Erythroblasts	
CD176	Ca	Stem cell subset	
CD177	M	Neutrophil subset	
CD178	Ck	Lymphoid cells	TNF
CD179a	B	Pro-B and early pre-B cells	IgSF
CD179b	B	Pro-B and early pre-B cells	IgSF
CD180	B	Mantle and marginal zone B cells, monocytes, dendritic cells	Toll-like receptor family
CD181-CD182		NA (reserved)	
CD183	Ck	T cells, plasmacytoid dendritic cells, subsets of NK and B-cells, eosinophils	Chemokine receptor
CD184	Ck	Broad in blood and tissue cells, CD34 stem cell subset	Chemokine receptor
CD185-CD194		NA (reserved)	
CD195	Ck	Monocytes, T cell subset	
CD196		NA (reserved)	
CDw197	Ck	Peripheral T and B cells, bone marrow and cord blood CD34$^+$ HPC, dendritic cells	Chemokine receptor
CD198-CD199		NA (reserved)	
CD200	X	Thymocytes, B cells, activated T cells	

Table 3.6 Human leukocyte differentiation antigens (*continued*)

CD no.	Session	Main antigen expression	Family
CD201	En	Endothelial cell subset	CD1/MHC super family
CD202b	En	Endothelial cells, stem cells	Tyrosine kinase receptor
CD203c	M	Basophils, mast cells	Ectoenzyme
CD204	M	Macrophages	
CD205	D	Dendritic cells	C-type lectin
CD206	D	Dendritic cells, macrophages	C-type lectin
CD207	D	Langerhans cells	C-type lectin
CD208	D	Interdigitating dendritic cells, mature dendritic cells	LAMP family
CD209	D	Dendritic cell subsets	C-type lectin
CDw210	Ck	T and B cells, NK cells, monocytes, macrophages	
CD211		*NA (reserved)*	
CD212	Ck	Activated T cells, activated NK cells	
CD213a1	Ck	Basophils, mastocytes	
CD213a2	Ck	B cells, monocytes	
CD214–CD216		*NA (reserved)*	
CDw217	Ck	Monocytes, erythroblasts	
CD218–CD219		*NA (reserved)*	
CD220	X	Broad	
CD221	X	Broad	
CD222	X	Broad	Lectins
CD223	X	Activated T cells, activated NK cells	
CD224	X	Vascular endothelium, peripheral blood macrophages, activated T cells, CD45RO+ T cells, B cell subset	Membrane-bound ectoenzyme
CD225	X	Broad	
CD226	T	NK cells, platelets, monocytes, subset of T cells, thymocytes	IgSF
CD227	X	Granular and ductal epithelial cells	Mucin
CD228	X	Melanoma cells, progenitor cells	Transferrin family, GPI anchor
CD229	X	T and B cells	
CD230	X	Broad	
CD231	X	T cell acute lymphoblastic leukemia, neuroblastoma cells	TM4SF

Table 3.6 Human leukocyte differentiation antigens (*continued*)

CD no.	Session	Main antigen expression	Family
CD232	X	Broad	
CD233	Q	Erythrocyte plasma membrane	Bicarbonate transporter
CD234	Q	Erythroid cells, endothelial cells, some epithelial cells	Chemokine receptor
CD235a	Q	Red blood cells, erythroid precursor cells	
CD235b	Q	Red blood cells, erythroid precursor cells	
CD235ab	Q	Red blood cells, erythroid precursor cells	
CD236	Q	Red blood cells, stem cell subset	
CD236R	Q	Red blood cells, stem cell subset	
CD237		*NA (reserved)*	
CD238	Q	Red blood cells, stem cell subset	Neutral endopeptidase
CD239	Q	Red blood cells, stem cell subset	IgSF
CD240CE	Q	Red blood cells	
CD240D	Q	Red blood cells	
CD240DCE	Q	Red blood cells	
CD241	Q	Red blood cells	
CD242	Q	Erythrocytes, lymphocytes	IgSF
CD243	S	Stem cells, NK cells, T cells, Tumor cells	ABC transporters
CD244	N	NK cells, T cell subset, monocytes, basophils	IgSF
CD245	T	T cell subset	
CD246	T	Anaplastic T cell leukemia	
CD247	T	T cells, NK cells	

Session keys:

Ad	Adhesion	M	Myeloid cells
B	B cells	N	NK cells
Ca	Carbohydrates/lectins	P	Platelets
Ck	Cytokines/chemokines	S	Stem cells
D	Dendritic cells	T	T cells
En	Endothelial cells	X	Blind panel
Er	Erythroid cells		

THE HLA DICTIONARY, 2001

4

A summary of HLA-A, -B, -C, -DRB1/3/4/5, -DQB1 alleles and their association with serologically defined HLA-A, -B, -C, -DR, and -DQ antigens

GMTh Schreuder and Associates*

- WHO NOMENCLATURE: ALLELE AND SEROLOGIC ASSIGNMENTS

- INTERNATIONAL CELL EXCHANGE, UCLA

- NATIONAL MARROW DONOR PROGRAM (NMDP)

- INFORMATION OBTAINED FROM OTHER SOURCES

- SEROLOGIC SPECIFICITIES LACKING OFFICIAL WHO NOMENCLATURE DESIGNATIONS

- FUTURE PLANS

Several years ago, the World Marrow Donor Association (WMDA) initiated a study aimed at identifying the serological types associated with each HLA-A, -B, -DRB allelic product. This work resulted in a 'serology to DNA equivalents dictionary' first published in 1997 with an update in 1999.[1,2] The dictionary is an attempt to aid searches for unrelated hematopoietic stem cell (hsc) donors in adult volunteer and umbilical cord blood banks. While most patients in need of hsc transplantation are HLA typed by DNA-based methods at medium or high resolution, substantial parts of the donor registries provide serologically based HLA typings only, at least for HLA class I. In this respect, the dictionary can help in the search for donors whose HLA phenotypes closely resemble that of the patient even though these typings are determined by different methods. Once identified, molecular class I typing of patient and selected donor can be performed to confirm the match. Since the appearance of the 1999 dictionary[2] based on the alleles listed in the 1998 WHO Nomenclature report,[3] a large number of additional alleles have received official allele designations.[4] Moreover, a substantial number of DNA-based typings have been added to the NMDP database

*GMTh Schreuder,[1,2] CK Hurley,[1,4] SGE Marsh,[1,2] M Lau,[1,3] M Maiers,[4] C Kollman,[4] HJ Noreen[4]

[1]World Marrow Donor Association Quality Assurance Working Group, Leiden, The Netherlands
[2]WHO Nomenclature Committee for Factors of the HLA system
[3]International Cell Exchange, UCLA, Los Angeles, California, USA
[4]US National Marrow Donor Program HLA testing laboratories and M. Setterholm
This Dictionary was published in parallel in *Tissue Antigens* 2001; 58: 109–140; *Human Immunology* 2001; 62: 826–849; and *European Journal of Immunogenetics* 2001; 28: 565–596. It is also available at http://www.worldmarrow.org.

prompting an update of the dictionary. As summarized in **Table 4.1**, serological equivalents are available for over 64 percent of the presently identified HLA-A, -B and -DRB1 alleles. For completeness, all known alleles as published in the Nomenclature Report 2000,[4] including null alleles, have been included in **Tables 4.2–4.7**. Although serological typing of HLA-C and -DQ antigens is notoriously difficult, often incomplete and unreliable, the information, when available for a donor, may help in the selection procedure. For that reason we have included HLA-C and -DQB1 serologic equivalents. The amount of data for these loci is substantially increased by the inclusion of data from the NMDP. Although HLA-DQ molecules are heterodimers combining polymorphic alpha and beta chains, the DQ serologic patterns correlate strongly with DQB1, but not with DQA1 polymorphisms. For that reason, **Table 4.7** gives equivalents between DQB1 alleles and DQ antigens.

WHO NOMENCLATURE: ALLELE AND SEROLOGIC ASSIGNMENTS

This update includes all presently identified alleles as included in the most recent nomenclature report.[4] Alleles are presented with their four-digit number designation, because silent substitutions (represented by the fifth and sixth digit) do not influence the antigenic expression. Non-expressed (N-(ull))-alleles are included for completeness. The serologic assignment (second column of **Tables 4.2–4.7**) is taken from the Nomenclature Report 2000 as well. It should be noted that the latter report now incorporates data from the previous dictionaries for well-defined antigens only. In the Nomenclature Report 2000, serologic specificities of newly identified alleles are only included when reported by the allele contributors as being identical or very similar to the standard specificity associated with the allele group (e.g., A1 associated with A*01 alleles). Likewise, the contributor may report an allele to be expressed as a serologically variant antigen. Such information is included

in the nomenclature report under the heading 'previous equivalents' and is included here in the comments column of **Tables 4.2–4.7**. Further information on the variant, if available, is included in **Table 4.8**. Some alleles have been assigned 'standard' WHO HLA specificities although they were known to be slightly different. For example, the antigen encoded by B*1524 was assigned B62, but known to express the Bw4 instead of the Bw6 antigenic determinant present on the 'standard' B62 (5).

INTERNATIONAL CELL EXCHANGE, UCLA

The HLA class I analysis included 88 samples, cells 977–1064, from healthy individuals, characterized between October 1998 and November 2000. Each cell was typed serologically by an average number of 201 laboratories, ranging from a low of 168 to a high of 229. The parallel typing for HLA-A, -B, -C alleles using DNA-based methods was performed by an average of 47 laboratories, ranging from 37 to 51 reporting monthly results. Some of the cells were re-examined in the DNA Extract-Class I Typing Exchange. Their high-resolution types were included in this update, together with the correlating serologically defined antigens of the initial typings. The number of participating labs typing the DNA extracts increased from 21 laboratories in 1997 to 60 laboratories in 2000. The average number was 54 laboratories during the time period used for this present update.

The HLA class II analysis included 44 B lymphoblastoid cell lines (TER229-272) typed during the period between October 1998 and October 2000. The previously published listings[1,2] have been updated with these alleles (**Tables 4.2–4.7**). During this time period, each cell was typed serologically by an average number of 33 laboratories, ranging from a low of 27 to a high of 44. The parallel typing for HLA-DRB1, -DRB3, -DRB4, -DRB5, and DQB1-alleles using DNA-based methods was performed by an average of 99 laboratories ranging from a low of 92 to a high of 103.

Table 4.1 Comparison of all presently known HLA alleles as listed in the WHO Nomenclature reports 1998[3] and 2000[4] and their known serologic equivalents as given in the previous[2] and the present 'Dictionary'*

	1999		2001	
Locus	WHO alleles[3] N	Serologic equivalents[2] N (%)	WHO alleles[4] N	Serologic equivalents N (%)
HLA-A	119	90 (76)	187	123 (65)
HLA-B	245	190 (77)	344	272 (79)
HLA-DRB1	201	145 (72)	243	155 (64)

*http://www.worldmarrow.org/Dictionary/Dict2001.html

Table 4.2 HLA-A alleles and their serologic designations

| HLA allele | WHO assigned | International cell exchange UCLA | | NMDP | | Comments |
		Cells tested	Assigned type [%]	Cells tested	Assigned type [%]	
A*0101	A1	14	A1 [99–100]	4920	A1 [96]	
A*0102	A1	—	—	122	A1 [92]	
A*0103	A1	1	A1 [99]	9	A1 [78]	A1[100][d]
A*0104N	Null	—	—	—	—	
A*0105N	Null	—	—	—	—	
A*0106	—	—	—	1	A1 [100][a]	
A*0107	A1	—	—	—	—	
A*0108	A1	—	—	—	—	
A*0201	A2	52	A2 [99–100]	3527	A2 [97]	
A*0202	A2	—	—	735	A2 [81]	
A*0203	A203	6	A2 [100]	361	A2 [62] A28 [1]	A2 [94] A203 [6][d]
A*0204	A2	—	—	16	A2 [69]	
A*0205	A2	2	A2 [100]	301	A2 [84]	
A*0206	A2	16	A2 [100]	276	A2 [67]	
A*0207	A2	4	A2 [100]	71	A2 [71]	
A*0208	A2	—	—	2	A2 [50][a]	
A*0209	A2	—	—	2	[b]	
A*0210	A210	—	—	2	A2 [51][a]	
A*0211	A2	—	—	191	A2 [73]	
A*0212	A2	—	—	3	A2 [100][a]	
A*0213	A2	—	—	2	A2 [51][a]	
A*0214	A2	—	—	7	A2 [71]	
A*0215N	Null	—	—	—	—	
A*0216	A2	—	—	18	A2 [50]	
A*0217	A2	—	—	33	A2 [68]	
A*0218	A2	—	—	1	A2 [100][a]	
A*0219	—	—	—	—	—	
A*0220	A2	—	—	3	A2 [68][a]	

Table 4.2 HLA-A alleles and their serologic designations (*continued*)

| HLA allele | WHO assigned | International cell exchange UCLA | | NMDP | | Comments |
		Cells tested	Assigned type [%]	Cells tested	Assigned type [%]	
A*0221	A2	—	—	—	—	
A*0222	A2	—	—	12	A2 [75]	
A*0224	A2	—	—	3	A2 [100][a]	
A*0225	A2	—	—	—	—	
A*0226	—	—	—	—	—	
A*0227	—	—	—	—	—	
A*0228	—	—	—	—	—	
A*0229	A2	—	—	—	—	
A*0230	—	—	—	1	[b]	
A*0231	A2	—	—	—	—	
A*0232N	Null	—	—	—	—	
A*0233	—	—	—	—	—	
A*0234	A2	—	—	—	—	
A*0235	—	—	—	—	—	
A*0236	—	—	—	1	A2 [100][a]	
A*0237	—	—	—	—	—	
A*0238	—	—	—	—	—	
A*0239	—	—	—	—	—	
A*0240	—	—	—	—	—	
A*0241	A2	—	—	—	—	(g)
A*0242	A2	—	—	—	—	
A*0243N	Null	—	—	—	—	
A*0244	—	—	—	—	—	
A*0245	—	—	—	—	—	
A*0246	A2	—	—	—	—	
A*0247	—	—	—	—	—	
A*0248	—	—	—	—	—	
A*0249	—	—	—	—	—	

Table 4.2 HLA-A alleles and their serologic designations (*continued*)

HLA allele	WHO assigned	International cell exchange UCLA		NMDP		Comments
		Cells tested	Assigned type [%]	Cells tested	Assigned type [%]	
A*0301	A3	22	A3 [100]	2807	A3 [95]	
A*0302	A3	1	A3 [100]	287	A3 [76]	
A*0303N	Null	—	—	—	—	
A*0304	A3	—	—	—	—	
A*0305	A3	—	—	1	blank[a]	
A*0306	—	—	—	—	—	
A*0307	—	—	—	—	—	
A*0308	—	—	—	—	—	
A*1101	A11	30	A11 [99–100]	1757	A11 [98]	
A*1102	A11	1	A11 [99]	23	A11 [88]	
A*1103	A11	—	—	13	A11 [92]	
A*1104	A11	2	A11 [63–69]	3	A11 [33] blank [67][a]	A11short
A*1105	A11	—	—	2	A11 [100][a]	A11var[c],
A*1106	—	—	—	—	—	
A*1107	A11	—	—	—	—	
A*1108	—	—	—	—	—	
A*1109	—	—	—	—	—	
A*2301	A23(9)	17	A23 [98–100]	4807	A23 [96] A24 [1]	
A*2302	—	—	—	2	A23 [100][a]	
A*2303	—	—	—	—	—	
A*2304	—	—	—	—	—	
A*2305	—	—	—	—	—	
A*2306	—	—	—	—	—	
A*2402	A24(9)	47	A24 [95–100]	2209	A24 [96] A2403 [1]	
A*2402102L	Low A24	—	—	—	—	A24Neg[d]
A*2403	A2403	7	A24 [44–67] A2403 [20–52]	239	A24 [50] blank [24] A9 [3] A2403 [3]	
A*2404	A24(9)	—	—	10	A24 [60] A2403 [10]	
A*2405	A24(9)	—	—	10	A24 [80] A23 [9]	
A*2406	A24(9)	—	—	3	A24 [33] blank [33] [a]	

Table 4.2 HLA-A alleles and their serologic designations (*continued*)

| HLA allele | WHO assigned | International cell exchange UCLA | | NMDP | | Comments |
		Cells tested	Assigned type [%]	Cells tested	Assigned type [%]	
A*2407	A24(9)	11	A24 [99]	232	A24 [61] blank [6]	
A*2408	A24(9)	—	—	33	A24 [61] blank [21]	A24var[c]
A*2409N	Null	—	—	—	—	
A*2410	A9	—	—	29	A24 [43] A9 [10] A2403 [3] blank [14]	A2403[e]
A*2411N	Null	—	—	—	—	
A*2413	A24(9)	—	—	1	[b]	
A*2414	A24(9)	—	—	11	A24 [65] blank [18]	
A*2415	—	—	—	—	—	
A*2416	—	—	—	—	—	A31var[c]
A*2417	—	—	—	—	—	
A*2418	—	—	—	—	—	A24x3[c]
A*2419	—	—	—	—	—	A9short[c]
A*2420	—	—	—	1	A24 [100][a]	
A*2421	—	—	—	—	—	A9var[c]
A*2422	A9	—	—	1	A24 [100][a]	
A*2423	A24(9)	—	—	—	—	
A*2424	—	—	—	—	—	
A*2425	—	—	—	—	—	
A*2426	—	—	—	—	—	
A*2427	A24(9)	—	—	—	—	
A*2428	—	—	—	—	—	
A*2501	A25(10)	5	A25 [94–99]	1226	A25 [95] A10 [2]	
A*2502	A10	—	—	2	A25 [100][a]	
A*2503	—	—	—	—	—	
A*2601	A26(10)	11	A26 [90–98]	1497	A26 [97] A10 [1]	
A*2602	A26(10)	—	—	20	A26 [75] A34 [5]	
A*2603	A26(10)	6	A26 [81–92]	31	A26 [65] A10 [10]	
A*2604	A26(10)	—	—	1	A26 [100][a]	

Table 4.2 HLA-A alleles and their serologic designations (*continued*)

| HLA allele | WHO assigned | International cell exchange UCLA | | NMDP | | Comments |
		Cells tested	Assigned type [%]	Cells tested	Assigned type [%]	
A*2605	A26(10)	—	—	10	A26 [90] A25 [10]	
A*2606	A26(10)	—	—	—	—	
A*2607	A26(10)	—	—	—	—	
A*2608	A26(10)	—	—	49	A26 [98] A66 [2]	A10short[c]
A*2609	A26(10)	—	—	2	A26 [100][a]	
A*2610	A10	—	—	—	—	A10var[c]
A*2611N	Null	—	—	—	—	
A*2612	—	—	—	—	—	
A*2613	—	—	—	—	—	
A*2614	—	—	—	—	—	
A*2615	—	—	—	—	—	
A*2616	—	—	—	—	—	
A*2617	—	—	—	—	—	
A*2901	A29(19)	—	—	94	A29 [94]	
A*2902	A29(19)	5	A29 [98–99]	1212	A29 [98]	
A*2903	—	—	—	3	A29 [100][a]	
A*2904	—	—	—	—	—	
A*3001	A30(19)	20	A30 [95–99]	2512	A30 [88] A31 [3] blank [6]	
A*3002	A30(19)	7	A30 [96–99]	1770	A30 [94] A31 [2]	
A*3003	A30(19)	—	—	6	A30 [67] blank [17]	
A*3004	A30(19)	—	—	50	A30 [93]	
A*3006	—	—	—	—	—	
A*3007	—	—	—	2	A30 [100][a]	
A*3008	—	—	—	—	—	
A*3009	—	—	—	—	—	
A*3101	A31(19)	21	A31 [97–99]	3209	A31 [90] A30 [3] A19 [1] blank [5]	
A*3102	—	—	—	—	—	
A*3103	—	—	—	1	A31 [100][a]	
A*3104	A31(19)	—	—	—	—	

Table 4.2 HLA-A alleles and their serologic designations (*continued*)

| HLA allele | WHO assigned | International cell exchange UCLA | | NMDP | | Comments |
		Cells tested	Assigned type [%]	Cells tested	Assigned type [%]	
A*3105	A31(19)	—	—	—	—	
A*3201	A32(19)	18	A32 [97–99]	3136	A32 [94]	
A*3202	A32(19)	—	—	3	A32 [67][a]	
A*3203	—	—	—	1	A32 [100][a]	
A*3204	—	—	—	3	A3 [33] blank [33][a]	A3[c]
A*3205	—	—	—	—	—	
A*3206	—	—	—	—	—	
A*3301	A33(19)	5	A33 [95–97]	540	A33 [87] A19 [3]	
A*3303	A33(19)	15	A33 [96–99]	670	A33 [94]	
A*3304	—	—	—	1	A33 [100][a]	
A*3305	—	—	—	—	—	
A*3401	A34(10)	9	A34 [87–95]	236	A34 [67] A26 [11] A10 [5] A66 [4] A33 [1] blank [8]	
A*3402	A34(10)	8	A34 [93–95]	900	A34 [90] A33 [3] A26 [2] A10 [2]	
A*3403	—	—	—	—	—	
A*3404	—	—	—	—	—	A34 [42] 34x31 [25] blank [33][dc]
A*3601	A36	3	A36 [91–93]	769	A36 [68] A1 [16] blank [8]	
A*4301	A43	—	—	8	A26 [39] A29 [12]	
A*6601	A66(10)	8	A'6601' [45–57] A66 [17–39] A26 [12–19]	712	A26 [61] A66 [20] A34 [8] A10 [4] A25 [1] A11 [1]	
A*6602	A66(10)	6	A'6602' [53–65] A66 [15–18] A34 [7–15]	180	A66 [41] A34 [24] A10 [11] A74 [7] A26 [3]	
A*6603	A10	—	—	25	A66 [36] A34 [16] A10 [9] A26 [8] A25 [4] A74 [4]	A10short[c]
A*6604	—	—	—	—	—	
A*6801	A68(28)	16	A68 [47–61] A28 [37–48]	2526	A28 [57] A68 [33]	
A*6802	A68(28)	4	A68 [55–63] A28 [36–44]	2012	A28 [70] A68 [21] A69 [1]	

Table 4.2 HLA-A alleles and their serologic designations (*continued*)

HLA allele	WHO assigned	International cell exchange UCLA		NMDP		Comments
		Cells tested	Assigned type [%]	Cells tested	Assigned type [%]	
A*6803	A68(28)	4	A68 [42–50] A28 [45–53]	125	A28 [62] A68 [28]	
A*6804	A68(28)	—	—	6	A28 [52] A68 [17]	
A*6805	A68(28)	—	—	46	A28 [49] A68 [33]	
A*6806	—	—	—	2	A28 [50] A68 [50][a]	
A*6807	—	—	—	—	—	
A*6808	A68(28)	2	A68 [38–45] A28 [53]	—	—	
A*6809	—	—	—	—	—	
A*6810	—	—	—	2	[b]	
A*6811N	Null	—	—	—	—	
A*6812	A28	—	—	2	[b]	A28short[c]
A*6813	—	—	—	—	—	
A*6814	—	—	—	—	—	
A*6815	—	—	—	—	—	
A*6816	A68(28)	—	—	—	—	
A*6817	—	—	—	—	—	
A*6818N	Null	—	—	—	—	
A*6819	—	—	—	—	—	
A*6901	A69(28)	1	A28 [49] A69 [40] A2 [8]	252	A28 [53] A69 [9] A2 [3] A68 [3]	
A*7401	A74(19)	5	A74 [77–85]	88	A74 [42] blank [28] A19 [8] A33 [3] A32 [2] A31 [2]	
A*7402	A74(19)	—	—	2	A74 [50] A33 [50][a]	
A*7403	A74(19)	3	A74 [86–89]	28	blank [67] A74 [11] A19 [7] A29 [4] A33 [4]	
A*7404	—	—	—	—	—	
A*7405	—	—	—	—	—	
A*8001	A80	7	A80 [69–81] A36 [4–5]	480	blank [70] A80 [10] A'X' [8] A36 [2] A1 [1]	

http://www.worldmarrow.org/Dictionary/Dict2001Table2.html
[a]Allele has been reported <6 times and/or serologically identified in <4 individuals.
[b]Allele identified but serology not informative.
[c]See remarks in Table 4.8.
[d]HLA-Club Cell Exchange: one sample typed by 17 laboratories and allele identified by at least three laboratories.
[e]Locally identified in Leiden.

Table 4.3 HLA-B alleles and their serologic designations

| HLA allele | WHO assigned | International cell exchange UCLA | | NMDP | | Comments |
		Cells tested	Assigned type [%]	Cells tested	Assigned type [%]	
B*0702	B7	20	B7 [96–99]	10283	B7 [98]	
B*0703	B703	—	—	14	B7 [93]	B703[c]
B*0704	B7	—	—	42	B7 [89]	
B*0705	B7	2	B7 [75–99]	59	B7 [94]	
B*0706	B7	—	—	8	B7 [100]	
B*0707	B7	—	—	12	B7 [81]	
B*0708	—	—	—	—	—	B7short[c]
B*0709	B7	2	B7 [97–98]	7	B7 [83]	
B*0710	—	—	—	—	—	
B*0711	B7	—	—	—	—	
B*0712	—	—	—	5	B7 [100][a]	
B*0713	—	—	—	3	blank [100][a]	
B*0714	—	—	—	1	B7 [100][a]	
B*0715	B7	—	—	1	B7 [100][a]	B7var[c]
B*0716	B7	—	—	—	—	B7short[c]
B*0717	—	—	—	—	—	
B*0718	—	—	—	—	—	
B*0719	—	—	—	—	—	
B*0720	—	—	—	—	—	B7short[c]
B*0721	—	—	—	—	—	
B*0722	—	—	—	—	—	
B*0723	—	—	—	—	—	
B*0724	B7	—	—	—	—	B7weak[c]
B*0725	—	—	—	—	—	
B*0726	—	—	—	—	—	
B*0801	B8	1	B8 [99]	6351	B8 [97]	
B*0802	B8	—	—	2	[b]	B8var[c]
B*0803	B8	—	—	—	—	B8var[c]
B*0804	—	3	B8 [51–71] B59 [6–7]	6	blank [25] B8 [17]	B8var[c]

Table 4.3 HLA-B alleles and their serologic designations (*continued*)

HLA allele	WHO assigned	International cell exchange UCLA			NMDP		Comments
		Cells tested	Assigned type [%]		Cells tested	Assigned type [%]	
B*0805	—	—	—		1	b	
B*0806	B8	—	—		1	b	
B*0807	B8	—	—		—	—	(10)
B*0808N	Null	—	—		—	—	
B*0809	B8	—	—		1	B8 [100]ᵃ	(7,10)
B*0810	B8	—	—		—	—	B8varᶜ
B*0811	—	—	—		—	—	
B*0812	—	—	—		—	—	
B*0813	—	—	—		—	—	
B*1301	B13	2	B13 [100]		210	B13 [97]	
B*1302	B13	3	B13 [99–100]		944	B13 [99]	
B*1303	—	—	—		1	B13 [100]ᵃ	B21varᶜ
B*1304	—	2	B21 [22–38] B49 [15] B15 [7–17] B15x21[0–15] B50 [5–11] B63 [0–8] B77 [4–7]		—	—	B15x21ᶜ B'X' [53] B15 [12] B21 [12]ᵈ
B*1306	—	—	—		—	—	
B*1307	Null	—	—		—	—	
B*1401	B64(14)	3	B14 [49–50] B64 [26–37] B65 [11–22]		649	B14 [69] B65 [14] B64 [9]	
B*1402	B65(14)	3	B65 [65–68] B14 [30–34]		1542	B14 [63] B65 [33] B64 [1]	
B*1403	B14	—	—		37	B14 [49] B65 [35] B64 [3]	
B*1404	—	—	—		2	B14 [50] B64 [50]ᵃ	
B*1405	—	—	—		2	blank [100]ᵃ	
B*1406	B14	—	—		—	—	B14weak(4)
B*1501	B62(15)	13	B62 [92–98]		1496	B62 [94] B15 [3]	
B*1502	B75(15)	12	B75 [83–88] B15 [5–10] B62 [6–7]		883	B75 [65] B62 [20] B15 [6]	
B*1503	B72(70)	12	B70 [50–59] B72 [28–37]		2295	B70 [60] B72 [10] B35 [2] B62 [1] blank [17]	
B*1504	B62(15)	—	—		21	B62 [71] B75 [10] B15 [2] blank [11]	
B*1505	B62(15)	1	B62 [88]		19	B62 [58] B15 [21] B75 [11]	
B*1506	B62(15)	—	—		10	B62 [40] B70 [20] B75 [11]	

Table 4.3 HLA-B alleles and their serologic designations (*continued*)

HLA allele	WHO assigned	International cell exchange UCLA		NMDP		Comments
		Cells tested	Assigned type [%]	Cells tested	Assigned type [%]	
B*1507	B62(15)	1	B62 [93]	85	B62 [78] B15 [1] B75 [1]	
B*1508	B75(15)	2	B75 [30–48] B15 [19–35] B62 [6–19] B'1508' [11–14]	42	B62 [21] B15 [17] B75 [17] B35 [10] B70 [7]	
B*1509	B70	—	—	63	B70 [51] B35 [3] B15 [2] blank [20]	
B*1510	B71(70)	5	B70 [60–67] B71 [22–33]	1051	B70 [46] B35 [4] B72 [2] B71 [2] B75 [1] B15 [1] blank [18]	
B*1511	B75(15)	3	B75 [31–51] B15 [21–38] B46 [6–12] B'1511' [0–19]	43	B75 [26] B62 [25] B46 [19] B15 [14] blank [9]	
B*1512	B76(15)	5	B76 [76–88] B45 [3–6]	9	B76 [22] B75 [11] B62 [11] B15 [11] B53 [11]	
B*1513	B77(15)	2	B77 [46–53] B63 [0–16]	91	B63 [24] B77 [19] B75 [15] B15 [13] B'X' [11] B62 [8] B76 [1]	B77 [59] B62 [29][d]
B*1514	B76(15)	1	B76 [23] B45 [20] B15 [20] B70 [6] B63 [5]	3	B62 [33][a]	
B*1515	B62(15)	4	B75 [51–59] B62 [15–29] B15 [13–27]	95	B62 [48] B75 [33] B15 [11] B70 [4]	B75 (11)
B*1516	B63(15)	7	B63 [87–91]	706	B63 [81] B57 [4] B15 [3] B62 [1] B'X' [2]	
B*1517	B63(15)	3	B63 [89–95]	651	B63 [88] B15 [3] B57 [1]	
B*1518	B71(70)	7	B70 [57–69] B71 [19–31]	491	B70 [49] blank [39] B71 [4] B72 [1]	
B*1519	B76(15)	1	B76 [29] B15 [19] B45 [11] B62 [8] B75 [7]	—	—	
B*1520	B62(15)	—	—	4	B62 [100][a]	
B*1521	B75(15)	6	B75 [78–86] B15 [12–13] B62 [4–8]	119	B75 [56] B62 [8] B15 [7] B35 [5] B'X' [5] B70 [3]	B75[76][d]
B*1522[e]	B35	1	B35 [79] B70 [12] B15 [5]	311	B35 [61] B70 [13] B15 [2] B53 [1] B78 [1]	
B*1523	—	2	B'1523' [17] B5 [14–15] B77 [7–11] B70 [7–9] B15 [5–8]	11	blank [64] B70 [18] B53 [9]	NM5, Bw4pos[c]
B*1524	B62(15)	4	B62 [45–48] B'1524' [23–24] B15 [21–22] B77 [4–6] B63 [2–3]	51	B62 [51] B'X' [7] B77 [4] B63 [2] B15 [2] blank [16]	B62Bw4[c]
B*1525	B62(15)	5	B62 [86–95]	84	B62 [76] B15 [9] B75 [6]	
B*1526N	Null	—	—	—	—	
B*1527	B62(15)	—	—	71	B62 [75] B15 [7]	
B*1528	B15	—	—	6	B62 [67] B15 [17] B75 [17]	
B*1529	B15	2	B70 [54–58] B15 [11–12] B71 [7–10]	20	blank [60] B70 [20] B35 [6]	

Table 4.3   HLA-B alleles and their serologic designations (*continued*)

HLA allele	WHO assigned	International cell exchange UCLA		NMDP		Comments
		Cells tested	Assigned type [%]	Cells tested	Assigned type [%]	
B*1530	B62(15)	3	B62 [63–86] B75 [4–13] B15 [12–19]	82	B62 [63] B15 [20] B75 [6]	
B*1531	B75(15)	—	—	8	B62 [63] B70 [13]	
B*1532	B62(15)	—	—	—	—	
B*1533	B15	—	—	—	—	
B*1534	B15	—	—	1	B62 [100][a]	
B*1535	B62(15)	5	B62 [88–95]	17	B62 [86] B15 [6]	
B*1536	—	—	—	3	B13 [100][a]	Bw4pos
B*1537	B70	2	B70 [46] B71 [11] B78 [9]	23	blank [44] B70 [18] B'X' [17] B15 [4] B71 [4] B78 [4]	B15short[c]
B*1538	—	—	—	4	B62 [75][a]	B62var[c]
B*1539	B62(15)	—	—	16	B62 [68] B15 [13]	
B*1540	—	—	—	22	blank [67] B70 [12] B62 [5]	B62short (11)
B*1542	—	—	—	—	—	
B*1543	—	—	—	—	—	Bw4pos
B*1544	—	—	—	—	—	
B*1545	B62(15)	—	—	16	B62 [46]	
B*1546	B72(70)	—	—	4	B50 [50]	
B*1547	—	—	—	15	blank [53] B35 [13] B70 [9] B15 [7]	
B*1548	B62(15)	—	—	—	—	
B*1549	—	—	—	—	—	
B*1550	—	—	—	—	—	
B*1551	B70	—	—	—	—	(10)
B*1552	—	—	—	—	—	
B*1553	—	—	—	—	—	
B*1554	—	—	—	—	—	
B*1555	B15	—	—	—	—	B15Bw6 (12)
B*1556	—	—	—	—	—	
B*1557	—	—	—	—	—	
B*1558	B62(15)	—	—	—	—	(9)
B*1559[e]	B35	—	—	—	—	B35 (13)

Table 4.3 HLA-B alleles and their serologic designations (*continued*)

HLA allele	WHO assigned	International cell exchange UCLA		NMDP		Comments
		Cells tested	Assigned type [%]	Cells tested	Assigned type [%]	
B*1560	—	—	—	—	—	
B*1561	—	—	—	—	—	
B*1562	—	—	—	—	—	
B*1563	—	—	—	—	—	
B*1564	—	—	—	—	—	
B*1801	B18	7	B18 [97–98]	1431	B18 [96]	
B*1802	B18	—	—	27	B18 [89]	
B*1803	B18	—	—	27	B18 [85] blank [15]	
B*1804	—	—	—	2	B18 [100][a]	
B*1805	B18	—	—	—	—	
B*1806	B18	—	—	—	—	B18var[c]
B*1807	—	—	—	1	[b]	
B*1808	—	—	—	1	B18 [100][a]	
B*1809	B18	—	—	—	—	B18Bw4[c]
B*1810	—	—	—	—	—	
B*1811	—	—	—	—	—	
B*1812	—	—	—	—	—	
B*1813	—	—	—	—	—	
B*2701	B27	—	—	6	B27 [100]	
B*2702	B27	1	B27 [97]	351	B27 [85] blank [7]	
B*2703	B27	1	B27 [97]	19	B27 [95]	
B*2704	B27	—	—	15	B27 [100]	
B*2705	B27	3	B27 [98–100]	1100	B27 [98]	
B*2706	B27	2	B27 [99–100]	21	B27 [95]	
B*2707	B27	—	—	6	B27 [100]	B27[100][d]
B*2708	B2708	3	B2708 [36–58] B7 [19–41] B27 [11–19]	13	B27 [46] B7 [15] blank [31]	B7Qui[c]
B*2709	B27	—	—	4	B27 [100][a]	
B*2710	B27	—	—	1	B27 [100][a]	

Table 4.3 HLA-B alleles and their serologic designations (*continued*)

HLA allele	WHO assigned	International cell exchange UCLA Cells tested	Assigned type [%]	NMDP Cells tested	Assigned type [%]	Comments
B*2711	B27	—	—	2	B27 [100][a]	B27var[c]
B*2712	—	—	—	14	blank [86]	BX [63] B40 [25] B61 [12][c,d]
B*2713	B27	—	—	—	—	
B*2714	—	—	—	4	B27 [75][a]	
B*2715	—	—	—	—	—	B'X'[c]
B*2716	—	—	—	—	—	
B*2717	B27	—	—	—	—	
B*2718	—	—	—	—	—	
B*2719	B27	—	—	—	—	
B*2720	B27	—	—	—	—	(9)
B*2721	—	—	—	—	—	
B*2722	B27	1	B27 [100]	—	—	(14)
B*2723	—	—	—	—	—	
B*3501	B35	13	B35 [94–99]	1252	B35 [96]	
B*3502	B35	2	B35 [94–97]	766	B35 [68]	
B*3503	B35	2	B35 [95–100]	904	B35 [96]	
B*3504	B35	—	—	15	B35 [93]	
B*3505	B35	7	B35 [96–99]	154	B35 [97]	
B*3506	B35	—	—	9	B35 [77]	
B*3507	B35	—	—	—	—	
B*3508	B35	4	B35 [95–99]	217	B35 [94]	
B*3509	B35	—	—	—	—	
B*3510	B35	—	—	10	B35 [80]	B35short (11)
B*3511	B35	—	—	17	B35 [67] B51 [10]	
B*3512	B35	2	B35 [72–80] B53 [11–16]	347	B35 [78]	
B*3513	B35	—	—	—	—	
B*3514	B35	—	—	15	B35 [87]	
B*3515	B35	—	—	5	B35 [40] blank [60]	B35var[c]
B*3516	B35	2	B35 [63] B70 [20]	33	B35 [37] B'X' [15] B70 [9] blank [27]	

Table 4.3 HLA-B alleles and their serologic designations (*continued*)

HLA allele	WHO assigned	International cell exchange UCLA		NMDP		Comments
		Cells tested	Assigned type [%]	Cells tested	Assigned type [%]	
B*3517	B35	3	B35 [98–99]	195	B35 [95]	
B*3518	B35	—	—	1	B35 [100][a]	
B*3519	B35	—	—	3	B35 [33] B21 [33] blank [33][a]	
B*3520	B35	—	—	16	B35 [39] B70 [6] blank [12]	
B*3521	—	—	—	7	B35 [22] B78 [14] blank [43]	
B*3522	—	—	—	1	B35 [100][a]	
B*3523	—	—	—	3	B35 [34] blank [66][a]	
B*3524	—	—	—	5	B35 [23] B78 [20][a]	
B*3525	—	—	—	1	[b]	
B*3526	—	—	—	—	—	
B*3527	B35	—	—	1	B35 [100][a]	
B*3528	—	—	—	4	B35 [50] blank [50][a]	
B*3529	B35	—	—	1	B35 [100][a]	(10)
B*3530	B35	—	—	—	—	
B*3531	—	—	—	3	[b]	Bfu/B40-like[c]
B*3532	B35	—	—	—	—	(10)
B*3533	—	—	—	—	—	
B*3534	—	—	—	—	—	
B*3535	B35	—	—	—	—	
B*3536	—	—	—	—	—	
B*3537	—	—	—	—	—	
B*3701	B37	3	B37 [98–99]	1622	B37 [93] blank [6]	
B*3702	—	—	—	4	B27 [50] B37 [25] B'X' [25][a]	B'blank'[c]
B*3703N	Null	—	—	—	—	
B*3801	B38(16)	3	B38 [97–98]	1874	B38 [93] B16 [1] B39 [1]	
B*3802	B38(16)	3	B38 [90–91]	641	B38 [90] B39 [2] B16 [2] blank [6]	
B*3803	B16	—	—	—	—	[c]
B*3804	—	—	—	—	—	
B*3805	B38(16)	—	—	—	—	

Table 4.3 HLA-B alleles and their serologic designations (*continued*)

HLA allele	WHO assigned	International cell exchange UCLA		NMDP		Comments
		Cells tested	Assigned type [%]	Cells tested	Assigned type [%]	
B*3806	—	—	—	—	—	
B*3807	—	—	—	—	—	
B*3901	B3901	4	B39 [99]	308	B39 [91] B3901 [6] B38 [1] B16 [1]	
B*3902	B3902	2	B39 [90–97] B67 [0–5]	71	B39 [85] blank [11]	
B*3903	B39(16)	—	—	16	B39 [78]	
B*3904	B39(16)	—	—	1	B39 [100]ᵃ	
B*3905	B16	6	B39 [69–90] B16 [3–11] B38 [3–9] B'3905' [3–8]	150	B39 [76] B38 [3] B16 [1] blank [7]	B39varᶜ
B*3906	B39(16)	3	B39 [99]	371	B39 [93]	
B*3907	B39(16)	—	—	3	B39 [36]ᵃ	
B*3908	B39(16)	1	B39 [48] B16 [12]	48	B39 [46] B'X' [12] blank [29]	B39 (11)
B*3909	B39(16)	—	—	4	B39 [79]ᵃ	
B*3910	B39(16)	2	B39 [95–99]	106	B39 [84] B16 [1] B38 [1] blank [14]	B39 (11)
B*3911	B39(16)	—	—	13	B39 [62] blank [23]	
B*3912	B39(16)	—	—	—	—	
B*3913	B39(16)	—	—	—	—	
B*3914	—	—	—	2	B39 [100]ᵃ	
B*3915	—	—	—	3	B39 [100]ᵃ	
B*3916	—	—	—	—	—	
B*3917	—	—	—	—	—	
B*3918	—	—	—	—	—	
B*3919	—	—	—	—	—	
B*3920	—	—	—	—	—	
B*3922	—	—	—	—	—	
B*3923	B39(16)	—	—	—	—	
B*3924	B39(16)	—	—	—	—	(15)
B*4001	B60(40)	23	B60 [96–99]	3285	B60 [93] B40 [3] B61 [2]	
B*4002	B61(40)	9	B61 [84–87]	1056	B61 [80] B60 [9] B40 [6]	
B*4003	B61(40)	—	—	23	B61 [62] B60 [22] blank [8]	
B*4004	B61(40)	—	—	43	B61 [35] B50 [16] B60 [12] B40 [5]	

Table 4.3 HLA-B alleles and their serologic designations (*continued*)

HLA allele	WHO assigned	International cell exchange UCLA		NMDP		Comments
		Cells tested	Assigned type [%]	Cells tested	Assigned type [%]	
B*4005	B4005	3	B4005 [41–57] B50 [28–36] B21 [5–9]	257	B50 [30] B4005 [13] blank [23] B'X' [4] B61[7] B70 [5] B21 [5] B60 [5]	BN21, B21var
B*4006	B61(40)	2	B61 [81–86] B60 [5–7]	413	B61 [74] B60 [5] B40 [5] blank [8]	
B*4007	B60(40)	—	—	12	B60 [83] B40 [8]	B'Fu'
B*4008	—	2	B61 [24–27] B40 [20–30] B48 [11–14] B'4008' [9]	127	blank [69] B61 [12] B40 [8] B48 [2]	B40x48var[c]
B*4009	B61(40)	—	—	4	B61 [75][a]	
B*4010	B60(40)	1	B60 [60] B40 [19] B61 [12] B48 [7]	2	B60 [100][a]	B60var[c]
B*4011	B40	—	—	8	B61 [41] B60 [14] B40 [12]	
B*4012	—	2	B48 [34] B70 [20] B72 [10]	8	B60 [50] B48 [13] blank [38]	B48x70[c]
B*4013	—	—	—	1	[b]	B40Bw4
B*4014	—	—	—	2	B60 [50] B61 [50][a]	
B*4015	—	—	—	—	—	
B*4016	B61(40)	—	—	3	B61 [67] B60 [33][a]	
B*4018	—	—	—	2	B61 [100][a]	
B*4019	—	—	—	—	—	Bw4pos
B*4020	—	—	—	4	B61 [50] B60 [27][a]	
B*4021	—	—	—	—	—	B40x15[c]
B*4022N	Null	—	—	—	—	
B*4023	—	—	—	—	—	
B*4024	—	—	—	—	—	
B*4025	—	—	—	—	—	B40short[c]
B*4026	B21	—	—	—	—	
B*4027	B61(40)	—	—	1	B61 [100][a]	
B*4028	—	—	—	—	—	
B*4029	B61(40)	—	—	—	—	
B*4030	—	—	—	—	—	
B*4031	B60(40)	—	—	—	—	
B*4032	—	—	—	—	—	

Table 4.3 HLA-B alleles and their serologic designations (*continued*)

HLA allele	WHO assigned	International cell exchange UCLA			NMDP		Comments
		Cells tested	Assigned type [%]		Cells tested	Assigned type [%]	
B*4033	—	—	—		—	—	
B*4034	B60(40)	—	—		—	—	
B*4035	—	—	—		—	—	
B*4101	B41	4	B41 [91–95]		479	B41 [86] blank [11]	
B*4102	B41	3	B41 [95–97]		390	B41 [87] blank [9]	
B*4103	B41	—	—		10	B41 [90]	
B*4104	—	—	—		—	—	
B*4105	—	—	—		—	—	
B*4201	B42	8	B42 [92–96]		926	B42 [93] B55 [1] B7 [1]	
B*4202	B42	5	B42 [87–96] B67 [0–12]		65	B42 [95]	
B*4402	B44(12)	6	B44 [99–100]		6374	B44 [97]	
B*4403	B44(12)	13	B44 [99–100]		2058	B44 [96]	
B*4404	B44(12)	—	—		30	B44 [58] B12 [3] B45 [3]	
B*4405	B44(12)	—	—		101	B44 [95]	B44[100][d]
B*4406	B44(12)	—	—		8	blank [50] B44 [37]	B12var[c]
B*4407	B44(12)	—	—		2	B44 [100][a]	
B*4408	B44(12)	—	—		1	B44 [100][a]	B44var[c]
B*4409	B12	—	—		6	B45 [83]	B12Bw6[c]
B*4410	B44(12)	—	—		32	B44 [78] B12 [3] B45 [3]	
B*4411	—	—	—		—	—	
B*4412	B44(12)	—	—		—	—	
B*4413	B44(12)	—	—		—	—	
B*4414	—	—	—		—	—	
B*4415	B12	—	—		—	—	B12Bw4[c]
B*4416	B47	1	B47 [52] B44 [12]		—	—	[c]
B*4417	—	—	—		—	—	B44 (17)
B*4418	—	—	—		—	—	
B*4419N	Null	—	—		—	—	
B*4420	—	—	—		—	—	
B*4421	—	—	—		—	—	

Table 4.3   HLA-B alleles and their serologic designations (*continued*)

HLA allele	WHO assigned	International cell exchange UCLA		NMDP		Comments
		Cells tested	Assigned type [%]	Cells tested	Assigned type [%]	
B*4422	—	—	—	—	—	
B*4423N	Null	—	—	—	—	
B*4424	—	—	—	—	—	
B*4501	B45(12)	4	B45 [96–99]	2073	B45 [94] B44 [1] B12 [1] B50 [1]	
B*4502	—	—	—	3	B45 [67][a]	
B*4503	—	—	—	—	—	
B*4504	—	—	—	—	—	
B*4601	B46	6	B46 [85–97]	1410	B46 [59] blank [35] B'X' [3] B62 [1]	
B*4602	B46	—	—	—	—	B46var[c]
B*4701	B47	2	B47 [90–94]	295	B47 [52] B27 [4] blank [34]	
B*4702	B47	—	—	6	B47 [33] B40 [34] blank [34]	B47Bw6[c]
B*4703	—	2	B47 [70–77] B60 [4–9] B61 [7–8] B40 [6–7]	3	B47 [33] B13 [33][a]	B47var[c]
B*4801	B48	6	B48 [92–96]	490	B48 [80] B60 [6] B61 [2] B40 [2] blank [5]	
B*4802	B48	—	—	39	blank [44] B35 [16] B70 [13] B48 [5]	
B*4803	B48	—	—	23	B48 [57] B40 [18] blank [13]	
B*4804	—	—	—	3	B48 [67][a]	
B*4805	B48	—	—	—	—	
B*4806	—	—	—	—	—	
B*4807	B48	—	—	1	B48 [100][a]	
B*4901	B49(21)	2	B49 [92–95]	2550	B49 [94] B50 [1] B21 [1]	
B*4902	—	—	—	—	—	B49 (32)
B*4903	—	—	—	—	—	
B*5001	B50(21)	7	B50 [87–99]	1527	B50 [84] B21 [3] B49 [2] B45 [2]	
B*5002	B45(12)	—	—	71	B45 [92] B50 [3]	
B*5004	B50(21)	—	—	—	—	
B*5101	B51(5)	22	B51 [92–98]	1425	B51 [96] B5 [2] B52 [1]	
B*5102	B5102	6	B51 [33–44] B5102 [28–31] B53 [14–21] B52 [2–5]	183	B51 [74] B53 [10] B52 [6] B5 [3] B5102 [3]	
B*5103	B5103	—	—	—	—	

Table 4.3   HLA-B alleles and their serologic designations (*continued*)

HLA allele	WHO assigned	International cell exchange UCLA		NMDP		Comments
		Cells tested	Assigned type [%]	Cells tested	Assigned type [%]	
B*5104	B51(5)	—	—	4	B51 [75] B5 [25][a]	
B*5105	B51(5)	—	—	24	B51 [54] B53 [21] B52 [8]	
B*5106	B5	1	B51 [43] B53 [32] B78 [11] B5102 [7] B5 [7]	34	B51 [62] B53 [15] B52 [9] B5 [3]	B5var[c]
B*5107	B51(5)	—	—	57	B51 [80] B52 [9]	
B*5108	B51(5)	—	—	131	B51 [77] B53 [4] B5 [2] B52 [2]	
B*5109	B51(5)	—	—	33	B51 [88] B5 [3] B52 [3] B53 [3] B78 [3]	
B*5110	—	—	—	3	[b]	
B*5111N	Null	—	—	—	—	
B*5112	—	—	—	—	—	B'blank'[c]
B*5113	—	—	—	3	B51 [100][a]	
B*5114	—	—	—	4	B51 [100][a]	
B*5115	—	—	—	—	—	
B*5116	B52(5)	—	—	—	—	
B*5117	B51(5)	—	—	—	—	
B*5118	B51(5)	—	—	—	—	
B*5119	—	—	—	—	—	
B*5120	—	—	—	—	—	
B*5121	—	—	—	—	—	
B*5122	—	—	—	—	—	
B*5123	—	—	—	—	—	
B*5124	B51(5)	—	—	—	—	(7)
B*5201	B52(5)	10	B52 [90–96]	2383	B52 [82] B51 [5] B5 [2]	
B*5202	—	—	—	—	—	
B*5301	B53	11	B53 [85–95]	2749	B53 [87] B35 [5]	
B*5302	—	—	—	3	B53 [67] B51 [33][a]	
B*5303	—	—	—	1	B35 [100][a]	
B*5304	—	—	—	—	—	B52/53[c]
B*5305	—	—	—	—	—	
B*5306	—	—	—	—	—	B51/53[c]

Table 4.3 HLA-B alleles and their serologic designations (*continued*)

| HLA allele | WHO assigned | International cell exchange UCLA | | NMDP | | Comments |
		Cells tested	Assigned type [%]	Cells tested	Assigned type [%]	
B*5307	—	—	—	—	—	B53x37[c]
B*5401	B54(22)	4	B54 [85–88]	590	B54 [68] blank [12] B55 [10] B56 [3] B22 [3] B'X' [2]	
B*5402	B54(22)	—	—	—	—	
B*5501	B55(22)	3	B55 [89–97]	314	B55 [97] B22 [2] B56 [1]	
B*5502	B55(22)	—	—	249	B55 [84] B54 [3] B22 [3]	
B*5503	—	—	—	1	B22 [100][a]	B55/67[c]
B*5504	B55(22)	—	—	3	B55 [67][a]	
B*5505	B22	—	—	—	—	
B*5507	B54(22)	—	—	2	B54 [50][a]	
B*5508	—	—	—	—	—	
B*5509	—	—	—	—	—	B22short[c]
B*5510	B55(22)	—	—	—	—	
B*5601	B56(22)	3	B56 [94–96]	581	B56 [73] B55 [3] B22 [3] blank [16]	
B*5602	B56(22)	—	—	5	B56 [80][a]	
B*5603	B22	—	—	13	B56 [46] B22 [8]	B22var[c]
B*5604	B56(22)	1	B56 [78] B22 [6]	8	B56 [63]	B56var[c]
B*5605	B56(22)	—	—	3	B7 [67][a]	
B*5606	B78	—	—	—	—	B78var[c]
B*5607	B56(22)	—	—	—	—	B56Bw4[c]
B*5701	B57(17)	6	B57 [97–98]	3335	B57 [93] B17 [4] B58 [2]	
B*5702	B57(17)	3	B57 [96–99]	52	B57 [75] B58 [10] B17 [8]	
B*5703	B57(17)	2	B57 [90–94]	289	B57 [83] B17 [9] B58 [6]	
B*5704	B57(17)	—	—	30	B57 [69] B17 [8]	
B*5705	—	—	—	2	B57 [100][a]	
B*5706	—	—	—	—	—	
B*5707	—	—	—	—	—	
B*5801	B58(17)	13	B58 [85–92]	1767	B58 [80] B57 [7] B17 [4]	
B*5802	B58(17)	1	B58 [82]	766	B58 [72] B57 [12] B17 [3]	

Table 4.3 HLA-B alleles and their serologic designations (*continued*)

HLA allele	WHO assigned	International cell exchange UCLA		NMDP		Comments
		Cells tested	Assigned type [%]	Cells tested	Assigned type [%]	
B*5804	—	—	—	—	—	
B*5805	—	—	—	—	—	
B*5806	—	—	—	—	—	
B*5901	B59	4	B59 [95–96]	118	B59 [44] blank [41] B38 [3] B55 [3]	
B*6701	B67	2	B67 [74] B39 [12] B42 [2–9] B55 [0–8]	76	B39 [41] B67 [31] B22 [11] blank [11]	
B*7301	B73	3	B73 [81–83]	99	blank [78] B73 [20]	
B*7801	B78	4	B78 [83–89] B35 [1–9] B51 [1–6]	182	B78 [43] B35 [13] B5 [14] B'X' [7] B70 [3] blank [15]	
B*7802	B78	—	—	—	—	
B*7803	—	—	—	—	—	
B*7804	—	—	—	2	B35 [51][a]	
B*7805	—	—	—	—	—	
B*8101	B81	9	B81 [79–89] B7 [13–19]	590	B7 [76] B81 [8] blank [6] B60 [1] B48 [1]	
B*8201	—	1	B'8201' [28] B22 [9]	140	blank [59] B'X' [10] B82 [4] B22 [6] B70 [4] B45 [4]	B22x45, B45var[c]
B*8202	—	—	—	—	—	B'8201'
B*8301	—	—	—	—	—	

http://www.worldmarrow.org/Dictionary/Dict2001Table3.html
[a]Allele has been reported <6 times and/or serologically identified in <4 individuals.
[b]Allele identified but serology not informative.
[c]See remarks in Table 4.8.
[d]HLA-Club Cell Exchange: one sample typed by 17 laboratories and allele identified by at least three laboratories.
[e]B*1522 and B*1559 renamed into B*3543 and B*3544 respectively.

Table 4.4 HLA-C alleles and their serologic designations

HLA alleles	WHO assigned	International cell exchange UCLA		NMDP		Comments
		Cells tested	Assigned type [%]	Cells tested	Assigned type [%]	
Cw*0102	Cw1	33	Cw1 [56–85]	292	Cw1 [92]	
Cw*0103	Cw1	—	—	1	Cw1 [100][a]	
Cw*0104	—	—	—	—	—	
Cw*0202	Cw2	17	Cw2 [58–87]	442	Cw2 [91]	
Cw*0203	—	—	—	—	—	
Cw*0204	—	—	—	—	—	
Cw*0302	Cw10(w3)	5	Cw3 [56–75] Cw10 [5–9]	40	Cw3 [86] Cw10 [5]	
Cw*0303	Cw9(w3)	23	Cw3 [64–74] Cw9 [5–10]	556	Cw3 [86] Cw9 [4]	
Cw*0304	Cw10(w3)	28	Cw3 [64–71] Cw10 [5–12]	798	Cw3 [86] Cw10 [3]	
Cw*0305	—	—	—	1	Cw3 [100][a]	
Cw*0306	—	—	—	1	Cw3 [100][a]	
Cw*0307	Cw3	—	—	—	—	
Cw*0308	—	—	—	—	—	
Cw*0309	Cw3	2	Cw3 [44–51] Cw10 [3–5]	—	—	
Cw*0310	Cw3	—	—	1	Cw3 [100][a]	
Cw*0311	—	—	—	—	—	
Cw*0312	—	—	—	—	—	
Cw*0401	Cw4	45	Cw4 [58–82] Cw6 [0–6]	998	Cw4 [91]	
Cw*0403	—	8	Cw6 [18–40] Cw4 [13–25] C'403' [2–5]	3	Cw6 [67][a]	Cw4NM, Cw4/6var
Cw*0404	—	—	—	2	[b]	
Cw*0405	—	—	—	1	Cw4 [100][a]	
Cw*0406	—	2	Cw6 [56–62] Cw4 [8–13]	—	—	
Cw*0407	—	—	—	—	—	
Cw*0408	—	—	—	—	—	
Cw*0501	Cw5	7	Cw5 [58–72]	869	Cw5 [82] blank [12]	
Cw*0502	Cw5	—	—	—	—	
Cw*0503	—	—	—	—	—	
Cw*0504	—	—	—	—	—	
Cw*0602	Cw6	18	Cw6 [52–74]	855	Cw6 [76] blank [18]	

Table 4.4 HLA-C alleles and their serologic designations (*continued*)

| HLA alleles | WHO assigned | International cell exchange UCLA | | NMDP | | Comments |
		Cells tested	Assigned type [%]	Cells tested	Assigned type [%]	
Cw*0603	—	—	—	1	Cw6 [100][a]	
Cw*0604	—	—	—	1	Cw6 [100][a]	
Cw*0605	Cw6	—	—	—	—	
Cw*0606	—	—	—	—	—	
Cw*0607	—	—	—	—	—	
Cw*0701	Cw7	17	Cw7 [51–83]	1227	Cw7 [69] blank [14]	
Cw*0702	Cw7	42	Cw7 [55–78]	1234	Cw7 [74] blank [12]	
Cw*0703	—	—	—	1	Cw7 [100][a]	
Cw*0704	Cw7	11	Cw7 [62–80]	104	Cw7 [65] blank [21]	
Cw*0705	—	—	—	—	—	
Cw*0706	Cw7	3	Cw7 [68–76]	3	Cw7 [80][a]	
Cw*0707	—	—	—	1	b	
Cw*0708	—	—	—	—	—	
Cw*0709	—	—	—	—	—	
Cw*0710	—	—	—	1	Cw7 [100][a]	
Cw*0711	—	—	—	—	—	
Cw*0712	—	—	—	—	—	
Cw*0713	—	—	—	—	—	
Cw*0714	Cw7	—	—	—	—	
Cw*0801	Cw8	25	Cw8 [48–64]	31	Cw8 [45] blank [45]	
Cw*0802	Cw8	9	Cw8 [35–65]	270	Cw8 [65] blank [27]	
Cw*0803	Cw8	1	Cw8 [8]	1	b	
Cw*0804	Cw8	3	Cw8 [43–59]	2	Cw8 [50]	
Cw*0805	—	—	—	—	—	
Cw*0806	—	—	—	—	—	
Cw*0807	—	—	—	—	—	
Cw*0808	—	—	—	—	—	
Cw*0809	—	—	—	—	—	
Cw*1202	—	7	blank[c]	82	blank [91]	
Cw*1203	—	12	Cw7 [2–14]	322	Cw7 [5] blank [61]	

Table 4.4   HLA-C alleles and their serologic designations (*continued*)

| HLA alleles | WHO assigned | International cell exchange UCLA | | NMDP | | Comments |
		Cells tested	Assigned type [%]	Cells tested	Assigned type [%]	
Cw*1204	—	—	—	2	b	
Cw*1205	—	—	—	—	—	
Cw*1206	—	—	—	—	—	
Cw*1207	—	—	—	—	—	
Cw*1301	—	—	—	—	—	
Cw*1402	—	11	blank[c]	102	blank [84] Cw1 [4]	
Cw*1403	—	1	blank[c]	8	blank [88]	
Cw*1404	—	—	—	—	—	
Cw*1502	—	12	Cw6 [0–11] Cw2 [0–7]	131	blank [78] Cw2 [6] Cw6 [3]	
Cw*1503	—	—	—	1	b	
Cw*1504	—	—	—	—	—	
Cw*1505	—	6	Cw2 [0–39]	28	blank [29] Cw7 [23]	
Cw*1506	—	—	—	1	b	
Cw*1507	—	—	—	—	—	
Cw*1508	—	—	—	—	—	
Cw*1509	—	—	—	1	b	
Cw*1510	—	—	—	—	—	
Cw*1601	—	22	Cw7 [0–13]	286	blank [92]	
Cw*1602	—	—	—	24	blank [81]	
Cw*1604	—	—	—	5	blank [78][a]	
Cw*1701	—	17	C'17' [33–34] Cw7 [16–24]	68	Cw7 [41] Cw2 [11] blank [35]	
Cw*1702	—	—	—	—	—	
Cw*1703	—	—	—	—	—	
Cw*1801	—	2	Cw6 [37–42]	8	Cw6 [62]	
Cw*1802	—	2	Cw6 [38–43]	—	—	

http://www.worldmarrow.org/Dictionary/Dict2001Table4.html
[a]Allele has been reported <6 times and/or serologically identified in <4 individuals.
[b]Allele identified but serology not informative.
[c]No Cw antigen reported by more than 5% of the laboratories.

Table 4.5 HLA-DRB1 alleles and their serologic designations

HLA allele	WHO assigned	International cell exchange UCLA		NMDP		Comments
		Cells tested	Assigned type [%]	Cells tested	Assigned type [%]	
DRB1*0101	DR1	12	DR1 [88–100]	9538	DR1 [97]	
DRB1*0102	DR1	2	DR1 [86–95]	2459	DR1 [95]	
DRB1*0103	DR103	1	DR103 [70] DR1 [17]	1666	DR1 [42] DR103 [33] blank [15]	
DRB1*0104	DR1	—	—	1	DR1 [100][a]	
DRB1*0105	—	—	—	1	DR1 [100][a]	
DRB1*0106	—	—	—	—	—	
DRB1*0107	—	—	—	—	—	
DRB1*0301	DR17(3)	22	DR17 [59–69] DR3 [28–41]	12867	DR17 [61] DR3 [36]	
DRB1*0302	DR18(3)	10	DR18 [62–67] DR3 [27–35]	370	DR18 [63] DR3 [28] DR17 [2]	
DRB1*0303	DR18(3)	—	—	7	DR18 [45] DR17 [29] DR3 [14]	
DRB1*0304	DR17(3)	1	DR3 [51] DR17 [47]	6	DR3 [67] DR17 [17]	
DRB1*0305	DR17(3)	—	—	—	—	DR3[d]
DRB1*0306	DR3	—	—	2	DR3 [50] blank [50][a]	
DRB1*0307	DR3	—	—	—	—	
DRB1*0308		—	—	—	—	
DRB1*0309	—	—	—	—	—	
DRB1*0310	DR17(3)	—	—	2	DR3 [50] DR17 [50][a]	
DRB1*0311	DR17(3)	—	—	2	DR3 [50] DR17 [50][a]	
DRB1*0312	DR3	—	—	—		(18)
DRB1*0313	—	—	—	—	—	
DRB1*0314	—	—	—	—	—	
DRB1*0315	—	—	—	—	—	
DRB1*0316	—	—	—	—	—	
DRB1*0317	—	—	—	—	—	
DRB1*0318	—	—	—	—	—	
DRB1*0401	DR4	5	DR4 [100]	10271	DR4 [97]	
DRB1*0402	DR4	4	DR4 [84–100]	1178	DR4 [91]	
DRB1*0403	DR4	7	DR4 [97–100]	960	DR4 [94]	
DRB1*0404	DR4	12	DR4 [95–100]	3003	DR4 [92]	

Table 4.5   HLA-DRB1 alleles and their serologic designations (*continued*)

| HLA allele | WHO assigned | International cell exchange UCLA | | NMDP | | Comments |
		Cells tested	Assigned type [%]	Cells tested	Assigned type [%]	
DRB1*0405	DR4	9	DR4 [94–100]	591	DR4 [94]	
DRB1*0406	DR4	2	DR4 [89–96]	84	DR4 [91]	
DRB1*0407	DR4	2	DR4 [97–98]	1154	DR4 [90]	
DRB1*0408	DR4	—	—	418	DR4 [89]	
DRB1*0409	DR4	—	—	16	DR4 [94]	
DRB1*0410	DR4	—	—	28	DR4 [86]	
DRB1*0411	DR4	1	DR4 [98]	74	DR4 [90]	
DRB1*0412	—	—	—	1	DR4 [100][a]	
DRB1*0413	DR4	—	—	6	DR4 [69]	
DRB1*0414	DR4	—	—	2	DR4 [100][a]	
DRB1*0415	DR4	2	DR4 [84–92] DR11 [24–49]	4	DR4 [75][a]	DR4x11
DRB1*0416	DR4	1	DR4 [100]	1	DR4 [100][a]	
DRB1*0417	DR4	—	—	1	DR4 [100][a]	
DRB1*0418	—	—	—	—	—	
DRB1*0419	DR4	—	—	—	—	
DRB1*0420	DR4	—	—	—	—	
DRB1*0421	DR4	—	—	1	DR4 [100][a]	
DRB1*0422	DR4	—	—	—	—	DR4x3
DRB1*0423	DR4	—	—	1	[b]	
DRB1*0424	DR4	—	—	—	—	
DRB1*0425	DR4	—	—	—	—	DR4short
DRB1*0426	DR4	2	DR4 [100]	—	—	
DRB1*0427	—	—	—	—	—	
DRB1*0428	DR4	—	—	—	—	
DRB1*0429	DR4	—	—	—	—	
DRB1*0430	—	—	—	—	—	
DRB1*0431	DR4	—	—	—	—	
DRB1*0432	DR4	1	DR4 [100]	—	—	(18)
DRB1*0433	—	—	—	—	—	

Table 4.5 HLA-DRB1 alleles and their serologic designations (*continued*)

HLA allele	WHO assigned	International cell exchange UCLA		NMDP		Comments
		Cells tested	Assigned type [%]	Cells tested	Assigned type [%]	
DRB1*0434	—	—	—	—	—	
DRB1*0435	—	—	—	—	—	
DRB1*0436	—	—	—	—	—	
DRB1*0437	—	—	—	—	—	
DRB1*0438	—	—	—	—	—	
DRB1*0701	DR7	15	DR7 [97–100]	13382	DR7 [99]	
DRB1*0703	DR7	—	—	—	—	
DRB1*0704	DR7	—	—	—	—	(19)
DRB1*0801	DR8	4	DR8 [95–100]	1975	DR8 [96]	
DRB1*0802	DR8	1	DR8 [99]	314	DR8 [97]	
DRB1*0803	DR8	11	DR8 [97–100]	285	DR8 [90] DR12 [3] blank [4]	
DRB1*0804	DR8	7	DR8 [88–99]	241	DR8 [95]	
DRB1*0805	DR8	—	—	11	DR8 [100]	
DRB1*0806	DR8	1	DR8 [100]	21	DR8 [86]	
DRB1*0807	DR8	—	—	4	DR8 [100][a]	
DRB1*0808		—	—	—	—	
DRB1*0809	DR8	—	—	1	DR8 [100][a]	DR8var
DRB1*0810	DR8	—	—	2	DR8 [100][a]	
DRB1*0811	DR8	—	—	9	DR8 [89]	
DRB1*0812	DR8	—	—	—	—	
DRB1*0813	—	—	—	—	—	
DRB1*0814	DR8	—	—	—	—	
DRB1*0815	—	—	—	—	—	
DRB1*0816	DR8	—	—	—	—	
DRB1*0817	DR8	—	—	—	—	
DRB1*0818	—	—	—	—	—	
DRB1*0819	—	—	—	—	—	
DRB1*0820	—	—	—	—	—	
DRB1*0821	—	—	—	—	—	
DRB1*0822	—	—	—	—	—	

Table 4.5 HLA-DRB1 alleles and their serologic designations (*continued*)

HLA allele	WHO assigned	International cell exchange UCLA		NMDP		Comments
		Cells tested	Assigned type [%]	Cells tested	Assigned type [%]	
DRB1*0823	—	—	—	—	—	
DRB1*0901	DR9	16	DR9 [97–100]	1869	DR9 [97]	
DRB1*1001	DR10	11	DR10 [94–95]	1490	DR10 [93] DR1 [2] blank [2]	
DRB1*1101	DR11(5)	8	DR11 [95–100]	4934	DR11 [93] DR5 [4]	
DRB1*1102	DR11(5)	5	DR11 [89–100]	527	DR11 [89] DR5 [3] DR6 [3]	
DRB1*1103	DR11(5)	2	DR11 [84–100]	553	DR11 [91] DR5 [2] DR13 [2]	
DRB1*1104	DR11(5)	11	DR11 [94–100]	2622	DR11 [89] DR5 [4]	
DRB1*1105	DR11(5)	—	—	5	DR11 [81][a]	
DRB1*1106	DR11(5)	—	—	16	DR5 [69] DR11 [25]	
DRB1*1107	DR11(5)	—	—	3	DR11 [100][a]	DR11x3
DRB1*1108	DR11(5)	—	—	—	—	
DRB1*1109	DR11(5)	—	—	5	DR11 [80] DR5 [20][a]	
DRB1*1110	DR11(5)	—	—	2	DR11 [52][a]	
DRB1*1111	DR11(5)	2	DR11 [87–90] DR13 [7–8] DR5 [4–5]	7	DR11 [78]	DR11x13
DRB1*1112	—	—	—	—	—	
DRB1*1113	DR11(5)	—	—	5	DR11[80] DR14 [20][a]	DR11x14
DRB1*1114	DR11(5)			3	DR11 [100][a]	
DRB1*1115	—	—	—	1	DR11 [100][a]	
DRB1*1116	—	—	—	—	—	DR11 [70] DR13 [30][c] DR11x13
DRB1*1117	—	—	—	1	DR14 [100][a]	
DRB1*1118	—	—	—	1	[b]	
DRB1*1119	DR11(5)	1	DR11 [100]	2	DR11 [100][a]	
DRB1*1120	DR11(5)	—	—	—	—	DR11x13
DRB1*1121	DR11(5)	1	DR11 [97]	1	DR11 [100][a]	
DRB1*1122	—	—	—	—	—	
DRB1*1123	DR11(5)	—	—	—	—	
DRB1*1124	—	—	—	2	DR11 [50] blank [50][a]	

Table 4.5 HLA-DRB1 alleles and their serologic designations (*continued*)

HLA allele	WHO assigned	International cell exchange UCLA			NMDP		Comments
		Cells tested	Assigned type [%]		Cells tested	Assigned type [%]	
DRB1*1125	DR11(5)	—	—		—	—	
DRB1*1126	DR11(5)	—	—		—	—	
DRB1*1127	DR11(5)	—	—		—	—	
DRB1*1128	—	—	—		1	DR11 [100][a]	
DRB1*1129	DR11(5)	—	—		—	—	
DRB1*1130	—	—	—		—	—	
DRB1*1131	—	—	—		—	—	
DRB1*1132	—	—	—		—	—	
DRB1*1133	—	—	—		—	—	
DRB1*1134	—	—	—		1	DR11 [100][a]	
DRB1*1135	—	—	—		—	—	
DRB1*1136					—	—	
DRB1*1137	—	—	—		—	—	
DRB1*1138	—	—	—		—	—	
DRB1*1139	—	—	—		—	—	
DRB1*1140	—	—	—		—	—	
DRB1*1141	—	—	—		—	—	
DRB1*1201	DR12(5)	8	DR12 [85–96] DR13 [5–12]		1399	DR12 [88] DR5 [3] DR11 [2]	
DRB1*1202	DR12(5)	15	DR12 [97–100]		303	DR12 [95]	
DRB1*1203	DR12(5)	—	—		4	DR12 [100][a]	
DRB1*1204	DR5	—	—		—	—	DR12x11
DRB1*1205	DR12(5)	—	—		—	—	
DRB1*1206	DR12(5)	—	—		2	DR12 [100][a]	
DRB1*1207	—	—	—		—	—	
DRB1*1301	DR13(6)	8	DR13 [88–90] DR6 [3–8]		5867	DR13 [84] DR6 [13]	
DRB1*1302	DR13(6)	15	DR13 [85–97] DR6 [7–9]		4646	DR13 [78] DR6 [13] DR14 [1]	
DRB1*1303	DR13(6)	10	DR13 [50–76] DR6 [10–21] DR'1303' [0–11] DR14 [0–17] DR3 [0–8]		1394	DR13 [75] DR6 [8] DR5 [2] DR11 [2] DR12 [2] DR14 [2] blank [4]	
DRB1*1304	DR13(6)	—	—		139	DR13 [86] DR5 [4] DR6 [2] blank [4]	

Table 4.5 HLA-DRB1 alleles and their serologic designations (*continued*)

HLA allele	WHO assigned	International cell exchange UCLA		NMDP		Comments
		Cells tested	Assigned type [%]	Cells tested	Assigned type [%]	
DRB1*1305	DR13(6)	3	DR13 [48–62] DR11 [11–34] DR6 [5–20] DR14 [3–8]	363	DR13 [61] DR6 [11] DR11 [7] DR12 [5] DR5 [3] DR14 [1]	
DRB1*1306	DR13(6)	—	—	8	DR13 [72] DR6 [25]	
DRB1*1307	DR13(6)	—	—	2	DR13 [100][a]	
DRB1*1308	DR13(6)	—	—	11	DR13 [47] DR6 [53]	
DRB1*1309	—	—	—	1	[b]	
DRB1*1310	DR13(6)	2	DR13 [84–85] DR6 [7–15]	9	DR13 [100]	
DRB1*1311	DR13(6)	—	—	—	—	
DRB1*1312	DR6	—	—	8	DR13 [37] DR11 [25] DR12 [13] DR6 [13]	DR13x12
DRB1*1313	DR13(6)	1	DR13 [83] DR6 [13]	—	—	DR52–neg (18)
DRB1*1314	DR13(6)	—	—	1	DR11 [100][a]	
DRB1*1315	—	—	—	1	[b]	
DRB1*1316	DR13(6)	—	—	—	—	
DRB1*1317	DR13(6)	—	—	—	—	DR13x8
DRB1*1318	DR13(6)	—	—	—	—	
DRB1*1319	DR13(6)	—	—	—	—	
DRB1*1320	DR13(6)	—	—	1	DR13 [100][a]	
DRB1*1321	—	—	—	1	DR13 [100][a]	
DRB1*1322	DR13(6)	—	—	3	DR13 [100][a]	DR13 [100][c]
DRB1*1323	—	—	—	—	—	
DRB1*1324	—	—	—	—	—	
DRB1*1325	—	—	—	—	—	
DRB1*1326	—	—	—	—	—	DRblank–DR52pos
DRB1*1327	DR13(6)	—	—	—	—	
DRB1*1328	—	—	—	—	—	
DRB1*1329	DR6	—	—	—	—	
DRB1*1330	—	—	—	—	—	
DRB1*1331	—	—	—	—	—	

Table 4.5 HLA-DRB1 alleles and their serologic designations (*continued*)

HLA allele	WHO assigned	International cell exchange UCLA		NMDP		Comments
		Cells tested	Assigned type [%]	Cells tested	Assigned type [%]	
DRB1*1332	—	—	—	—	—	
DRB1*1333	—	—	—	—	—	
DRB1*1334	—	—	—	—	—	
DRB1*1335	—	—	—	—	—	
DRB1*1336	DR13(6)	—	—	—	—	
DRB1*1337	—	—	—	—	—	
DRB1*1338	—	—	—	—	—	
DRB1*1339	—	—	—	—	—	
DRB1*1340	—	—	—	—	—	
DRB1*1341	—	—	—	—	—	
DRB1*1342	DR13(6)	—	—	—	—	
DRB1*1343	—	—	—	—	—	
DRB1*1344	—	—	—	—	—	
DRB1*1345	—	—	—	—	—	
DRB1*1346	—	—	—	—	—	
DRB1*1347	—	—	—	—	—	
DRB1*1401	DR14(6)	15	DR14 [90–92] DR6 [5–7]	2654	DR14 [78] DR6 [13] DR13 [2]	
DRB1*1402	DR14(6)	6	DR14 [46–75] DR6 [12–23] DR17 [8–15] DR13 [0–8]	234	DR14 [65] DR6 [11] DR13 [9] DR5 [6] blank [6]	
DRB1*1403	DR1403	—	—	46	DR14 [66] DR6 [12] DR13 [7]	
DRB1*1404	DR1404	8	DR14 [84–89] DR1404 [5–8] DR6 [3–7] DR8 [0–10]	116	DR14 [73] DR6 [9] DR1404 [7] DR12 [2] DR8 [2]	
DRB1*1405	DR14(6)	2	DR14 [74–85] DR6 [11–15] DR3 [0–13]	63	DR14 [84] DR6 [11] DR13 [3]	
DRB1*1406	DR14(6)	1	DR13 [28] DR14 [26] DR6 [19] DR3 [11]	130	DR14 [46] DR6 [15] DR13 [13] DR5 [11] blank [8]	
DRB1*1407	DR14(6)	—	—	15	DR14 [93]	
DRB1*1408	DR14(6)	1	DR14 [94]	1	DR14 [100][a]	DR14[d]
DRB1*1409	—	—	—	—	—	
DRB1*1410	DR14(6)	1	DR14 [76] DR6 [10]	5	DR14 [100][a]	
DRB1*1411	DR14(6)	1	DR14 [59] DR6 [12] DR11 [12] DR'14x11' [9]	2	DR14 [100][a]	DR14x11

Table 4.5 HLA-DRB1 alleles and their serologic designations (*continued*)

HLA allele	WHO assigned	International cell exchange UCLA		NMDP		Comments
		Cells tested	Assigned type [%]	Cells tested	Assigned type [%]	
DRB1*1412	DR14(6)	—	—	—	—	
DRB1*1413	DR14(6)	1	DR14 [59] DR6 [17] DR13 [10]	—	—	
DRB1*1414	DR14(6)	1	DR14 [89] DR13 [5]	—	—	
DRB1*1415	DR8	1	DR8 [50] DR14 [26] DR13 [8]	—	—	DR8+DR52
DRB1*1416	DR6	—	—	4	DR14 [75] DR13 [25][a]	DR14x11
DRB1*1417	DR6	—	—	1	DR14 [100][a]	DR14x13
DRB1*1418	DR6	—	—	—	—	
DRB1*1419	DR14(6)	—	—	—	—	DR14–DQ7[d]
DRB1*1420	DR14(6)	—	—	1	[b]	
DRB1*1421	DR6	1	DR14 [49] DR13 [23] DR6 [19] DR17 [6]	1	DR6 [100][a]	DR13[d]
DRB1*1422	DR14(6)	—	—	—	—	
DRB1*1423	—	—	—	—	—	
DRB1*1424	—	—	—	—	—	
DRB1*1425	—	—	—	—	—	
DRB1*1426	DR14(6)	—	—	—	—	
DRB1*1427	DR14(6)	—	—	—	—	
DRB1*1428	—	—	—	—	—	
DRB1*1429	DR14(6)	—	—	—	—	
DRB1*1430	—	—	—	—	—	
DRB1*1431	—	—	—	—	—	
DRB1*1432	—	—	—	—	—	
DRB1*1433	—	—	—	—	—	
DRB1*1434	—	—	—	—	—	
DRB1*1435	—	—	—	—	—	
DRB1*1436	—	—	—	—	—	
DRB1*1437	—	—	—	—	—	
DRB1*1438	—	—	—	—	—	
DRB1*1439	—	—	—	—	—	

Table 4.5 HLA-DRB1 alleles and their serologic designations (*continued*)

| HLA allele | WHO assigned | International cell exchange UCLA | | NMDP | | Comments |
		Cells tested	Assigned type [%]	Cells tested	Assigned type [%]	
DRB1*1440	—	—	—	—	—	
DRB1*1501	DR15(2)	19	DR15 [78–91] DR2 [7–22]	9701	DR15 [76] DR2 [22]	
DRB1*1502	DR15(2)	11	DR15 [87–90] DR2 [7–10]	816	DR15 [75] DR2 [18] DR16 [1]	
DRB1*1503	DR15(2)	6	DR15 [86–94] DR2 [3–10]	352	DR15 [84] DR2 [11]	
DRB1*1504	DR15(2)	—	—	1	DR15 [100][a]	
DRB1*1505	DR15(2)	—	—	2	DR15 [50][a]	
DRB1*1506	DR15(2)	—	—	1	DR15 [100][a]	
DRB1*1507	DR15(2)	—	—	—	—	(19)
DRB1*1508	DR2	—	—	—		
DRB1*1509	—	—	—	—	—	
DRB1*1510	—	—	—	—	—	
DRB1*1511	—	—	—	—	—	
DRB1*1601	DR16(2)	1	DR16 [31] DR15 [38] DR2 [31]	1150	DR16 [51] DR2 [27] DR15 [13]	
DRB1*1602	DR16(2)	6	DR16 [63–85] DR2 [11–17] DR15 [7–8]	290	DR16 [54] DR2 [26] DR15 [15]	
DRB1*1603	DR2	—	—	—	—	
DRB1*1604	DR16(2)	—	—	2	DR16 [100][a]	
DRB1*1605	DR2	—	—	1	DR16 [100][a]	DR16[d]
DRB1*1607	—	—	—	—	—	
DRB1*1608	—	—	—	—	—	

http://www.worldmarrow.org/Dictionary/Dict2001Table5.html

[a]Allele has been reported <6 times and/or serologically identified in <4 individuals.
[b]Allele identified but serology not informative.
[c]HLA-Club Cell Exchange: one sample typed by 17 laboratories and allele identified by at least 3 laboratories.
[d]Locally identified in Leiden.

Table 4.6 HLA-DRB3/4/5 alleles and their serologic designations

HLA alleles	WHO assigned	Cells tested	International cell exchange UCLA Assigned type [%]	Comments
DRB3*0101	DR52	53	DR52 [91–100]	
DRB3*0102	DR52	1	DR52 [99]	
DRB3*0103	—	—	—	
DRB3*0104	—	—	—	
DRB3*0105	—	—	—	
DRB3*0106	DR52	—	—	
DRB3*0107	DR52	—	—	
DRB3*0201	DR52	2	DR52 [97–99]	
DRB3*0202	DR52	72	DR52 [89–100]	
DRB3*0203	DR52	1	DR52 [100]	
DRB3*0204	—	—	—	
DRB3*0205	—	—	—	
DRB3*0206	—	—	—	
DRB3*0207	DR52	—	—	
DRB3*0208	DR52	—	—	
DRB3*0209	DR52	—	—	
DRB3*0210	DR52	—	—	
DRB3*0211	DR52	—	—	
DRB3*0212	—	—	—	
DRB3*0213	—	—	—	
DRB3*0301	DR52	28	DR52 [85–100]	
DRB3*0302	DR52	—	—	
DRB3*0303	—	—	—	
DRB4*0101	DR53	33	DR53 [100]	
DRB4*0102	DR53	—	—	
DRB4*0103	DR53	28	DR53 [90–100]	
DRB4*0103102N	Null	7	DR53 [42–100]	
DRB4*0104	—	—	—	
DRB4*0105	DR53	—	—	
DRB4*0201N	Null	—	—	

Table 4.6   HLA-DRB3/4/5 alleles and their serologic designations (*continued*)

HLA alleles	WHO assigned	Cells tested	International cell exchange UCLA Assigned type [%]	Comments
DRB4*0301N	Null	—	—	
DRB5*0101	DR51	22	DR51 [82–87]	
DRB5*0102	DR51	6	DR51 [86–89]	
DRB5*0103	—	—	—	
DRB5*0104	—	—	—	
DRB5*0105	—	—	—	
DRB5*0106	—	—	—	
DRB5*0107	DR51	—	—	
DRB5*0108N	Null	—	—	
DRB5*0109	—	—	—	
DRB5*0110N	Null	—	—	
DRB5*0202	DR51	6	DR51 [86–92]	
DRB5*0203	—	—	—	
DRB5*0204	—	—	—	
DRB5*0205	—	—	—	

http://www.worldmarrow.org/Dictionary/Dict2001Table6.html

HLA serologic equivalents have been analysed in heterozygous samples only if the other HLA allelic product would not interfere with the serologic recognition. For example, B*4010 expression was not analysed on a cell also expressing B*4001 (cell 914 = 998). The associated serologic specificities as reported by the participating laboratories are given with the lowest and highest percentage recognition.

The percentage recognition of the serological specificities has increased considerably. Therefore the percentage range as observed in the update analysis of 88 cells for Class I and 44 cells for Class II are now presented in the tables.

The International Cell Exchange (UCLA) provided up to now serologic equivalents for 39 HLA-A, 99 HLA-B, 29 HLA-Cw, 58 HLA-DRB1, 6 HLA-DRB3, 3 HLA-DRB4, 3 HLA-DRB5, and 17 HLA-DQB1 alleles.

NATIONAL MARROW DONOR PROGRAM (NMDP)

Cells from bone marrow transplant patients and unrelated donors were HLA typed using serological and molecular methods by typing laboratories participating in a project to retrospectively type transplant pairs[6] and by laboratories associated with NMDP-affiliated donor or transplant centers as outlined in the previous report.[2] The NMDP dataset is based on the typing information from 35 102 individuals for HLA-A; 56 718 individuals for HLA-B; and 65 752 individuals HLA-DRB1. Analysis was extended to HLA-C (4749 individuals) and HLA-DQB1 (36 297 individuals). **Tables 4.2–4.5** and **Table 4.7** present the number of individuals carrying a certain allele, together with the most frequent serologic type(s) associated with that allele. For example (**Table 4.2**), there were 4920 individuals typed as A*0101 and most (96 percent) received the HLA-A1 serologic assignment. Other alleles appear with several common assignments. For example, B*5108 was observed in 131 individuals of which 77 percent were serologically typed as B51, 4 percent as B53, 2 percent as B5 and 2 percent as B52.

The EM algorithm, used to obtain these data, has been described previously.[2] If an allele appeared five times or less, the serologic assignment(s) should be considered as only an approximation of the serologic reactivity of the

Table 4.7 HLA-DQB1 alleles and their serologic designations

HLA allele	WHO assigned	International cell exchange UCLA		NMDP		Comments
		Cells tested	Assigned type [%]	Cells tested	Assigned type [%]	
DQB1*0201	DQ2	7	DQ2 [93–100]	10258	DQ2 [97]	
DQB1*0202	DQ2	5	DQ2 [90–94]	882	DQ2 [81] DQ1 [6]	
DQB1*0203	DQ2	—	—	2	DQ2 [50][a]	
DQB1*0301	DQ7(3)	21	DQ7 [70–80] DQ3 [16–27] DQ8 [0–8]	14154	DQ7 [81] DQ3 [12] DQ8 [1]	
DQB1*0302	DQ8(3)	13	DQ3 [47–74] DQ8 [22–56] DQ7 [3–7] DQ9 [0–6]	7814	DQ8 [62] DQ3 [28] DQ7 [3]	
DQB1*0303	DQ9(3)	4	DQ3 [45–60] DQ9 [35–42]	3110	DQ9 [64] DQ3 [16] DQ2 [10] DQ8 [1] DQ7 [1]	
DQB1*0304	DQ7(3)	1	DQ7 [60] DQ3 [34]	120	DQ3 [44] DQ7 [37] DQ8 [8]	
DQB1*0305	DQ8(3)	—	—	60	DQ8 [36] DQ3 [34]	DQ7negDQ8neg[c]
DQB1*0306	DQ3	—	—	1	DQ8 [100][a]	
DQB1*0307	—	—	—	—	—	
DQB1*0308	—	—	—	—	—	
DQB1*0309	—	—	—	—	—	
DQB1*0310	DQ8(3)	—	—	—	—	
DQB1*0401	DQ4	1	DQ4 [88]	124	DQ4 [77] DQ3 [7] DQ8 [4] DQ7 [3]	
DQB1*0402	DQ4	7	DQ4 [81–91]	1197	DQ4 [87] DQ3 [3] blank [6]	
DQB1*0501	DQ5(1)	10	DQ1 [51–71] DQ5 [29–41] DQ6 [0–7]	6537	DQ5 [69] DQ1 [20] DQ6 [1]	
DQB1*0502	DQ5(1)	7	DQ1 [54–64] DQ6 [13–27] DQ5 [0–24]	997	DQ5 [50] DQ1 [24] DQ6 [14]	
DQB1*0503	DQ5(1)	3	DQ1 [59] DQ6 [24] DQ5 [14]	1540	DQ5 [60] DQ1 [22] DQ6 [4]	
DQB1*0504	DQ5(1)	1	DQ1 [66] DQ5 [23] DQ6 [8]	63	DQ5 [55] DQ1 [23] DQ6 [2]	
DQB1*0601	DQ6(1)	2	DQ1 [49–54] DQ6 [40]	674	DQ6 [66] DQ1 [23] DQ5 [1]	
DQB1*0602	DQ6(1)	7	DQ1 [48–57] DQ6 [38–48]	6390	DQ6 [67] DQ1 [26]	
DQB1*0603	DQ6(1)	3	DQ1 [42–58] DQ6 [34–56]	3587	DQ6 [62] DQ1 [23] DQ5 [2]	
DQB1*0604	DQ6(1)	1	DQ1 [50] DQ6 [44]	2065	DQ6 [60] DQ1 [25] DQ5 [2]	
DQB1*0605	DQ6(1)	—	—	379	DQ6 [76] DQ1 [15] DQ5 [1]	
DQB1*0606	—	—	—	1	DQ5 [100][a]	
DQB1*0607	—	—	—	4	DQ1 [100][a]	

Table 4.7 HLA-DQB1 alleles and their serologic designations (*continued*)

HLA allele	WHO assigned	International cell exchange UCLA			NMDP		Comments
		Cells tested	Assigned type [%]		Cells tested	Assigned type [%]	
DQB1*0608	—	—	—		8	DQ6 [39] DQ1 [28]	DQ6 [c]
DQB1*0609	DQ6(1)	1	DQ6 [54] DQ1 [43]		150	DQ6 [47] DQ1 [36] DQ5 [2]	
DQB1*0610	—	—	—		—	—	
DQB1*0611	DQ1	—	—		—	—	
DQB1*0612	DQ1	—	—		—	—	
DQB1*0613	—	—	—		1	[b]	
DQB1*0614	DQ6(1)	—	—				
DQB1*0615	—	—	—		—	—	
DQB1*0616	—	—	—		—	—	
DQB1*0617	—	—	—		—	—	

http://www.worldmarrow.org/Dictionary/Dict2001Table7.html
[a] Allele has been reported <6 times and/or serologically identified in <4 individuals.
[b] Allele identified but serology nor informative.
[c] Locally identified in Leiden.

resultant antigen. These cases are indicated in the tables. The NMDP data presented in the tables lacks information on the actual serologic reaction patterns, which might vary from the standard assignment.

INFORMATION OBTAINED FROM OTHER SOURCES

The comments column in the tables indicates references to recent publications that have included remarks on the serologic expression of newly described alleles. Expression of alleles tested in the HLA-Club Cell Exchange is indicated in the tables as well. Also, personal communications to and local observations in the Leiden laboratory are included, but it should be realized that such observations are often based on very few samples tested in only one laboratory.

SEROLOGIC SPECIFICITIES LACKING OFFICIAL WHO NOMENCLATURE DESIGNATIONS

As shown in the tables, a number of alleles appear with several common serologic assignments, indicating that their serologic reaction patterns are not well characterized. For example, the molecule encoded by B*5306 exhibits both B5 and B53 reactivity.[7] **Table 4.8** provides a description of

some of these variant reaction patterns for HLA class I alleleic products.

A number of officially named HLA specificities cannot readily be identified by the majority of typing laboratories. Both NMDP and the UCLA cell exchange data demonstrate that relatively new specificities such as A203, A2403, B5102, and B703 are rarely used. They are usually designated by their associated or broad specificities (A2, A24, B51, and B7, respectively). Even specificities such as A66, B64, and B65 are poorly identified. Therefore, the percentage recognition in the included tables can be helpful in selecting alternative antigens as search determinants.

FUTURE PLANS

For many alleles, serologic equivalents have now been identified and represent between 64 and 79 percent of the WHO designated alleles (**Table 4.1**). It will be important to include samples with poorly defined serologic types into serologic testing programs such as the International Cell Exchange, UCLA, or other locally organized cell exchange programs. Laboratories are also encouraged to participate in the 13th International Histocompatibility Workshop serology component, which will study rare and newly identified HLA-A and -B alleles by serological methods.[8] Therefore,

Table 4.8   HLA Class I alleles expressed as HLA-A and -B variants for which no official nomenclature is available and/or with known variation in serologic reaction pattern

Allele	Serotype	Description of serological pattern	Reference
A*1104	A11short	Reactive with A11, A11+3(+1) sera; negative with A11+6601 sera	
A*1105	A11var	Only 3 out of 31 A11 antisera reactive	
A*2408	A24var	Similar to A2403; A24 sera positive, A9 sera negative	(20)
A*2416	A31var	A31-like. A*3101 with 3'exon2 partly homologous to A*24 alleles	
A*2418	A24x3	Reactive with most A3, A9 and A24 sera	
A*2419	A9short	Very short A9, shorter than A2403	
A*2421	A9var	A9 sera positive, A24 and A23 sera negative	(20)
A*2608	A10short	A26 and A66 specific sera negative	(11)
A*2610	A10var	A26 short	
A*3204	A3	A19 and A32 sera negative. All A3 sera positive. One A32 mAb pos	[b]
A*3404	A34x31	Reactive with A34 and some of the A31 sera	[c]
A*6603	A10short	A26 and A66 specific sera negative	(11)
A*6812	A28short	Most A28+2 sera negative, A28+33+34+sera positive	(21)
B*0703	B703, BPOT	Short B7 with B7+42 sera positive, but B7+22 sera negative	
B*0708	B7short	Very few B7 sera reactive; similar to B*0703	[d]
B*0715	B7(Bw6var)	B7 with aberrant Bw6 epitope. Some Bw6 sera non-reactive	(22)
B*0716	B7short	Very few B7 sera reactive; pattern similar to B703	(23)
B*0720	B7short	B7+42 sera reactive; B7 specific sera weak/negative	(30)
B*0724	B7weak	B7 sera weak, B7+27(+40) sera positive	(24)
B*0802	B8var(Bw4+)	Reactive with Bw4, B8 and B8+59 sera; negative with Bw6 sera	
B*0803	B8var(Bw4+)		(5)
B*0804	B8neg(Bw6+)	Some B8+59 and B49 sera positive; most B8 sera negative	
B*0810	B8var(Bw6+)	Short B8; part of B8 specific sera non-reactive	(25)
B*1303	B21var	B21 like specificity; also some B15 sera reactive; Bw4 positive	
B*1304	B15x21 (Bw4+)	Similar to B*1303; Some B15 and B21 sera reactive; UCLA Int. cell exchange TER 847; HLA-Club cell exchange	(11)
B*1523	'NM5'	B5/B53/B77; B5 CREG sera positive; B70-Bw4	(11)
B*1524	B62(Bw4+)	Regular B15 and B62 sera positive; Bw6 negative, Bw4 positive	
B*1537	B15short	Very few B15 sera reactive, also typed as B70; Bw6 associated	
B*1538	B62var	B62+B52, B62, B15 and B52 sera positive	

Table 4.8 HLA Class I alleles expressed as HLA-A and -B variants for which no official nomenclature is available and/or with known variation in serologic reaction pattern (*continued*)

Allele	Serotype	Description of serological pattern	Reference
B*1806	B18(Bw6-)	B18 which is non-reactive with Bw4 and Bw6 sera	
B*1809	B18(Bw4+)	B18 with Bw4 epitope	(22)
B*2708	B2708, B7Qui	Most B7 sera positive, but B27 sera mostly negative; Bw6 associated	
B*2711	B27var	B27 with some B40 sera positive	
B*2712	B'X'(Bw6+), B40var	Reactive with few B40 and B27+7, B7CREG sera; B27 and B7 specific sera negative	d
B*2715	B'X'(Bw6+)	B27 sera negative, reactive with some Bw4 sera	(31)[a]
B*3515	B35var	Only part of B35 sera reactive. B75 CREG sera reactive	
B*3531	Bfu-var	Bfu-B40-like pattern	(9)
B*3702	B'37x27'	Some B37 and some B27 sera reactive; Bw4 associated	
B*3803	B16(Bw4+)	Some B16 sera reactive; B38 and B39 sera negative	
B*3905	B39var	Previously called ST-16; some B38 reactivity and some B16/B39 sera non-reactive	
B*4008	B40x48var	Only some B40+48(+7) CREG sera reactive; tested on TER-969	
B*4010	B60var	Some B60 sera reactive; B48-like	
B*4012	B48x70	Additional to B40+48 also some B15+B70 CREG sera positive	(26)
B*4019	B40(Bw4+)	B*4002 sequence with Bw4 epitope	(16)
B*4021	B40x15	Reactive with B40+48, B15+57 and some other B15 sera; neg with B40+13+47 sera	(27)
B*4025	B40short	Only 2 out of 15 B40 sera reactive	(10)
B*4406	B12var/ 'blank'	Few B12 sera reactive; can easily be missed	
B*4408	B44short	Reactive with all B44, most B12 and most B62+63+57 sera; detected in Welsh population but extremely rare	
B*4409	B12(Bw6+)	B*4402 allele with Bw6 sequence; will be typed as B45	
B*4415	B44var	B*4501 allele with Bw4 sequence: will be typed as B44	(15)
B*4416	B47/44	B47 sera positive, few B12 and B44 sera positive; tested on TER-1017	
B*4602	B46var	B46+75 sera negative, B46+Cw3 sera reactive	(28)
B*4702	B47(Bw6+)	Reactive with B40+47+13 sera; B27+47 sera negative	d
B*4703	B47var	Strongly reactive with B40+47+13, but not with B27+47 sera; aberrant Bw4 epitope, reactive with both Bw4 and Bw6 antisera; tested on TER-977	(29)
B*4802	B48?	B70-B71-like	(11)[a]
B*5106	B5	B51 specific sera negative	
B*5112	B'blank'	Possibly not expressed	
B*5304	B52x53	B5+53 sera reactive	(10)

Table 4.8 HLA Class I alleles expressed as HLA-A and -B variants for which no official nomenclature is available and/or with known variation in serologic reaction pattern (*continued*)

Allele	Serotype	Description of serological pattern	Reference
B*5306	B51x53	B51 and B53 specific sera positive, some B5 sera negative	(7)
B*5307	B53x37	B53 with mAb 12W069 (anti B37+52+18+39) positive, but most allo B37 sera negative	d
B*5503	B55xB67var	Maybe less reactive with Bw6 sera; Cw1 associated	
B*5509	B22short	B55 and B56 specific sera negative, only few broad B22 sera reactive	a
B*5603	B22var	B22 with some B15 features; reactive with B22 sera and with Bw6; detected in Orientals	
B*5604	B56var	Reactive with part of B56 and B22 specific sera; infrequent in Korean population	
B*5606	B78var	Some B78 sera positive, all B22 sera negative	(7)
B*5607	B56(Bw4+)	Bw56 with Bw4 epitope	(22)
B*8201	B45x56var	Some B22 and B45 specific sera reactive; B12 sera negative	(11)

http://www.worldmarrow.org/Dictionary/Dict2001Table8.html
[a]EM van den Berg-Loonen, personal communication.
[b]A de Smet, Antwerp, Belgium, personal communication.
[c]HLA-Club Cell Exchange: one sample typed by serology in 17 laboratories.
[d]Local observation in Leiden.

laboratories that encounter rare or poorly identified alleles during typing procedures, should store typing material, preferably viable lymphocytes or B cell lines for further testing.

CONCLUSION

The serologic equivalents of 123 HLA-A, 272 HLA-B, and 155 HLA-DRB1 alleles are presented here, covering over 64 percent of the presently identified HLA-A, -B, and -DRB1 alleles. This dictionary is an update of the one published in 1999 and also includes equivalents for HLA-C, DRB3, DRB4, DRB5, and DQB1 alleles. The data summarize information obtained by the WHO Nomenclature Committee for Factors of the HLA System, the International Cell Exchange (UCLA), the National Marrow Donor Program (NMDP), and individual laboratories. In addition, alleles are listed that are expressed as antigens with serologic reaction patterns that differ from the well-established HLA specificities. The equivalents provided will be useful in guiding searches for unrelated hematopoietic stem cell donors in which patients and/or potential donors are typed by either serology or DNA-based methods. These equivalents will also serve typing and matching procedures for organ transplant programs where HLA typings from donors and from recipients on waiting lists represent mixtures of serologic and molecular typings. The tables with HLA equivalents

and a questionnaire for submission of serologic reaction patterns for poorly identified allelic products will also be available on the WMDA web page: www.worldmarrow.org.

ACKNOWLEDGMENTS

We thank Prof Dr EM van den Berg-Loonen (Maastricht, The Netherlands) and Dr C Darke (Cardiff, UK) for their personal contributions. The efforts of the participants of the UCLA International Cell Exchange are gratefully acknowledged. We specifically thank the typing laboratories in the Netherlands, Belgium, and Luxembourg which participate in the HLA-Club Cell Exchange, for submitting, typing, and commenting on the samples with new and rare HLA alleles.

This work was supported in part by the JA Cohen Institute for Radiopathology and Radioprotection (IRS), by the Dutch National Reference Center for Histocompatibility, and by the US Office of Naval Research N00014-96-2-0016, N00014-99-2-0006, to the National Marrow Donor Program, and N00014-99-1-0551 to the CW Bill Young Marrow Donor Recruitment and Research Program.

KEY REFERENCES

1 Hurley CK, Schreuder GMT, Marsh SGE, Lau M, Middleton D, Noreen H (1997) The search for HLA-matched donors: a summary of HLA-A*, -B*, -DRB1/3/4/

5* alleles and their association with serologically defined HLA-A,-B,-DR antigens. *Tissue Antigens* 50: 401–18

2 Schreuder GMTh, Hurley CK, Marsh SGE, Lau M, Maiers M, Kollman C, Noreen H (1999) The HLA dictionary 1999: a summary of HLA-A, -B, -C, -DRB1/3/4/5, -DQB1 alleles and their association with serologically defined HLA-A, -B, -C, -DR and -DQ antigens. *Tissue Antigens* 1999: 54: 409–37

3 Bodmer JG, Marsh SGE, Albert ED *et al.* (1999) Nomenclature for factors of the HLA system, 1998. *Tissue Antigens* 1999: 53: 407–46

4 Marsh SGE, Bodmer JG, Albert ED *et al.* (2001) Nomenclature for factors of the HLA system, 2000. *Tissue Antigens* 57: 236–83

5 Marsh SGE, Parham P, Barber LD (2000) *The HLA FactsBook.* Academic Press, London

6 Hurley CK, Baxter-Lowe L, Begovich AB, Fernandez-Vina M, Noreen H, Ng J, Hartzman RJ, Hegland J *et al* (2000) The extent of HLA class II allele level disparity in unrelated bone marrow transplantation: Analysis of 1259 National Marrow Donor Program donor-recipient pairs. *Bone Marrow Transplant* 25: 385–93

7 Anholts JDH, Kemps-Mols B, Verduijn W, Oudshoorn M, Schreuder GMT (2001) Three newly identified HLA-B alleles: B*5124, B*5306, B*5307 and confirmation of B*0809 and B*5606. *Tissue Antigens* 58: 38–41

8 Fernandez-Vina M, Schreuder GMT, Juji T, Holdsworth R, Ferrara GB, Moraes ME, Strothman R, Alosco S, Cannady B, Lee JH *et al* (2000) Progress report of the serology component of the 13th International Histocompatibility Workshop [abstract] *Hum Immunol*: 61: Suppl 2: S32

9 Kim MA, Kang SJ, Han KS, Park MH (2000) Five new HLA alleles (A*0241, B*1558, B*2720, B*3531, DRB1*1339) found in Koreans [abstract] *Hum Immunol* 61: Suppl 2: S50

10 Kennedy CT, Dodd R, Le T, Wallace R, Ng G, Greville WD, Kennedy A, Taverniti A, Moses JH, Clow N, Watson N, Dunckley H (2000) Routine HLA-B genotyping with PCR-SSO detects eight new alleles: B*0807, B*0809, B*1551, B*3529, B*3532, B*4025, B*5304 and B*5508. *Tissue Antigens* 55: 266–70

11 Gutierrez M, Araujo HA, Cao K, Burdett L, Osowski L, Fernandez-Vina MA. Serological reactivity of novel HLA-A and-B alleles [abstract] *Hum Immunol* 2000: 61: Suppl 2: S34

12 Hoddinot MA, Cox ST, McWhinnie AJ, Marsh SGE, Ogilvie H, Madrigal JA, Little AM (2000) A new allele, HLA-B*1555, identified in an African patient awaiting bone marrow transplantation. *Tissue Antigens* 55: 463–6

13 Magira E, Beznik B, Monos D (2000) HLA-B*1559: a hybrid allele including exon 2 of B35 and exon 3 of B15 and serologically typed as B35. *Tissue Antigens* 56: 460–62

14 Witter K, Lau M, Zahn R, Scholtz S, Albert ED (2001) Identification of a novel HLA-B*2722 allele from a Filipino cell. Tissue Antigens 58: 263–8

15 Williams F, Curran MD, Leheny WA, Daar AS, Middleton D (2000) Characterization of HLA-B*3921 and confirmation of HLA-B*4415, two variant HLA-B alleles identified in Omani population. *Tissue Antigens* 56: 376–9

16 Pimtanothai N, Rizzuto GA, Slack R, Steiner NK, Kosman CA, Jones PF, Koester R, Ng J, Hartzman RJ, Hurley

CK(2000) Diversity of alleles encoding HLA-B40: relative frequencies in US populations and description of five novel alleles. *Hum Immunol* 61: 808–15

17 Elsner H, Schmitz G, Ballas M, Lenhard V, Blaszczyk R (2000) Characterization of the novel allele HLA-B*4417: implication for bone marrow transplantation. *Tissue Antigens* 56: 463–6

18 Anholts JDH, Verduijn W, Drabbels J, Mulder A, Doxiadis IIN, Schreuder GMT (2000) Identification of two new alleles HLA-DRB1*0312, DRB1*0432 and of a DRB3- negative DRB1*1313-positive haplotype. *Tissue Antigens* 56: 87–9

19 Darke C, Guttridge MG, Street J, Thompson J, Thomas M (2000) Molecular, serologic and genetic studies on two new HLA-DRB1 alleles: DRB1*0704 and DRB1*1507. *Tissue Antigens* 56: 467–9

20 Lee KW, Cho HC (2000) Serologic heterogeneity of HLA-A24 correlates with allelic types in the Korean population. *J Korean Med Sci* 13: 623–6

21 Fae I, Gleissner B, Kriks D, Leitner D, Fischer GF (2000) Serological and nucleotide sequence analysis of HLA-A*6812. *Tissue Antigens* 56: 276–8

22 Voorter CEM, van der Vlies S, Kik M, van den Berg-Loonen EM (2000) Unexpected Bw4 and Bw6 reactivity patterns in new alleles. *Tissue Antigens* 56: 363–70

23 Dunn PPJ, Carter V, Dunn A, Day S, Fuggle SV, Ross J, Cavanagh G (2000) Identification of an HLA-B7 serological variant and its characterization by sequencing based typing. *Tissue Antigens* 55: 71–3

24 Middleton D, Curran MD, Anholts JDH, Reilly ER, Schreuder GMT (2001) Characterization of a new HLA-B allele, HLA-B*0724. *Tissue Antigens* 57: 471–3

25 Ward JJ, Bowerman CJ, Day S, Younie M, Dunn PPJ (2000) Two novel HLA-B alleles, B*0810 and B*51014, identified by discrepancies between serological and DNA tissue types. Abstract. *Hum Immunol* 61: Suppl 2: S127

26 Elsner HA, O'Brien BJ, Bryan CF, Meyer JA, Diederich DA, Peirce GE, Lau M, Terasaki PI, Blaszczyk R (2000) HLA-B*4012: a new allele with unique serological features. *Tissue Antigens* 56: 180–3

27 Lee KW (1999) HLA-B*4021: hybrid linking the B15 and B40 families. *Tissue Antigens* 54: 625–7

28 Akesaka T, Kashiwase K, Shimamura M, Ishikawa Y, Tanaka H, Fuji M, Akaza T, Hando K, Yuasa S, Takahashi T, Juji T (2000) Identification of a novel HLA-B46 allele, B*4602, in Japanese. *Tissue Antigens* 55: 460–2

29 Darke C, Guttridge MG, Street J, Thompson J, Thomas M (1999) HLA-B*4703: sequence confirmation, serology and distribution. *Tissue Antigens* 53: 586–90

30 Ligeiro D, Elsner HA, Sancho MR, Almeida AP, Trindade H, Blaszczyk R (2001) Serological and molecular characterization of the new allele HLA-B*0720. *Eur J Immunogen* 28: 352

31 Voorter CEM, vd Berg-Loonen EM (2001) Molecular diagnostics of B*27. *Eur J Immunogen* 28: 231

32. Poli F, Longhi E, Frison S, Scalamogna M, Bianchi P, Sirchia G (2001). Identification of a novel HLA-B allele – HLA-B*4902. *Tissue Antigens*. 57: 173–74.

5

NOMENCLATURE FOR FACTORS OF THE HLA SYSTEM, 2002

SGE Marsh and Associates*

- **NAMING OF ADDITIONAL GENES WITHIN THE HLA REGION**

- **NAMING OF ADDITIONAL ALLELES**

- **RENAMING OF ALLELES AND REMOVAL OF INCORRECT ALLELES**

- **EXTENSION OF HLA ALLELE NAMES**

- **NAMING OF ALLELES WITH ABERRANT EXPRESSION**

- **NAMING OF HLA-G ISOFORMS**

- **KILLER IMMUNOGLOBULIN-LIKE RECEPTOR (KIR) GENE AND ALLELE NOMENCLATURE**

- **THE IMGT/HLA SEQUENCE DATABASE**

- **NEW COMMITTEE MEMBERS**

The WHO Nomenclature Committee for Factors of the HLA System met in Victoria, Canada in May 2002 after the 13th International Histocompatibility Workshop to consider additions and revisions to the nomenclature of HLA specificities following the principles established in previous reports.[1–16]

The main subjects discussed were:

- Naming of additional genes within the HLA region
- Naming of additional alleles
- Renaming of alleles and removal of incorrect alleles
- Extension of HLA allele names

*SGE Marsh,[1] ED Albert,[2] WF Bodmer,[3] RE Bontrop,[4] B Dupont,[5] HA Erlich,[6] DE Geraghty,[7] JA Hansen,[8] B Mach,[9] WR Mayr,[10] P Parham,[11] EW Petersdorf,[12] T Sasazuki,[13] GMTh Schreuder,[14] JL Strominger,[15] A Svejgaard,[16] PI Terasaki[17]
[1]Anthony Nolan Research Institute, London, UK
[2]Policlinic for Children, University of Munich, Germany
[3]Cancer Research UK, Oxford, UK
[4]Biomedical Primate Research Centre, Rijswijk, The Netherlands
[5]Sloan-Kettering Institute for Cancer Research, New York, USA
[6]Roche Molecular Systems, Alameda, USA
[7]Fred Hutchinson Cancer Center, Seattle, USA
[8]Fred Hutchinson Cancer Center, Seattle, USA
[9]University of Geneva, Geneva, Switzerland
[10]University of Vienna, Vienna, Austria
[11]Stanford University School of Medicine, Stanford, USA
[12]Fred Hutchinson Cancer Center, Seattle, USA
[13]Kyushu University, Fukuoka, Japan
[14]Leiden University Medical Center, Leiden, The Netherlands
[15]Harvard University, Cambridge, USA
[16]State University Hospital, Copenhagen, Denmark
[17]Los Angeles, USA
The report on which this chapter is based was originally published in *Tissue Antigens* 2002; 60: 407–464.

- Naming of alleles with aberrant expression
- Naming of HLA-G isoforms
- Killer Immunoglobulin-like Receptor (KIR) gene and allele nomenclature
- The IMGT/HLA Sequence Database
- New committee members

NAMING OF ADDITIONAL GENES WITHIN THE HLA REGION

A number of class I and II gene fragments within the HLA region have been previously described but had yet to be named. Official designations were given to these gene fragments. Three class I gene fragments, previously called HLA-30,[17] HLA-17[17] and HLA-X[18] are now named HLA-N, HLA-S, and HLA-X respectively. An HLA class I gene fragment located within the HLA class II region previously called HLA-Z1 has been officially named HLA-Z.[19]

An HLA class II pseudogene found centromeric to the pseudogene HLA-DPB2 and most closely related to HLA-DPA2 has been named HLA-DPA3.[20]

The names LMP2 and LMP7 used previously for the two proteasome genes in the HLA class II region have been renamed by the Human Genome Nomenclature committee (HGNC) PSMB9 and PSMB8 respectively.[21] After discussion with the HGNC it was decided to keep the names TAP1 and TAP2 as the official names for the two transporter genes and the names ABCB2 and ABCB3 as aliases for these genes. More information can be found on the HGNC's website (www.gene.ucl.ac.uk/nomenclature/).

The list of those genes in the HLA region considered by the WHO Nomenclature Committee is given in **Table 5.1**.

NAMING OF ADDITIONAL ALLELES

Conditions for acceptance of new allele sequences

As emphasized in previous reports, there are required conditions for acceptance of new sequences for official names.

1 Where a sequence is obtained from cDNA, or where PCR products are subcloned prior to sequencing, several clones should have been sequenced.
2 Sequencing should always be performed in both directions.
3 If direct sequencing of PCR amplified material is performed, products from at least two separate PCR reactions should have been sequenced.
4 In individuals who are heterozygous for a locus, and where one of the alleles is novel, the novel allele must be sequenced in isolation from the second allele. Thus an allele sequence which is derived using a sequence-based typing (SBT) methodology, where both alleles of a heterozygous individual are sequenced together, is insufficient evidence for assignment of an official designation.

5 Sequence derived solely from the primers used to amplify an allele should not be included in the submitted sequence.
6 Where possible, a novel sequence should be confirmed by typing of genomic DNA using a method such as PCR-SSOP or PCR-SSP. Where a new sequence contains either a novel mutation or a previously unseen combination of nucleotides (sequence motif), this must be confirmed by a DNA typing technique. This may require the use of newly designed probes or primers to cover the new mutation; these reagents should also be described.
7 An accession number in a databank should have been obtained. Sequences may be submitted to the databases online at the following addresses:
    - EMBL: www.ebi.ac.uk/Submissions/index.html;
    - GenBank: www.ncbi.nlm.nih.gov/Genbank/index.html
    - DDBJ: www.ddbj.nig.ac.jp/sub-e.html
8 Full-length sequences are preferable though not essential; the minimum requirements are exons 2 and 3 for an HLA class I sequence and exon 2 for an HLA class II sequence.
9 Where possible, a paper in which the new sequence is described should have been submitted for publication.
10 DNA or other material, preferably cell lines, should, wherever possible, be made available in a publicly accessible repository or, alternatively, at least in the originating laboratory. Documentation on this will be maintained by the WHO Nomenclature Committee.
11 Submission of a sequence to the WHO Nomenclature Committee should be performed using the online submission tool available at www.ebi.ac.uk/imgt/hla/subs/submit.html. Researchers are expected to complete a questionnaire relating to the sequence and provide a comparison of their new sequence with known related alleles. If the sequence cannot be submitted using the online web tools, researchers should contact hladb@ebi.ac.uk directly for details of alternative submission methods.

Although at present it is only a recommendation that full-length sequences of the coding region of novel alleles be submitted it was widely felt that this should become in the future a requirement for submission. Such a requirement would remove many of the currently encountered ambiguities in the assignment of names to alleles for which partial sequences have been submitted and should not be burdensome as sequencing techniques have improved substantially

Table 5.1   Names for genes in the HLA region considered by the WHO Nomenclature Committee

Name[a]	Previous equivalents	Molecular characteristics	References
HLA-A	—	Class I α-chain	
HLA-B	—	Class I α-chain	
HLA-C	—	Class I α-chain	
HLA-E	E, '6.2'	Associated with class I 6.2 kB Hind III fragment	
HLA-F	F, '5.4'	Associated with class I 5.4 kB Hind III fragment	
HLA-G	G, '6.0'	Associated with class I 6.0 kB Hind III fragment	
HLA-H	H, AR, '12.4', HLA-54	Class I pseudogene associated with 5.4 kB Hind III fragment	
HLA-J	cda12, HLA-59	Class I pseudogene associated with 5.9 kB Hind III fragment	
HLA-K	HLA-70	Class I pseudogene associated with 7.0 kB Hind III fragment	
HLA-L	HLA-92	Class I pseudogene associated with 9.2 kB Hind III fragment	
HLA-N	HLA-30	Class I gene fragment associated with a 1.7 kb Hind III fragment	(17)
HLA-S	HLA-17	Class I gene fragment associated with a 3.0 kb Hind III fragment	(17)
HLA-X	HLA-X	Class I gene fragment	(18)
HLA-Z	HLA Z1	Class I gene fragment located within the HLA Class II region	(19)
HLA-DRA	DRα	DR α chain	
HLA-DRB1	DRβI, DR1B	DR β1 chain determining specificities DR1, DR2, DR3, DR4, DR5 etc.	
HLA-DRB2	DRβII	Pseudogene with DR β-like sequences	
HLA-DRB3	DRβIII, DR3B	DR β3 chain determining DR52 and Dw24, Dw25, Dw26 specificities	
HLA-DRB4	DRβIV, DR4B	DR β4 chain determining DR53	
HLA-DRB5	DRβIII	DR β5 chain determining DR51	
HLA-DRB6	DRBX, DRBσ	DRB pseudogene found on DR1, DR2 and DR10 haplotypes.	
HLA-DRB7	DRBψ1	DRB pseudogene found on DR4, DR7 and DR9 haplotypes.	
HLA-DRB8	DRBψ2	DRB pseudogene found on DR4, DR7 and DR9 haplotypes.	
HLA-DRB9	M4.2 β exon	DRB pseudogene, isolated fragment	
HLA-DQA1	DQα1, DQ1A	DQ α chain as expressed	
HLA-DQB1	DQβ1, DQ1B	DQ β chain as expressed	
HLA-DQA2	DXα, DQ2A	DQ α chain-related sequence, not known to be expressed	
HLA-DQB2	DXβ, DQ2B	DQ β chain-related sequence, not known to be expressed	
HLA-DQB3	DVβ, DQB3	DQ β chain-related sequence, not known to be expressed	
HLA-DOA	DNA, DZα, DOα	DO α chain	
HLA-DOB	DOβ	DO β chain	

Table 5.1 Names for genes in the HLA region considered by the WHO Nomenclature Committee (*continued*)

Name[a]	Previous equivalents	Molecular characteristics	References
HLA-DMA	RING6	DM α chain	
HLA-DMB	RING7	DM β chain	
HLA-DPA1	DPα1, DP1A	DP α chain as expressed	
HLA-DPB1	DPβ1, DP1B	DP β chain as expressed	
HLA-DPA2	DPα2, DP2A	DP α chain-related pseudogene	
HLA-DPA3	DPA3	DP α chain-related pseudogene	(20)
HLA-DPB2	DPβ2, DP2B	DP β chain-related pseudogene	
TAP1	ABCB2, RING4, Y3, PSF1	ABC (ATP Binding Cassette) transporter	
TAP2	ABCB3, RING11, Y1, PSF2	ABC (ATP Binding Cassette) transporter	
PSMB9	LMP2, RING12	Proteasome-related sequence	
PSMB8	LMP7, RING10	Proteasome-related sequence	
MICA	MICA, PERB11.1	Class I chain-related gene	
MICB	MICB, PERB11.2	Class I chain-related gene	
MICC	MICC, PERB11.3	Class I chain-related pseudogene	
MICD	MICD, PERB11.4	Class I chain-related pseudogene	
MICE	MICE, PERB11.5	Class I chain-related pseudogene	

[a]Gene names given in bold type have been assigned or changed since the 2000 Nomenclature report.

in the time since the submission conditions were first devised.

It should be noted with some caution that cells from which only partial sequences have been obtained may later be shown to have different or novel alleles when further sequencing is performed. This is of particular importance in cases where partial sequences of what appears to be the same allele have been obtained from several different cells. In such cases, all cells studied have been listed in this report.

Current practice is that official designations will be promptly assigned to newly described alleles in periods between Nomenclature Committee meetings, provided the submitted data and its accompanying description meet the criteria outlined above. A list of the newly reported alleles is published each month in nomenclature updates in the journals *Tissue Antigens*, *Human Immunology*, and the *European Journal of Immunogenetics*. The listing of references to new sequences does not imply priority of publication. The use of numbers or names for alleles, genes or specificities which pre-empt assignment of official designations by the Nomenclature Committee is strongly discouraged.

New allele sequences

A total of 209 HLA alleles have been named since the last report.[16] The newly named alleles are shown in bold typeface in **Tables 5.2–5.10**. For HLA class I, 42 HLA-A, 79 HLA-B, and 19 HLA-C alleles were named, making a total of 881 class I alleles with official names. For HLA class II, 52 HLA-DRB, one DRA, one DQA1, eight DQB1, one DPA1, and six DPB1 alleles were named, making a total of 611 class II alleles with official names. Three MICA alleles were named bringing their total to 54 (**Table 5.11**). The total number of alleles at each locus assigned with official names as of 31 July 2002 is given in **Table 5.12**.

As the database of HLA allele sequences has expanded, it has become increasingly difficult to maintain consistent linkage between allele names assigned on the basis of nucleotide sequences and the serological profiles of the encoded proteins. These difficulties are in part technological and part due to the inherent biological properties of the HLA system. In the first category is the increasing emphasis on DNA technology and consequent lack of a serological description for many newly discovered HLA alleles.

Table 5.2 Designations of HLA-A alleles

HLA alleles[a]	HLA specificity	Previous equivalents	Individual or cell line from which the sequence was derived	Accession number	References or Submitting author(s)
A*010101	A1	—	LCL721, MOLT4, PP	X55710, M24043, Z93949, AJ278305	
A*010102	—	—	GN00348	AF248059, AF248060	
A*0102	A1	—	DAUDI	U07161	
A*0103	A1[c]	—	04VC, UCLA 144, BONIFACE, FU-GP, JF-GP, BR-GP	Y12469, Y12470, AJ002528, AJ002529, AF098160	
A*0104N	Null	A*01N, A*01N-Ca	PELa, PEFr, PEPi, PEPa, CAFL, CB1280	Z93776, Z97027, AJ011125, AJ011126, AJ011127	(24)[b]
A*0106	—	A*0101V	GN00280	AF143231, AF143232	
A*0107	A1	BLP-N	BLP-N	AF219632, AF219633	
A*0108	A1	A*01	34040	AJ277792	
A*0109	—	—	T110	AJ315641	A-M Little
A*020101	A2	A2.1	LCL721, JY, GM637, GRC138, T5-1, JD	K02883, M84379, X02457	
A*020102	A2	—	CHI564, CHI557	Y14624, Y14625	
A*020103	A2	A*02DKP	DKP, 19673946	AF108449, AF108450, AF255333, AF190713, AF190714, AF190715	
A*020104	A2	A*02New	NM4a189	AF139832, AF139833	
A*020105	A2	A*02AR	32711	AJ277793	
A*020106	—	A0201V3	JCB11458	AB032595, AB048347	
A*0202	A2	A2.2F	M7, 951314	M17566, M17568, X94566	
A*0203	A203	A2.3	DK1, 951315	U03863, M17567, M19670, X94567	
A*0204	A2	—	RML, AN, 951316	X57954, M86404, X94568 AJ297476	
A*0205	A2	A2.2Y	WT49, AM, SUS-NF, 951317	U03862, L76290, X94569	
A*0206	A2	A2.4a	CLA, T7526, 951318	M24042, X94570	
A*0207	A2	A2.4b	KNE, KTO	D50458	
A*0208	A2	A2.4c	KLO	X94571	
A*0209	A2	A2-OZB	OZB	AJ249241	
A*0210	A210	A2-LEE	XLI-ND, 951322	Z23071, X94572	
A*0211	A2	A2.5	KIME, GRC138, 951366	X60764, M84377, X94573	
A*0212	A2	—	KRC033, KRC005	M84378	
A*0213	A2	A2SLU	SLUGEO	Z27120	
A*0214	A2	A2'1S'	1S, ML1260	Z30341, AF305699	(25)[b]
A*0215N	Null	HLA-Anull	TSU	D38525	
A*0216	A2	A2'TUB'	TUBO	Z46633	

Table 5.2   Designations of HLA-A alleles (*continued*)

HLA alleles[a]	HLA specificity	Previous equivalents	Individual or cell line from which the sequence was derived	Accession number	References or Submitting author(s)
A*021701	A2	A*New	AMALA, LZL, C.S.	U18930, L43526, L43527, X89707, X89708	
A*021702	A2	—	H.K	Y13267	
A*0218	A2	A*2K	ENDO	D83515	
A*0219	—	A-02X09	TOB-81	L76936	
A*022001	A2	—	BI	X96724	
A*022002	A2	A*02New	MT-SN	AJ276069	
A*0221	A2	A206W331R	W331R	U56825	
A*0222	A2	A-02x28	TER-109, OCA1/4	U76398, U76399, Y11441	
A*0224	A2	A*02JG	11952547, 13041452, RP122	Y11201, Y11202, AF036921, AF001956, AF001957	
A*0225	A2	—	NP814, 970551	U70863, Y13028	
A*0226	—	—	C.C	AF008933, U90138, U90139	
A*0227	—	A*02TK	TRK	AJ001269	
A*0228	—	—	NM3298	AF041365, AF041366	
A*0229	A2	—	RAG	AF053479, AF053480, AF012766	
A*0230	—	A*02WP	NM332, CL154, WP	AF101162, AF101163, AF116215, AF133091, AF133092	
A*0231	A2	A*02011V	19703222	AF113923, AF113924	
A*0232N	Null	A*02xxN	NDS-AN	AF117228	
A*0233	—	A*0201New	CL-PPA	AF140506	
A*0234	A2	A*AAT	AAT	AF129429, AF129430, AF129431	
A*0235	—	A*0201V	GN00279, GN00300	AF140600, AF140601, AF157310, AF157311	
A*0236	—	A*02011V	GN00297	AF157308, AF157309	
A*0237	—	A*0212Variant	GN00303	AF157563, AF157564	
A*0238	—	A*0213V	GN00260, GN00286, GN00346	AF135542, AF135543, AF181101, AF181102, AF232705, AF232706	
A*0239	—	A*02011V	GN00308, 99-2203	AF173873, AF173874, AF198352, AF198353	
A*0240	—	A*CB2406	CB2406, CB2406(MUM)	AF194531, AF194532	
A*0241	A2	A*02CIS	KMP01-636	AF170580, AF170581	
A*0242	A2	A0201V2	JCB6898	AB032594	
A*0243N	Null	A*02ROUB	ROUB	AJ251960	
A*0244	—	—	GN00337	AF226834, AF226835	
A*0245	—	—	1998-302-2581	AF251354, AF251355	

Table 5.2 Designations of HLA-A alleles (*continued*)

HLA alleles[a]	HLA specificity	Previous equivalents	Individual or cell line from which the sequence was derived	Accession number	References or Submitting author(s)
A*0246	A2	A*02COL	COL	AJ289156	
A*0247	—	—	GN00378	AF291839, AF291840	
A*0248	—	—	GN00381	AF299250, AF299251	
A*0249	—	A*02new	22697	AJ291697, AJ291698	
A*0250	A2	A*02X68	A02X68	AF162678, AF162679	P Stastry
A*0251	—	—	2000-7-206, Taramahara31	AF372047, AF372048, AJ457988	CK Hurley, A-M Little
A*0252	—	—	JSILV	AF417237, AF417238	H Dunckley
A*0253N	Null	—	Yanli, VTIS25793	AF416455, AF479485, AF479486	(26), BD Tait
A*0254	—	—	2000-084-3329	AF440104, AF440105	CK Hurley
A*0255	—	—	1PFA8	AY045739, AY045740	WH Hildebrand
A*0256	—	—	MYTCZA-A202x	AJ430523, AJ430524, AJ431714, AJ431715	(27)
A*0257	—	—	Taramahara35	AJ457989	A-M Little
A*0258	—	—	RL*D	AY100700, AY100701	L-A Baxter-Lowe
A*030101	A3	A3.1	JG, JD, PP, AP630	X00492, U32184	
A*030102	A3	DT18-A*0301v	DT18	AF053128, AF053129	
A*030103	A3	A*03NJ	12244015, NM4a227	Y17000, Y17001, AF146365, AF146366	
A*0302	A3	A3.2	E1B2, R69772, CL183	U56434, U56435, AF217561	
A*0303N	Null	A3blank	MMK	L77702	
A*0304	A3	—	CTM-2983694	AF015930	
A*0305	A3	A*03011V	GN00262, GN00309, 99-2197, CS, 34507	AF135546, AF135547, AF173877, AF173878, AF190718, AF190719, AJ252283, AJ252284, AJ252285, AJ401085, AJ401086, AJ401087	
A*0306	—	A*03011V	GN00341	AF226842, AF226843	
A*0307	—	A*03011New	NM5A488	AF268399, AF268400	
A*0308	—	A*03011v	GN00375	AF288047, AF288048	
A*0309	—	—	BY00016	AF372049, AF372050	CK Hurley
A*110101	A11	A11E, A11.1, A11	CJO-A, K.LIE, MMU, YMU, THA-DCH412, THA-DCH926, THA-DCH1093	M16007, M16008, X13111, D16841, AF030899, AF030900, AF030901, AF030902, AF030897, AF030898	

Table 5.2 Designations of HLA-A alleles (*continued*)

HLA alleles[a]	HLA specificity	Previous equivalents	Individual or cell line from which the sequence was derived	Accession number	References or Submitting author(s)
A*110102	A11	A*1101new	UCLA201	AJ238608, AJ238609, AJ238610	
A*1102	A11	A11K, A11.2	K.LIE, KOK, CTA, THA-DCH538, THA-DCH639	X13112, D16842, AF030903, AF030904, AF030905, AF030906	
A*1103	A11	—	AMAD	X91399	
A*1104	A11	87A	HM, I65, 87A, THA-DCH7672, THA-DCH7673	U50574, U59701, U59702, U88250, AF017309, AF030907, AF030908, AF030909, AF030910	
A*1105	A11	—	KH, GN00302, HOATWAY	Y15223, AF147454, AF147455, AJ306733	P Dunn[b]
A*1106	—	A*1101V	GN00259	AF135540, AF135541	
A*1107	A11	A*11	CMC1	AF165065	
A*1108	—	A11nou	VPH-IM0002135	AF284443	
A*1109	—	—	9315466	AF260828, AF260829	
A*1110	A11	A11v	VTIS38035	AF329874, AF329875	BD Tait
A*1111	—	—	2001-26-469	AF440108, AF440109	CK Hurley
A*1112	A11	—	103201, 103195, 106843, 106844	AF439511	(28)
A*1113	A11	A11v	B5997	AB073216, AB073217	H Ikeda
A*2301	A23(9)	—	SHJO, ELON	M64742, L76288	
A*2302	—	A*2301V	GN00274	AF137079, AF137080	
A*2303	—	A*2301 variant	GN00250	AF102571, AF102572	
A*2304	—	A*2301V	GN00263	AF135548, AF135549	
A*2305	—	A*2301New	GN00284, NM5A405	AF140859, AF140860, AF255718, AF255719	
A*2306	—	A*2301New	GM14672	AJ271340	
A*2307N	Null	A*23MATSi	MATSi	AJ306634	A Dormoy
A*2308N	Null	—	SH38	AY028848, AY028849, AY028850	A Smith
A*2309	—	—	MAWE0816AN	AJ426561	A-M Little
A*24020101	A24(9)	A24, A2402	SHJO, 32/37, KRC032, KRC110, THA-DCH538	M64740, L47206, Z72423, AF030911, AF03091	
A*24020102L	Low A24	A2402LOW, APET, A24L-LACC	6319, PAn, PMa, Pmi, LACC	L76291, Z72422, Z97370	
A*240202	A24(9)	—	NM426	AF101160, AF101161	
A*240203	A24(9)	—	KBM-2	AY121128	KW Lee

Table 5.2 Designations of HLA-A alleles (*continued*)

HLA alleles[a]	HLA specificity	Previous equivalents	Individual or cell line from which the sequence was derived	Accession number	References or Submitting author(s)
A*240301	A2403	A9.3	APA, KPE, THA-DCH412, THA-DCH8151, THA-DCH8152	M64741, AF030913, AF030914, AF030915, AF030916, AF030917, AF030918	
A*240302	A2403	A*2403 Variant	GN00247	AF102565, AF102566	
A*2404	A24(9)	A24AK,	ITOU, KJRAID5	D26550, L43532, L43533	
A*2405	A24(9)	—	DST, FST	X82161, X82189	
A*2406	A24(9)	A*24YM	YM29	U18987, U19733	
A*2407	A24(9)	A#46	PICH, A#46, K92068, THA-DCH507, THA-DCH522, THA-DCH1109, THA-DCH5342	U25971, U36914, L43530, L43531, AF030921, AF030922, AF030919, AF030920, AF030923, AF030924, AF030925, AF030926	
A*2408	A24(9)	A*9HH	K62098, HIRH	L43528, L43529, D83516	
A*2409N	Null	A24Null	SUS-NF, WAG	L47231, AJ251621	
A*2410	A2403[c]	A*24JV	JV1458, KM315, CH121, THA-DCH611, THA-DCH639, THA-DCH1109	U37110, U37111, U59699, U59700, Y10695, AF030927, AF030928, AF030929, AF030930, AF030931	
A*2411N	Null	A*24LM	LUME	L76289	
A*2413	A24(9)	A*24YM2	YM81	U37112, U37113	
A*2414	A24(9)	A*24SA	SBD6380	U37114, U37115	
A*2415	—	—	NM3469	AF042666, AF042667	
A*2417		A*2402v, A*VB	NDS-NH, VB-ARCBS, 0234	AF067436, AF067437, AF117764, AF117765, AJ239035, AJ239036	
A*2418	—	A*2403v	3362	AF065401, AF065402	
A*2419	—	—	HP-CV	Y17292, Y17291	
A*2420	—	—	SW36, 21833843, JCBB26794	Y16948, Y16949, AF190716, AF190717, AB032596	
A*2421	—	A*24Var	DHL, JSL	AF106688, AF106689	
A*2422	A9	A*2403New, A9v	CL153, GN00272	AF116214, AF137081, AF137082	
A*2423	A24(9)	A*24021New, A24v(9)	EA31, 26586	AF128537, AF128538, AJ278667	
A*2424	—	—	GN00275	AF140723, AF140724	
A*2425	—	—	10296952, NM5A251	AF190708, AF190709, AF255716, AF255717	
A*2426	—	—	12318945	AF190710, AF190711, AF190712	
A*2427	—	A*24Mall, ?A24(9)	MALL	AJ271626	
A*2428	—	—	GN00359	AF266519, AF266520	

Table 5.2   Designations of HLA-A alleles (*continued*)

HLA alleles[a]	HLA specificity	Previous equivalents	Individual or cell line from which the sequence was derived	Accession number	References or Submitting author(s)
A*2429	—	A*2402V	GN00379	AF291843, AF291844	
A*2430	—	A*2402V	GN00380	AF291841, AF291842	
A*2431	—	—	2000-35-513	AF298583, AF298584	
A*2432	—	—	GN00390, 07-S-0025#0100	AF359393, AF359394, AY038075	CK Hurley, S Grams
A*2433	A2403	—	17933/00, 17670/00	AF363678, AF363679, AF363680	A Smith
A*2434	—	—	2000-182-2100	AF443283, AF443284	CK Hurley
A*2435	—	—	PR45	AY045731, AY045732	WH Hildebrand
A*2436N	Null	—	022659718	AF486832	(29)
A*2501	A25(10)	—	BM92	M32321	
A*2502	A10	A66var	M54672, TW	X97802, AJ238524	
A*2503	—	A*2501V	GN00273, GN00301	AF137075, AF137076, AF148897, AF148898	
A*2504	—	—	BY0019	AY042682, AY042683	CK Hurley
A*2601	A26(10)	A26.1, A26.3	GM637, O2BN5, MGAR, N.M., MIY-2, MIY-3	M24095, U03697, D16843, D32130, D32131	
A*2602	A26(10)	A26.2, A26.1	KT14, Y.I., E.K.	M98453, D14350	
A*2603	A26(10)	A26.4	T.M., S.M., MIY-1	D14351, D32129	
A*2604	A26(10)	A10SA	Y.S.	D14354	
A*2605	A26(10)	A26KY	SAJ022, K91089, K93022	D50068, L43536, L43537	
A*2606	A26(10)	—	KHB102	L43534, L43535	
A*2607	A26(10)	A26mic	MIC-ND	L48341	
A*2608	A26(10)	A26RMH, A*26new-66A	MI108, W652D, M.McL, 66A	U45480, U52429, X99733, U43334, AF017310	
A*2609	A26(10)[c]	—	GN00158	U90242, U90243	
A*2610	A10	—	034-SEA-HK	AF001553, AF001554	
A*2611N	Null	A26Null	JBO13900	AB005048	
A*2612	—	A*2601V	NM1183, CS3, GN00249	AF042186, AF042187, AF065486, AF065487	
A*2613	—	A*2601V	GN00271	AF139766, AF139767	
A*2614	—	A*MJUL	MJUL	AF194529, AF194530	
A*2615	—	A*26FONT	FONT, FRED, FRED(-1), 22663, 6A29	AJ271225, AJ291695, AJ291696, AY045729, AY045730	WH Hildebrand[b]

Table 5.2   Designations of HLA-A alleles (*continued*)

HLA alleles[a]	HLA specificity	Previous equivalents	Individual or cell line from which the sequence was derived	Accession number	References or Submitting author(s)
A*2616	—	—	2000-7-951	AF303952, AF303953	
A*2617	—	—	GN0384	AF310142, AF310143	
A*2618	—	—	GN00399	AY050205, AY050206	CK Hurley
A*29010101	A29(19)	A2901W652R	JOE, W652R, AKB96676	M23739, U83415, AJ303359	R Blasczy[b]
A*29010102N	Null	A*GBnu29	GBnu29	AJ293507	(30)
A*2902	A29(19)	A29.2	LAM	X60108	
A*2903	—	—	CMD004AN	Y09218, AJ000661	
A*2904	—	—	NM3234	AF042188, AF042189	
A*2905	—	—	BY0020	AY042684, AY042685	CK Hurley
A*2906	—	—	GN00412	AY062005, AY062006	CK Hurley
A*3001	A30(19)	A30.3, A30RSH	LBF, RSH	M30576, M28414, U07234	
A*3002	A30(19)	A30.2	CR-B, T.B.B.	X61702, AF148862	
A*3003	A30(19)	A30JS	JS, HT	M93657	
A*3004	A30(19)	A*30AD, A30W7, A30JW	AD7563, W7(CC), ASE	U18988, U19734, Z34921, X83770, X83771	
A*3006	—	—	CS48	AF028713, AF028714	
A*3007	—	—	318-409	AF065642, AF065643	
A*3008	—	—	I3753	AJ249308, AJ249309, AJ249310, AJ249311, AJ249312, AJ249313, AJ249314, AJ249315	
A*3009	—	A*3002V	99-2196, GN00351	AF198350, AF198351, AF266529, AF266530	
A*3010	—	—	E249	AF323494, AF323495, AF323496	D Adorno
A*3011	A30(19)	A*30New	19302, 23031	AJ308423, AJ308424	(31)
A*3012	—	—	0995970	AF480841	F Garcia Sanchez
A*310102	A31(19)	—	KRC033, TB, KRC110, JHAF, KT12, 0229	M30578, M28416, M84375, M86405, L78918, AJ239045, AJ239046	
A*3102	—	—	NM2492	AF041369, AF041370	
A*3103	—	A*3101v	NDS-MA	AF067438, AF067439	
A*3104	A31(19)	A31V	NMDP#013528641, NMDP#012891701, NMDP#012797924, T.B.B.	AF105027, AF105028, AF148863	

Table 5.2   Designations of HLA-A alleles (*continued*)

HLA alleles[a]	HLA specificity	Previous equivalents	Individual or cell line from which the sequence was derived	Accession number	References or Submitting author(s)
A*3105	A31(19)	A3101V1, A31v(19)	JCBT1569	AB032597	
A*3106	—	—	2000-133-482	AF440106, AF440107	CK Hurley
A*3107	—	—	BY00041	AY094132, AY094133	CK Hurley
A*3108	—	A*19New, A31CT, A*2416	DD3, CRT	AF053481, AF053482, AF012767, AJ011699 AJ011700	L Gebuhrer, (32)
A*3201	A32(19)	—	AM	P10314	
A*3202	A32(19)	—	MP	X97120	
A*3203	—	—	023-8001, VTIS70350	AF072761, AF072762, AF517561, AF517562	BD Tait[b]
A*3204	—	A*0301V	GN00277, GN00278	AF139891, AF139892, AF137077, AF137078	
A*3205	—	A*32New	CL183	AF217560	
A*3206	—	A*3201V	GN00338	AF226836, AF226837	
A*3207	—	—	GN00388	AF359389, AF359390	CK Hurley
A*3301	A33(19)	Aw33.1, A3301W776R	JOE, LWAGS, LCL80, W776R	M30580, M28415, U18989, U19735, X83004-5, U83416	
A*3303	A33(19)	A33NC, A33MK	CTM4955926, GAO801, LCL82, HOR, IT	U09740, U18990, U19736, X83002-3, L06440	
A*3304	—	—	NM2442	AF041367, AF041368	
A*3305	A33(19)	A*33DU	DU, NM5A679, Leiden-QC1504	AF108447, AF108448, AF268401, AF268402, AJ251541	K Witter[b]
A*3306	—	A33 variant	ASM	AF234539, AF234540, AF234541	
A*3401	A34(10)	—	ENA	X61704	
A*3402	A34(10)	—	WWAI	X61705	
A*3403	—	A*3402V	1998-302-1407, GN00377	AF251352, AF251353, AF315685, AF315686	
A*3404	—	A34new	ATG	AJ297499, AJ297500	
A*3601	A36	—	MASCH	X61700	
A*3602	—	A*3601V	GN00347	AF244504, AF244505	
A*3603	A36	—	HC030101, F.G.	AF384666	MGJ Tilanus

Table 5.2 Designations of HLA-A alleles (*continued*)

HLA alleles[a]	HLA specificity	Previous equivalents	Individual or cell line from which the sequence was derived	Accession number	References or Submitting author(s)
A*4301	A43	—	CC, GN00174	X61703, AF008305, AF008306	
A*6601	A66(10)	—	25/1506, TEM, GU5175	X61711, U17571	
A*6602	A66(10)	—	CR-B, MALS, HUT102	X61712, X51745	
A*6603	A10	A66KA	AKI	X96638	
A*6604	—	—	BY00015	AF321832, AF321833	
A*680101	A68(28)	Aw68.1	LB, 10063349	X03070, X03071, AF106692, AJ315642	A-M Little[b]
A*680102	A68(28)	Aw68.1	GRC187	L06425	
A*6802	A68(28)	Aw68.2	PA, TO	U03861	
A*680301	A28	A*68new-69A, A68N	AA859, PIME, 69A, FC	U41057, U56436, U56437, U43336, AF017311, U89946	
A*680302	A28	A68N2	GP	U89947	
A*6804	A68(28)[c]	A*68new-65A	65A	U41844, AF017312	
A*6805	A68(28)[c]	A*68new-67A	67A	U43335, AF017313	
A*6806	—	A*6801Var	GN00156	U91627, U91628	
A*6807	—	—	NM2514	AF041371, AF041372	
A*6808	A68(28)	A68V	TER#934	AJ223972	
A*6809	—	—	262-492	AF072769, AF072770	
A*6810	—	A*68011Variant	346-00642	AF108430, AF108431	
A*6811N	Null	A68Null	HP2, OV	AF101046	
A*6812	A28	A*68New	KE-GF	AJ238362, AJ238363, AJ238364	
A*6813	—	A*68KM	FAH	AJ238523, AJ238151, AJ238152, AJ238153	
A*6814	—	A*68xx	NMDP0247-8661-2	AF145954, AF145955	
A*6815	—	A*6802V	GN00261, GN00299	AF135544, AF135545, AF181103, AF181104	
A*6816	A68(28)	A68PA	PA87	AF144013	
A*6817	—	A*68Dan	K45, NM5A815	AJ245567, AF268397, AF268398	
A*6818N	Null	A*68BLA	BLA-Fab	AJ278501	
A*6819	—	A*68012V	GN00376, GN00410	AF288049, AF288050, AF408168, AF408169	CK Hurley[b]
A*6820	—	—	GN00389	AF359391, AF359392	CK Hurley
A*6821	—	—	2001-7399	AF479818, AF479819	J Wu

Table 5.2 Designations of HLA-A alleles (*continued*)

HLA alleles[a]	HLA specificity	Previous equivalents	Individual or cell line from which the sequence was derived	Accession number	References or Submitting author(s)
A*6822	—	—	K83467	AJ420528	J Crowley
A*6901	A69(28)	—	IDF, ZM, BJ	X03158, X03159	
A*7401	A74(19)	—	CC, PDAV, ATUR, GU2037, GU2040	X61701, U17569, U17570	
A*7402	A74(19)	A*74dc	DCH-HLA05, BT2358	X95409, AJ223060	
A*7403	A74(19)[c]	A*74pb	PEB, JB-R.B.	X95561, AJ002678	
A*7404	—	A*74New	U3765	AJ249370	
A*7405	—	—	NM5A142	AF255720, AF255721	
A*7406	A74(19)	—	VTIS23531	AF329872, AF329873	BD Tait
A*7407	—	—	BY0021	AY050187, AY050188	CK Hurley
A*7408	—	—	2001-40-660	AF440110, AF440111	CK Hurley
A*8001	A80	AX"BG", A-new	VH, 35020, 35841, 32511, CODI, MIKA, LADA, CTM3953540, CTM1953541	M94880, L18898, L19403, U03754	

[a]Allele names given in bold type have been assigned since the 2000 Nomenclature report.
[b]This reference is to a confirmatory sequence.
[c]HLA specificity provided from the HLA dictionary (33, 34).

Table 5.3 Designations of HLA-B alleles

HLA alleles	HLA specificity	Previous equivalents	Individual or cell line from which the sequence was derived	Accession number	References or Submitting author(s)
B*070201	B7	B7.2, B*07L	JY, PP, RD105U, RD105, L5, L7, GN00105, 383008	M16102, M32317, P01889, U29057, L47338, U49904, U49905, AJ292075, AJ309047	A-M Little[b] SGE Marsh[b]
B*070202	B7	B*0702V, B*07AD	HGW12327, DZA10	Y13567, AJ002675	
B*070203	B7	B*07N	RN1373B	AF002273, AF017314	
B*0703	B703	BPOT	POT71, BPot	X64454, U21053	
B*0704	B7	B7E	10243	U04245	
B*0705	B7	B*07ZEL	GEE018, ZEL, CF	L33922, U18661, U21052	
B*0706	B7	B7-L79	L7901	X91749	
B*0707	B7	—	DAPO	Z70315	
B*0708	—	—	A.McG	X99735	
B*0709	B7	B*07ML, B*07DKDC	TER#939, DKDC, 011147550	AJ003063, AF106043, AF106044, AF106045, AF132018, AF132019, AF132020	
B*0710	—	B*07AE	A.E.	AJ223602	
B*0711	B7	B-0702v	001524990	AF056481, AF056482	
B*0712	B7	—	GN00216, GN00232	AF061865, AF061866, AF072443, AF072444	
B*0713	—	—	346-808	AF065646, AF065647	
B*0714	—	B*0707Var	012774733, NM4B169, GN00330	AF127806, AF127807, AF132491, AF165854, AF165855, AF205532, AF205533	
B*0715	B7	B*07021Var	NM4B274, 4344PL	AF148809, AF148810, AJ243371, AJ243372	
B*0716	—	B*0703Variant, ?B7	CT-VC	AJ237594, AJ237595	
B*0717	B7	—	R99171035G	AF173936	
B*0718	—	—	CL183	AF189017	
B*0719	—	B*0704V	GN00323, GN00335	AF198648, AF198649, AF226689, AF226690	
B*0720	—	B*0702V, B*07MSB	CU26, SMB7N, MHH-000773	AJ251770, AJ251771, AF244146, AF244147, AJ278043, AJ278044	
B*0721	—	B*07021new	NM5b91	AF255714, AF255715	
B*0722	—	B*07021variant	10009909	AJ400823	
B*0723	—	B*07021V	GN00368	AF279113, AF279114	
B*0724	B7	B*07021var	BEL-LEI	AJ401222	

Table 5.3   Designations of HLA-B alleles (*continued*)

HLA alleles	HLA specificity	Previous equivalents	Individual or cell line from which the sequence was derived	Accession number	References or Submitting author(s)
B*0725	—	B*CBU138	CBU138	AF313415, AF313416	
B*0726	B7	B*07BJ	BSF, 14622	AF317496, AF317497, AJ311257	S Vidal[b]
B*0727	—	B*KHOLM	KHOLM	AF343000, AF343001	H Dunckley
B*0728	—	B*ALTHO	ALTHO	AF402322, AF402323	H Dunckley
B*0729	—	—	BY0029	AF443285, AF443286	CK Hurley
B*0730	—	—	D25857	AB073300, AB073668	H Ikeda
B*0731	—	—	VTIS87843	AY124570, AY124571	BD Tait
B*0801	B8	—	LCL721, MF, CGM1, HECO, 12506397	M59841, M24036, M28204, L76093, AJ295294	
B*0802	B8	B8JON, B8V	20015, 19315	U04244	
B*0803	B8	B*08NR	NR	U28759	
B*0804	—	B*08New-UW	BLB, JS, PF	U67330, U67331, U74386	
B*0805	—	rn083B	rn083B	U88254, AF017315	
B*0806	B8	B-08v	009048430	AF056483, AF056484	
B*0807	B8	B*NV	BM1 101910	AF105226	
B*0808N	Null	B8Null	STRIJOHN, RS	Y1855	
B*0809	B8	B*08HO, B*MW	H.O., MW-ARCBS, HM-ARCBS, GN00244, GN00287, ANO	AJ131852, AJ131853, AF117768, AF117769, AF127247, AF127248, AF102559, AF102560, AF176073, AF176074, AJ276994	
B*0810	B8	B*0801Var	R.E	AJ133101, AJ133102	
B*0811	—	—	NMDP -ID#035343375	AF213681, AF213682	
B*0812	—	B*0801V	GN00344, G3543, GN00371	AF226150, AF226151, AJ276427, AF279674, AF279675	
B*0813	—	—	2000-21-622-7	AF310144, AF310145	
B*0814	—	—	GN00386	AY016211, AY016212	(35)
B*0815	—	—	VTIS37741	AY057398, AY057399	BD Tait
B*0816	—	—	026575043	AF468046, AF468047	TM Williams
B*1301	B13	B13.1	HE, SDI, YTY, TAC	M24075, D50290	
B*1302	B13	B13.2, B13N	LBF, TO, HJB, PKM, TAC, L7901	M19757, M24041, D50291, AJ295278	
B*1303	—	B New	CTM4956865, CTM2956866	U14943	

Table 5.3 Designations of HLA-B alleles (*continued*)

HLA alleles	HLA specificity	Previous equivalents	Individual or cell line from which the sequence was derived	Accession number	References or Submitting author(s)
B*1304	—	B*15X21	TER847, 27B, 76002	U75533, U88248, AF017316, Y12378, Y12379	
B*1306	—	B*1301V	GN00336	AF226691, AF226692	
B*1307N	Null	B1301V	JCB13747	AB032598	
B*1308	—	—	PACO	AJ295279	V Carter
B*1309	—	—	2000-112-197	AY034808, AY034809	CK Hurley
B*1310	—	—	2001-7709	AF461046, AF461047	TM Williams
B*1401	B64(14)	—	MRWC, 32367, W6106, WT51	M24040, X94574	
B*1402	B65(14)	—	BB, CGM1, CM1402, 10038822	M59840, M24032, U90558, AJ301657	
B*1403	B14	B*1402v	DT16, DT3, E210	U91330, U91331, AF015271, AF015272, AF279664	
B*1404	—	B*14N	RN1429B	AF002275, AF017317	
B*1405	—	—	S18, 012867131	AF031142, AF031143, AF110259, AF110260, F110261	
B*140601	B14	Sofh3713, wk B14	FLi	AJ131193, AJ131194	
B*140602	B14	B*1402 Variant	GN00248	AF102567, AF102568	
B*15010101	B62(15)	—	MF, HA, BCK, OLGA (OLL)ᶜ, KT17, PP, FUR, YAG, BA3, BA4, BA5	M28203, M83193, U03859, D50292, L48400, AJ295140	
B*15010102N	Null	BM1947	BEL-13-JA	Y17110	
B*150102	B62(15)	B*1501Var1	PUSPAT, BWH56458, NMDP#015329287, NMDP#015329535, NMDP#015329246, NMDP#015329097, NMDP#015329436	Y17063, Y17168, AF053999, AF054000, AF106626, AF106627	
B*150103	B62(15)	B*15New	AG-SP	AF109724, AF109725	
B*150104	B62(15)	B*15SRE	ET79538	AJ297940, AJ297941	
B*1502	B75(15)	B15N, B*1502	APA, LW, CAY, DCH4060, DCH4061, DCH3086, 12WDCH018, 12WDCH017, 12WDCH002, 12WDCH003, 12WDCH016	M75138, M83192, D50293, AF014769, AF014770, AF014771, AF014772, AF014773, AF014774, AF014775, AF014776, AF014777, F014778, AF014779, AF014780, AF014781, AF014782, AF014783, AF014784	

Table 5.3 Designations of HLA-B alleles (*continued*)

HLA alleles	HLA specificity	Previous equivalents	Individual or cell line from which the sequence was derived	Accession number	References or Submitting author(s)
B*1503	B72(70)	—	CC, 26931, 31708	X61709	
B*1504	B62(15)	Bw62-G	GRC138, KG, GRC187, GRC-150	M84382, AJ292970	
B*1505	B62(15)	Bw62.1	VB	M83191	
B*1506	B62(15)	Bw62.4	WI	M83194	
B*1507	B62(15)	Bw62.5	SB	M83195	
B*1508	B75(15)	B62variant	KHAGNI, LATIF, DAN723	L11666	
B*1509	B70	B70.1	34863	L11571	
B*1510	B71(70)	B70.2	25514, 19014, GU373, GU2092, GU2037, GU5175	L11570, U11262, U11264, U11269	
B*151101	B75(15)	B15variant, B75v	LEE743, AZ195, AZ319	L11604, D50294	
B*151102	75(15)	B1511V1	JCBT2513	AB036051	
B*1512	B76(15)	B76	THAI742	L11603	
B*1513	B77(15)	B77	RSA-ND, CAM020, PETCH, 12WDCH009, 12WDCH010, 12WDCH011, 12WDCH028	L15005, D50295, U90424, U90425, U90422, U90423, U90420, U90121, U90418, U90419	
B*1514	B76(15)	B76	SS713	L19937	
B*1515	B62(15)	B62s	MLH727, LDM	L22027, L49343	
B*1516	B63(15)	B63.1, 8W66	DOP-ND, 21909, 31133	L09735	
B*15170101	B63(15)	B63	JAP-NF, PARMG	U01848, U35431, AJ300181	A-M Little[b]
B*15170102	B63(15)	B*1517 var	Terasaki EXT#95	AJ308397	A-M Little
B*1518	B71(70)	B*7901, B"X"-HS, B71	HS, GU2739, GU2760, MSU, ML108, ML108U	U11266, U11268, D50296, U57966	
B*1519	B76(15)	B76	GEE018	U03027	
B*1520	B62(15)	—	OLGA (OLL), KRC110	U06862	
B*1521	B75(15)	B15Ab	BJ, HWY, 14247373, 12WDCH022	L32862, D44500, U32678, U91332, U91333	
B*1523	—	B'NM5'	TK765	L37881	
B*1524	B62(15)	B*15ZEL, 1501-B4a, B*1501-Bw4	ZEL, SF94-140	U16309, L42146	
B*1525	B62(15)	B*15AOH, B*1525	WON, M, HM, BY0007, 12WDCH012, 12WDCH023, 12WDCH025, DCH3258, DCH1109	U18660, U50710, U52177, U52178, U91336, U91337, U91334, U91335, AF014785, AF014786, AF014787, AF014788, AF014789, AF014790	
B*1526N	Null	B-null	K.I.	D49824	

Table 5.3 Designations of HLA-B alleles (*continued*)

HLA alleles	HLA specificity	Previous equivalents	Individual or cell line from which the sequence was derived	Accession number	References or Submitting author(s)
B*1527	B62(15)	—	PELE	L42144, L40182	
B*1528	B62(15)ᶜ	B15v1	YTR	D44499	
B*1529	B15	B15v3	DKA	D44501	
B*1530	B62(15)	B*1501V1	EFTO, GN00104, GN00108	42296, U49900, U49901, U52171, U52172	
B*1531	B75(15)	B*1502V	ALDE, GN00110	L42145, U52173, U52174	
B*1532	B62(15)	—	DCH036, 12WDCH038, 12WDCH027	X95410, U83580, U83581	
B*1533	B15	—	GN00103	U49898, U49899	
B*1534	B15	—	GN00105	U49902, U49903	
B*1535	B62(15)ᶜ	—	GN00106	U52167, U52168	
B*1536	—	B*15MD	MD674	U58315, U58316	
B*1537	B70ᶜ	—	11112331, CTM1984782	U55022, U55023, AF016641	
B*1538	—	—	#10	U95084, U95085	
B*1539	B62(15)ᶜ	ZA016, B*15MZH	ZA016, GN00177, T228, NM3906	AF016302, AF009681, AF017080, AF017081, AF033501, AF033502, AF060504, AF060505	
B*1540	—	—	GN00181, GN00206	AF028597, AF028598, AF054003, AF054004	
B*1542	—	B*15/55Var	PB(16962)	Y15841	
B*1543	—	B*1501Var2	GN00211	AF054011, AF054012	
B*1544	—	B*1521Var	GN00212	AF061857, AF061858	
B*1545	B62(15)	B*15JL	J.L, GN00219	AJ007605, AJ007606, AF071765, AF071766	
B*1546	B72(70)	B*15UL, B1501V2	S.Z., 97-02707, JCBB13806	AJ007603, AJ007604, AF110250, AF110251, AF110252, AB036049	
B*1547	—	—	346-516	AF07265, AF072266	
B*1548	B62(15)	—	009326174/HR1858	AF072377, AF072378	
B*1549	—	B*1503V	NMDP#016220287	AF105029, AF105030	
B*1550	—	B*1501Variant	121-08035	AF108424, AF108425	
B*1551	B70	B*NO = B*27New	NO-ARCBS	AF117766, AF117767	
B*1552	—	B*15 Variant	01223584, UCLA01203301, GN00288, 99-2200, GN00328, GN00343	AF127810, AF127811, AF132488, AF172869, AF172870, AF176075, AF176076, AF189248, AF189249, AF189250, AF202451, AF202452, AF226152, AF226153	

Table 5.3 Designations of HLA-B alleles (*continued*)

HLA alleles	HLA specificity	Previous equivalents	Individual or cell line from which the sequence was derived	Accession number	References or Submitting author(s)
B*1553	—	B*15 Variant	012436002	AF129296, AF129297, AF132487	
B*1554	—	B*1503v	GN00257, E3541	AF135536, AF135537, AJ245869	
B*1555	B15	B*1531new	T2059	AJ249316, AJ249317, AJ249318, AJ249319, AJ249320, AJ249321, AJ249322	
B*1556	—	B*1501V2	GN00315	AF181846, AF181847	
B*1557	—	B*15New	NDS-758	AF188885, AF188886, AF188887	
B*1558	B15	B*15KSW, ?B62(15)	99-2202, KSW	AF190278, AF190279, AF190280, AF184607, AF184608	
B*1560	—	B1501V4	JCBT1283	AB036050	
B*1561	—	B*1503V	1999-158-3366	AF251356, AF251357	
B*1562	—	—	GN00363	AF266527, AF266528	
B*1563	—	B*1545V	Toba44, GN00364	AF275626, AF275627, AF281150, AF281151	
B*1564	—	B*1518V	GN00367	AF279111, AF279112	
B*1565	—	B*CB3654	CB3654	AF335310, AF335311	H Dunckley
B*1566	—	—	UCB-163-1999	AJ308399	(36)
B*1567	—	—	MCH104, ML1777	AF335547	(37)
B*1568	—	B*15/48	13365831	AY033429, AY033430, AY033431	(38)
B*1569	—	B*15var	BHCP	AJ298282, AJ298289	(39)
B*1570	B62(15)	—	285D	AY057402, AY057403	BD Tait
B*1571	B62(15)	—	FH66, FH67	AY065827, AY065828, AY065829	A Smith
B*1572	—	—	FH60	AY065830, AY065831, AY065832	A Smith
B*1573	B62(15)	—	11470, 28580	AJ459483, AJ489936, AJ489937	S Vidal, T Gervais
B*180101	B18	—	SGAR, F24, MM1801, VEN	M24039, U90559, AJ310507	A-M Little[b]
B*180102	B18	—	6ABC124	AY045737, AY045738	WH Hildeband
B*1802	B18	B18PE	PETCH	D25275	
B*1803	B18	B1803	BM66, GSW002, T36121	X94480, Y07824, AJ309979	P Dunn[b]
B*1804	—	B*18IM	IMM348	U38792, U38793	
B*1805	B18	B*18GSW	GSW001, DZA1	Y07710, AJ002676	
B*1806	B18	—	CTM-9985836	AF033351	
B*1807	—	B*MF	GN00210, MF-ARCBS	AF054009, AF054010, AF117774, AF117775	

Table 5.3 Designations of HLA-B alleles (*continued*)

HLA alleles	HLA specificity	Previous equivalents	Individual or cell line from which the sequence was derived	Accession number	References or Submitting author(s)
B*1808	—	B*1801New	NM4b448	AF148636, AF148637	
B*1809	B18	B18OP	6259OP, GN00345	AJ243374, AJ243376, AF274500, AF274501	
B*1810	—	B*1801V	GN00324	AF198650, AF198651	
B*1811	—	—	GN00362	AF266525, AF266526	
B*1812	—	B*1801V	GN00366	AF275716, AF275717	
B*1813	—	—	2000-56-617	AF310138, AF310139	
B*1814	—	—	2000-224-257, 00-809	AY042672, AY042673, AF403249	CK Hurley, LA Baxter-Lowe
B*1815	—	—	2000-084-2159	AY042686, AY042687	CK Hurley
B*1817N	—	—	WVAN, AVAN	AF416771	(40)
B*1818	—	—	28626	AJ489938, AJ489939	T Gervais
B*2701	B27	27f	LH, PIL-139	L76935	
B*2702	B27	27e, 27K, B27.2	BRUG, NV, KSH	X03664, X03667, L38504, U18659	
B*2703	B27	27d, 27J	CH (CHI)	M54883	
B*2704	B27	27b, 27C, B27.3	WEWAK 1, DH, DEW-ND	U27608	
B*270502	B27	27a, 27W, B27.1	CD, HC, MRWC, KCA, MVL, LG2, BRUG, BTB	X03945, M12967, L20086, M14013, M12678, AJ420238	A-M Little[b]
B*270503	B27	27MW	HHE	X83727, X83737	
B*270504	B27	FMVB27	20836	AJ250630, AJ250631, AJ250632	
B*270505	B27	—	8998871, 6998872	AF480612	F Garcia Sanchez
B*2706	B27	27D, B27.4	LIE, PAR, TER Cell#995	X73578, U35734, AJ292971	(41)[b]
B*2707	B27	B27-HS	HS	M62852	
B*2708	B2708	B7Qui	19418, BCK	L19923	
B*2709	B27	B27-ci	Ci	Z33453	
B*2710	B27	B2705v	KRICO, NMDP0392-7903-9	L76095, AJ308990, AJ310147	M Bengtsson[b]
B*2711	B27	B27KH	K.H.	D83043	
B*2712	—	WIS1/C846	RW, MT3, RK, CTM4896	U90244, U90245, Y14582, AF022783	
B*2713	B27	B27052W496D	W496D	AF026218	
B*2714	—	—	65-90810, 01168999	AF072763, AF072764, AF110256, AF110257, AF110258	

Table 5.3 Designations of HLA-B alleles (*continued*)

HLA alleles	HLA specificity	Previous equivalents	Individual or cell line from which the sequence was derived	Accession number	References or Submitting author(s)
B*2715	—	B"X"-Bw6	KC	Y16637, Y16638	
B*2716	—	B*27052 Variant	GN00246	AF102563, AF102564	
B*2717	B27	B27TO	4388TO	AJ243373, AJ243375	
B*2718	—	—	99-2198	AF189012, AF189013, AF189014	
B*2719	B27	—	BFLR	AF190146, AF190147	
B*2720	B27	B*27CHN	KMP01-1379	AF170578, AF170579	
B*2721	—	B*2706V	GN00334	AF218578, AF218579	
B*2723	—	B*27IG	30733VTIS, 35520	AF305196, AF305197, AJ298262	M Guttridge[b]
B*2724	—	—	2000-161-3004	AY042670, AY042671	CK Hurley
B*2725	—	—	2000-119-979	AF408160, AF408161	CK Hurley
B*350101	B35	—	HS, KT17, GU2739, CMM, KT12	M28109-12, U11265, L63544, AJ420239	A-M Little[b]
B*350102	B35	—	GN00356	AF260977, AF260978	
B*3502	B35	—	DL, 388	M63454, U90563	
B*3503	B35	—	C1R, HMY2, 12405, 13159, 093	M81798, D50299, U90564	
B*3504	B35	—	AN, RB22, 12.36JK	M86403, U30936, L47986	
B*3505	B35	B35-G	GRC212, KRC032, TOB-115	M84385, L76930	
B*3506	B35	B35-K	KRC032	M84381	
B*3507	B35	—	#20073	L04695	
B*3508	B35	B35TL	#22338, TL	L04696, Z22651	
B*350901	B35	—	MA9, 30	U17107, U90565	
B*350902	B35	—	WIC-54	L76932	
B*3510	B35[c]	—	JK1.2, JK5.13, JK14.41	L36979	
B*3511	B35	B35v	GRC-187	L40599	
B*3512	B35	B-3504v	BAON, FEME, PNS	L42281, L76094, L49342	
B*3513	B35	2993	RCE80, THA-DCH 0654, THA-DCH 9675	X87268, AF208430, AF208431, AF208432, AF208433	
B*3514	B35	B*35M	JLG, JGS	S83195, S83196	
B*3515	B35	—	PARMG	U30904	
B*3516	B35[c]	B*35GAR	GAR	U29880	

Table 5.3 Designations of HLA-B alleles (*continued*)

HLA alleles	HLA specificity	Previous equivalents	Individual or cell line from which the sequence was derived	Accession number	References or Submitting author(s)
B*3517	B35	B35V1, B*35PNS, B-3505v	JM (G2744), PNS, AMYE	U34618, L49341, L75941	
B*3518	B35	B-3508v	TOB-137	L75942	
B*3519	B35	B-40X35	WIC-54, VTIS43878	L76933, AF387905, AF387906	BD Tait[b]
B*3520	B35	B-3501V	TER-135	U76392, U76393	
B*3521	—	B-3511H	TER-109	U76390, U76391	
B*3522	—	M001B	M001B	AF017327, AF009685	
B*3523	—	MA080B	MA080B	AF016301, AF009680	
B*3524	—	MA086B	MA086B	AF016300, AF009679	
B*3525	—	—	GN00215	AF061863, AF061864	
B*3526	—	B15/35 7-1 clone 24	NMDP#027669746	AF105031, AF105032	
B*3527	B35	B*35JAC	JAC	Y18288, Y18289	
B*3528	—	B*3510Variant	304-00651, 016696205	AF108428, AF108429, AF127808, AF127809, AF132486	
B*3529	B35	B*KG	KG-ARCBS, GN00289	AF117770, AF117771, AF176077, AF176078	
B*3530	B35	B*3517Variant	GN00242	AF110504, AF110505	
B*3531	—	B*35/40	KYR, KKW, MOV	AF138164, AF138165, AF170577, AJ278744	
B*3532	B35	B*TMUL	BM1 139852	AF134866, AF134867	
B*3533	—	B*35New	0000-3034-6	AJ238411, AJ238412	
B*3534	—	—	GN00329	AF205530, AF205531, AF201762	
B*3535	B35	B3501V1, B35v	JCBT1635	AB032093	
B*3536	—	B*3503V	GN00353	AF282765, AF282766	
B*3537	—	B*35KM	DZA1999-16/MHH994949	AJ243737, AJ243738	
B*3538	—	—	BSB620, BSB620-MO	AJ312287	K Witter
B*3539	—	—	2000-140-1975	AY042688, AY042689	CK Hurley
B*3540N	—	—	IBTC-B35N	AJ418040	(42)
B*3541	B35	—	2HT21, WAC1087870, CAP13	AY045735, AY045736, AF480613, AF497262	WH Hildebrand, F Garcia Sanchez, D Smith
B*3542	B35	—	MS21871	AJ316289, AJ426469, AJ426468, AJ417680, AJ417669	EM van den Berg Loonen

Table 5.3 Designations of HLA-B alleles (*continued*)

HLA alleles	HLA specificity	Previous equivalents	Individual or cell line from which the sequence was derived	Accession number	References or Submitting author(s)
B*3543	B35	B15UW1, B35V2, B*1522, B15/35 7-1 clone 27	1274, B503, JC (G2997), FFAJ, NMDP#027669746	U14756, L42506, U34619, U80945, AF106630, AF106631	
B*3544	B35	B*1559	013221023	AF206514, AF206515	
B*3701	B37	—	KAS011, MG, GU2760	M32320, U11267	
B*3702	—	B27-37	CTM-8958127	U31971	
B*3703N	Null	B*37OMI	OMI	AJ277845	
B*3704	—	—	GN00382, H156H2	AF303101, AF303102, AF389378	(43)[b]
B*3705	—	—	CMC2	AF284826, AF284827, AF284828	
B*3801	38(16)	B16.1	Z, JAP-NF, YAR, JBUSH, TEM, WDV, ELON, LB96-SAR	M29864, L36591, U40498	
B*380201	B38(16)	—	RSA-ND, Terasaki EXT#58, 32764	L22028, AJ297317, AJ308991, AJ308992	A-M Little[b], M Guttridge[b]
B*380202	B38(16)	—	GN00155, GN00416	U90240, U90241, AY094134, AY094135	CK Hurley[b]
B*3803	B16	—	CTM-4786786	AF081275, AF081276	
B*3804	—	—	49-TA	AF181857, AF181858	
B*3805	B38(16)	B*38New	CTM-1095139	AF218802, AF218803, AF218804	
B*3806	—	—	GN00357, GN00372	AF262960, AF262961, AF282769, AF282770	
B*3807	—	B*3801New	MCB4	AF281053, AF281054	
B*3808	—	B*SSHAM	SSHAM	AF402320, AF402321	H Dunckley
B*390101	B3901	B39.1, B16.2	S, JC	M94052, M29865	
B*390103	B3901	B39.1J	IT, #591	M94051	
B*390104	B3901	B*39011New	NM4B380, JCB11331	AF165852, AF165853, AB032096	
B*390201	B3902	B39.2	YAM	M94053	
B*390202	B3902	B39.2	CL170	U04243	
B*3903	B39(16)	—	AUCA#19, VTIS46155	L20088, AF387907, AF387908	BD Tait[b]
B*3904	B39(16)	B39N	TO ?KO	L22649	
B*3905	B16	ST-16, B*39UW1, B*39JAI	11, HGOM, 12.35JK, 12.63JK	U15638, L36318, L36980	

Table 5.3 Designations of HLA-B alleles (*continued*)

HLA alleles	HLA specificity	Previous equivalents	Individual or cell line from which the sequence was derived	Accession number	References or Submitting author(s)
B*390601	B39(16)	B*39UW2	15, HAA, BA1, TER-102	U15639, L42024, L76640, L76639, U76396, U76397	
B*390602	B39(16)	B*39DBU, B39G	DBU, GVA, CVL, RD105, NAVAJO	U16298, L40562, U29083, U32660	
B*3907	B39(16)[c]	B*39UW3	1276	U15640	
B*3908	B39(16)	—	822	L42280	
B*3909	B39(16)	B39-143.2	143.2, XAV-50, 072	U29480, L76088, U90580	
B*3910	B39(16)	B39.ZU47	Zu47, GN00110, GB32, MA-31750	U56246, U52175, U52176, Y09058, AJ237703	
B*3911	B39(16)[c]	—	KUNA 20	U74387	
B*3912	B39(16)	B-3901V	TER-103	U76394, U76395	
B*3913	B39(16)	—	MCDS	AJ223282	
B*3914	—	—	GN00217	AF061867, AF061868	
B*3915	—	—	178-260	AF065640, AF065641	
B*3916	—	BA-39V	BAKA	AF098266, AF098267	
B*3917	—	B*39Var	010760981	AF110262, AF110263, AF110264	
B*3918	—	B*39011V	GN00310	AF173875, AF173876	
B*3919	—	B*3901V	GN00293	AF176081, AF176082	
B*3920	—	B*3910V	GN00317	AF184216, AF184217	
B*3922	—		GN00332	AF205536, AF205537	
B*3923	B39(16)	B3902V1	JCB12110	AB032097	
B*3924	B39(16)	B*CB2261, B*3903V	NDS-IH, CBu 10474, POHS-397, OC311, OC350, OC311, OC350, C183	AF220288, AF220289, AF231101, AF231102, AF293020, AF293021, AF293022, AJ251768, AJ251769, AJ251768, AJ251769, AF428252	C Vilches[b]
B*3925N	—	—	13W09502	AF363012, AF363013, AF363014	A. Smith
B*3926	—	—	2000-333-343	AF408162, AF408163	CK Hurley
B*400101	B60(40)	—	LB	P01890, U03698	
B*400102	B60(40)[c]	B60Ut	Ut-m, JD, #W7079	M95530, L41628	
B*400103	B60(40)	B*40(93090)	93090	AJ309573	A Dormoy
B*4002	B61(40)	B40*	SWEIG, CALOGERO, YUKI, 19014, TOB-105	L09736, D14343, L76089	
B*4003	B61(40)[c]	B40-G1	GRC138	M84383	
B*4004	B61(40)[c]	B40-G2	GRC212, TOB-0087	M84384, L76090	

Table 5.3 Designations of HLA-B alleles (*continued*)

HLA alleles	HLA specificity	Previous equivalents	Individual or cell line from which the sequence was derived	Accession number	References or Submitting author(s)
B*4005	B4005	BN21	00136	M84694	
B*40060101	B61(40)	B61	Ot-s	M95531, AJ300180	A-M Little[b]
B*40060102	B61(40)	B*4006new	Terasaki EXT#58	AJ292253	A-M Little
B*4007	B60(40)[c]	B'Fu'	MSU, FTA, KTA	D31816	
B*4008	—	—	4008	L41353	
B*4009	B61(40)	B-4003V	PIL-117	L76934	
B*4010	B60(40)	B*40MD, B*40Var, B*40011Var, B40New	MD676, GN00160, 10PNG, PK, NMDP#019350966	U58643, U58644, U93915, U93916, Y15840, Y16636, Y16639, AF106628, AF106629	
B*4011	B40	B*40N	098, UCLA160	U75864, U75865, AF016299, AF009682	
B*4012	—	B*40x15	TER-914, TE914, 015740137/467	Y13029, AF017334, AF017335, AF132492, AF132493, AF132494	
B*4013	—	—	NBER	U96942	
B*4014	—	—	104B	AF002274, AF017318	
B*4015	—	—	M008B	AF002268, AF002269	
B*4016	B61(40)	—	EW, CS25, CS48, 98-00101	Y14606, AF017022, AF017023, AF027296, AF027297, AF110253, AF110254, AF110255	
B*4018	—	RN988B	RN988B	AF017332, AF017333	
B*4019	—	—	329-8016	AF065644, AF065645	
B*4020	—	—	290-596, 010818557	AF065648, AF065649, AF127812, AF127813, AF132017	
B*4021	—	B*15Var	CBP, #6749	AF106686, AF106687	
B*4022N	Null	B40VN	40FC	AF129291, AF129292	
B*4023	—	B*40Var, B*CB2880	011743051, 702502, CB2880	AF129298, AF129299, AF132489, AJ278749, AJ278750, AF335312, AF335313	H Dunckley[b]
B*4024	—	B*4018 Variant	GN00251	AF102573, AF102574	
B*4025	—	B*BM	BM1 131485	AF134864, AF134865	
B*4026	B21	B40Var	Akbasaim	AJ243433, AJ243434	
B*4027	B61(40)	B*4002V1	JC12323, GN00316	AB030575, AF181471, AF181472	
B*4028	—	B*4004V	GN00313	AF181842, AF181843	
B*4029	B61(40)	B4002V2, B61v(40)	JC16904	AB032599	
B*4030	—	B*40011V	GN00340, GN00352, GN00373	AF226840, AF226841, AF257507, AF257508, AF282767, AF282768	

Table 5.3 Designations of HLA-B alleles (*continued*)

HLA alleles	HLA specificity	Previous equivalents	Individual or cell line from which the sequence was derived	Accession number	References or Submitting author(s)
B*4031	B60(40)	B*40RG	33692	AJ271160	
B*4032	—	B*4016V	GN00361	AF266523, AF266524	
B*4033	—	B*40011V	GN00369	AF279115, AF279116	
B*4034	B60(40)	B*40var	386619	AJ404846	
B*4035	—	—	ZFI	AJ290949, AJ290950	
B*4036	—	B*RRACH	RRACH	AY034093, AY034094	H Dunckley
B*4037	—	B*4002V	2000-343-446, 2000-343-785	AY034806, AY034807, AY042676, AY042677	CK Hurley
B*4038	—	—	VTIS39243	AF387901, AF387902	BD Tait
B*4039	—	—	BUMC-40v	AY040540	D Smith
B*4040	—	—	BY0018, BY0025, BY0022	AY042680, AY042681, AY050193, AY050194, AY050189, AY050190	CK Hurley
B*4042	—	—	2000-350-252	AF408164, AF408165	CK Hurley
B*4043	—	—	BY00040	AF494281, AF494282	CK Hurley
B*4044	—	—	GN00417	AY094136, AY094137	CK Hurley
B*4101	B41	—	SGAR, CM4101, BM21	M24035, U90560, AJ309193	A-M Little[b]
B*4102	B41	B41.2	SBD4, GU5175, BM2684	X81363, U17572, X86704	
B*4103	B41[c]	—	GN00182, GN00245	AF028595, AF028596, AF102561, AF102562	
B*4104	—	—	99126462S	AF258782	
B*4105	—	B*4101V	GN00370	AF279117, AF279118	
B*4106	—	—	UC-B434, 09-S-0029#0001	AJ308547, AY033291, AY033292	(44), S Grams
B*4201	B42	—	BB, BJ	M24034, AJ309194	A-M Little
B*4202	B42	B42ANDO, 71B	E-117, E-119, 71B, 31-650, DZA9	D50709, U88249, AF017319, U88407, AJ002677	
B*4204	—	—	BY0027	AY050197, AY050198	CK Hurley
B*44020101	B44(12)	B44.1, B44.2, B44021	FMB, BAU, RG-BR	M24038, M15470, AJ309936	P Dunn[b]
B*44020102S	—	—	PIO	AF384095	L Gebuhrer
B*440202	B44(12)	B*4402V	GN00350	AF253326, AF253327, AF386759	D Adorno[b]
B*440203	B44(12)	—	2000-238-831	AY034810, AY034811	CK Hurley

Table 5.3 Designations of HLA-B alleles (*continued*)

HLA alleles	HLA specificity	Previous equivalents	Individual or cell line from which the sequence was derived	Accession number	References or Submitting author(s)
B*440301	B44(12)	B44.1:New	PITOUT, F24, MM44031	X64366, U90561	
B*440302	B44(12)	—	OBH, SHCHA, CAUC44032	L42282, U58469, U58470, AF056981	
B*4404	B44(12)	B44.4	TAN, BEB	X75953, X78426, X78427	
B*4405	B44(12)	B44WJG, B44KB	WJG, KB, 14-AS-0013#0001	X78849, X78850, L31798, AF288472, AF288473	
B*4406	B44(12)	—	GIJM, KARY	X83400, X83401-3, L42345	
B*4407	B44(12)	B*44GB	GB92	X90391	
B*4408	B44(12)	B44bo, B*44DM	19662, DM	U64801, AJ132659, AJ132660	
B*4409	B12	B4409	S.A., RG-BR	X99734, AJ309937	P Dunn[b]
B*4410	B44(12)[c]	—	S32	U63559, U63560	
B*4411	—	—	GN00220	AF071767, AF071768	
B*4412	B44(12)	B*4402Var	MOV002AN	AJ133267	
B*4413	B44(12)	B*44New1	AMI005AN	AJ131118	
B*4414	B12	B44IP	IP	AJ238702	
B*4415	B12	B45New, B*45V	ML1805, 3880, SMN44	AJ133471, AJ133472, AJ251766, AJ251767, AF215918, AF215919	
B*4416	B47	B*4402New	10000009	AF190446, AF190447	
B*4417	B44(12)	B*44SR	B1268	AJ249724, AJ249725	
B*4418	—	—	99-2201	AF190275, AF190276, AF190277	
B*4419N	Null	B44N	ALBA	AJ251593	
B*4420	—	—	GN00331	AF205534, AF205535	
B*4421	—	B*TBAL	GN00333, TBAL	AF205538, AF205539, AF231098, AF231099	
B*4422	—	—	15-S-0032#0102	AY003906, AY003907	
B*4423N	Null	B*44MP	12506397, FH33	AJ278766, AJ295293, AF363681, AF363682, AF363683	A Smith[b]
B*4424	—	—	GN00383	AF310140, AF310141	
B*4425	—	B*CB2913	CB2913	AF335308, AF335309	(45)
B*4426	—	—	MCH48	AF349440	(37)
B*4427	B44(12)	—	E487, FH50, FH48	AF329843, AF329845, AF419293, AF419294, AF419295	(46), A Smith
B*4428	—	—	GN00396, GN00397	AY050199, AY050200, AY050201, AY050202	CK Hurley
B*4429	—	—	GN00406	AY050212, AY050213	CK Hurley

Table 5.3 Designations of HLA-B alleles (*continued*)

HLA alleles	HLA specificity	Previous equivalents	Individual or cell line from which the sequence was derived	Accession number	References or Submitting author(s)
B*4430	—	—	2000-301-424	AF408158, AF408159	CK Hurley
B*4431	B44(12)	—	AKAR	AJ297942, AJ297043	(47)
B*4432	—	—	VBD25061	AY057404, AY057405	BD Tait
B*4501	B45(12)	—	OMW, CM4501	X61710, U90562	
B*4502	—	—	GN00214	AF061861, AF061862	
B*4503	—	B*4501New	O3499	AJ275937	
B*4504	—	—	PMF	AJ278944	
B*4505	—	—	GN00387	AY016213, AY016214	CK Hurley
B*4506	—	—	013969175	AF469652, AF469653	TM Williams
B*4601	B46	—	T7527, THAI742, T7526	M24033, AJ310508	A-M Little[b]
B*4602	B46	B46V	JCB15113	AB032091	
B*47010101	B47	—	PLH	M19756, AJ295141	A-M Little[b]
B*47010102	B47	—	383008	AJ308398	A-M Little
B*4702	B47[c]	—	CAL	Y09118	
B*4703	—	B*47RG, B*47TAIB	DT-32, 29182, TAIB, GN00218, VELT	AF016842, AF016843, Y17193, Y1919, AJ006978, AF071763, AF071764, AJ251003	
B*4704	—	—	05-S-0012#1001	AY033293, AY033294	(48)
B*4801	B48	—	KRC103, HS67, CM4801, 26/27	M84380, U66250, AJ309139	A-M Little[b]
B*4802	B48	—	AUCA#18	L20089	
B*4803	B48[c]	B-48.3	TOB-115	L76931	
B*4804	B48	0328	0328, JC20008	AF017328, AF017329, AB063626, AB063627, AB063628	T Noda[b]
B*4805	B48	B*40Var	GLAD, 011837630/48	AF096631, AF096632, AF127805, AF129293, AF132490	
B*4806	—	B*4801Variant	234-01069	AF108426, AF108427	
B*4807	B48	B*4801Var	30007, GN00258	AF136393, AF136394, AF135538, AF135539	
B*4901	B49(21)	—	AM, GU2092	M24037, U11263, AJ311600	A-M Little[b]

Table 5.3 Designations of HLA-B alleles (*continued*)

HLA alleles	HLA specificity	Previous equivalents	Individual or cell line from which the sequence was derived	Accession number	References or Submitting author(s)
B*4902	B49(21)	B*4901V	MC2918, GN00358	AJ269496, AJ269497, AJ269498, AF262958, AF262959	
B*4903	—	B*RA	29037	AJ288980	
B*5001	B50(21)	—	SH.JO, JD, GU2037	X61706, U11261	
B*5002	B45(12)	B*50IM, B*45v, B*45ZJ	IMM754, WM1366C, CTM-1983039, GN00173, UBM13129406	U58317, U58318, Y08995, AF006634, AF008926, AF008927, Y14205	
B*5004	B50(21)	—	3011	AF136397, AF136398	
B*510101	B51(5)	—	LKT-2, TO, BM92, CD, LCL721, KRC110, KRC005, BA1, BA6	M32319, M22786, M22787-M22788, M28205, Z46808, L47985	
B*510102	B51(5)	B*51V	GN00106, 12WDCH010, 12WDCH028, UCB-1999-163	U52169, U52170, U90611, U90612, U90613, U90614, AJ278903	
B*510103	B51(5)	B*51011V	GN00264	AF135550, AF135551	
B*510104	B51(5)	—	DLM	AJ249937, AJ249938	
B*510105	B51(5)	—	MS22035	AJ426462, AJ426465, AJ426466, AJ426463, AJ426464	EM van den Berg-Loonen
B*510201	B5102	B5.35	UM, 02627	M68964	
B*510202	B5102	—	MY823, 12WDCH011	L41925, U90615, U90616	
B*5103	B5103	BTA	30-BY3	M80670	
B*5104	B51(5)	—	GRC150	Z15143	
B*5105	B51(5)	B51v	LK, 10030381	U06697, AJ297934	
B*5106	B51(5)ᶜ	—	GN097, GN088	U31334, U32661	
B*5107	B51(5)	B5101v	RCE55	X94481	
B*5108	B51(5)	B*51FA, B*51GAC	F.A., GN00109, NDS-DG, AS7235	X96473, U52815, U52816, Y08994, Y10031, Y11228, Y11229	
B*5109	B51(5)	B*51IM, B*51N	IMM721, NMDP-0004, RN285B, GN00178, GN00205, GN00204, NM4B437	U58319, U58320, U76400, U76401, AF002272, AF017320, AF028599, AF028600, AF054001, AF054002, AF165848, AF165849	
B*5110	—	HLA-B*51like, B-51v	KUNA 14, 009041674	AF004370, AF056479, AF056480	
B*5111N	Null	B*51N	HGW6178	Y13566	
B*5112	—	B51Va	RTCV	AF023442, AF023443	

Table 5.3 Designations of HLA-B alleles (*continued*)

HLA alleles	HLA specificity	Previous equivalents	Individual or cell line from which the sequence was derived	Accession number	References or Submitting author(s)
B*511301	—	B*51vK60	K60	AJ002151	
B*511302	—	B*51011V	GN00140	AF135534, AF135535	
B*5114	—	—	GN00207, GN00208	AF054005, AF054006, AF054007, AF054008	
B*5115	—	—	GN00183	AF072445, AF072446	
B*5116	B52(5)	DT51v	DTEC	AF098264, AF098265	
B*5117	B51(5)	—	3010	AF136395, AF136396	
B*5118	B51(5)	B*51New	MEFG	AJ133773, AJ133814	
B*5119	—	—	TN01/1210	AJ238971, AJ238972	
B*5120	—	B*5108V	GN00285	AF140861, AF140862	
B*5121	—	B*51011V	GN291	AF176079, AF176080	
B*5122	—	B*51011V	GN00349, GN00355	AF248061, AF248062, AF260975, AF260976	
B*5123	—	B*5102V	GN00342	AF226844, AF226845	
B*5124	B51(5)	B*51New	46643	AJ276995	
B*5126	—	—	GN00385	AY016209, AY016210	(49)
B*5127N	—	—	5761	AF363789, AF363790	M Kamoun
B*5128	B51(5)	—	VTIS40888	AY057400, AY057401	BD Tait
B*5129	B51(5)	—	FH59, FH38	AY056451, AY056452, AY056453	A Smith
B*520101	B52(5)	—	MT, LK707, E4181324	M22793-9, AJ420240	A-M Little[b]
B*520102	B52(5)	—	AUCA#2, TOB-137, BA8	L20090, L76091, L47984	
B*520103	B52(5)	B*52011V	GN00339	AF226838, AF226839	
B*5202	—	B*52012V	GN00314	AF181844, AF181845	
B*5203	—	B*52012V	GN00365	AF281152, AF281153	
B*5204	B52(5)	—	MS23477	AJ316288, AJ426470, AJ426467, AJ417684, AJ417673	EM van den Berg Loonen
B*5301	B53	—	AMAI, AM, 046	M58636, U90566, AJ311599	A-M Little[b]
B*5302	—	—	S15(28)	U63561, U63562	
B*5303	—	—	GN00231	AF071769, AF071770	
B*5304	—	B*CD	CD-ARCBS	AF117772, AF117773	
B*5305	—	B*5301V	GN00325, 24961vtis	AF198652, AF198653, AF304002, AF304003	

Table 5.3   Designations of HLA-B alleles (*continued*)

HLA alleles	HLA specificity	Previous equivalents	Individual or cell line from which the sequence was derived	Accession number	References or Submitting author(s)
B*5306	—	B*51/53New	SIA	AJ276996	
B*5307	B53	B*53/37	49716	AJ293856, AJ293857	
B*5308	—	—	2000-077-189	AY034802, AY034803	CK Hurley
B*5309	—	—	BY0023	AY050191, AY050192	CK Hurley
B*5401	B54(22)	—	LKT-3, TTL	M77774	
B*5402	B54(22)	B5401V1	JCBB18561	AB032095	
B*5501	B55(22)	—	VEN	M77778, AJ310509	A-M Little[b]
B*5502	B55(22)	—	APA	M77777	
B*5503	B55(22)[c]	B5501v	RCE70	X94482	
B*5504	B55(22)	B-4201v, B55.2	TAGO, 11840Kane, KIW	L76225, D85761, D89333, D89334	
B*5505	B22	B5501 W669R	B55W669R	U63653	
B*5507	B54(22)	—	8138, 9070	AF042289, AF042290	
B*5508	B56(22)	B*ER	DIA2 98629, VTIS31300	AF091343, AF091344, AF304004, AF304005	
B*5509	—	S-PB55	13215	AJ250628, AJ250629	
B*5510	B55(22)	B5502V1, B55v	JCBB1366, BY0028	AB032094, AF408166, AF408167	CK Hurley[b]
B*5511	—	—	2000-259-501	AY042674, AY042675	CK Hurley
B*5512	—	—	10002057	AJ420106	A-M Little
B*5601	B56(22)	—	VOO	M77776	
B*5602	B56(22)	—	ENA	M77775	
B*5603	B22	B22N, B56/46	15630Naka, 01300, 01094, NPC-4	D85762, U67746, U67747, U67749, U73113	
B*5604	B56(22)	B*5602Var	5227, 5274	U93911, U93912, U93913, U93914	
B*5605	B56(22)	B56v	234-1047, CBC11028	AF072767, AF072768, AB030574	
B*5606	—	B*7801New, ?B78	20598, AFM	Y18542, Y18543, AJ276993	
B*5607	B56(22)	B*New B56-Bw4	20193, VTIS45561	Y18544, Y18545, AF387903, AF387904	BD Tait[b]
B*5608	—	—	1PF6	AY045733, AY045734	WH Hildebrand

Table 5.3 Designations of HLA-B alleles (*continued*)

HLA alleles	HLA specificity	Previous equivalents	Individual or cell line from which the sequence was derived	Accession number	References or Submitting author(s)
B*570101	B57(17)	—	WIN, MOC, MOLT4	X55711, M32318	
B*570102	B57(17)	—	GN00398	AY050203, AY050204	CK Hurley
B*5702	B57(17)	Bw57.2	32/32	X61707	
B*570301	B57(17)	B*57SAU	SAU, MAME, GB32	U18790, U39088, Y09157	
B*570302	B57(17)	B*57New	E187	AF279663	
B*5704	B57(17)	B-5702v	OPOU	L76096	
B*5705	—	—	GN00213	AF061859, AF061860	
B*5706	—	B*57New	CTM2988653	AF130734	
B*5707	—	—	GN00327	AF202449, AF202450	
B*5708	B57(17)	—	35980	AJ409214	M Guttridge
B*5709	—	—	2000-245-285	AY034804, AY034805	CK Hurley
B*5801	B58(17)	—	WT49, DAUDI, GN00107, 1075011, HGN, KBM	M11799, U52813, U52814, U65395, U65396, AB008102, AJ420241	A-M Little[b]
B*5802	B58(17)	B58v	DAUDI, RCE56, CR-30609	L33923, X86703, AJ133780, AJ133781	
B*5804	—	—	99-2199	AF189245, AF189246, AF189247	
B*5805	—	B*5801V	GN00322	AF201474, AF201475	
B*5806	—	B*5802V	GN003714	AF288046	
B*5901	B59	—	AT, KY, MAS	L07743, D50300	
B*670101	B67	—	HS67, #591, #W7079, PVR	L17005, L76252	
B*670102	B67	B*67LAV	LAV	U18789	
B*6702	—	—	BY00014, BY0026, JH66203	AF321834, AF321835, AY050195, AY050196, AF487379	CK Hurley[b], MS Leffell[b]
B*7301	B73	—	LK707, LE023, HL	U04787, X77658, L24373, AJ311601	A-M Little[b]
B*7801	B78	B'SNA',Bx1	SNA, 32/32, Terasaki Ext#69	X61708, M33573, AJ309192	A-M Little[b]
B*780201	B78	—	RC654	L41214	

Table 5.3   Designations of HLA-B alleles (*continued*)

HLA alleles	HLA specificity	Previous equivalents	Individual or cell line from which the sequence was derived	Accession number	References or Submitting author(s)
B*780202	B78	B78Hen	Hen	X96534, X96533	
B*7803	—	—	GN00209	AF061855, AF061856	
B*7804	—	B*78New	COH#1058	AJ012471, AJ132713, AJ132714	
B*7805	—	B52 variant	B5859	AB051357	
B*8101	B81	B'DT', B*7x48GB, B56b	AP630, GB92, 56B	L37880, X90390, U34810	
B*8201	—	B22x45, B45v, B*82new-64B	MAME, MAMA, MAPA, RB22, VWAR, 64B	U29241, U38800, U36492, U43337, AF017321	
B*8202	—	B*8201New	CEK008AN, VTIS68967	AJ251755, AF525409, AF525410	BD Tait[b]
B*8301	—	B*5603V	GN00298, GN00298	AF176083, AF176084, AF275748, AF275749	

[a]Allele names given in bold type have been assigned since the 2000 Nomenclature report.
[b]This reference is to a confirmatory sequence.
[c]HLA specificity provided from the HLA dictionary (33, 34).

Table 5.4 Designations of HLA-C, -E, -F, -G alleles

HLA alleles	HLA specificity	Previous equivalents	Individual or cell line from which the sequence was derived	Accession number	References or Submitting author(s)
Cw*0102	Cw1	Cw1.2, C1J1	T7527, AP, LCL721, KRC005, TTY, BRUG, LCL721	M84171, Z46809, D50852, M16272, AJ420242	A-M Little[b]
Cw*0103	Cw1	C1J2	ITOU	D64145	
Cw*0104	—	Cw*01/12	J.V	AJ133100	
Cw*0105	—	Cw*01variant	607990	AJ300765, AJ300766	(50)
Cw*0106	—	—	SWC231	AJ418708, AJ418709	(51)
Cw*0107	—	—	VTIS67160	AF525405, AF525406	BD Tait
Cw*020201	Cw2	Cw2.2	MVL	M24030	
Cw*020202	Cw2	Cw2.2	SWEIG, BDG, BRUG, SWEIG007	M26712, D83029, M16273, AJ420243	A-M Little[b]
Cw*020203	Cw2	—	KACD	Z72007	
Cw*020204	Cw2	Cw2.4	HEL299, NM155, NM233, NM239, NM303, NM366, NM72, MAN527, 19215	U88838, U88839, U97346, U97347, Z96924, AJ011881, Y18660, Y18661, Y18144, Y18145	
Cw*020205	—	—	1177	AY028705, AY028706	M Bunce
Cw*0203	—	—	NM3340	AF037449, AF037450	
Cw*0204	—	—	PRC32	AF281055, AF281056	
Cw*0205	—	—	1206	AY028707, AY028708	M Bunce
Cw*030201	Cw10(w3)	—	AP, JG	M84172, AJ011884	
Cw*030202	Cw10(w3)	—	DAUDI	AJ318865	A-M Little
Cw*030301	Cw9(w3)	C3J1	GRC150, SJK	M99390, D50853	
Cw*030302	Cw9(w3)	—	NM2688, NM3499	AF036554, AF036555	
Cw*030303	Cw9(w3)	—	TER#1054	AJ298837	
Cw*030401	Cw10(w3)	C3J2	KRC110, JD, SKA, JG	M99389, D64150, U44064, U31372, U31373	
Cw*030402	Cw10(w3)	—	NM233, NM303, NM366, ML1805	U97344, U97345, AJ133473, AJ133474	
Cw*0305	—	MA083C, Cw*03MAC	MA083C, NM3214, NM3222, PAM	AF016303, AF009683, AJ005199	
Cw*0306	—	—	NM133, NM627, NM2203, NM2415, NM2616	AF003283, AF003284	
Cw*0307	Cw3	—	CTM-7980718	AF039198	

Table 5.4   Designations of HLA-C, -E, -F, -G alleles (*continued*)

HLA alleles	HLA specificity	Previous equivalents	Individual or cell line from which the sequence was derived	Accession number	References or Submitting author(s)
Cw*0308	—	—	NM1931, TER0171	AF037074, AF037075, Y16411, Y16412, Y18656, Y16411, Y16412, Y18142, Y18143	
Cw*0309	Cw3[c]	—	NM4305	AF037076, AF037077	
Cw*0310	Cw3	Cw*03041New	NM4C187, DKM	AF138276, AF138277, AF147701, AF147702	
Cw*0311	—	Cw*03xx	NMDP0187-1868-4	AF145466, AF145467	
Cw*0312	—	—	UCLA022679917	AF172867, AF172868	
Cw*0313	—	Cw*03031var	10050195	AJ298116	(52)
Cw*0314	—	Cw*KCULL	KCULL	AF335314, AF335315	H Dunckley
Cw*0315	—	—	N322	AY078078, AY078079	S Chapple
Cw*04010101	Cw4	C4J1, BeWo C.1	C1R, KSE, BeWo, CJO-A	M84386, X58536, D83030, AJ238694, AJ292559, M26432	
Cw*04010102	—	—	Tersaki EXT40	AJ278494	
Cw*040102	Cw4	Cw*04N	RN1238C	AF002271, AF017322	
Cw*0403	—	Cw4NM, Cw4x6	KWO010	L54059	
Cw*0404	—	rn126C, Cw*0401new	rn126C, NM157, NM187	U88251, AF017323, U96786, U96787	
Cw*0405	—	Cw*0401New	NM2602	AF036556, AF036557	
Cw*0406	—	TREC1, Cw4x6	DM4, MP3	AF062587, AF062588, AF076476	
Cw*0407	—	Cw*0401Variant	ML1805	AJ133475, AJ133476	
Cw*0408	—	Cw*04new	NMDP-0196-1628-3	AF284582, AF284583	
Cw*0409N	Null	Cw4New	CTM6991383, LCL13W09501	AF196489, AF405691	(53), ZC Wang
Cw*0410	Cw4	—	VTIS64141	AF525407, AF525408	BD Tait
Cw*0501	Cw5	Cw5N	QBL, RC, JME, QBL, LB129-SCLC	M58630, L24491, D64148, D83742, AJ010748, Y18146, AJ420244	A-M Little[b]
Cw*0502	Cw5	Cw5New	CTM-5957411	AF047366, AF047367	
Cw*0503	—	Cw*05DZ	BB90-MEL	AF168611	
Cw*0504	—	Cw5New	CTM-4990904	AF173007, AF173008	
Cw*0505	—	—	609648	AJ440717, AJ440718	M Bengtsson
Cw*0602	Cw6	Cw6(W), C6J1	MS, G088, DJS, JOE, JD, TTU	M28206, X70857, Z22752-4, M28160, D64147	

Table 5.4 Designations of HLA-C, -E, -F, -G alleles (*continued*)

HLA alleles	HLA specificity	Previous equivalents	Individual or cell line from which the sequence was derived	Accession number	References or Submitting author(s)
Cw*0603	—	—	NM779	AF019567, AF019568	
Cw*0604	—	Cw6V	MA43, MA95	AB008136	
Cw*0605	Cw6	Cw*06NF	NF	AF105240, AF105241	
Cw*0606	—	—	675/99	AJ277100, AJ277101, AJ277102, AJ277103	
Cw*0607	—	Cw*06DKM	DEDKM	AJ293511	
Cw*070101	Cw7	—	MF, LCL721	M28207, Z46810, Y16418	
Cw*070102	Cw7	Cw*07New	19323	Y18499, Y18533, Y18534, Y18535, Y18536	
Cw*07020101	Cw7	JY328, Cw7J1, Cw7.5	JY, TID, KOK, WEHO	D38526, Z49112, AJ293016	
Cw*07020102	Cw7	—	Terasaki EXT48	AJ293017	
Cw*0703	—	HLA-4	?	M11886	
Cw*070401	Cw7	Cw7/8v	LB33-MEL, KRO3/4, SSA, 40C, 10050195	U09853, X83394, D49552, U38976, AJ291815	A-M Little[b]
Cw*070402	Cw7	—	NDS-HM	AF220290, AF220291, AY064404	M Bunce
Cw*0705	—	39C	39C	U38975	
Cw*0706	Cw7	Cw*07GB	GB92	X97321	
Cw*0707	—	Cw7v	HAUP	Z79751	
Cw*0708	—	RN2157C	RN2157C	AF017330, AF017331	
Cw*0709	—	—	NM388	AF015556, AF015557	
Cw*0710	—	—	NM1279	AF038573, AF038574	
Cw*0711	—	Cw*0704x	LB129-SCLC	AJ010749	
Cw*0712	—	Cw-0704N	TER#877, TER#878, TER#857	U60217, U60218	
Cw*0713	—	Cw*JFOR	JFOR, PFOR	AF144664, AF144665	
Cw*0714	Cw7	—	14783D3	AJ242661	
Cw*0715	—	—	500900	AF316035, AF316036	TM Williams
Cw*0716	—	—	NY00000850	AF480614	F Garcia Sanchez
Cw*080101	Cw8	C8J1	02627, KNM, SFK, HTS, 26/27	M84174, D64151, AJ420246	A-M Little[b]
Cw*080102	Cw8	—	SWN8, PU03	AJ438882, AJ438883, AF510721	(54), M Lin-Chu

Table 5.4 Designations of HLA-C, -E, -F, -G alleles (*continued*)

HLA alleles	HLA specificity	Previous equivalents	Individual or cell line from which the sequence was derived	Accession number	References or Submitting author(s)
Cw*0802	Cw8	—	CGM1, LWAGS, WT51	M59865, M84173	
Cw*0803	Cw8	C8J2	KRC103, SSK	Z15144, D50854	
Cw*0804	Cw8[c]	—	NM313, NM914, C03, TER#876	U96784, U96785, AF016304, AF009684, U60321, U60322	
Cw*0805	—	Cw*08Var	NEQ2A10/97	Y15842	
Cw*0806	—	—	EC22	AF082800, AF082801	
Cw*0807	—	Cw*CCAI	CCAI	AF179631, AF179632	
Cw*0808	—	Cw*0801V	CSR	AF245437	
Cw*0809	—	—	Kolla 34	AJ278509	
Cw*120201	—	Cb-2	MT	M28172	
Cw*120202	—	Cw*1202gyp, C12J1	G085, MSU, AKIBA, E4181324	X70856, D64152, D83741, M21963, D12471, D12472, AJ420247	A-M Little[b]
Cw*120203	—	Cw*PBAG	PBAG	AF189725, AF189726	
Cw*120301	—	Cw12New, C12J12	D0208915, WDV, YAR, GB002, HNT, JBUSH	U06695, U06696, X82122, D64146, AJ420248	A-M Little[b]
Cw*120302	—	—	PI151	AF289031	
Cw*120401	—	Sy/9-2	M.H(9-2)	X99704	
Cw*120402	—	Cw*12JD,	NDS-JD, NM2018	Y11843, AF015558, AF015559	
Cw*1205	—	Cw12x16	ANDP	Z80228, Z83247	
Cw*1206	—	—	NM1699	AF036552, AF036553	
Cw*1207	—	—	Atuwagu, Atuwaogu	AJ249163, AJ249164	
Cw*1208	—	Cw*12new	10030006	AJ304496	(52)
Cw*140201	—	—	LUY, TC106, LKT2	U06487, Z47377, U41386, D49820, M28171,	
Cw*140202	—	—	NM1991	AF015554, AF015555	
Cw*1403	—	Cx4	TID, DK1	D31817, AJ420249	A-M Little[b]
Cw*1404	—	—	CTM-1986765	AF104218, AF104219	
Cw*1405	—	Cw*1402v	NMDP0121-0146-5	AJ306617, AJ306618	(50)
Cw*150201	—	C*X, Cw*6.2, Cl.9, Cw15J	AUCA#2, G085, G088, KUE, GM637, BOB	L20091, X67818, D83031, M24096, AJ420250	A-M Little[b]
Cw*150202	—	Cw*1502new	NM4C376	AF139727, AF139728	

Table 5.4 Designations of HLA-C, -E, -F, -G alleles (*continued*)

HLA alleles	HLA specificity	Previous equivalents	Individual or cell line from which the sequence was derived	Accession number	References or Submitting author(s)
Cw*1503	—	—	GRC150	M99388	
Cw*1504	—	Cw*15Sp	C047	X73518	
Cw*150501	—	Cw*15v	LE023	X78343	
Cw*150502	—	Cw*1505v	L7901	X87841	
Cw*1506	—	Cw*15N	M001C, NM2732, JF	AF002270, AF017324, AF036550, AF036551, Y15745, Y15746, AJ011882 Y15746, Y15745, Y18140, Y18141	
Cw*1507	—	—	PUSPAN	Y17064, Y17065	
Cw*1508	—	Cw*15P	Peru-15	AJ010322, AJ010323	
Cw*1509	—	Cw*1504New	NM4C159	AF165850, AF165851	
Cw*1510	—	—	SLGJ	AF302133, AF302134	
Cw*1511	—	Cw*KDILL	CBM2598	AF335316, AF335317	H Dunckley
Cw*1601	—	Cl.10	GM637, TC106, PITOUT	M24097, U41420, U56259, U56260, AJ420251	A-M Little[b]
Cw*1602	—	Cw*16v	C073	X76189	
Cw*160401	—	rn183C, wt30L	BOJ, rn183C, wt30C, NM290, NM633, 4136	Z75172, U88252, AF017326, U88253, AF017325, U96788, U96789, AJ011883, Y18657, Y18658, Y18659, Y18139	
Cw*1701	—	Cw16New	RSH, GB86, BM21	U06835, X98742, Y10520, AJ420252	A-M Little[b]
Cw*1702	—	Cw17N	KSU	D64149	
Cw*1703	—	Cw*17New	17767	Y18537, Y18538, Y18539, Y18540, Y18541	
Cw*1801	—	Cw*04GB, Cw4x6	GB92, DIJL, TERASAKI926	X96582, Z80227, AJ420253	A-M Little[b]
Cw*1802	—	Cw*18GB	GB32	Y09156	
E*0101	—	JTW15	JT, YN, HF, SPAARN70	M20022, L78934	
E*0102	—	HLA-6.2	LCL721	M21533	
E*010301	—	M32507, E*01C230	MT, MH, TK, SPAARN70, CHI009, JFE, CR	M32507, L78455, X87678, X87679, L78455, AJ002533, AJ002534	
E*010302	—	E*01T230	MSC, CHI004, 17771	X87680, X87681, L79943	

Table 5.4   Designations of HLA-C, -E, -F, -G alleles (*continued*)

HLA alleles	HLA specificity	Previous equivalents	Individual or cell line from which the sequence was derived	Accession number	References or Submitting author(s)
E*010303	—	—	CD	AJ293263	
E*0104	—	M32508	KS	M32508	
F*0101	—	HLA-5.4	LCL721.144	X17093	
G*010101	—	HLA-6.0, G*I, GCO1	LCL721.144, ASR53, MOU, SPO010, YRK, HT68	J03027, X17273, L27836, L27837, D77998, D77999, D78000, U76216, U76217	
G*010102	—	BeWo G7, G*II, GJ2, GCO2	BeWo, COX, DHIF, WT47, STK, HT43, TB250	M32800, X60983, L07784, L41392, D85032, D67009, D67010, D67011, U65245, U65246, U88244	
G*010103	—	G*IV, GJ4, GCO5	BeWo, KKH, HT147	L07784, L20777, L41363, D67003-5, D85033, U65235, U65236	
G*010104	—	G*0101d, GCO4	HT180	U65233, U65234	
G*010105	—	CEPH G1	1305	U58024	
G*010106	—	CEPH G5	2702	U58027	
G*010107	—	CEPH G6	3101	U58028	
G*010108	—	CEPH G7	3102	U58029	
G*0102	—	Ice 6.23-5.4H	ICE 6	S69897	
G*0103	—	G*III, GCO9	LWAGS, HT59	L20777, U65241, U65242	
G*010401	—	GJ3, GCO7, CEPH G2	KMR, CHI525, HT98, 1302	D67006, D67007, D67008, L78072, U65237, U65238, U58025	
G*010402	—	CEPH G3	2701	U58094	
G*010403	—	CEPH G4	2701	U58026	
G*0105N	—	G*1.5	DCH027	L78073	
G*0106	—	—	050900cA537	AF312697	

[a]Allele names given in bold type have been assigned since the 2000 Nomenclature report.
[b]This reference is to a confirmatory sequence.
[c]HLA specificity provided from the HLA dictionary (33, 34).

Table 5.5 Designations of HLA-DR alleles

HLA alleles	HLA-DR serological specificities	HLA-D-associated (T cell defined) specificities	Previous equivalents	Individual or cell line from which the sequence was derived	Accession number	References or Submitting author(s)
DRA*0101	—	—	DRα, PDR-α-2	JY, RAJI, F.G.	J00194, J00196, J00203	
DRA*010201	—	—	DR-H	JY	J00201, AF481359	(55)[b]
DRA*010202	—	—	—	HSF7	Z84814	S Williams
DRB1*010101	DR1	Dw1	—	45.1, LG2, JSA, DRH, CHG	X03069, M11161, AF029288	
DRB1*010102	—	—	—	9380965	AF479570	J Wu
DRB1*010201	DR1	Dw20	DR1-NASC	NASC, 1568, MUM	AF029293	
DRB1*010202	DR1	Dw20	DRB1*01DMT	TO0973	Z50871	
DRB1*0103	DR103	Dw'BON'	DR1-CETUS, DRB1*BON	RAI, BG, BON	M33600	
DRB1*0104	DR1	—	DRB1*01New	L.R., LAUTH J	X70261, X99896	
DRB1*0105	—	—	DRB1*0101V1	JC10218	AB015184	
DRB1*0106	—	—	—	MGM14106	AJ089723	
DRB1*0107	—	—	DRB1*New	ZAE, IOL Gae, IOL Ire	AJ276206, AJ303118	A Dormoy[b]
DRB1*0108	—	—	DR1-BCN	HSP934010	AY034875	(56)
DRB1*030101	DR17(3)	Dw3	dJ93N13	RAJI, AVL, WT49, DM24, DM28, DM29, CMCC, HSF7, APR, ALL, MVJ, MUR, U-STH	M17379, X04054, Z84489, AF029265, AF152843	
DRB1*030102	DR17(3)	Dw3	DRB1*IMR	21, M.R.	M91807, L07767	
DRB1*030201	DR18(3)	Dw'RSH'	—	2041, 1563, 24A1	M27689, AF029266	
DRB1*030202	DR18(3)	Dw'RSH'	—	GN055, GMONT	U29342, U82403	
DRB1*0303	DR18(3)	—	—	RBL B25	M81743	
DRB1*0304	DR17(3)	—	03MIT	MIT3758, 35919	X75441, AJ409216	M Guttridge[b]
DRB1*030501	DR17(3)	—	DR3New	U-HFI, TTO5607	L29807, U26557	
DRB1*030502	—	—	—	LAHRE	AF335318	(57)
DRB1*0306	DR3	—	—	JV1094	X90644	

Table 5.5 Designations of HLA-DR alleles (*continued*)

HLA alleles	HLA-DR serological specificities	HLA-D-associated (T cell defined) specificities	Previous equivalents	Individual or cell line from which the sequence was derived	Accession number	References or Submitting author(s)
DRB1*0307	DR3[c]	—	—	GN073	U37433	
DRB1*0308	—	—	—	GN090	U47028	
DRB1*0309	—	—	—	D438	X93315	
DRB1*0310	DR17(3)[c]	—	—	PMR	U65585	
DRB1*0311	DR17(3)[c]	—	—	UWE02	U79028	
DRB1*0312	DR3	—	DRB1*03AGC	WVN	Y17274	
DRB1*0313	—	—	—	DELA	AJ012424	
DRB1*0314	DR3	—	DR'KW'	KW	Y17863	
DRB1*0315	—	—	DRB1*0301A	DKMS 585607	AJ237899	
DRB1*0316	—	—	—	09343336	AF169240	
DRB1*0317	—	—	DRB1*13KM	SMS202-147-KerHut	AJ238154	
DRB1*0318	—	—	DRB1*03XX	RSA036575, MSA058812	AJ279010	
DRB1*0319	—	—	—	GCASS	AF343002	H Dunckley
DRB1*0320	—	—	DRB1*03011var	NT0022	AF352294	CK Hurley
DRB1*0321	—	—	—	Patient#17839	AJ297266	(58)
DRB1*0322	—	—	—	MAWE0816AN	AJ420288	A-M Little
DRB1*0323	—	—	—	DNA6060	AY116505	A Reil
DRB1*040101	DR4	Dw4	—	WT51, PRIESS, MJ4, BOLETH, LTC	K02776, M17381, M20548-50, AF029267	
DRB1*040102	DR4	Dw4	—	MC	X96851	
DRB1*0402	DR4	Dw10	—	FS, DM24, MMCC, LPB, YAR	M15068, AF029268, AJ245881, J297586	SGE Marsh[b]
DRB1*040301	DR4[c]	Dw13	DR4 Dw13A, 13.1	SSTO, TAS, NBP	AF029269	
DRB1*040302	DR4	Dw13	DRB1*SD	BM1 116040, 32891	AF112876, AJ295845	
DRB1*0404	DR4	Dw14	DR4 Dw14A, 14.1	BIN40, LS40, DM29, RGR	X02902, M15069, M15073, M15074, AF029270	
DRB1*040501	DR4	Dw15	—	KT3, JML, AHC, CRP, DOS	M15070, L13875, AF029271	

Table 5.5   Designations of HLA-DR alleles (*continued*)

HLA alleles	HLA-DR serological specificities	HLA-D-associated (T cell defined) specificities	Previous equivalents	Individual or cell line from which the sequence was derived	Accession number	References or Submitting author(s)
DRB1*040502	DR4	Dw15	DRB1*KOM	KOM	D50889, D49952	
DRB1*040503	DR4	—	DRB1*JVASA	JVASA	AF450094	(59)
DRB1*040504	—	—	—	GN00419	AY094139	CK Hurley
DRB1*0406	DR4	Dw'KT2'	—	KT2, 43A3	AF029272	
DRB1*040701	DR4	Dw13	DR4 Dw13B, 13.2	JHF, R88, JRR	M37771, AF029273	
DRB1*040702	DR4		DRB1*0407var	NT0019	AF352291	(60)
DRB1*0408	DR4	Dw14	DR4-CETUS, Dw14B, 14.2	M36, RA1, SUDNA0254, RGR	M37770, L78169, AF029274	
DRB1*0409	DR4	—	—	R80	M64794	
DRB1*0410	DR4	—	DR4.CB	CB, ABCC60, EGR	M81670, M80192, AF029275	
DRB1*0411	DR4	—	DR4.EC EC,	HV846, HAA, JMJ	M81700, M55615, L42143, L79973	
DRB1*0412	—	—	AB2	ABO1078	M77672	
DRB1*0413	DR4	—	DRB1*LEV	LEV	M94460	
DRB1*0414	DR4	—	DR4 Dw10.2	VK	X65031	
DRB1*0415	DR4	—	—	NIC, HOU	X68272	
DRB1*0416	DR4	—	DR4-BELF	BEL5GB	X70788	
DRB1*0417	DR4	—	DRB1*04SAM	TOB-0070	L14481	
DRB1*0418	—	—	DRB1*04.N	AI7, AI8, 74DR	X71610, U38974	
DRB1*0419	DR4	—	DR4FK	FK	L21985	
DRB1*0420	DR4	—	DRB1*04MC	AD-7863, BM29/92	L27217	
DRB1*0421	DR4	—	DR4New	SMH	X80288	
DRB1*0422	DR4	—	DR4New	D18002	U17014	
DRB1*0423	DR4	—	—	MAG	Z68503	
DRB1*0424	DR4	—	DRB1*Mi	Mi	Z71541	
DRB1*0425	DR4	—	DRB1*04ISA	RI, HB	Y09211	
DRB1*0426	DR4	—	DRB1*04CMT	T010148	AJ001252	
DRB1*0427	—	—	—	NOR03	AF030439	
DRB1*0428	DR4	—	DRB1*0405V1	JC4772	AB007635	
DRB1*0429	DR4	—	DRB1*0405V2	JC7616	AB007636	

Table 5.5   Designations of HLA-DR alleles (*continued*)

HLA alleles	HLA-DR serological specificities	HLA-D-associated (T cell defined) specificities	Previous equivalents	Individual or cell line from which the sequence was derived	Accession number	References or Submitting author(s)
DRB1*0430	—	—	DRB1*0405V3	JC9227	AB015185	
DRB1*0431	DR4[c]	—	DRB1*04New	GE47192	AJ009755	
DRB1*0432	DR4[c]	—	DRB1*04-A	NIE	Y17273	
DRB1*0433	—	—	DRB1*04_7468	WBD7468	AF023153	
DRB1*0434	—	—	DRB1*04new	CB1653	AJ133492	
DRB1*0435	—	—	DRB1*04New	NT0009	AF242355	
DRB1*0436	—	—	—	BN61	AF240637	
DRB1*0437	—	—	DRB1*04nv	MDPH0002764	AY007565	
DRB1*0438	—	—	—	SLTA, VTIS72428	AF235034, AF489510	M Varney[b]
DRB1*0439	—	—	DRB1*04031var	NT0024	AF352296	(60)
DRB1*0440	—	—	DRB1*0404var	NT0020	AF352292	(60)
DRB1*0441	—	—	DRB1*04031var	NT0021	AF352293	(60)
DRB1*0442	DR4	—	—	JH71321	AF304866	(61)
DRB1*0443	—	—	—	OORCH18	AY042678, AF349316	(60), J Tang
DRB1*0444	—	—	—	satt44124	AF497643	I Humphreys
DRB1*070101	DR7	Dw17, Dw'DB1'	—	BURKHARDT, MANN, LBF	M16941, M17384, U09201	
DRB1*070102	DR7	—	DRB1*07New	CBM500	AJ243327	
DRB1*0703	DR7	—	DRB1*07RMT	ED01436	Y13785	
DRB1*0704	DR7[c]	—	DRB1*07ROS	12827878	Y16224	
DRB1*0705	—	—	—	NT0012	AF327742	(62)
DRB1*0706	—	—	—	13765	AJ311892	JHM Cohen
DRB1*080101	DR8	Dw8.1	DRB1*0801	MADURA, SUDNA0140, U-STH, BM9, MTP1 134873, MULRe, 1823-T, BTB	M17386, L78166, AF144105, AF121971, AJ249626, AF278701, AY028514, AY028515, AY028516, AY028517, AY028518, AY028519	(63)[b]
DRB1*080102	—	—	—	GN00415	AF491843	CK Hurley
DRB1*080201	DR8	Dw8.2	DRw8-SPL	SPL, 24A2	AF029277	

Table 5.5 Designations of HLA-DR alleles (*continued*)

HLA alleles	HLA-DR serological specificities	HLA-D-associated (T cell defined) specificities	Previous equivalents	Individual or cell line from which the sequence was derived	Accession number	References or Submitting author(s)
DRB1*080202	DR8	Dw8.2	DRw8b	OLL, C-78	AF029278	
DRB1*080203	DR8	—	—	NT0014	AF327743	CK Hurley
DRB1*080302	DR8	Dw8.3	DRw8-TAB	KT, FO, POPE, TAB089	M27511, AJ001094	
DRB1*080401	DR8	—	RB1066-1,DR8-V86	1066, 1127, PM, MTR	M84446, M34315, AF029279	
DRB1*080402	DR8	—	—	CAY3, CAY5, CAY92, CAY96	L10402	
DRB1*080403	DR8	—	—	UWEH03	U88135	
DRB1*080404	DR8	—	—	NT0016	AF330103	CK Hurley
DRB1*0805	DR8	—	DR8-A74	MS	M84357	
DRB1*0806	DR8	—	DR8.6	RBL B24, RBL B124, SET, BOU, ALG, C.R., SUDNA0095	M87543, M86590, Z32685, L78165	
DRB1*0807	DR8	—	DR8BZ	AG, RG, L2, L4, TIC03, TIC04, TIC06	L22341, L28096	
DRB1*0808	—	—	08New	ETH3754	X75443	
DRB1*0809	R8	—	DR8.7, DRB1*8.2V	BRI-10, JB44585	L23987, D45046, AB046526	
DRB1*0810	DR8	—	LP10-1	K.R., R.R., TH10559	L19054, X82553	
DRB1*0811	DR8	—	DR8TL, DR8New	ARA016, ARAC25, JR	L29082, L32810	
DRB1*0812	DR8	—	DRB#52	4390, DRB#52	X88854, U36836	
DRB1*0813	—	—	DRB#47	DRB#47, 29168	U36571, AJ495001	T Gervais[b]
DRB1*0814	DR8	—	DR8WE	WE, KE	U24179	
DRB1*0815	—	—	DRB1*08Taree	TDS-023	U63802	
DRB1*0816	DR8	—	DRB1*08JST	ML0273, 24131	X99840, AJ309930	M Guttridge[b]
DRB1*0817	DR8	—	DRB1*08LRT	RV0253	Y09665	
DRB1*0818	—	—	HLAAL1, HLA-DR8.5va	DKM379804, dJAE-0173, DU32971	U96926, Z99006, AJ223124	
DRB1*0819	—	—	DRB1*08YF, DRB1*08BL	VBD21599B, RP-BL046	AF016225, AF028011	
DRB1*0820	—	—	DRB182624	82624	AJ000927	

Table 5.5   Designations of HLA-DR alleles (*continued*)

HLA alleles	HLA-DR serological specificities	HLA-D-associated (T cell defined) specificities	Previous equivalents	Individual or cell line from which the sequence was derived	Accession number	References or Submitting author(s)
DRB1*0821	—	—	—	ROD01	AF049875	
DRB1*0822	—	—	DRB1*08New	R9846, R9028	AJ276711	
DRB1*0823	—	—	DRB1*08032V1	JCB13444	AB049829	
DRB1*0824	—	—	DRB1*08022var	GN00391	AF363728	L Burdett
DRB1*090102	DR9	Dw23	—	DKB, 09012, PMR, ISK	M17387, U66826, D89917	
DRB1*0902	—	—	—	J69	AY043181	(64)
DRB1*100101	DR10	—	—	RAJI, NASC	M20138	
DRB1*100102	DR10	—	DRB1*10New	AW10-LCL	AF225565	
DRB1*110101	DR11(5)	Dw5	DRw11.1	SWEIG	M11867	
DRB1*110102	DR11(5)	Dw5	—	1180, 1249	M34316	
DRB1*110103	DR11(5)	Dw5	DR11.MD, DRB1*11DCT	DR11MDA, DR11MDB, BV3402	X86803, Y07590	
DRB1*110104	DR11(5)	—	—	NT0015	AF329281	(62)
DRB1*1102	DR11(5)	Dw'JVM'	DRw11.2	JVM, LTI	M17382, AF029280	
DRB1*1103	DR11(5)	—	DRw11.3	UA-S2	M21966, M22047-49	
DRB1*110401	DR11(5)	Dw'FS'	—	FPA (FPF), 34A2, FPF	AF029281, AJ297587	SGE Marsh[b]
DRB1*110402	DR11(5)	—	—	2094, 17A1	M34317, AF029282	
DRB1*1105	DR11(5)	—	—	DBUG	M84188	
DRB1*1106	DR11(5)	—	DR11.CCY, 11PMH	CCY, PMH161	M98436, D14352	
DRB1*1107	DR11(5)[c]	—	DR11+3	BEL6KG, RMS21	X73027, X82507	
DRB1*110801	DR11(5)	—	DR11JL	JL	L21984	
DRB1*110802	DR11(5)	—	DR11HW	HW	L21983	
DRB1*1109	DR11(5)	—	DRB1*MON	BEL7MON	X75347	
DRB1*1110	DR11(5)[c]	—	DR11.5	BRI-6	L23986	
DRB1*1111	DR11(5)[c]	—	DR11.6, DR11BRA	BRI-7, 1082	L23990, L26306	

Table 5.5 Designations of HLA-DR alleles (*continued*)

HLA alleles	HLA-DR serological specificities	HLA-D-associated (T cell defined) specificities	Previous equivalents	Individual or cell line from which the sequence was derived	Accession number	References or Submitting author(s)
DRB1*111201	—	—	DR11.7	BRI-9, 008	L23988, AF234175	
DRB1*111202	DR11(5)	—	—	SWP71	AJ251984	
DRB1*1113	DR11(5)	—	DR11-14, DR11+14	PAL-6117, 30251, EmKa, SB, BV0595, JOK	X76194, L29081, U09200, U03291, Z37162, X87677	
DRB1*1114	DR11(5)	—	F1363, 115T, 94-09865	BRI-11, HN0605, DJB, BEN, 12762	U08932, Z37161, U25639, Z50187, AJ245714	
DRB1*1115	—	—	DR1101v	Z.S., Z.Z., Z.Z.V., GN041, GN037	Z34824, U17380	
DRB1*1116	DR11(5)ᶜ	—	DRB1*OULA, DR11+13	OULA, HB7542AKG	U13009, X87200	
DRB1*1117	—	—	UCSF-D3152, DR11-14N, 0104D0335	D3152, D3153, GN032, 950104-D0335	X77776, U17379, U33474	
DRB1*1118	—	—	RMS16	RMS16	X82211	
DRB1*1119	DR11(5)ᶜ	—	RMS117, DR11Loel	RMS117, MB, KBD	X82210, Z47353, U26558	
DRB1*1120	DR11(5)	—	—	CV	U25442	
DRB1*1121	DR11(5)	—	—	MUL	X86976	
DRB1*1122	—	—	—	ZL3096	Z49113	
DRB1*1123	DR11(5)	—	DRB1*11OS	YAS	D49468	
DRB1*1124	—	—	7CGCE	JB, DZA95-7C	X89193, Z50746	
DRB1*1125	DR11(5)	—	DR11x08	SimE, TAR	X91823, X97291	
DRB1*1126	DR11(5)	—	DRB.W11	WAN	X94350	
DRB1*112701	DR11(5)	—	2166/1018	M.K.	X95656	
DRB1*112702	DR11(5)	—	DRB1*11New	E404, E405, E434, NMDP0361-0724-1	AF186407, AF186408, AJ401148	
DRB1*1128	—	—	DRB1*11Var	LELIEAM, 980102	X97722, AF047350	
DRB1*1129	DR11(5)	—	DRB1*11PBT	CL1281, 21690	X99841, AJ245715	
DRB1*1130	—	—	—	GN00153	U79027	
DRB1*1131	—	—	DRB1*VIC	CTM4065412	U72064	
DRB1*1132	—	—	MANDRAY Arlette	MA96401984	AF011786	

Table 5.5 Designations of HLA-DR alleles (*continued*)

HLA alleles	HLA-DR serological specificities	HLA-D-associated (T cell defined) specificities	Previous equivalents	Individual or cell line from which the sequence was derived	Accession number	References or Submitting author(s)
DRB1*1133	—	—	DR11New, DRB1*JG	DU13673, BM1 101910	AF034858, AF112877	
DRB1*1134	—	—	—	GN00236	AF081676	
DRB1*1135	—	—	DRB1*TG	DIA3 128504	AF112878	
DRB1*1136	—	—	DRB1*1102v	NT0001	AF144081	
DRB1*1137	—	—	DRB1*11LF	LIFU	AJ249726	
DRB1*1138	—	—	DRB1*CB3202	CB3202	AF247534	
DRB1*1139	—	—	DRB1*CB1801	CB1801, DKM649157	AF267639, AJ404618	
DRB1*1140	—	—	DRB1*11MMK	TO05334	AJ289124	
DRB1*1141	—	—	DRB1*1103v	NT0011	AF280436	
DRB1*1142	—	—	—	FPO	AJ306404	A Dormoy
DRB1*1143	—	—	DRB1*CB4551	CB4551	AF450093	(59)
DRB1*120101	DR12(5)	Dw'DB6'	DRB1*EBROW	HERLUF, FO, HK, POPE, SWS53, EBROW	M27635, M27509, S48645, AJ293695, AJ293696, AF335319, AF335320	(63)[b]
DRB1*120102	DR12(5)	—	—	BS464263	AJ293725, AJ302075	(65)
DRB1*120201	DR12(5)	—	DRw12b	KI	M27510	
DRB1*120202	DR12(5)	—	DRB1*1202X	BP-9, BP-21	L34353	
DRB1*120302	DR12(5)	—	DRB1*12JBT	T00341	X83455	
DRB1*1204	DR5[c]	—	MHT#12v	MHT#918	U39087	
DRB1*1205	DR12(5)	—	—	JC2862	D86503	
DRB1*1206	DR12(5)	—	DRB1*12XX	K-KT	U95989, AF017439	
DRB1*1207	—	—	DRB1*TCOX	TCOX	AF315825, AF316619	
DRB1*1208	—	—	DRB1*12 variant	13365831	AY033428	(38)
DRB1*130101	DR13(6)	Dw18	DRw6a I, DR1301Var	HHKB, APD, W468R, W468D	M17383, X04056, U83583	
DRB1*130102	DR13(6)	—	DRB1*13new	19783VO	AJ271206	
DRB1*130201	DR13(6)	Dw19	DRw6c I, DR1302Var	WT46, CMCC, AS, W556R, W556D	L76133, U83584	

Table 5.5 Designations of HLA-DR alleles (*continued*)

HLA alleles	HLA-DR serological specificities	HLA-D-associated (T cell defined) specificities	Previous equivalents	Individual or cell line from which the sequence was derived	Accession number	References or Submitting author(s)
DRB1*130202	DR13(6)	—	DRB1*RMAY, FM99/810	RMAY, FM99/810	AF176834, AF217961	
DRB1*130301	DR13(6)	Dw'HAG'	—	HAG, MRS, EGS, OSC, MGA, JRS, 1181, 1183, 2708, IH, JS, MD, SK	X52451, X16649, M59798, M57599	
DRB1*130302	DR13(6)	Dw'HAG'	—	11118-CMN, 22127-EC	U41634, U34602	
DRB1*1304	DR13(6)	—	RB1125-14	1124, 1125	M59803	
DRB1*1305	DR13(6)	—	DRw6'PEV',	TA, JP, HS, BP, DES.DI, SUDNA0165, 17A2	M57600, L78167, AF029283	
DRB1*1306	DR13(6)	—	DRB1*13.MW	MW	M81343	
DRB1*130701	DR13(6)	—	DRB1*JJY, DRB1*SHN	JJY, SHN, SLIR1-13	L06847, D13189, AF305212	
DRB1*130702	DR13(6)	—	—	GN00185	AF036944	
DRB1*1308	DR13(6)	—	—	THA	L03531	
DRB1*1309	—	—	DRB1*YUN	MJD	L23534	
DRB1*1310	DR13(6)	—	13NEW	ARA, 13345532, 13976036	X75442, AJ245716, AJ409215	M Guttridge[b]
DRB1*1311	DR13(6)	—	1303-Like	H108, HER-2698, 1083933x	X74313, X75445, AJ243898	
DRB1*1312	DR13(6)[c]	—	DR13BRA, DR13.7	650, 651, 681, BRI-8, N170, CC75, AD-6168, DNAQC012, RMS103	L25427, L23989, D29836, L27216, X82508	
DRB1*1313	DR13(6)[c]	—	DRB1*13/8	NORH01, NORH02, XX406	U79025, U79026, Y17272	
DRB1*131401	DR13(6)	—	1101A58, 13New	BRI-12, YAS, 11684232	U08274, X82239, AJ245717	
DRB1*131402	DR13(6)	—	DRB1*13MJ	31854	AJ243897	
DRB1*1315	—	—	83-7601	BRI-14, GN070	U08276, U32325	
DRB1*1316	DR13(6)	—	DRB1*D86	BRI-15, JA	U08277, U25638	
DRB1*1317	DR13(6)	—	RB1194 13/12	RB	U03721	
DRB1*1318	DR13(6)	—	DRB1*13HZ	K27418, TH10913, ZAN FR	Z36884, X82549, Z48631	

Table 5.5 Designations of HLA-DR alleles (*continued*)

HLA alleles	HLA-DR serological specificities	HLA-D-associated (T cell defined) specificities	Previous equivalents	Individual or cell line from which the sequence was derived	Accession number	References or Submitting author(s)
DRB1*1319	DR13(6)[c]	—	DR1308V	GN033	U17381	
DRB1*1320	DR13(6)	—	DRB1*13VHT, DRB1*13PL	SR0300, 10843566	Z48803, Y17695	
DRB1*1321	—	—	DR13TAS	ATAS	L41992	
DRB1*1322	DR13(6)	—	—	GvdP, LI3936	X86326, X87886	
DRB1*1323	—	—	—	GN079	U36827	
DRB1*1324	—	—	—	GN039	U36825	
DRB1*1325	—	—	—	MRN5981	X93924	
DRB1*1326	—	—	DRB1*16WIL, DRB1*14/16New	WIL3966, B.A-B	X96396, Y11462	
DRB1*1327	DR13(6)	—	DRB1*13MS, DRB1*13NW	NVE 802	Z71289, U59691, X97601	
DRB1*1328	—	—	—	DU25503	X97407	
DRB1*1329	DR6	—	—	JC6267	D87822	
DRB1*1330	—	—	DRB1*13DAS	DAS 094	U72264	
DRB1*1331	—	—	—	GN00133, GN00138	U88133, U88134	
DRB1*1332	—	—	DR13MC	AD-2111	U97554	
DRB1*1333	—	—	DRB1*13TMT	OTO1567	AJ001254	
DRB1*1334	—	—	—	974770	AF048688	
DRB1*1335	—	—	DRB1*13Var	GN00266-FV2397	AF136155	
DRB1*1336	DR13(6)	—	DR'RD', DRB1*JSMA	RD-DJ, AA-DJ, JSMA, 30638	AF089719, AF195786, AJ293898	
DRB1*1337	—	—	DRB1*13New	GN00256, NT0003	AF169238, AF164346	
DRB1*1338	—	—	DRB1*13New	031188956	AF169239	
DRB1*1339	—	—	DRB1*13PSB	KMDP01-415	AF170582, AF104018	
DRB1*1340	—	—	DRB1*13JP	NE3114, NE3005	AJ237964	
DRB1*1341	—	—	DRB1*Laton	Laton	AJ249591	
DRB1*1342	DR13(6)	—	DRB1*1318V	NT0010, AN3SP6	AF243537, AF288212	
DRB1*1343	—	—	DRB1*14New	GN00221	AF243538	

Table 5.5 Designations of HLA-DR alleles (*continued*)

HLA alleles	HLA-DR serological specificities	HLA-D-associated (T cell defined) specificities	Previous equivalents	Individual or cell line from which the sequence was derived	Accession number	References or Submitting author(s)
DRB1*1344	—	—	DRB1*GDES	GDES	AF247533	
DRB1*1345	—	—	—	SG606319	AJ276873	
DRB1*1346	—	—	DRB1*AHAW	AHAW	AF306862	
DRB1*1347	—	—	DRB1*1307V1	JCB12184	AB049459	
DRB1*1348	—	—	—	20281	AJ401236	A Moine
DRB1*1349	—	—	DRB1*1312var	NT0023	AF352295	(62)
DRB1*1350	—	—	—	1DM4038S1	AY048687	Y-J Lee
DRB1*1351	—	—	—	LPC14	AF441789	M Lin-Chu
DRB1*1352	—	—	—	R.171	AF499445	SG Rodriguez Marino
DRB1*140101	DR14(6)	Dw9	DRw6b I	4/w6, TEM, 15B1	X04057, AF029284, AJ297582	(66)[b]
DRB1*140102	DR14(6)	—	DRB1*14ML	BV17214	AJ289123	
DRB1*1402	DR14(6)	Dw16	—	AMALA (LIA,AZL)[e], 15B3	AF029285, AJ297583	(66)[b]
DRB1*1403	DR1403	—	JX6	MI	AJ297584	(66)[b]
DRB1*1404	DR1404	—	DRB1*LY10, DRw6b.2	CEPH-137502, KGU	M58632, AJ297585	(66)[b]
DRB1*1405	DR14(6)	—	DRB1*14c	36M, 38M, SUDNA0503, GN00402, GN00404	M60209, L78168, AY050209, AY050210	(67)[b]
DRB1*1406	DR14(6)	—	DRB1*14.GB, 14.6	GB, SAS5041, SAS9080, SUDNA0164, 24A3, GN00405, GN00407	M63927, M74032, L78164, AF029286, AY050211, AY050214	(67)[b]
DRB1*140701	DR14(6)	—	14.7	PNG141, PNG196, 43A1, GN00400, GN00401	M74030, AF029287, AY050207	CK Hurley[b]
DRB1*140702	DR14(6)	—	—	GN00403	AY052549	CK Hurley
DRB1*1408	DR14(6)[c]	—	AO1,14.8	HV178, PNG198, PNG202, GN00409	M77673, M74031, AY052550	(67)[b]
DRB1*1409	—	—	AB4	1103	M77671	
DRB1*1410	DR14(6)	—	AB3	ABCC31	M77670	

Table 5.5 Designations of HLA-DR alleles (*continued*)

HLA alleles	HLA-DR serological specificities	HLA-D-associated (T cell defined) specificities	Previous equivalents	Individual or cell line from which the sequence was derived	Accession number	References or Submitting author(s)
DRB1*1411	DR14(6)	—	DRw14x11	MARBrun, MARMari, MARMarg	M84238	
DRB1*1412	DR14(6)	—	DRB1*YOS	YOS	D16110	
DRB1*1413	DR14(6)	—	—	GRC138	L21755	
DRB1*1414	DR14(6)	—	DRB1*14N	AD-2927, AD-3798, IHL AD036	L17044	
DRB1*1415	DR8	—	DRB1*14af	D.M.	U02561	
DRB1*1416	DR6	—	DR13+14	FVA-0166	X76195	
DRB1*1417	DR6	—	1412T	#15310-LN	X76938	
DRB1*1418	DR6	—	81-4641	BRI-13, TH6994, DR14BBD	U08275, X82552, U37264	
DRB1*1419	DR14(6)	—	DRB1*14MA, DRB.14a	MA-TE, AKKAL	Z38072, X86973	
DRB1*1420	DR14(6)	—	DRB.14o	OND-52971	X86974	
DRB1*1421	DR14(6)[c]	—	DRB.14t	TGI	X86975	
DRB1*1422	DR14(6)[c]	—	DRB1*BA	LS005, BA	Z50730, Z71275	
DRB1*1423	—	—	DRB1*14	#66820, SAR	X91640, Z84375	
DRB1*1424	—	—	BY14V, BRAVOG, DRB1*14Pal	BY00002, HDB, PALT, SERL	U41489, AJ000900, AF052574	
DRB1*1425	—	—	HL14V	HL.BWH, MF.BWH	U41490, U41491	
DRB1*1426	DR14(6)	—	—	JC1980	D86502, D50865	
DRB1*1427	DR14(6)	—	—	MO52	D86504	
DRB1*1428	—	—	DRB1*14DKT	TO4138	X99839	
DRB1*1429	DR14(6)	—	—	JC6094	D88310	
DRB1*1430	—	—	DRB1*14CB	CB-254	U95115	
DRB1*1431	—	—	DRB1*14JV	RP-JV129	AF028010	
DRB1*1432	—	—	DRB1*14JW	GAIB	AJ010982	
DRB1*1433	—	—	DRB1*LAM	CB1 116643	AF112879	
DRB1*1434	—	—	—	R98-333250Q	AF172071	
DRB1*1435	—	—	DRB1*SDAV	SDAV	AF177215	
DRB1*1436	—	—	DRB1*New	IHL	AJ242985	

Table 5.5 Designations of HLA-DR alleles (*continued*)

HLA alleles	HLA-DR serological specificities	HLA-D-associated (T cell defined) specificities	Previous equivalents	Individual or cell line from which the sequence was derived	Accession number	References or Submitting author(s)
DRB1*1437	—	—	DRB1*1309New	SWP43	AJ251985	
DRB1*1438	—	—	DRB1*1401V1	JCB14069	AB049830	
DRB1*1439	—	—	DRB1*1401V2	JCB15932	AB049831	
DRB1*1440	—	—	DRB1*1403V2	JCB24742	AB049832	
DRB1*1441	—	—	—	04RCH28	AY050186, AF339884	CK Hurley, J Tang
DRB1*1442	—	—	—	GN00411	AY054375	CK Hurley
DRB1*1443	—	—	—	P87043M1	AF400066	M Lin-Chu
DRB1*150101	DR15(2)	Dw2	DR2B Dw2	PGF, ROF-NL	M17378, M16957, M20430	
DRB1*150102	DR15(2)	Dw2	DRB1*15MT	LD0797	Z48359	
DRB1*150103	—	—	DRB1*15011var	BY00017	AF363727	L Burdett
DRB1*150104	—	—	—	R24489	AJ431718	J Mytilineos
DRB1*150201	DR15(2)	Dw12	DR2B Dw12	BGE, DHO, 20A1	M16958, M30180, M28584, AF029289	
DRB1*150202	DR15(2)	Dw12	DR2MU	CMURD	L23964	
DRB1*150203	DR15(2)	—	DRB1*15JMT	HN08729	AJ001253	
DRB1*1503	DR15(2)	—	—	G247, M851, M848, 20A2	M35159, AF010142, AF029290	
DRB1*1504	DR15(2)	—	DR2DAI	D13, D53, HM	L23963, L34025	
DRB1*1505	DR15(2)	—	DRB1*15KY	K.W.	D49823	
DRB1*1506	DR15(2)	—	—	JB317836, RP, CANSIN009, INDRAN001, INDRAN003	D63586, U45999, X98256	
DRB1*1507	DR15(2)[c]	—	DRB1*15LJM	UBM12218693	Y15404	
DRB1*1508	DR2	—	DRB1*15021V	JC3399	AB007634	
DRB1*1509	—	—	—	R98-903841B	AF172070	
DRB1*1510	—	—	—	98-2028, 98-2500, GN00320	AF191104, AF243536	
DRB1*1511	—	—	—	NR-GLW	AJ293861	
DRB1*1512	—	—	—	VTIS24502	AF373015	M Varney
DRB1*1513	—	—	DRB1*TT68	TT68	AF239244	R Holdsworth

Table 5.5 Designations of HLA-DR alleles (*continued*)

HLA alleles	HLA-DR serological specificities	HLA-D-associated (T cell defined) specificities	Previous equivalents	Individual or cell line from which the sequence was derived	Accession number	References or Submitting author(s)
DRB1*160101	DR16(2)	Dw21	DR2B Dw21	AZH, MN-2, FJO, W692D, W738D, 20A3	M16959, M30179, M28583, U56640, AF029291	
DRB1*160102	DR16(2)	Dw21	—	GN00150	U59686	
DRB1*160201	DR16(2)	Dw22	DR2B Dw22	REM (RML), 20A4	M20504, AF029292	
DRB1*160202	DR16(2)	Dw22	DRB1*16MADANG	MAD009	U38520	
DRB1*1603	DR2	—	—	JWR	L02545	
DRB1*1604	DR16(2)	—	DRB1*16x8	BONA, FORE	L14852	
DRB1*1605	DR16(2)c	—	16PRET	E.H.B., PRET4149	X74343, X75444	
DRB1*1607	—	—	DR2Mut	USH	U26659	
DRB1*1608	—	—	DRB1*(Gi+Pi)	Gi, Pi	Z72424	
DRB2*0101	—	—	—	AVL	M86691, M86694, M16274, M16275	
DRB3*010101	DR52	Dw24	DR3 III, DRw6a III	AVL, HHKB, DM28, DM29, CMCC, U-STH	X04055, X04058, AF152844	
DRB3*01010201	DR52	Dw24	dJ172K2, DRB3*01012	PMR, HSF7, W461R	U66825, Z84814, AF000448	
DRB3*01010202	DR52	Dw24	—	GN00199, 23054638	AF092089, AF092176, AF199236	
DRB3*010103	DR52	Dw24	DRB3*MOBD	MO, BD	X99771	
DRB3*010104	DR52	—	DRB3*01BTT	TO02021	Y10553	
DRB3*0102	DR52	—	DRB3*N409	409/96-UKN	Y08063	
DRB3*0103	—	—	DRB3*DF	DF	U94590	
DRB3*0104	—	—	—	GN00139	AF026467	
DRB3*0105	—	—	—	GN00234	AF081677	
DRB3*0106	DR52	—	DRB3*01EGT	EG-OT	AJ242860	
DRB3*0107	DR52	—	DRB3*01ABT	AB-OT	AJ242862	
DRB3*0108	—	—	—	1507-33405	AF361865	C Löliger
DRB3*0109	—	—	—	GN00394	AY042679	CK Hurley

Table 5.5 Designations of HLA-DR alleles (*continued*)

HLA alleles	HLA-DR serological specificities	HLA-D-associated (T cell defined) specificities	Previous equivalents	Individual or cell line from which the sequence was derived	Accession number	References or Submitting author(s)
DRB3*0110	—	—	DRB3*01MGT	CL06453	AJ315477	S Tavoularis
DRB3*0201	DR52	Dw25	DRw6b III	4/w6, DM24	M17380, V00522	
DRB3*020201	DR52	Dw25	pDR5b.3	SWEIG, WT49, U-STH	X99690, AF152845	
DRB3*020202	DR52	—	DRB3*02CVT	CV-OT	AJ242861	
DRB3*020203	DR52	—	DRB3*SSOM	SSOM	AF177216	
DRB3*020204	DR52	—	—	GN00418	AY094138	CK Hurley
DRB3*0203	DR52	—	DRB3.02p	POS	X86977	
DRB3*0204	—	—	—	SCHT	X91639	
DRB3*0205	—	—	DRB3*02-03v	GN068	U36826	
DRB3*0206	—	—	DRB3*02MT	BV1661	X95760	
DRB3*0207	DR52	—	DRB3 new	BML	Y10180	
DRB3*0208	DR52	—	DRB3*02HMT	BV02755	AJ001255	
DRB3*0209	DR52	—	DRB3*02New	p1454/bg287, Orietta Q.C.16/98	AF148518, AF132810	
DRB3*0210	DR52	—	DRB3*02KM	SMS145263 Diakon, CTM-9991295, NMDP#0236-9013-4	AJ238155, AF192259, AB035378	
DRB3*0211	DR52	—	DRB3*02NEW-A	CTM-9991121	AF192258	
DRB3*0212	—	—	DRB3*JWOO	JWOO	AF208484	
DRB3*0213	—	—	DRB3*HMAR	HMAR	AF208485	
DRB3*0214	—	—	—	00F03, 00F10, 00F13	AJ290395	A Moine
DRB3*0215	—	—	—	VTIS45001, VTIS45004	AF427138, AF427139	M Varney
DRB3*0216	—	—	—	74356	AF455114	MS Leffell
DRB3*0217	—	—	DRB3*VNGAZ	VNGAZ, emanuela, PB-MID 65347, FR-MID 65690, FR-MID 65691	AF461431, AY033875, AJ441058	H Dunckley, A Malagoli, F. Poli
DRB3*030101	DR52	Dw26	—	WT46, CMCC	—	

Table 5.5 Designations of HLA-DR alleles (*continued*)

HLA alleles	HLA-DR serological specificities	HLA-D-associated (T cell defined) specificities	Previous equivalents	Individual or cell line from which the sequence was derived	Accession number	References or Submitting author(s)
DRB3*030102	DR52	Dw26	DRB3*KL044	RP-KL044	AF242306	
DRB3*0302	DR52	—	DRB3*03KLT	SJ00198	Y13715	
DRB3*0303	—	—	DRB3*03SM	RP-SM073	AF028012	
DRB4*01010101	DR53	—	—	MANN, LBF, DKB, BURKHARDT, KT3, PRIESS, FS, DM24, DM29, MMCC	M16942, M17385, M17388, M15071, K02775	
DRB4*0102	DR53	—	DRB4*ICML	C.M.L., CML	L08621, D89879	
DRB4*01030101	DR53	—	dJ93N13	MJ4, BOLETH, HSF7, G081	M15178, M20555, M19556, Z84477, AF361548	(68)[b]
DRB4*01030102N	Null	—	DRB4 null	DBB	D89918	
DRB4*010302	DR53	—	DRB4W778R	W778R	AF048707	
DRB4*010303	DR53	—	DRB4GL	MG-CV, FOA2362, G081	AJ242833, AJ297503, AF207709, AF361549	(68)[b]
DRB4*010304	—	—	—	14242	AJ292564	ME Fasano
DRB4*0104	—	—	DRB4*CR210	69-218, 76-394	X92712	
DRB4*0105	DR53	—	DRB4New	17345	Y09313	
DRB4*0106	—	—	—	MKOST	AF450316, AF450317	H Dunckley
DRB4*0201N	Null	—	DRB4*VI	GN016	U50061, U70543, U70544	
DRB4*0301N	Null	—	DRB4*v2	GN017	U70542	
DRB5*010101	DR51	Dw2	DR2A Dw2	PGF, ROF-NL	M17377, M16954, M20429	
DRB5*010102	DR51	Dw2	—	GN00152	U66721	
DRB5*0102	DR51	Dw12	DR2A Dw12	BGE, DHO	M16955, M30182, M16086	
DRB5*0103	—	—	DRB5.Oli	IND-24, IND-59, NT0002	X86978, AF122887	
DRB5*0104	—	—	DRB5*0101V	GN045	U31770	
DRB5*0105	—	—	—	CP5570	X87210	

Table 5.5 Designations of HLA-DR alleles (*continued*)

HLA alleles	HLA-DR serological specificities	HLA-D-associated (T cell defined) specificities	Previous equivalents	Individual or cell line from which the sequence was derived	Accession number	References or Submitting author(s)
DRB5*0106	—	—	DRB5*New	ZL4062	Z83201	
DRB5*0107	DR51	—	DRB5*01CBT	WI01846	Y09342	
DRB5*0108N	Null	—	—	ES	Y10318, Y17819	
DRB5*0109	—	—	DRB5*01ART	BV08663	Y13727	
DRB5*0110N	Null	—	DRB5*0102Null, DRB5*CB848	JAS, CB848	AF097680, AF314541	H Dunckley[b]
DRB5*0202	DR51	Dw21, Dw22	DR2A Dw21, DR2A Dw22	REM (RML), FJO, MN-2, AZH	M16956, M30181, M20503, M15992, M32578, X99939	
DRB5*0203	—	—	DRB5*HK	HK55	M91001	
DRB5*0204	—	—	—	GN00151	U59685	
DRB5*0205	—	—	DRB5*02 variant	TT030822	AJ271159	
DRB6*0101	—	—	DRBσ*0101, DRBX11	BAC, BRO-2, HOM-2, KAS116, MZ070782, HON, SAS6211	X53357, M83892	
DRB6*0201	—	—	DRBX21, DRBVI	PGF, D0208915, CGG, BA, E4181324	M77284-7, X53358, M83893	
DRB6*0202	—	—	DRBσ*0201, DRBX22, DRB6III	RML, KAS011	M83204, M83894	
DRB7*010101	—	—	DRBψ1	BOLETH, BH13	K02772-4, L31617	
DRB7*010102	—	—	—	PITOUT	L31618	
DRB8*0101	—	—	DRBy2	BOLETH	M20556, M20557	
DRB9*0101	—	—	M4.2 b exon	MOU	M15563	

[a]Allele names given in bold type have been assigned since the 2000 Nomenclature report.
[b]This reference is to a confirmatory sequence.
[c]HLA specificity provided from the HLA dictionary (33, 34).

Table 5.6   Designations of HLA-DQA1, -DQB1 alleles

HLA alleles	HLA-DR serological specificities	HLA-D-associated (T cell defined) specificities	Previous equivalents	Individual or cell line from which the sequence was derived	Accession number	References or Submitting author(s)
DQA1*010101	—	Dw1	DQA 1.1, 1.9	LG2, BML, KAS116	L34082	
DQA1*010102	—	—	DQA1*0101new	MZ070782, LWAGS, PMG075	AF322867, AF322868, AF322869	ML Ashdown
DQA1*010201	—	Dw2, w21, w19	DQA 1.2, 1.19, 1.AZH	PGF, LB, CMCC, AZH, WT46, DRA, ROF-NL, EMJ	M20431, L34083	
DQA1*010202	—	Dw21	—	KAS011	L34084	
DQA1*0103	—	Dw18, w12, w8, Dw'FS'	DQA 1.3, 1.18, DRw8-DQw1	APD, TAB, FPF, WVB, 2012, E4181324	M59802, L34085	
DQA1*010401	—	Dw9	—	1183, 2013, 2012, 2708, 31227ABO, EK, KOSE, DEK, REN	M95170, L34086	
DQA1*010402	—	—	DQA1*new	KGU	AJ296091, AJ296092	
DQA1*0105	—	—	—	AK93007, 1183, 2708	L42625, L46877	
DQA1*0106	—	—	183DQA1	183		
DQA1*0201	—	Dw7, w11	DQA 2, 3.7	LG-10, BEI, DM24, DM28, DM29, MOU	L34087	
DQA1*030101	—	Dw4, w10, w13, w14, w15	DQA 3, 3.1, 3.2	MMCC, JY, NIN, BML, DM24, DM29, BOLETH	M29613, M29616, L34088	
DQA1*0302	—	Dw23	DQA 3, 3.1, 3.2, DR9-DQw3	ISK, DKB, YT	M11124, L34089	
DQA1*0303	—	—	—	YT	L34089, L46878	
DQA1*0401	—	Dw8, Dw'RSH'	DQA 4.2, 3.8	ARC, 2041, MADURA, SPL (SPACH)[e]	M33906, L34090	
DQA1*050101	—	Dw3, w5, w22	DQA 4.1, 2, dJ93N13	RAJI, CMCC, VAVY, HSF7, SWEIG	X00370, K01160, L34091, Z84489	
DQA1*050102	—	Dw5	DQA 4.1, 2	MG3	—	
DQA1*0502	—	—	—	EMA	U03675	
DQA1*0503	—	Dw16	—	AMALA	L34093	
DQA1*0504	—	—	DQA1*05YD, DQA05MC	YD-069, AD-YM23	U85035, U97555	
DQA1*0505	—	Dw5, Dw22	DQA 4.1, 2	BM21, REM (RML), BM16	AB006908, M20506, L34092	
DQA1*060101	—	Dw8	DQA 4.3	LUY	L34094	

Table 5.6 Designations of HLA-DQA1, -DQB1 alleles (*continued*)

HLA alleles	HLA-DR serological specificities	HLA-D-associated (T cell defined) specificities	Previous equivalents	Individual or cell line from which the sequence was derived	Accession number	References or Submitting author(s)
DQA1*060102	—	—	—	RV	Y09968	
DQB1*0201	DQ2	Dw3	DQB 2	WT49, CMCC, QBL, MZ, LD, VW, MOR, JNP, DM24, DM28, DM29, BEI, VAVY	K02405, M65043, M81140, L40179	
DQB1*0202	DQ2	Dw7	DQB 2	BURKHARDT, BH, MOU	M81141, U07848, L34095	
DQB1*0203	DQ2	—	DQB1*02DL, DQB1*GHA30	RAQ, CAUCA254, CAUCA288, DL-13, GHA30	Z35099, U33329, U39089, U39090, AB002468	
DQB1*030101	DQ7(3)	Dw4, w5, w8, w13	DQB 3.1, DQ0301W515R	SWEIG, DQB37, NIN, JHA, JR, JME, DC, JGL, LUY, BML, DM23, MG3, AMALA, W515R, CjAr, CaAr, 06-006	M65040, L34096, U83582, M25325	
DQB1*030102	DQ7(3)	—	DQB1*03GPT	HM00214	AJ001256, Y10428	
DQB1*0302	DQ8(3)	Dw4, w10, w13, w14	DQB 3.2	BOLETH, FS, BIN40, WT51, DM24, DM29, JS, MMCC, VW, JNP, JOP, Priess, BrEh, 145b, DaHa	M65038, K01499, L34097, M25326	
DQB1*03032	DQ9(3)	Dw23, w11	DQB 3.3	DBB, KOZ, 5112.103, DKB, 06-006	M65039, M60028, L34098, M25328	
DQB1*03033	DQ9(3)	—	DQB1*03New	G.C.	AF093815	
DQB1*0304	DQ7(3)	—	DQB1*03HP, *03new	HP, RG, M.M.	M74842, M83770, X76553	
DQB1*030501	DQ8(3)	—	DQB1*03KC	G.P., M.A.	X69169, X76554	
DQB1*030502	—	—	—	00L53	AJ290396	A Moine
DQB1*0306	DQ3	—	DQB1*MAT	MAT	D78569	
DQB1*0307	—	—	DQB1*D4	D4	Z49215	
DQB1*0308	—	—	—	97-459#1	AJ003005	
DQB1*0309	—	—	DQ3 Var	W469D, W469R	U66400	
DQB1*0310	DQ8(3)	—	DQB1*03new	CTM-8991127	AF195245	
DQB1*0311	—	—	—	VBALA	AF439338	H Dunckley
DQB1*0312	—	—	—	216305	AF469118	K Schwarz

Table 5.6   Designations of HLA-DQA1, -DQB1 alleles (*continued*)

HLA alleles	HLA-DR serological specificities	HLA-D-associated (T cell defined) specificities	Previous equivalents	Individual or cell line from which the sequence was derived	Accession number	References or Submitting author(s)
DQB1*0313	—	—	—	10993426	AF479569	J Wu
DQB1*0401	DQ4	Dw15	DQB 4.1, Wa	KT3, YT	M13279, L34099	
DQB1*0402	DQ4	Dw8, Dw'RSH'	DQB 4.2, Wa, E1448	ARC, OLN, MZ, 2041, SPL (SPACH), MADURA, RPET01	M33907, M65042, L34100, Z80898	
DQB1*050101	DQ5(1)	Dw1	DQB 1.1, DRw10-DQw1.1	LG2, 45.1, BML, MVL, JR, MDR, WG, DC, KAS116	X03068, M65044, L34101	
DQB1*050102	DQ5(1)	—	DQB1*05COT	COT.DA	Y17290	
DQB1*050201	DQ5(1)	Dw21	DQB 1.2, 1.21	AZH, FJO, KAS011	L34102	
DQB1*050202	—	—	—	J16	AF463516	(69)
DQB1*050301	DQ5(1)	Dw9	DQB 1.3, 1.9, 1.3.1	WT52, HU129, HU128, EK	M65047, L34103, L40180	
DQB1*050302	DQ5(1)	Dw9	DQB 1.3, 1.9, 1.3.2	AP106, AP109, AP110, AP115	—	
DQB1*0504	DQ5(1)	—	DQB 1.9	DG, R.F.	M65046, M94773	
DQB1*060101	DQ6(1)	Dw12, w8	DQB 1.4, 1.12	AKIBA, BGE, TAB, E4181324, B.H., B.S.	L34104, X89194, L40181	
DQB1*060102	DQ6(1)	Dw12, w8	DQB1*0601var.	Sk, Rb	M86740	
DQB1*060103	DQ6(1)	—	DQ06W649R	W649R	AF000447	
DQB1*0602	DQ6(1)	Dw2	DQB 1.5, 1.2	PGF, VYT, 2041, ROF NL, AMAI, CjAr, CaAr	M20432, M65048, L34105, M25327	
DQB1*0603	DQ6(1)	Dw18, Dw'FS'	DQB 1.6, 1.18	WVB, APD, FPF, 2012, OMW	M65050, M34322, L34106	
DQB1*060401	DQ6(1)	Dw19	DQB 1.7, 1.19	CMCC, DAUDI, DM23, LD, WG, EMJ	M65051, L34107	
DQB1*060402	DQ6(1)	—	DQB1*0604-Variant	GN015	AF113250, U63321	
DQB1*060501	DQ6(1)	Dw19	DQB 1.8, DQBSLE, 1.19b, 2013-24	CI, KT, MR, 2013	M36472, M59800, M65052	
DQB1*060502	DQ6(1)	Dw19	DQB1*MDvR-1	BEN53	L26325	
DQB1*0606	—	—	DQB1*WA1	LINE66	M86226	

Table 5.6 Designations of HLA-DQA1, -DQB1 alleles (*continued*)

HLA alleles	HLA-DR serological specificities	HLA-D-associated (T cell defined) specificities	Previous equivalents	Individual or cell line from which the sequence was derived	Accession number	References or Submitting author(s)
DQB1*0607	—	—	DQB1*06BRI1	08-2779-0, BN151	M87041, AF112463	
DQB1*0608	DQ6(1)[c]	—	DQB1*06BRI2	R.W., BM675	M87042, AF112464	
DQB1*0609	DQ6(1)	—	DQB1*06AA	HO301, TRACHT, N076, AK93022	L19951, L27345, D29918, L42626	
DQB1*0610	—	—	DQB1MC	M.M., M.G., N205, L13, L90	X86327, Z75044	
DQB1*061101	DQ1	—	UNM-95-228	#MUD0130-14998	U39086	
DQB1*061102	DQ1	—	DQB1*06new1	6658K	AJ012155	
DQB1*0612	DQ1	—	DQB1*06GB	GB002	X96420	
DQB1*0613	—	—	DQB1*0602V	BB-(2)	U77344	
DQB1*0614	DQ6(1)	—	DQB1*06EMT	OG00018	AJ001257	
DQB1*0615	—	—	DQB1*06new2	T890	AJ012156	
DQB1*0616	—	—	052DQB1	052	AF087939	
DQB1*0617	—	—	99-3039	15427-00/01/02	AF181983	
DQB1*0618	—	—	DQB1*06nou	IM0000053	AY026349	(70)
DQB1*0619	—	—	DQB1*0602-Variant	ACAR	AF091305	S Bowman
DQB1*0620	—	—	—	CB846		H Dunckley

[a]Allele names given in bold type have been assigned since the 2000 Nomenclature report.
[b]This reference is to a confirmatory sequence.
[c]HLA specificity provided from the HLA dictionary (33, 34).

In the second category is the finding that a newly defined antigen does not comfortably place within any known serological grouping. This is especially true of the DRB1*03, *11, *13, *14, and *08 family of alleles, for which the description of new alleles has revealed a continuum of allelic diversity rather than five discrete subfamilies. It should be stressed that, although a goal is to indicate the serological grouping into which an allele will fall, this is not always possible. Most importantly, the allele name should be seen as no more than a unique designation.

RENAMING OF ALLELES AND REMOVAL OF INCORRECT ALLELES

There was discussion on the renaming of several HLA class I alleles, stimulated by the more extensive nucleotide sequence information obtained subsequent to the official naming of the alleles based upon partial sequences. For three alleles it was agreed that they had been named inappropriately and a decision was therefore made to rename them as follows: A*2416 becomes A*3108, B*1522 becomes

Table 5.7 Designations of HLA-DPA1 and DPB1 alleles

HLA alleles[a]	Associated HLA-DP specificities	Previous equivalents	Individual or cell line from which the sequence was derived	Accession number	References or Submitting author(s)
DPA1*010301	—	DPw4α1	BOLETH, 3.1.0, LG2, PRIESS, LB	X03100, X82390, X82392, X82389	
DPA1*010302	—	DPA1	933-302-2	AF074848	
DPA1*0104	—	01New	SK	X78198, X81348, X82391	
DPA1*0105	—	DPA1*RK	DNA-RK	X96984	
DPA1*0106	—	DPA1*Indian-024	I024	U87556	
DPA1*0107	—	DPA1*0103New	#913	AF076284	
DPA1*0108	—	—	936-563-6	AF346471	(71)
DPA1*020101	—	DPA2, pDAα13B	DAUDI, AKIBA	X82394, X82393, X78199	
DPA1*020102	—	DPA1*TF	A371, L67, LB0410278	L31624, X83610	
DPA1*020103	—	DPA1-CAM024, DPA1*Cameroon2	CAM024, CAM241, #63	U94838, AF015295, AF076285	
DPA1*020104	—	DPA1	533-2929, 922-485-8	AF074847	
DPA1*020105	—	DPA1*PERR	CC109	AF098794	
DPA1*020106	—	—	A.L.	AF165160	
DPA1*020201	—	2.21	CB6B	M83906, L11642, X79475, X80482, X79479	
DPA1*020202	—	2.22	LKT3, KT17, WI-L2 NS, CT46, EsSm, GIWh	M83907, L11641, X79476, X80484, X79480	
DPA1*020203	—	DPA1*0202New	#904	AF092049	
DPA1*0203	—	DPA1*TC48	TC48	Z48473	
DPA1*0301	—	3.1	AMAI	M83908, X79477, X81347, X79481	
DPA1*0302	—	DPA1*Cameroon	CAM48, CAM59, CAM66, CAM100, CAM151	AF013767	
DPA1*0401	—	4.1	T7526	M83909, L11643, X79478, X80483, X78200	
DPB1*010101	DPw1	DPB1,DPw1b	LUY, RSH, P0077, FB11	M83129, M83664, M62338, X72070	
DPB1*010102	DPw1	DPB1*WA6	LINE 101, AH1457	L19220, L27662	
DPB1*020102	DPw2	DPB2.1	45.1, WJR076, LB, JY	M62328, X03067, X99689	
DPB1*020103	DPw2	DPB2.1	CJ	X94078	
DPB1*020104	DPw2	—	CQ930-SEQ1643	AF326565	(72)
DPB1*020105	DPw2	—	27D	AF462072	M Varney
DPB1*020106	DPw2	—	UCLA-344	AF517128	M Tilanus

Table 5.7 Designations of HLA-DPA1 and DPB1 alleles (*continued*)

HLA alleles[a]	Associated HLA-DP specificities	Previous equivalents	Individual or cell line from which the sequence was derived	Accession number	References or Submitting author(s)
DPB1*0202	DPw2	DPB2.2	QBL, DUCAF, 99101422	M62329, X72071, AF492642	M Luo[b]
DPB1*030101	DPw3	DPB3	SLE, PRIESS, ETH9-0226	M62334, X02964, X03023, X78044	
DPB1*030102	—	DPB1*03var	POHS-161	AF234538	
DPB1*0401	DPw4	DPB4.1, DPw4a	HHKB, BOLETH, PRIESS, LC11, KAS011	M62326, M23675, K01615, M23906-8, L29174, X03022, X030025-8, X02228, X72072	
DPB1*0402	DPw4	DPB4.2, DPw4b	APD, BURKHARDT	M62327, M21886	
DPB1*0501	DPw5	DPB5	HAS, LKT3, 99101467	M62333, X72073, AF492638	M Luo[b]
DPB1*0601	DPw6	DPB6	JMOS, FB11	M62335, X72074	
DPB1*0801	—	DPB8	PIAZ	M62331	
DPB1*0901	—	DPB9, DP'Cp63'	TOKUNAGA, 99100402	M62341, X72075, AF492637	M Luo[b]
DPB1*1001	—	DPB10	BM21, SAVC, 99101332	M85223, M62342, X72076, AF492640	M Luo[b]
DPB1*110101	—	DPB11	CRK, AVE G	M62336, X78046	
DPB1*110102	—	—	AH696	L23399	
DPB1*1301	—	DPB13	NB, KAS116	M62337, X72077	
DPB1*1401	—	DPB14	8268, KAS011	M31778, M62343, X72078	
DPB1*1501	—	DPB15	PLH, 99100835	M31779, M62339, X72079, AF492636	M Luo[b]
DPB1*1601	—	DPB16	JRA, WT46, 99101659	M31780, M62332, X72080, AF492641	M Luo[b]
DPB1*1701	—	DPB17	JRAB, LBUF, 99101046	M31781, M62344, X72082, AF492643	M Luo[b]
DPB1*1801	—	DPB18	JCA	M62340	
DPB1*1901	—	DPB19	CB6B, 99101467	M62330, X72081, AF492639	M Luo[b]
DPB1*200101	—	Oos, DPB-JA	OOS, ARENT, BEL8-CC	M58608, M63508	
DPB1*200102	—	DPB1*BRI6	NT	M97685	
DPB1*2101	—	DPB-GM, DPB30, NewD	GM, PEI52, PEI74, C1, T7527	M77659, M83915, M84621, M80300	
DPB1*2201	—	DPB1*AB1, NewH	HV152, HV385, SAS60103, SAS60106	M77674, M83919	
DPB1*2301	—	DPB32, NewB	D0208915, UK3082, UK5496, PT35, IT22, I132	M83913, M84014	
DPB1*2401	—	DPB33, NewC	UK7430	M83914	
DPB1*2501	—	DPB34, NewE	PEI46	M83916	
DPB1*260101	—	DPB31, WA2	LINE70	M86229	
DPB1*260102	—	DPB1*WA8	4-BEN NO2	L24387	

Table 5.7 Designations of HLA-DPA1 and DPB1 alleles (*continued*)

HLA alleles[a]	Associated HLA-DP specificities	Previous equivalents	Individual or cell line from which the sequence was derived	Accession number	References or Submitting author(s)
DPB1*2701	—	DPB23, WA3	LINE92, H033	M84619, M86230	
DPB1*2801	—	DPB21, JAVA2	I57, I147, JOG1489	M84617, L00599	
DPB1*2901	—	DPB27, NewG	RBLB66, NG105, NG113, PNG112, PNG177, SCZ244	M84625, M83918, L01467	
DPB1*3001	—	DPB28	AH1377, EB5, ETH-0245	M84620, X78045	
DPB1*3101	—	DPB22, NewF, JAVA1	I68, I147, I6, PEI03, JOG1427, JOG1471	M84618, M83917, L00598	
DPB1*3201	—	DPB24, NewI	NG78, PNG167	M84622, M85222	
DPB1*3301	—	DPB25	HO23	M84623	
DPB1*3401	—	DPB26	HO26, DH67	M84624	
DPB1*3501	—	DPB29	AH1450, AH521	M84626	
DPB1*3601	—	New A, SSK2	SASBE41, THM1, KT	M83912, D10479, D10882	
DPB1*3701	—	DPB1*WA4	LINE41	M87046	
DPB1*3801	—	SSK1	THKK	D10478	
DPB1*3901	—	DPB1*BRI4	EM, ETH-0203	M97686, X78043	
DPB1*4001	—	DPB1*BRI5, WA5	5D, LINE103, LINE105, LINE116, LINE117, LINE119, EB39, HO62	M97684, L19219, L23400	
DPB1*4101	—	DPB2.3	HT	D13174	
DPB1*4401	—	STCZ	SCZ259, SCZ244	L01466	
DPB1*4501	—	DPB1*NM	C212	L09236	
DPB1*4601	—	DPB1*NIB	V.E.C., R130	L07768, L31817	
DPB1*4701	—	DPB1*02KY, *SUT	SAJ008, SAJ119, SUT	D14344, D10834	
DPB1*4801	—	—	SE107	L17314	
DPB1*4901	—	—	HO21	L17313	
DPB1*5001	—	—	DIEDE	L17311	
DPB1*5101	—	DPB1*WA7, *EA1, *JYO	C2#3, 15-BEN, NMDP#00800-2553-8, JYO	L17310, L19219, L27073, D28809	
DPB1*5201	—	—	HO82	L22076	
DPB1*5301	—	—	EB26	L22077	
DPB1*5401	—	DPB1 New2	ETH-0222	X78042	
DPB1*5501	—	DPB1 New3, DPBGUY	ETH-0271, J.M.	X78041, X80331	

Table 5.7 Designations of HLA-DPA1 and DPB1 alleles (*continued*)

HLA alleles[a]	Associated HLA-DP specificities	Previous equivalents	Individual or cell line from which the sequence was derived	Accession number	References or Submitting author(s)
DPB1*5601	—	DPB1-R90	R90	L31816	
DPB1*5701	—	DPBMYT4220	H.R.	X80752	
DPB1*5801	—	DPB1newAW	HAM006	X82123, X85966	
DPB1*5901	—	—	GA Au, HBO1242, HBO1243, HBO1244, 0000-5922-0	Z47806, U29534, U59422	
DPB1*6001	—	—	JN, BPN	U22313	
DPB1*6101N	Null	—	ZN, Nel., Nan	U22312, AJ002530	
DPB1*6201	—	—	LE, CT	U22311	
DPB1*6301	—	DPB1*IsOr	IsOr	U34033	
DPB1*6401N	Null	DPB1*IsAr	IsAr	U34032	
DPB1*6501	—	—	E.L.	X91886	
DPB1*6601	—	DPB1*BR	DNA-128	X96986	
DPB1*6701	—	DPB1*TF	DNA-TF	X96985	
DPB1*6801	—	DPB1*BAC	BAC1283, 902-258-3	Z70731, U59440	
DPB1*6901	—	—	SBD3497	X97406	
DPB1*7001	—	—	900-132-2	U59441	
DPB1*7101	—	—	905-967-6, I045	U59438	
DPB1*7201	—	—	0014-3022-2	U59439	
DPB1*7301	—	—	0076-0684-1	U59437	
DPB1*7401	—	DPB1-512ld	512ld	U94839	
DPB1*7501	—	0402-GA	U73	Y09327	
DPB1*7601	—	DPB1*14new	19835	Z92523	
DPB1*7701	—	DPBnewBR	U.R.	Y14230	
DPB1*7801	—	DPBNew	M541	Y13900	
DPB1*7901	—	DPB1New	1197	Y16095	
DPB1*8001	—	DPB1	18055285	AF074845	
DPB1*8101	—	DPB1*dre	009340662, dre	AF074846, AJ245640	
DPB1*8201	—	DPB1*04New	19045	Y18498	
DPB1*8301	—	—	GM-CV	AJ238005	
DPB1*8401	—	DPB1*PERR	CC109	AF077015	
DPB1*8501	—	DPB1*27New	MGD, UCLA212	AF184168, AF211979	
DPB1*8601	—	DP New	605861, 606165	AJ271373	

Table 5.7 Designations of HLA-DPA1 and DPB1 alleles (*continued*)

HLA alleles[a]	Associated HLA-DP specificities	Previous equivalents	Individual or cell line from which the sequence was derived	Accession number	References or Submitting author(s)
DPB1*8701	—	DPB1*2001new	#014738363	AF288354	
DPB1*8801	—	DPB1*3701new	#009519430	AF288355	
DPB1*8901	—	DPB1*MO	MOP	AJ297820	
DPB1*9001	—	DPBnew	608050	AJ292074	73
DPB1*9101	—	DP14New	VTIS20927	AY029777	M Varney
DPB1*9201	—	—	VTIA71787	AF489518	M Varney

[a]Allele names given in bold type have been assigned since the 2000 Nomenclature report.
[b]This reference is to a confirmatory sequence.

Table 5.8 Designations of HLA-DOA, -DOB alleles

HLA alleles	Previous equivalents	Individual or cell line from which the sequence was derived	Accession number
DOA*010101	DZα, DNA1.2a	JG, MANN, DBB	X02882, Z81310, AB005994
DOA*01010201	pII-α-6, DNA1.1b	SPL, TOK	M26039, AB005992
DOA*01010202	PGDZ1, DNA1.1a	PGF, SA	M31525, AB005991
DOA*01010203	DNA1.1c	SPO101	AB005993
DOA*010103	DNA1.2b	DKB	AB005995
DOA*01010401	DNA1.3a	U937	AB005996
DOA*01010402	DNA1.3b	U937	AB005997
DOA*010105	DNA1.4	COX	AB005998
DOB*01010101	DO, pII-b-9	45.1, SPL, SA, LCL721	X03066, M26040, AB035249
DOB*01010102	—	WT100BIS, LCL721	AB035250
DOB*010102	DOB1.6	SR117	AB035254
DOB*010201	DOB	BOLETH	L29472
DOB*010202	DOB1.3	AKIBA	AB035251
DOB*0103	HA14	MANN	X87344
DOB*01040101	DOB1.4	PEA	AB035252
DOB*01040102	DOB1.5	SPO010	AB035253

Table 5.9   Designations of HLA-DM alleles

HLA alleles	Previous equivalents	Individual or cell line from which the sequence was derived	Accession number
DMA*0101	RING6	JY, MANN	X62744
DMA*0102	DMA-Ile 140	AZL	Z24753
DMA*0103	DMA3.2	HOM-2	U04878
DMA*0104	DMA3.4	BM21	U04877
DMB*0101	RING7	JY, MANN	Z23139
DMB*0102	DMB-Glu 143	YAR	Z24750
DMB*0103	DMB-Thr 179	BM16	Z24751
DMB*0104	DMB3.4	CEPH 23-01	U00700
DMB*0105	HY595, DMB*KV1	HY595, H.S.K.	D32055, U16762
DMB*0106	DMB*PERR	CC44	AF134890, AF072680

Table 5.10   Designations of TAP alleles

HLA alleles	Previous equivalents	Individual or cell line from which the sequence was derived	Accession number
TAP1*0101	RING4, PSF(Y3), TAP1A	U937, LCL721.45, HB00028, HB00032	X57522, X57521, L21204
TAP1*0102N	TAP1*0101Null	KMW	AB012644, AB012645
TAP1*020101	TAP1B	CK	L21206
TAP1*020102	TAP1E	HEH	L21205
TAP1*0301	TAP1C	JT	L21208
TAP1*0401	TAP1D	HB00031	L21207
TAP2*0101	RING11A, TAP2A	CEM-CCRF	M84748
TAP2*0102	TAP2E	JY	Z22936
TAP2*0103	TAP2F	S-2	U07844
TAP2*0201	RING11B, TAP2B	DX3	Z22935

Table 5.11 Designations of MICA alleles

MICA alleles	Previous equivalents	Individual or cell line from which the sequence was derived	Accession number	References or Submitting author(s)
MICA*001	MICA001, PERB11.1-18.2, MICA-EIBA	IMR90, EJ32B, DUCAF, EVA,SP	L14848, U56940, L29406, U69965, AF085059, AF085060, AF085061, AF085062, AF336085, AF336086	(74)[b]
MICA*00201	MICA002, MICA-BEBF	YAR, AMAI, WT49, TEM, JBUSH, 9-2,ZR75-1	U56941, AF085043, AF085044, AF085045, AF085046	
MICA*00202	MICA-BEE, MICA042	Individual1	AF011877, AF011878, AF011879	
MICA*004	MICA004, MICA-AJCD	MOU, BM15, PF97387, MANN, RSH, Individual2	U56943, X92841, AF085031, AF085032, AF085033, AF085034	
MICA*005	MICA005	U373	U56944	
MICA*006	MICA006, MICA-ADCD	KAS116	U56945, AF085023, AF085024, AF085025, AF085026, AF336065, AF336066	(74)[b]
MICA*00701	MICA007, MICA-CEEA	JESTHOM, BM92, WT24	U56946, AF085047, AF085048, AF085049, AF085050	
MICA*00702	MICA-CEB, MUC-22, MICA023	A34, B27-ci, SchS(child1)-MUC	AF011880, AF011881, AF011882, Y16805	
MICA*00801	MICA008, PERB11.1-44.1, PERB11.1-8.1, PERB 11.1-60.3, PERB11.1-47.1, MICA-AAAC	SCHU, MGAR, SAVC, LB, JY, R90/7379, REE,GD, EMJ, PLH, DKB, LBF, WT8, APD, MADURA	U56947, U69624, U69967, L29409, U69977, U69628, L29411, U69625, U69970, U69976, AF085015, AF085016, AF085017, AF085018, AF336067, AF336068	(74)[b]
MICA*00802	MICA-AAD, MICA-AN23, MUC-26, MICA026, MICA-silent B	Individual3, GUA-ND, BrI(f)-MUC, MLA-MUC, BrID(child1)-MUC, Thai-DCH019, 01083208, 01065930, 0183074	AF011883, AF011884, AF011885, AJ250499, AJ250500, Y16809, AF106650, AF106651, AF106652	
MICA*00803	MICA-silent C, MICA054	01083082	AF106653, AF106654, AF106655	
MICA*00901	MICA009, PERB11.1-52.1, MICA-ABCD	RML, AKIBA, HARA, BOB, C1R, JHAF, LUY, Individual2, E4181324	U56948, U69626, U69971, AF085019, AF085020, AF085021, AF085022, AF336069	(74)[b]
MICA*00902	MICA-AFC, MICA-TAND, MICA020, MUC-20	MANIKA, TAA, AE(F)-MUC, AS(Child2)-MUC, DZA 97-19	AF011886, AF011887, AF011888, AF097419, AF079420, AF079421, AF079406, Y16803, AY029762, AY029763	A Kimura[b]
MICA*010	MICA010, PERB11.1-62.1, PERB11.1-46.1, MICA-DGAB, MUC-18	AMALA, BOLETH, T7526, BSM, KAS011, TAB089, EM(M)-MUC, EM(Child1)-MUC EK-MUC, ES-MUC, T7526	U56949, U69629, U69974, L29408, U69969, AF085055, AF085056, AF085057, AF085058, Y16801, AF336071, AF336072	(74)[b]
MICA*011	MICA011, PERB11.1-65.1, MICA-BCGE	LWAGS, T47D	U56950, U69630, U69975, AF085035, AF085036, AF085037, AF085038 AF336073, AF336074	(74)[b]
MICA*01201	MICA012, PERB11.1-54.1	LKT3, HOKKAIDO, TA94	U56951, U69627, U69972, AF336081, AF336082	(74)[b]

Table 5.11 Designations of MICA alleles (*continued*)

MICA alleles	Previous equivalents	Individual or cell line from which the sequence was derived	Accession number	References or Submitting author(s)
MICA*01202	MICA-silent A, MICA053	01082123	AF106647, AF106648, AF106649	
MICA*013	MICA013	PAR1	U56952	
MICA*014	MICA014	PAR2	U56953	
MICA*015	MICA015, MICA-39	OMW	U56954, AF136157, AF136158, AF136159, AF264738, AF264739, AF264740	(74)[b]
MICA*016	MICA016, PERB11.1-35.1, MICA-AGFB, MUC-19	J0528239, FPAF, Q85/8086, NR(M)-MUC, NR(Child1)-MUC, NM(Child2)-MUC, TISI	U56955, U69623, U69966, AF085027, AF085028, AF085029, AF085030, Y16802, AF336075, AF336076	A Kimura[b]
MICA*017	MICA-KMCE, MICA017, MUC-27, MICA-AN31	KSM, DBB, DEU, WJR076, DEM, FD(F)-MUC, FM(child1)-MUC, HF(M)-MUC, HS(child1)-MUC, HT(child2)-MUC, Thai-DCH013, Thai-DCH020, Thai-DCH024	AF079413, AF079414, AF079415, AF097403, AJ250803, Y16810, AF264735, AF264736, F264737	
MICA*018	MICA-EEBA, MICA-GKIT, MICA018, MUC-23, MICA-AN22	31227ABO, BM16, CBA, DO208915, SE(F)-MUC, KU(F)-MUC, KF(child1)-MUC, Thai-DCH036, DZA 97-8, DZA 97-18, DZA 97-20, BM16, DO208915	AF011874, AF011875, AF011876, AF093116, AF079425, AF079426, AF079427, AF097404, Y16806, AJ250805, AF336077	(74)[b]
MICA*019	MICA-AMW, MICA AGAB, MICA-DPCA, MICA019, MICA-AN26	SSA, HSB27, OLL, WEWAK1, DPCA, CF996, DHIF, WT51	AB015600, AF011835, AF011836, AF011837, AF093113, AF079416, AF079417, AF079418, AF097405, AJ250804, AF336079, AF336080	(74)[b]
MICA*020	MICA-AN33	25/1506	AJ249394	
MICA*021	MUC-17, MICA021	AA-MUC, AM(child1)-MUC, AS(child2)-MUC	Y18110	
MICA*022	MICA-BGA, MUC-21, MICA022	Individual10, Thai-DCH021	AF011856, AF011857, AF011858, Y16804	
MICA*023	MICA-BEBC	WDV	AF085039, AF085040, AF085041, AF085042	
MICA*024	MICA-AAC, MUC-24, MICA024	BT594, Individual7, DZA 97-17	AF011832, AF011833, AF011834, Y16807	
MICA*025	MICA-DEB, MUC-25, MICA025	BT20, Thai-DCH032	AF011853, AF011854, AF011855, Y16808	
MICA*026	MICA-CEED	HOM-2	AF085051, AF085052, AF085053, AF085054	
MICA*027	MICA-AAAB, MICA-AN21	SWEIG007, HSB27	AF085011, AF085012, AF085013, AF085014, AJ250802	
MICA*028	MICA-AABC, MUC-29, MICA028	DKB, KUR-MUC	AF011829, AF011830, AF011831, AF093115, Y18111	

Table 5.11 Designations of MICA alleles (*continued*)

MICA alleles	Previous equivalents	Individual or cell line from which the sequence was derived	Accession number	References or Submitting author(s)
MICA*029	MUC-30, MICA-AN27, MICA029	DZA 97-08, MFO-ND	Y18112, AJ250503, AJ250504	
MICA*030	MICA-KWHT, MICA036	WKD	AF079422, AF079423, AF079424	
MICA*031	MICA-AIB, MICA037	MCF7	AF011838, AF011839, AF011840	
MICA*032	MICA-AKB, MICA038	CAR, NS2TA, NS2TA1, S2T2	AF011841, AF011842, AF011843	
MICA*033	MICA-ALAB, MICA039, MICA-AN24	WEWAK1	AF011844, AF011845, AF011846, AF093114, AJ250505	
MICA*034	MICA-BCC, MICA040	Individual18	AF011847, AF011848, AF011849	
MICA*035	MICA-BEA, MICA041	SK-BR3	AF011850, AF011851, AF011852	
MICA*036	MICA-BHB, MICA043	EHM	AF011859, AF011860, AF011861	
MICA*037	MICA-CEA, MICA044	AVE G, GRE G, LS40, LH, IHL, AD031, Individual10, B7Qui, 8TB	AF011862, AF011863, AF011864	
MICA*038	MICA-CEC, MICA045	Individual12, Individual14	AF011865, AF011866, AF011867	
MICA*039	MICA-CEF, MICA046	Individual13	AF011868, AF011869, AF011870	
MICA*040	MICA-CIB, MICA047	A34	AF011871, AF011872, AF011873	
MICA*041	MICA-AN25, MICA048, MICA-newA	M7, 01083098, 01083208, 01081318, 01065894	AJ271789, AF106632, AF106633, AF106634	
MICA*042	MICA-newB, MICA049	01065869	AF106635, AF106636, AF106637	
MICA*043	MICA-AN32, MICA050, MICA-newC	RB22, 01084383	AJ250990, AJ250991, AF106638, AF106639, AF106640	
MICA*044	MICA-newD, MICA051	01083114	AF106641, AF106642, AF106643	
MICA*045	MICA-AN30, MICA052, MICA-newE	DEW-ND, 01083268, 01065876	AJ250506, AJ250507, AF106644, AF106645, AF106646	
MICA*046	MICA-AN28	M7	AJ250501, AJ250502	
MICA*047	MICA-055D	COYA3408, KM	AJ295250, AJ295251, AF286732	(75, 76)
MICA*048	—	TA21	AF264741, AF264742, AF264743	(74)
MICA*049	—	LUY	AF264744, AF264746, AF264747	(74)

[a]Allele names given in bold type have been assigned since the 2000 Nomenclature report.
[b]This reference is to a confirmatory sequence.

Table 5.12 Numbers of alleles
with official names at each locus
by 31 July 2002

Locus	Number of alleles
HLA-A	250
HLA-B	490
HLA-C	119
HLA-E	6
HLA-F	1
HLA-G	15
HLA-DRA	3
HLA-DRB1	315
HLA-DRB2	1
HLA-DRB3	38
HLA-DRB4	12
HLA-DRB5	15
HLA-DRB6	3
HLA-DRB7	2
HLA-DRB8	1
HLA-DRB9	1
HLA-DQA1	22
HLA-DQB1	53
HLA-DPA1	20
HLA-DPB1	99
HLA-DOA	8
HLA-DOB	8
HLA-DMA	4
HLA-DMB	6
TAP1	6
TAP2	4
MICA	54

B*3543, and B*1559 becomes B*3544. These three examples vividly illustrate the problems inherent in naming sequences just consisting of exons 2 and 3 of HLA class I alleles. Determination of longer sequences, full coding sequences or more, should avoid future assignment of inappropriate names and lead to more accurate and interesting interpretation of the sequence data.

It has been accepted that the nucleotide sequence designated as the Cw*1301 allele was in error and so this designation has been deleted. There is also some doubt as to the validity of certain of the HLA-E allele sequences, which is currently being investigated. A comprehensive list of all the allele names that have been deleted is given in **Table 5.13**.

EXTENSION OF HLA ALLELE NAMES

The convention of using a four digit code to distinguish HLA alleles that differ in the proteins they encode was first implemented in the 1987 Nomenclature Report.[8] In 1990 a fifth digit was added to permit the distinction of sequences differing only by synonymous (non-coding) nucleotide substitutions within the exons.[10] When these conventions were adopted it was anticipated that the nomenclature system would accommodate all the HLA alleles likely to be sequenced. Unfortunately that is not proving to be the case, as the number of alleles for certain genes is fast approaching the maximum possible with the current naming convention.

In particular there are three problem areas; firstly the fifth digit, used for synonymous substitutions, can distinguish only nine variants of an allele. Already there are six named variants of the A*0201 allele – A*02011 to A*02016 – and eight variants of the G*0101 – G*01011 to G*01018. The second problem area concerns the third and fourth digits used to distinguish up to 99 variants within the allele families defined by the first and second digits. The first allele family to exceed 99 named alleles is likely to be the B*15 family for which 73 variants have been named to date, soon followed by the A*02 and DRB1*13 families for which over 50 allele variants have already been named. The most immediate problem concerns the DP genes, for which the decision was taken in 1989 to name all alleles which differ by non-synonymous (coding) substitutions with different combinations of the first two digits, a system that can only accommodate 99 alleles.[9] The most recently assigned name was DPB1*9201, so that once an additional eight coding sequences have been reported there will be no capacity left in this system for naming newly discovered DPB1 alleles.

There was much discussion of this topic at the meeting. Several different options were considered, including the splitting up of the allele names into discrete fields separated by colons or semi-colons. This option while it would have

Table 5.13   List of allele names which have been deleted

Old name now deleted	New name	Reason for change
A*0105N	A*0104N	Sequence shown to be in error
A*0223	A*0222	Sequence named in error
A*2401	—	Sequence shown to be in error
A*2412	A*2408	Sequence shown to be in error
A*2416	A*3108	Sequence renamed
A*3005	A*3004	Sequence shown to be in error
A*31011	A*310102	Sequence shown to be in error
A*3302	A*3303	Sequence shown to be in error
B*0701	—	Sequence shown to be in error
B*1305	B*1304	Sequence submitted with errors
B*1522	B*3543	Sequence renamed
B*1541	B*1539	Sequence named in error
B*1559	B*3544	Sequence renamed
B*1816	B*1814	Sequence named in error
B*27051	B*270502	Sequence shown to be in error
B*2722	B*2706	Sequence shown to be in error
B*39012	B*390101	Sequence shown to be in error
B*3921	B*3924	Sequence submitted with errors
B*4017	B*4016	Sequence named in error
B*4041	B*4040	Sequence named in error
B*4203	B*4202	Name never officially assigned
B*4401	B*4402	Sequence shown to be in error
B*5003	B*5002	Sequence named in error
B*5125	B*5122	Sequence named in error
B*5506	B*5504	Sequence submitted with errors
B*5803	—	Name never officially assigned
B*7901	B*1518	Sequence renamed
Cw*0101	Cw*0102	Sequence shown to be in error
Cw*0201	Cw*020202	Sequence shown to be in error
Cw*0301	Cw*0304	Sequence shown to be in error
Cw*0402	Cw*040101	Sequence shown to be in error

Table 5.13 List of allele names which have been deleted (*continued*)

Old name now deleted	New name	Reason for change
Cw*0601	Cw*0602	Sequence shown to be in error
Cw*1101	—	Sequencing artefact
Cw*1201	Cw*120202	Sequence shown to be in error
Cw*1301	—	Sequence shown to be in error
Cw*1401	Cw*1402	Sequence shown to be in error
Cw*1501	Cw*1502	Sequence shown to be in error
Cw*1603	Cw*1403	Sequence shown to be in error
Cw*16042	Cw*160401	Sequence submitted with errors
Cw*1605	Cw*160401	Sequence submitted with errors
DRB1*0702	DRB1*0701	Sequence shown to be in error
DRB1*08031	DRB1*080302	Sequence shown to be in error
DRB1*09011	DRB1*090102	Sequence shown to be in error
DRB1*12031	DRB1*1201	Sequence shown to be in error
DRB1*1606	DRB1*1605	Sequence shown to be in error
DRB4*0101102N	DRB4*01030102N	Sequence named in error
DRB5*0201	DRB5*0202	Sequence shown to be in error
DQA1*03012	DQA1*0302	Sequence shown to be in error
DQA1*05013	DQA1*0505	Additional coding polymorphism detected
DQB1*03031	DQB1*030302	Sequence shown to be in error
DPA1*0101	DPA1*0103	Sequence shown to be in error
DPA1*0102	DPA1*0103	Sequence shown to be in error
DPB1*02011	DPB1*020102	Sequence shown to be in error
DPB1*0701	—	Name never assigned
DPB1*1201	—	Name never assigned
DPB1*4201	DPB1*3101	Sequence shown to be in error
DPB1*4301	DPB1*2801	Sequence shown to be in error
MICA*003	—	Name never assigned

no limit to the number of names available, was in the end considered by the committee to be too radical and disruptive a solution for the problems at hand. It was therefore decided to seek solutions with minimal change to the existing format of the alleles, so as to limit the changes that would have to be made to existing database structure. The following decisions were taken to solve the three major problems.

1 To introduce an extra digit between the current fourth and fifth digit, to allow for up to 99 synonymous variants of each allele. This expands the full allele name to eight digits, the first two digits defining the allele family and where possible corresponding to the serological family, the third and fourth digits describing coding variation, the fifth and sixth digits describing synonymous variation, and the seventh and eighth digits describing variation in introns or $5'$ or $3'$ regions of the gene.

2 In cases where the total number of coding variants exceeds 99, a second number series will be used to extend the first one. For example, for the B*15 family of alleles, the B*95 series will be reserved and used to code for additional B*15 alleles. Consequently, the next B*15 allele to be named following B*1599 will be B*9501. Likewise, the A*92 series will be reserved as a second series for the A*02 allele family.

3 For HLA-DPB1 alleles, it was decided to assign new alleles within the existing system, hence once DPB1*9901 has been assigned, the next allele would be DPB1*0102, followed by DPB1*0203, DPB1*0302 etc.

The introduction of the additional digit for synonymous variation will take place immediately and all allele names that are currently five digits or above will be renamed accordingly, as shown in **Tables 5.2–5.10**. The other changes will only be implemented when necessary, as dictated by submission of novel allele sequences.

NAMING OF ALLELES WITH ABERRANT EXPRESSION

The use of an optional 'N' or 'L' suffix to an allele name to indicate either 'Null' or 'Low' expression was introduced in previous Nomenclature Reports.[12–14] At this committee meeting there was discussion on the introduction of additional suffixes and concern that some alleles which had previously been given an 'N' suffix should be reconsidered in light of new data indicative of some type of protein expression.

Three new suffixes will be introduced. An 'S' to denote an allele specifying a protein which is expressed as a soluble 'Secreted' molecule but is not present on the cell surface; a 'C' to indicate an allele product which is present in the 'Cytoplasm' but not at the cell surface; an 'A' to indicate 'Aberrant' expression where there is some doubt as to whether a protein is expressed. The first example of a secreted only molecule is that encoded by the newly assigned B*44020102S allele which by virtue of a single intronic mutation fails to express the transmembrane domain and is therefore produced in a secretory form only. A comprehensive reanalysis of all the alleles that have previously been assigned the Null status will be undertaken and alleles found to fit better into these new categories will be reassigned.

NAMING OF HLA-G ISOFORMS

There is evidence of differential splicing of HLA-G that leads to the production of both membrane-bound and soluble forms of the same allele. It was felt that while different naming conventions are already being used by researchers in the field, it would be unnecessarily complex to assign official names to all different isoforms produced by expression of a single allele. The committee recommended the use of a lower case 's' or 'm' to indicate 'soluble' or 'membrane'-bound as a prefix to the HLA-G allele name. Thus the soluble or membrane bound forms of the HLA-G*0101 allele, would be described as sHLA-G*0101 and mHLA-G*0101, respectively.

KILLER IMMUNOGLOBULIN-LIKE RECEPTOR (KIR) GENE AND ALLELE NOMENCLATURE

Discussion took place on the naming of the Killer Immunoglobulin-like Receptor (KIR) genes and alleles. While the naming of the genes remains under the remit of the HGNC, it was decided to establish a subcommittee comprising Drs Bo Dupont (New York, USA), Daniel Geraghty (Seattle, USA), Peter Parham (Stanford, USA), Derek Middleton (Belfast, UK) Steven Marsh (London, UK), and John Trowsdale (Cambridge, UK) who will put forward a set of recommendations for the naming of KIR alleles and haplotypes. The recommendations of this subcommittee will be published in a separate report.

THE IMGT/HLA SEQUENCE DATABASE

The IMGT/HLA Sequence Database is the official repository for HLA sequences named by the WHO Nomenclature Committee for Factors of the HLA System.[22,23] The database contains sequences for all HLA alleles officially recognized by the WHO Nomenclature Committee for Factors of the HLA System and provides users with online tools and facilities for their retrieval and analysis. These include allele reports, alignment tools, and detailed descriptions of the source cells. The online IMGT/HLA submission tool allows both new and confirmatory sequences to be submitted directly to the WHO Nomenclature Committee. New releases of the database are made quarterly with the latest version (release 1.15.0 July 2002) containing 1482 HLA alleles derived from over 3980 component sequences from the EMBL/GenBank/DDBJ databases. The database may be accessed via the World Wide Web at www.ebi.ac.uk/imgt/hla.

The IMGT/HLA Database is currently supported by the following organizations: the American Society for Histocompatibility and Immunogenetics (ASHI), the Anthony Nolan Trust (ANT), Biotest, Dynal, European Federation for Immunogenetics (EFI), Forensic Analytical, Innogenetics, the National Marrow Donor Program (NMDP), and Orchid Biosciences. Initial support for the IMGT/HLA database project was from the Imperial Cancer Research Fund and an EU Biotech grant (BIO4CT960037).

NEW COMMITTEE MEMBERS

The following individuals have been invited to serve on the WHO Nomenclature Committee for Factors of the HLA System: Daniel Geraghty (Seattle, USA), John Hansen (Seattle, USA), Carolyn Hurley (Washington DC, USA), Effie Petersdorf (Seattle, USA) and John Trowsdale (Cambridge, UK).

ACKNOWLEDGMENTS

The work of James Robinson and Matthew Waller with the IMGT/HLA Sequence Database and their help in the preparation of tables is gratefully acknowledged. Also thanked are Dr Peter Stoehr and the staff at the European Bioinformatics Institute for their continued support of the IMGT/HLA Database, and Biotest for their support of the WHO Nomenclature Committee meeting.

KEY REFERENCES

1 WHO Nomenclature Committee (1968) *Bull WHO* 39: 483

2 WHO Nomenclature Committee (1970) WHO terminology report. In: PI Terasaki, ed. *Histocompatibility Testing, 1970.* Copenhagen: Munksgaard, p 49

3 WHO Nomenclature Committee (1972) *Bull WHO* 47: 659

4 WHO Nomenclature Committee (1975) *Bull WHO* 52: 261

5 WHO Nomenclature Committee (1978) *Bull WHO* 56: 461

6 WHO Nomenclature Committee (1980) In: PI Terasaki, ed. *Histocompatibility Testing, 1980.* Los Angeles: UCLA Tissue Typing Laboratory, pp. 18–20.

7 WHO Nomenclature Committee (1985) Nomenclature for factors of the HLA system 1984. In: ED Albert, MP Baur, WR Mayr, eds. *Histocompatibility Testing, 1984.* Berlin: Springer-Verlag, 4–8

8 Bodmer WF, Albert E, Bodmer JG, Dupont B, Mach B, Mayr WR, Sasazuki T, Schreuder GMT, Svejgaard A, Terasaki PI (1989) Nomenclature for factors of the HLA system, 1987. In: B. Dupont, ed. *Immunobiology of HLA.* New York: Springer-Verlag, pp 72–799

9 Bodmer JG, Marsh SGE, Parham P, Erlich HA, Albert E, Bodmer WF, Dupont B, Mach B, Mayr WR, Sasasuki T, Schreuder GMT, Strominger JL, Svejgaard A, Terasaki PI (1990) Nomenclature for factors of the HLA system, 1989. *Tissue Antigens* 35: 1–8

10 Bodmer JG, Marsh SGE, Albert E, Bodmer WE Dupont B, Erlich HA, Mach B, Mayr WR, Parham P, Sasasuki T, Schreuder GMT, Strominger JL, Svejgaard A, Terasaki PI (1991) Nomenclature for factors of the HLA system, 1990. *Tissue Antigens* 37: 97–104

11 Bodmer JG, Marsh SGE, Albert E, Bodmer WF, Dupont B, Erlich HA, Mach B, Mayr WR, Parham P, Sasasuki T, Schreuder GMT, Strominger JL, Svejgaard A, Terasaki PI (1992) Nomenclature for factors of the HLA system, 1991. In: T Tsuji, M Aizawa, T Sasazuki, ed. *HLA 1991.* Oxford: Oxford University Press, p 17–31

12 Bodmer JG, Marsh SGE, Albert E, Bodmer WF, Dupont B, Erlich HA, Mach B, Mayr WR, Parham P, Sasasuki T, Schreuder GMT, Strominger JL, Svejgaard A, Terasaki PI (1994) Nomenclature for factors of the HLA system, 1994. *Tissue Antigens* 44: 1–18

13 Bodmer JG, Marsh SGE, Albert E, Bodmer WT, Bontrop RE, Charron D, Dupont B, Erlich HA, Mach B, Mayr WR, Parham P, Sasasuki T, Schreuder GMT, Strominger JL, Svejgaard A, Terasaki PI (1995) Nomenclature for factors of the HLA system, 1995. *Tissue Antigens* 46: 1–18

14 Bodmer JG, Marsh SGE; Albert ED, Bodmer WF, Bontrop RE, Charron D, Dupont B, Erlich HA, Fauchet R, Mach B, Mayr WR, Parham P, Sasasuki T, Schreuder GMT, Strominger JL, Svejgaard A, Terasaki PI (1997) Nomenclature for factors of the HLA system, 1996. *Tissue Antigens* 49: 297–321

15 Bodmer JG, Marsh SGE, Albert ED, Bodmer WF, Bontrop RE, Dupont B, Erlich HA, Hansen JA, Mach B, Mayr WR, Parham P, Petersdorf EW Sasazuki T, Schreuder GM, Strominger JL, Svejgaard A, Terasaki PI (1999) Nomenclature for factors of the HLA system, 1998. *Tissue Antigens* 53: 407–446

16 Marsh SGE, Bodmer JG, Albert ED, Bodmer AT, Bontrop RE, Dupont B, Erlich HA, Hansen JA, Mach B, Mayr WK Parham P, Petersdorf EW, Sasazuki T, Schreuder GM, Strominger JL, Svejgaard A, Terasaki PI (2001) Nomenclature for factors of the HLA system, 2000. *Tissue Antigens* 57: 236–283

17 Geraghty DE, Koller BH, Hansen JA, Orr HT (1992) The HLA class I gene family includes at least six genes and twelve pseudogenes and gene fragments. *J Immunol* 149: 1934–1946

18 Vernet C, Ribouchon MT, Chimini G, Jouanolle AM, Sidibe I, Pontarotti P (1993) A novel coding sequence belonging to a new multicopy gene family mapping within the human MHC class I region. *Immunogenetics* 38: 47–53

19 Beck S, Abdulla S, Alderton RP, Glynne RJ, Gut IG, Hosking LK, Jackson A, Kelly A, Newell WR, Sanseau P, Radley E, Thorpe KL, Trowsdale J (1996) Evolutionary dynamics of non-coding sequences within the class II region of the human MHC. *J Mot Biol* 255: 1–13

20 Stephens R, Horton R, Humphray S, Rowen L, Trowsdale J, Beck S (1999) Gene organisation, sequence variation and isochore structure at the centromeric boundary of the human MHC. *J Mot Biol* 291: 789–799

21 Wain HM, Bruford EA, Lovering RC, Lush MJ, Wright MW Povey S (2002) Guidelines for human gene nomenclature. *Genomics* 79: 464–470

22 Robinson J, Malik A, Parham P, Bodmer JG, Marsh SGE (2000) IMGT/HLA database – a sequence database for the human major histocompatibility complex. *Tissue Antigens* 55: 280–287

23 Robinson J, Waller MJ, Parham P, Bodmer JG, Marsh SGE (2001) IMGT/HLA Database – a sequence database for the human major histocompatibility complex. *Nucleic Acids Res* 29: 210–213

24 Poli F, Bianchi P, Scalamogna M, Crespiatico L, Ghidoli N, Puglisi G, Sirchia G (1999) A nucleotide deletion in exon 4 is responsible for an HLA-A null allele (A*0105N). *Tissue Antigens* 54: 300–302

25 Luscher MA, MacDonald KS, Bwayo JJ, Plummer FA, Barber BH (2001) Sequence and peptide-binding motif for a variant of HLA-A*0214 (A*02142) in an HIV-1-resistant individual from the Nairobi Sex Worker cohort. *Immunogenetics* 53: 10–14

26 Wu G-G, Cheng L-H, Deng Z-H, Wang L-X, Wei T-L, Yang C-S, Zhao T-M (2002) Cloning and *complete sequence of* a novel HLA-A null *allele, A*0253N, with a termination codon generated by a C to G mutation in exon 2. *Tissue Antigens* 59: 328–330

27 Czachurski D, Rausch M, Scherer S, Opelz G, Mytilineos J (2002) Characterisation of a new HLA-A *allele, A*0256, identified in Caucasian individual. *Tissue Antigens* 60: 181–183

28 Garino E, Belvedere M, Berrino M, Bertola L, M DOA, Mazzola G, Rossetto C, Carcassi C, Lai S, Gay E, Ricotti M, Curtoni ES (2002). A new HLA-A*11 allele, A*1112, identified by Sequence-Based-Typing. *Tissue Antigens* 60: 84–87

29 Li J, den Hollander N, van Oosterwijk A, Ng C-M, Singal DP (2002) Identification of a new HLA-A null allele A*2436N. *Tissue Antigens* 60: 184–185

30 Elsner H-A, Bernard G, Bernard A, EizVesper B, Blaszczyk R (2002) Non-expression of HLA-A*2901102N is caused by a nucleotide exchange in the mRNA splicing site at the beginning of intron 4. *Tissue Antigens* 59: 139–141

31 Voorter CEM, van den Berg-Loonen EM (2002) Identification of 2 new HLA-A alleles, A*2419 and A*3011, by sequence-based typing. *Tissue Antigens* 59: 136–138

32 Binder T, Heym J, Horn B, Blaszczyk R (2000) HLA-A*2416: a new allele of the HLA-A19 lineage. *Tissue Antigens* 55: 178–181

33 Hurley CK, Schreuder GMT, Marsh SGE, Lau M, Middleton D, Noreen H (1997) The search for HLA-matched donors: a summary of HLA-A*, -B*, -DRBl/3/4/5* alleles and their association with serologically defined HLA-A, -B, -DR antigens. *Tissue Antigens* 50: 401–418

34 Schreuder GMT, Hurley CK, Marsh SGE, Lau M, Maiers M, Kollman C, Noreen H (1999) The HLA Dictionary 1999: a summary of HLA-A, -B, -C, -DRBl/3/4/5, -DQBl alleles and their association with serologically defined HLA-A, -B, -C, -DR and -DQ antigens. *Tissue Antigens* 54: 409–437

35 Steiner NK, Gans CP, Kosman C, Baldassarre LA, Edson S, Jones PF, Rizzuto G, Pimtanothai N, Koester R, Mitton W, Ng J, Hartzman RJ, Hurley CK (2001) Novel HLA-B alleles associated with antigens in the 8C CREG. *Tissue Antigens* 57: 373–375

36 Cox ST, Prokupek B, Baker F, Holman R, Leung VTC, Wong ASW, Madrigal JA, Little A-M (2002) Identification of HLA-B*1566. *Tissue Antigens* 59: 424–425

37 Luo M, Embree J, Ramdahin S, Ndinya-Achola J, Njehga S, Bwayo JB, Pan S, Mao X, Stuart T, Brunham RC, Plummer F (2002) HLA-A and HLA-B in Kenya, Africa: allele frequencies and identification of HLA-B*1567 and B*4426. *Tissue Antigens* 59: 370–380

38 Sheldon. MH, Bunce M, Dunn PPJ, Day S, Lee GD, Park Y-J, Bang BK, Kim BK, Oh E-J (2002) Identification of two new alleles in a single Korean individual: HLA-B*1568 and DRBl*1208. *Tissue Antigens* 59: 430–432

39 Smillie DM, Smith PA, Day S, Dunn PPJ. Identification of a new HLA-B*15 allele: HLA-B*1569. *Tissue Antigens* 59: 151–153.

40 den Hollander N, Li J, van Oosterwijk A, Ng C-M, Singal DP (2002) Identification of a new HLA-B null, B*1817N, allele in a family. *Tissue Antigens* 59: 341–343

41 Witter K, Lau M, Zahn R, Scholz S, Albert ED (2001) Identification of a novel HLA-B*2722 allele from a Filipino cell. *Tissue Antigens* 58: 263–268

42 Dunne C, Little A-M, Cox ST, Masson D, Crowley J, Barnes T, Marsh SGE, Rooney G, Hagan R, Lawlor E, Madrigal JA (2002) Identification and nucleotide sequence of a new null allele, HLA B*3540N. *Tissue Antigens* 59: 522–524

43 Estefania E, Gamez-Lozano N, de Pablo R, Moreno ME, Vilches C (2002) Complementary DNA sequence of the novel HLA-B*3704 allele. *Tissue Antigens* 59: 142–144

44 Vidal S, Morante MP, Mosquera AM, Moga E, Querol S, Garcia J, Rodriguez-Sanchez JL. Identification and sequencing of the HLA-B*4106 allele. *Tissue Antigens* 58: 343–344

45 Kennedy CT, Greville WD, Dodd R, Le T, Taverniti A, Chapman G, Wallace R, Kennedy A, Strickland J, Dunckley H (2002) Six new HLA Class I alleles detected by PCR-SSO genotyping. *Tissue Antigens* 59: 320–324

46 Canossi A, Papola F, Liberators G, Del Beato T, Piancatelli D, Tessitore A, Vicentini MT, Maccarone D, Aureli A, Cervelli C, Di Rocco M, Casciaru CU, Adorno D (2002) Identification of the novel allele B*4427 and a confirmatory sequence (B*44022). *Tissue Antigens* 59: 331–334

47 Przemeck M, Elsner H-A, Sel S, Pastucha LT, Blaszczyk R (2002) HLA-B*4431, a new HLA-B variant with a complex ancestral history involving HLA-B*44, B*40 and B*07 alleles. *Tissue Antigens* 60: 91–94

48 Grams SE, Moonsamy PV, Mano C, Oksenberg JR, Begovich AB (2002) Two new HLA-B alleles, B*4422 and B*4704, identified in a study of families with autoimmunity. *Tissue Antigens* 59: 338–340

49 Steiner NK, Kosman C, Jones PF, Gans CP, Rodriguez-Marino SG, Rizzuto G, Baldassarre LA, Edson S, Koester

R, Sese D, Milton W, Ng J, Hartzman RJ, Hurley CK (2001) Twenty-nine new HLA-B alleles associated with antigens in the 5C CREG. *Tissue Antigens* 57: 481–485

50 Bengtsson M, Danielsson F, Egle Jansson I, Jidell E, Johansson U (2002) Two new HLA-Cw* alleles, Cw*0105 and Cw*1405, detected by sequence based typing. *Tissue Antigens* 59: 226–228

51 Wu S, Lai C-Y, Lai S-M, Chen S-P, Chou FC, Shiao YM, Huang C-S (2002) Point mutation in the a helix of the HLA-C a2 domain generates a novel HLA-C allele (HI,A-Cw*0106) in a Han Chinese individual in Taiwan. *Tissue Antigens* 59: 433–435

52 Cox ST, Ogilvie H, Bohan EM, Holman R, Prentice G, Potter M, Madrigal JA, Little AM (2002) Sequence of two new HLA-C alleles: HLA-Cw*0313 and HLA-Cw*1208. *Tissue Antigens* 59: 49–51

53 Balas A, Santos S, Aviles MJ, GaraaSanchez F, Lillo R, Alvarez A, Villar-Guimerans LM, Vicario JL (2002) The elongation of the cytoplasmic domain due to a point deletion at exon 7 results in a HLA-C null allele (Cw*0409N). *Tissue Antigens* 59: 95–100

54 Wu S, Lai C-Y, Chou F-C, Lai S-M, Chen SP, Shiao YM (2002) Point mutation in the (3 sheet of the HLA-C a2 domain generates a novel HLA-C allele (HI.A-Cw*08012) in a Puyuma aboriginal individual in Taiwan. *Tissue Antigens* 2002; 60: 333–335

55 Kralovicova J, Marsh SGE, Waller MJ, Hammarsfrom L, Vorechovsky I (2002) The HLA-DRA*0102 allele: correct nucleotide sequence and associated HLA haplotypes. *Tissue Antigens* 2002; 60: 266–267

56 Martinez-Gallo M, Martinez-Lostao L, Martinez-Carretero MA, Gago I, Maroto P, Rodriguez-Sanchez JL, de la Calls-Martin O (2002) Identification of a novel HLA-DRBl allele (DRBl*0108) by sequence-based DRB typing in two siblings. *Tissue Antigens* 59: 350–351

57 Greville WD, Chapman G, Hogbin J-P, Kennedy A, Dunckley H (2002) Novel HLA-DRBl alleles discovered during routine sequencing based typing: DRBl*03052, DRBl*04032, DRBl*1139 and DRBl*1346. *Tissue Antigens* 59: 154–156

58 Gagne K, Bonneville F, Cheneau M-L, Herry P, Pinson M -J, Herfray P, Cesbron-Gautier A, FoMa G, Bignon J-D (2002) Identification of a new HLA-DRB1 allele, DRBl*0321. *Tissue Antigens* 2002; 60: 268–270

59 Greville WD, Chapman G, Hogbin J-P, Dunckley H (2002) Novel HLA-DRBl alleles revealed by sequencing based typing: DRBl*04053 and DRBl*1143. *Tissue Antigens* 59: 347–349

60 Chen DS, Tang TF, Pulyaeva H, Slack R, Tu B, Wagage D, Li L, Perlee L, Ng J, Hartzman RJ, Hurler CK (2002) Relative HLA-DRBl*04 allele frequencies in five United States populations found in a hematopoietic stem cell voluteer donor registry and seven new DRBl*04 alleles. *Tissue Antigens* 2002; 63: 665–672

61 Dunn TA, Iglehart BA, Leffell MS (2001) Identification of a novel DR4 allele, DRBl*0442. *Tissue Antigens* 58: 198–200

62 Tang TF, Lin YS, Robbins FM, Li L, Sintasath D, Coquillard G, Huang A, Heine U Ng J, Hartzman RJ,

Hurler CK (2002) Description of fourteen new DRB alleles found in a stem cell donor registry. *Tissue Antigens* 59: 63–65

63 Greville WD, Dunckley H (2002) Identification of sequence errors in HLA-DRBl*0801 and HLA-DRBl*12011. *Tissue Antigens* 59: 52–54

64 Lin J, Liu Z, Chen W, Jia Z, Pan D, Fu Y, Zhang F, Xu A (2002) A novel DRBl*09 allelic sequence in Jing ethnic minority of China. *European Journal of Immunogenetics* 29: 335–336

65 Zanone R, Bettens F, Tiercy J-M (2002) Sequence of a new DR12 allele with two silent mutations that affect PCR-SSP typing. *Tissue Antigens* 59: 165–167

66 Corell A, Cox ST, Soteriou B, Ramon D, Madrigal JA, Marsh SGE (2002) Complete cDNA sequences of the HLA-DRBl*14011, *1402, *1403 and *1404 alleles. *Tissue Antigens* 59: 66–69

67 Gans CP, Tang TF, Slack R, Ng J, Hartzman RJ, Hurler CK (2002) DRBl*14 diversity and DRB3 associations in four major population groups in the United States. *Tissue Antigens* 59: 364–369

68 De Pablo R, Solis R, Balas A, Vilches C (2002) Specific amplification of the HLA-DRB4 gene from c-DNA. Complete coding sequence of the HLA alleles DRB4*0103101 and DRB4*01033. *Tissue Antigens* 59: 44–46

69 Fu Y, Chen W, Liu Z, Lin J, Jia Z, Pan D, Xu A (2002) A novel HLA-DQBl allele isolated from Jing ethnic group in Southwest China. *Tissue Antigens* 60: 102–103

70 Casamitjana N, Gil J, Campus E, Santos M, Nogues N, Ribera A, Palou E (2001) Identification of a novel HLA-DQBl*06 allele, DQBl*0618. *Tissue Antigens* 58: 269–271

71 Grams SE, Wu J, Noreen HJ, Mangaccat J, Cognato MA, Johnson S, Scgall M, Williams TM, Begovich AB (2001) Three new DP alleles identified in a study of 800 unrelated bone marrow donor-recipient pairs. *Tissue Antigens* 58: 272–275

72 Curcio M, Lapi S, Italia S, Rizzo G (2002) Identification of a novel HLA-DPBl allele, DPBl*02014, by sequence- based typing. *Tissue Antigens* 59: 58–59

73 Bengtsson M, Danielsson F, Egle Jansson I, Johansson U (2002) Identification of a new HLA-DPBl allele, HLA-DPBl*9001. *Tissue Antigens* 59: 344–346

74 Obuchi N, Takahashi M, Nouchi T, Satoh 1VI, Arimura T, Ueda K, Akai J, Ota M, Naruse T, Inoko H, Numano F, Kimura A (2001) Identification of MICA alleles with a long Leu-repeat in the transmembrane region and no cytoplasmic tail due to a frameshift-deletion in exon 4. *Tissue Antigens* 57: 520–535

75 Perez-Rodriguez M, Raimondi E, Marsh SGE, Madrigal JA (2002) Identification of a new MICA allele, MICA*047. *Tissue Antigens* 59: 216–218

76 Zhang Y, Lazaro AM, Mirbaha F, Lavingia B, Vorhaben R, Stastny P (2002) Characterization of a novel MICA allele: MICA*047. *Tissue Antigens* 59: 308–310

77 Marsh SGE, Albert ED, Bodmer WF, Bontrop RE, Dupont B, Erlich HA, Geraghty DE, Hansen JA, Mach B, Mayr WR, Parham P, Petersdorf EW, Sasazuki T, Schreuder GMTh, Strominger JL, Svejgaard A, Terasaki PI (2002) Nomenclature for factors of the HLA system, 2002. *Tissue Antigens* 60: 407–464

i

KEY DATABASES

- IMGT/HLA Sequence Database: Robinson J, Malik A, Parham P, Bodmer JG, Marsh SGE (2000) IMGT/HLA Database – sequence database for the Human Major Histocompatibility Complex. *Tissue Antigens* 55: 280–287
- Robinson J, Waller MJ, Parham P, de Groot N, Bontrop R, Kennedy LJ, Stoehr P, Marsh SGE (2003) IMGT/HLA and IMGT/MHC – sequences databases for the study of the Major Histocompatibility Complex. *Nucleic Acid Research* 31: 311–314
- Website address for the IMGT/HLA Sequence Database www.ebi.ac.uk/ imgt/hla
- Website address for the Anthony Nolan Research Institute <www.anthonynolan.org.uk/hig/>

Nomenclature for Factors of the Dog Major Histocompatibility System (DLA), 2000

6

LJ Kennedy and Associates*

- **NAMING OF GENES WITHIN THE DLA REGION**
- **NAMING OF MHC ALLELES FROM OTHER CANIDAE**
- **NAMING OF NEW ALLELES**
- **UPDATED SEQUENCE ALIGNMENTS**

The Major Histocompatibility Complex (MHC) of the dog and other Canidae appears to be highly polymorphic, and alleles of these genes are likely to be functionally relevant in regulating the immune response and the susceptibility/ resistance to immune-mediated diseases. Considerable effort has recently been made in characterizing the extent of the polymorphisms in DLA class II genes.

NAMING OF GENES WITHIN THE DLA REGION

Class I

Following the first Nomenclature Committee report,[1,2] studies by JL Wagner (unpublished) have identified locus-specific primers for four transcribed class I genes. Thus the previously described alleles[3] can be assigned unequivocally to particular loci. In order to avoid the suggestion that any DLA class I genes were homologs of particular HLA class I genes, the Nomenclature Committee decided that numbers rather than letters would be used at present to name DLA

class I genes. The updated list of DLA genes is shown in **Table 6.1a**, while **Table 6.1b** lists other genes which have yet to be confirmed and do not have official names. Since the first DLA nomenclature report, the DLA region has been mapped to dog chromosome 12,[4] and it has been confirmed that DLA-79 maps to a separate region on chromosome 18. Although DLA-79 and C1pg-26 are orphan genes not located in the DLA region, the Committee considered that naming such genes fell within its remit.

Class II

No new class II gene names were assigned by the Committee in its second report.

NAMING OF NEW ALLELES

The Committee reaffirmed the published conditions for naming new alleles.[1,2]

Since the first nomenclature report, 48 DLA-88, 16 DLA-DRB1, 6 DLA-DQA1, and 15 DLA-DQB1 alleles have been

*Lorna J Kennedy,[1] John M Angles,[2] Annette Barnes,[1] Stuart D Carter,[1] Olga Francino,[3] John A Gerlach,[4] George M Happ,[5] William ER Ollier,[1] Wendy Thomson,[1] John L Wagner[6]

[1]Mammalian Immunogenetics Research Group, Veterinary Clinical Sciences, University of Liverpool, UK
[2]Small Animal Clinical Studies, University College Dublin, Ireland
[3] Facultat de Veterinaria, Universitat Autonoma de Barcelona, Spain
[4]Medical Technology Program, Department of Medicine, Michigan State University, East Lansing, Michigan, USA
[5]Institute of Arctic Biology, University of Alaska, Fairbanks, Alaska, USA
[6]Thomas Jefferson University, Department of Medicine, Philadelphia, PA, USA
This chapter is based on the second report of the ISAG DLA Nomenclature Committee, which was originally published in *Tissue Antigens* 58: 55–70.

Table 6.1a   Genes in the DLA complex

Official name	Previous equivalents	Molecular characteristics	References
DLA-79	DLA-79	Non-classical class I gene associated with 7.9 kb *Hin*d III fragment. Not in DLA region	3,5,6
DLA-88	DLA-88	Class I gene associated with 8.8 kb *Hin*d III fragment	3,6
DLA-12	DLA-12	Non-classical class I gene associated with 12 kb *Hin*d III fragment	3,6
DLA-64	DLA-64	Non-classical class I gene associated with 6.4 kb *Hin*d III fragment	3,6
DLA-DRA1	DRA	DR alpha chain	7,8
DLA-DRB1	DRB1 DRBB1	DR beta chain	7,9, 10,11
DLA-DRB2	DRB2, DRBB2	DRB pseudogene	7,12, 10,11
DLA-DQA1	DQA1	DQ alpha chain	7,10
DLA-DQB1	DQB1	DQ beta chain	7,13
LMP2	LMP2		Gerlach, personal communication

Table 6.1b   Other unconfirmed genes associated with the DLA complex

Name	Molecular characteristics	References
DLA-A	Unknown	9
DLA-12a	Class I pseudogene associated with 12 kb *Hin*d III fragment	6
C1pg-26	Class I processed gene associated with 2.6 kb *Hin*d III fragment. Not in DLA region	6
DLA-53	Class I pseudogene associated with 5.3 kb *Hin*d III fragment	6
DQB2	?Pseudogene	7,13
DPA	DP alpha chain	7
DPB1	DP beta chain	7
DPB2	DP beta chain	7
DOB	DO beta chain	7

Table 6.2   Accession numbers and references for DLA88 alleles

Allele	Previous equivalents	Accession numbers (Exon 2, Exon 3)	Reference
DLA-88*00101	dla88-01	AF100567, AF101486	(3)
DLA-88*00201	dla88-02	AF100568, AF101487	(3)
DLA-88*00301	dla88-03	AF100569, AF101488	(3)
DLA-88*00401	dla88-04	AF100570, AF101489	(3)
DLA-88*00402	dla88-12	AF100578, AF101497	(3)
DLA-88*00501	dla88-05	AF100571, AF101490	(3)
DLA-88*00601	dla88-06	AF100572, AF101491	(3)
DLA-88*00701	dla88-07	AF100573, AF101492	(3)
DLA-88*00801	dla88-08	AF100574, AF101493	(3)
DLA-88*00901	dla88-09	AF100575, AF101494	(3)
DLA-88*01001	dla88-10	AF100576, AF101495	(3)
DLA-88*01301	dla88-13	AF100579, AF101498	(3)
DLA-88*01401	dla88-14	AF100580, AF101499	(3)
DLA-88*01501	dla88-15	AF100581, AF101500	(3)
DLA-88*01601	dla88-16	AF100582, AF101501	(3)
DLA-88*01602	dla88-40	AF100606, AF101525	(3)
DLA-88*01701	dla88-17	AF100583, AF101502	(3)
DLA-88*01801	dla88-18	AF100584, AF101503	(3)
DLA-88*01901	dla88-19	AF100585, AF101504	(3)
DLA-88*02001	dla88-20	AF100586, AF101505	(3)
DLA-88*02201	dla88-22	AF100588, AF101507	(3)
DLA-88*02301	dla88-23	AF100589, AF101508	(3)
DLA-88*02401	dla88-24	AF100590, AF101509	(3)
DLA-88*02501	dla88-25	AF100591, AF101510	(3)
DLA-88*02601	dla88-26	AF100592, AF101511	(3)
DLA-88*02701	dla88-27	AF100593, AF101512	(3)
DLA-88*02801	dla88-28	AF100594, AF101513	(3)
DLA-88*02802	dla88-29	AF100595, AF101514	(3)
DLA-88*03001	dla88-30	AF100596, AF101515	(3)
DLA-88*03101	dla88-31	AF100597, AF101516	(3)
DLA-88*03401	dla88-34	AF100600, AF101519	(3)

Table 6.2   Accession numbers and references for DLA88 alleles (*continued*)

Allele	Previous equivalents	Accession numbers (Exon 2, Exon 3)	Reference
DLA-88*03501	dla88-35	AF100601, AF101520	(3)
DLA-88*03701	dla88-37	AF100603, AF101522	(3)
DLA-88*03801	dla88-38	AF100604, AF101523	(3)
DLA-88*03901	dla88-39	AF100605, AF101524	(3)
DLA-88*04101	dla88-41	AF100607, AF101526	(3)
DLA-88*04201	dla88-42	AF100608, AF101527	(3)
DLA-88*04301	dla88-43	AF100609, AF101528	(3)
DLA-88*04401	dla88-44	AF100610, AF101529	(3)
DLA-88*04801	dla88-48	AF218299, AF218300	(14)
DLA-88*50101	dla88-11	AF100577, AF101496	(3)
DLA-88*50201	dla88-33	AF100599, AF101518	(3)
DLA-88*50301	dla88-36	AF100602, AF101521	(3)
DLA-88*50401	dla88-46	AF218297, AF218298	(14)
DLA-88*50501	dla88-32	AF100598, AF101517	(3)
DLA-88*50601	dla88-47	AF218303, AF218304	(14)
DLA-88*50701	dla88-49	AF218301, AF218302	(14)
DLA-88*50801	dla88-21	AF100587, AF101506	(3)

named. Although there is some evidence for 2, 3, and 4 alleles respectively for DLA-12, DLA-64, and DLA-79,[3] no sequence alignments have been published, and no alleles have been lodged in GenBank to date. Therefore no allele names have been assigned by the Committee for these class I loci. **Tables 6.2–6.5** list all the named alleles at the following loci: DLA-88, DLA-DRB1, DLA-DQA1, and DLA-DQB1. The tables show new alleles in bold type, and include previous names, accession numbers, and for class II, the Canidae in which each allele has been found to date.

The class I alleles for the DLA-88 locus have been named according to the rules defined in the previous nomenclature report. Some of these alleles have an additional amino acid at codon 156, and these alleles have received names starting at DLA-88*50101. The sequential numbering for the nucleotides and the codons therefore includes codon 156, although many alleles are missing that codon.

NAMING OF MHC ALLELES FROM OTHER CANIDAE

The principles for naming the MHC genes and alleles in different Canidae (dog, grey wolf, red wolf, and coyote lineages), and how to apply them to sub-species and hybrids, were considered. The principles established here for the Canidae group may have applications in other animals that have been domesticated.

At this point in time it is not considered possible to distinguish class II alleles from domestic dogs, grey wolves, red wolves, and coyotes. As can be seen from **Tables 6.3– 6.5**, there is some overlap (especially for DLA-DQA1) in the occurrence of alleles in these different Canidae, at all three of the class II loci studied to date. Given that the sample sizes are small for some of the Canidae presently examined, it would seem likely that the degree of overlap will increase. As full-length genomic sequences are not as

Table 6.3 Accession numbers and references for DLA-DRB1 alleles and their distribution in different Canidae

Allele	Previous equivalents	Accession numbers	Ref	Dog (n >800)	Grey wolf (n = 50)	Red wolf (n = 2)	Mexican wolf (n = 5)	Coyote (n = 4)
DRB1*00101	Dw4, D1	M57529	15	+				
DRB1*00102	Dw3, D3	M57528, S76138	15, 19	+				
DRB1*00201	Dw1, D2	M57537	15	+				
DRB1*00202	D2a	U44777	11	+				
DRB1*00301	0902	AJ003012	17	+				
DRB1*00401	D4, D4m	M57532	15, 11	+				
DRB1*00501	D24, 2302	AJ003017, AF098496	17, [a]	+				
DRB1*00601	D6, D6m	M57534	15, 11	+	+		+	
DRB1*00701	D7	M57533	15	+				
RB1*00801	D8, D8m	M57535	15, 11	+				
DRB1*00802		AJ012456	18	+				
DRB1*00901	D9	M57531	15	+	+			
DRB1*010011	D25	AF016910	[b]	+				
DRB1*010012	Cafa-10, 1102	X93572	16	+				
DRB1*01101	2102, Cafa-11, 1112	X93573	16	+				
DRB1*01201	1902	AJ003015	17	+	+			
DRB1*01301	D13	U44778	11	+				
DRB1*01401	D14	U44779	11	+				
DRB1*01501	D15/Dw8, D15m, D24(partial seq)	M57536, AF016912	15, 11, [b]	+				
DRB1*01502	1502	AJ003013	17	+				
DRB1*01503	1503	AJ003014	17	+				
DRB1*01601		AJ012454	18	+				
DRB1*01701	D17	U44780	11	+				
DRB1*01801	D18	U44781	11	+				
DRB1*01901	D19	U44782	11	+				
DRB1*02001	D20	U58684	11	+				

Table 6.3   Accession numbers and references for DLA-DRB1 alleles and their distribution in different Canidae (*continued*)

Allele	Previous equivalents	Accession numbers	Ref	Dog (n >800)	Grey wolf (n = 50)	Red wolf (n = 2)	Mexican wolf (n = 5)	Coyote (n = 4)
DRB1*02101	D21	U44783	11	+				
DRB1*02201	D22	U58685	11	+				
DRB1*02301	2301	AJ003016	17	+				
DRB1*02401	2401	AJ003018	17	+				
DRB1*02501	2501	AJ003019	17	+				
DRB1*02601	2601	AJ003020	17	+				
DRB1*02701	drb 26	AF061039	20	+				
DRB1*02801	drb 25	AF061038	20	+				
DRB1*02901		AJ012455	18	+	+			
DRB1*03001	D23	AF016911	b	+				
Partial sequence	1-Dob-A	M30129	c	+				
Partial sequence	1-Dob-B	M30130	c	+				
Partial sequence	2-Dob	M30131	c	+				
Partial sequence	3-Lab	M30132	c	+				
Partial sequence	4-Pood	M30133	c	+				
DRB1*03101		AF336108	e		+			
DRB1*03201		AY009941	f	+				
DRB1*03301		AF343737	f	+				
DRB1*03501		AF336109	e		+			
DRB1*03601		AF336110	e		+			
DRB1*03701		AF343738	d		+			
DRB1*03801		AF343739	d		+		+	
DRB1*03901		AF343740	d, g		+	+		
DRB1*04001		AF343741	d	+				
DRB1*04101		AF343742	d, g		+			
DRB1*04201		AF343743	d, g					+
DRB1*04301		AF343744	d, g				+	
DRB1*04401		AF343745	d, g		+			
DRB1*04501		AF343746	d, g		+			+

Table 6.3 Accession numbers and references for DLA-DRB1 alleles and their distribution in different Canidae (*continued*)

Allele	Previous equivalents	Accession numbers	Ref	Dog (n >800)	Grey wolf (n = 50)	Red wolf (n = 2)	Mexican wolf (n = 5)	Coyote (n = 4)
DRB1*04601		AF343747	d	+				
DRB1*04701		AF343748	d	+				

Additional references:

[a]Wagner, GenBank 1998, unpublished
[b]Francino, GenBank 1997, unpublished
[c]Motoyama, GenBank 1996, unpublished

[d]LJ Kennedy, unpublished
[e]LJ Kennedy and JM Angles, unpublished
[f]JM Angles, unpublished
[g]LJ Kennedy and P Hedrick, unpublished

Table 6.4 Accession numbers and references for DLA-DQA1 alleles and their distribution in different Canidae

Allele	Previous equivalents	Accession numbers	Ref	Dog (n >800)	Grey wolf (n = 50)	Red wolf (n = 2)	Mexican wolf (n = 5)	Coyote (n = 4)
DQA1*00101	0101, Dqa2	M74907, U44786	21, 23	+	+		+	+
DQA1*00201	0201, Dqa9	M74909, U75455	21, [a]	+	+		+	
DQA1*00301	0301	Y07944	22	+			+	
DQA1*00401	0203, Dqa4	Y07943, U44788	22, 23	+	+			
DQA1*005011	0202, Dqa3	M74910, U44787	21, 23	+	+		+	
DQA1*005012	Dqa5	U44789	23	+				
DQA1*00601	0103, Dqa6	Y07942, U44790	22, 23	+	+			
DQA1*00701	Dqa7	U44842	23	+	+	+		
DQA1*00801	Dqa8	U61400	[a]	+				
DQA1*00901	0102, Dqa1	M74908, U44785	21, 23	+			+	+
DQA1*01001		AJ130870	18	+	+			
DQA1*01101		AF343733	[b]		+			
DQA1*01201		AF343734	[c]	+	+			+
DQA1*01301		AF343735	[b]		+			
DQA1*01401		AF336107	[b]	+	+			
DQA1*01501		AF343736	[c]	+				

Additional references:

[a]Wagner, GenBank 1996, unpublished
[b]LJ Kennedy and JM Angles, unpublished
[c]LJ Kennedy unpublished

Table 6.5   Accession numbers and references for DLA-DQB1 alleles and their distribution in different Canidae

Allele	Previous equivalents	Accession numbers	Ref	Dog (n = 200)	Grey wolf (n = 40)	Red wolf (n = 2)	Mexican wolf (n = 7)	Coyote (n = 0)
DQB1*00101	0101, dqb2, dqb0102	M90802, AF043147, AF016905	24, 13, a	+				
DQB1*00201	0201, dqb3, dqb0203	M90803, AF043148, AF016908	24, 13, a	+				
DQB1*00301	0301, dqb6	M90804, AF043151	24, 13	+				
DQB1*00401	0401, dqb5	M90805, AF043150	24, 13	+				
DQB1*00501	0501, dqb12	Y07947, AF043157	22, 13	+				
DQB1*00502		AF336111	e	+				
DQB1*00701	0701, dqb4, dqb1001	Y07949, AF043149, AF016907	22, 13, a	+	+		+	
DQB1*008011	0801, dqb1	AF043492, AF043167	b, 13	+	+		+	
DQB1*008012		AF336112	e					
DQB1*00802		AF343731	d	+				
DQB1*01101	dqb1101	AF016904	a	+				
DQB1*01201		AY009942	e	+				
DQB1*01301	dqb13	AF043158	13	+				
DQB1*01302	dqb14	AF043159	13	+				
DQB1*01303	qb7, dqb0901	AF043152, AF016906	13, a	+			+	
DQB1*01401		AF343732	c		+			
DQB1*01501	dqb15	AF043160	13	+		+		
DQB1*01601	dqb16	AF043161	13	+				
DQB1*01701	dqb17	AF043162	13	+	+			
DQB1*01801	dqb18	AF043163	13	+				
DQB1*01901	dqb9	AF043154	13	+				
DQB1*02001	dqb20, dqb23	AF043165, AF113705	13, 25	+				
DQB1*02002	dqb19	AF043164	13	+	+			
DQB1*02101	dqb11	AF043156	13	+				
DQB1*02201	dqb10	AF043155	13	+				
DQB1*02301	dqb8, dqb0303	AF043153, AF016909	13, a	+	+			
Partial sequence	0302	Y07946	22	+				

Table 6.5   Accession numbers and references for DLA-DQB1 alleles and their distribution in different Canidae (*continued*)

Allele	Previous equivalents	Accession numbers	Ref	Dog (n = 200)	Grey wolf (n = 40)	Red wolf (n = 2)	Mexican wolf (n = 7)	Coyote (n = 0)
Partial sequence	0601	Y07948	22	+				
Partial sequence	0202	Y07945	22	+				
DQB1*02401		AY009940	e		+			
DQB1*02601	dqb22	F113704	25	+				
DQB1*02701	dqb24	AF113706	25	+				
DQB1*02801		AF343730	d	+				
DQB1*03001	dqb26	AF241781	25	+				
DQB1*03101	dqb27	AF241529	25	+				
DQB1*03201		AJ311104	c		+			
DQB1*03301		AJ311105	c		+			
DQB1*03401		AJ311106	c		+			
DQB1*03501		AJ311107	c	+	+			

Additional references:
[a]Francino, GenBank, 1997 unpublished
[b]Polvi, GenBank 1998, unpublished
[c]LJ Kennedy and JM Angles, unpublished
[d] LJ Kennedy, unpublished
[e]JM Angles, unpublished

yet available, there are no data as to whether the other exons and the intron sequences are also identical. Compelling evidence of MHC identity exists between these Canidae, since several three locus haplotypes have been shown to be shared between dogs, wolves and coyotes.

DLA class II alleles found in any of these Canidae will be named in a common series, until such time as they can be shown to be different.

UPDATED SEQUENCE ALIGNMENTS

Class I sequence alignments

Figure 6.1 and **Figure 6.2** show the updated comparative nucleotide and amino acid sequence alignments for the class I gene DLA-88. The hypervariable regions are highlighted on each alignment.

Class II sequence alignments

Figures 6.3–6.8 show the updated comparative nucleotide and amino acid sequence alignments for the class II genes: DLA-DRB1, DLA-DQA1, and DLA-DQB1. The hypervariable regions are highlighted on each alignment.

ACKNOWLEDGMENTS

The conclusions reported here were reached after discussion involving input from Philip Hedrick, Steve O'Brien, Peter Parham, Ronald Bontrop, Bob Wayne, Steven Marsh, George Russell, Shirley Ellis, and the DLA Nomenclature Committee.

The help of SGE Marsh and J Robinson in preparing the sequence alignments in this report is gratefully acknowledged. Comparative sequence alignments were generated by Format_Aln.cgi, which was written by James Robinson for the IMGT/HLA Database.

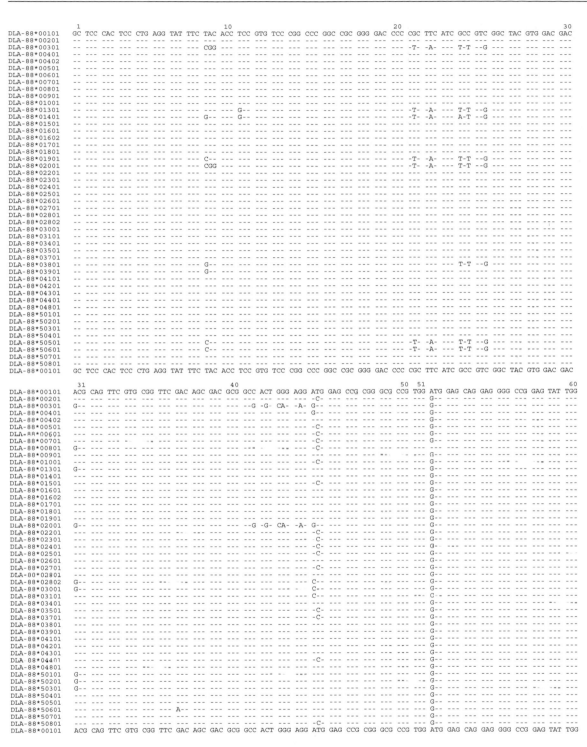

Figure 6.1   DLA-88 nucleotide alignment

```
                    |---------------------------------HVR 1-----------------------------------|
                                                           70                  75                  80                                               90
DLA-88*00101    GAC CGG GAG ACG CGG ACC GCC AAG GAG ACC GCA CAG AGG TAC CGA GTG GAC CTG GAC ACC CTG CGC GGC TAC TAC AAC CAG AGC GAG GCC
DLA-88*00201    --- --- C-- --- --- --- TT- --- --- --- --- --- GT- --- --- --- --- --- --- --- --- --- --- --- --- --- --- --- --- ---
DLA-88*00301    --- --- --- --- --- --- T-- --- --- -A- --- --- GT- --- --- --- --- --- --- --- --- --- --- --- --- --- --- --- --- ---
DLA-88*00401    --- -C- C-- --- --- --- AT- --- --- --- --- -G- -CT -T- --- --- --- --- --- --- --- --- --- --- --- --- --- --- --- ---
DLA-88*00402    --- -C- C-- --- --- --- AT- --- --- --- --- -G- -CT -T- --- --- --- --- --- --- --- --- --- --- --- --- --- --- --- ---
DLA-88*00501    --- --- C-- --- --- -A- TT- --- --- --- --- -G- -AT -T- --- --- --- --- --- --- --- --- --- --- --- --- --- --- --- ---
DLA-88*00601    --- -C- C-- --- --- --- AT- --- --- --- --- --- CT- --- --- --- --- --- --- --- --- --- --- --- --- --- --- --- --- ---
DLA-88*00701    --- --- --- --- --- -A- TT- --- --- --- --- -G- -AT -T- --- --- --- --- --- --- --- --- --- --- --- --- --- --- --- ---
DLA-88*00801    --- --- --- --- --- --- --- --- --- --- --- --- --- --- --- --- --- --- --- --- --- --- --- --- --- --- --- --- --- ---
DLA-88*00901    --- -C- C-- --- --- G-- AT- --- --- --- --- -G- -CT -T- --- --- --- --- --- --- --- --- --- --- --- --- --- --- --- ---
DLA-88*01001    --- --- C-- --- --- --- AT- --- --- --- --- -G- -CT -T- --- --- --- --- --- --- --- --- --- --- --- --- --- --- --- ---
DLA-88*01301    --- G-- --- --- --- -A- -T- --- --- --- --- -G- -CT -T- --- --- A-- --- --- --- --- --- --- --- --- --- --- --- --- ---
DLA-88*01401    --- G-- --- --- --- -AA -T- --- --- --- --- -G- -CT -T- --- --- AG- --- --- --- --- --- --- --- --- --- --- --- --- ---
DLA-88*01501    --- --- C-- --- --- --- TT- --- --- --- --- -G- -CT -T- --- --- --- --- --- --- --- --- --- --- --- --- --- --- --- ---
DLA-88*01601    --- -C- C-- --- --- --- AT- --- --- --- --- --- --- --- --- --- AG- --- --- --- --- --- --- --- --- --- --- --- --- ---
DLA-88*01602    --- -C- C-- --- --- --- AT- --- --- --- --- --- --- --- --- --- AG- --- --- --- --- --- --- --- --- --- --- --- --- ---
DLA-88*01701    --- -C- C-- --- --- --- AT- --- --- --- --- --- --- --- --- --- AG- --- --- --- --- --- --- --- --- --- --- --- --- ---
DLA-88*01801    --- -C- C-- --- --- --- AT- -G- --- --- --- --- --- --- --- --- AG- --- --- --- --- --- --- --- --- --- --- --- --- ---
DLA-88*01901    --- G-- C-- --- --- -T- T-- --- --- -A- --- --- CT- --- --- --- --- --- --- --- --- --- --- --- --- --- --- --- --- ---
DLA-88*02001    --- --- --- --- --- --- T-- --- --- -A- --- --- GT- --- --- --- --- T-- --- --- --- -A- --- --- --- --- --- --- --- ---
DLA-88*02201    --- -C- C-- --- --- --- AT- --- --- --- --- -G- -CT -T- --- --- --- --- --- --- --- --- --- --- --- --- --- --- --- ---
DLA-88*02301    --- G-- C-- --- --- -AA -T- --- --- --- --- --- CT- --- --- --- --- --- --- C-- --- --- --- --- --- --- --- --- --- ---
DLA-88*02401    --- --- C-- --- --- --- AT- --- --- --- --- -GA G-- --- --- --- -G- --- --- --- --- --- --T --- --- --- --- --- --- ---
DLA-88*02501    --- --- --- --- --- -A- TT- --- --- --- --- -GA G-- --- --- --- --- --- --- --- --- --- --- --- --- --- --- --- --- ---
DLA-88*02601    --- --- --- --- --- -A- TT- --- --- --- --- -GA G-- --- --- --- -G- --- --- --- --- --- --- --- --- --- --- --- --- ---
DLA-88*02701    --- --- C-- --- --- -A- TT- --- --- --- --- -GA G-- --- --- --- -G- --- --- --- --- --- --- --- --- --- --- --- --- ---
DLA-88*02801    --- --- --- --- --- -A- TT- --- --- --- --- --- GT- --- --- --- -G- --- --- --- --- --- --- --- --- --- --- --- --- ---
DLA-88*02802    --- --- --- --- --- -A- TT- --- --- --- --- --- GT- --- --- --- -G- --- --- --- --- --- --- --- --- --- --- --- --- ---
DLA-88*03001    --- --- --- --- --- -A- CT- --- --- --- --- --- GT- --- --- --- -G- --- --- --- --- --- --- --- --- --- --- --- --- ---
DLA-88*03101    --- -C- C-- --- --- --- TT- --- --- --- --- --- CT- --- --- --- --- --- --- --- --- --- --- --- --- --- --- --- --- ---
DLA-88*03401    --- G-- --- --- --- -AA -T- --- --- --- --- --- CT- --- --- --- --- --- --- --- --- --- --- --- --- --- --- --- --- ---
DLA-88*03501    --- --- C-- --- --- --- AT- --- --- --- --- -G- -CT -T- --- --- --- --- --- --- --- --- --- --- --- --- --- --- --- ---
DLA-88*03701    --- --- C-- --- --- --- TT- --- --- --- --- --- GT- -T- --- --- --- --- --- --- --- --- --- --- --- --- --- --- --- ---
DLA-88*03801    --- --- C-- --- --- --- T-- --- --- -A- --- --- GT- -G- --- --- --- --- --- --- --- --- --- --- --- --- --- --- --- ---
DLA-88*03901    --- G-- --- --- --- -AA -T- --- --- --- --- --- CT- --- --- --- --- --- --- --- --- --- --- --- --- --- --- --- --- ---
DLA-88*04101    --- G-- --- --- --- -A- CT- --- --- --- --- --- GT- --- --- --- --- --- --- --- --- --- --- --- --- --- --- --- --- ---
DLA-88*04201    --- --- C-- --- --- --- TT- --- --- --- --- -GA G-- --- --- --- -G- --- --- --- --- --- --- --- --- --- --- --- --- ---
DLA-88*04301    --- --- --- --- --- --- --- --- --- --- --- --- GT- --- --- --- --- --- --- --- --- --- --- --- --- --- --- --- --- ---
DLA-88*04401    --- -C- C-- --- --- --- AT- --- --- --- --- --- CT- --- --- --- --- --- --- --- --- --- --- --- --- --- --- --- --- ---
DLA-88*04801    --- -C- C-- --- --- --- AT- --- --- --- --- -G- -CT -T- --- --- --- --- --- --- --- --- --- --- --- --- --- --- --- ---
DLA-88*50101    --- -C- C-- --- --- --- AT- --- --- --- --- -G- -CT -T- --- --- --- --- --- --- --- --- --- --- --- --- --- --- --- ---
DLA-88*50201    --- -C- C-- --- --- --- AT- --- --- --- --- -G- -CT -T- --- --- --- --- --- --- --- --- --- --- --- --- --- --- --- ---
DLA-88*50301    --- --- C-- --- --- --- AT- --- --- --- --- -G- -CT -T- --- --- --- --- --- --- --- --- --- --- --- --- --- --- --- ---
DLA-88*50401    --- -C- C-- --- --- --- AT- --- --- --- --- -G- -CT -T- --- --- --- --- --- --- --- --- --- --- --- --- --- --- --- ---
DLA-88*50501    --- --- C-- --- --- -T- T-- --- --- -A- --- --- GT- --- --- --- --- --- --- --- --- --- --- --- --- --- --- --- --- ---
DLA-88*50601    --- -C- C-- --- --- -T- T-- --- --- -A- --- --- GT- --- --- --- AG- --- --- --- --- --- --- --- --- --- --- --- --- ---
DLA-88*50701    --- G-- --- --- --- -AA -T- --- --- --- --- --- GT- --- --- --- --- --- --- --- --- --- --- --- --- --- --- --- --- ---
DLA-88*50801    --- G-- --- --- --- -AA -T- --- --- --- --- --- GT- --- --- --- --- --- --- --- --- --- --- --- --- --- --- --- --- ---
DLA-88*00101    GAC CGG GAG ACG CGG ACC GCC AAG GAG ACC GCA CAG AGG TAC CGA GTG GAC CTG GAC ACC CTG CGC GGC TAC TAC AAC CAG AGC GAG GCC
```

```
                |-----------------------------------HVR 2-----------------------------------------------------|
                 91                                  100                                     110                                     120
DLA-88*00101    GGG TCT CAC ACC CGC CAG ACC ATG TAC GGC TGT GAC CTG GGG CCC GGC GGG CGC CTC CTC CGC GGG TAC AGT CAG GAC GCC TAC GAC GGC GCC
DLA-88*00201    --- --- --- --- --- --- --- -C- --- --- --- --- --- --- --- --- --- --- --- --- --- --- --- --- --- --- --- --- --- --- ---
DLA-88*00301    --- --- --- --- --- --- --- -C- -TT --- --- --- --- --- --- --- --- --- --- --- --- --- --- T-G --- --- --- --- --- --- ---
DLA-88*00401    --- --- --- --- --- --- --- --- --- --- --- --- --- --- --- --- --- --- --- --- --- --- --- --G --- --- --- --- --- --- ---
DLA-88*00402    --- --- --- --- --- --- --- --- --- --- --- --- --- --- --- --- --- --- --- --- --- --- --- --G --- --- --- --- --- --- ---
DLA-88*00501    --- G-- --- --- -T- --- TGG --- --- --- --- --- --- --- --- --- --- --- --- --- --- --- --- T-G --- --- --- --- --- --- ---
DLA-88*00601    --- --- --- --- --- --- --- --- --- --- --- --- --- --- --- --- --- --- --- --- --- --- --- --- --- --- --- --- --- --- ---
DLA-88*00701    --- G-- --- --- -T- --- TGG --- --- --- --- --- --- --- --- --- --- --- --- --- --- --- --- T-G --- --- --- --- --- --- ---
DLA-88*00801    --- --- --- --- --- --- --- --- --- --- --- --- --- --- --- --- --- --- --- --- --- --- --- --- --- --- --- --- --- --- ---
DLA-88*00901    --- --- --- --- --- --- --- --- --- --- --- --- --- --- --- --- --- --- --- --- --- --- --- --G --- --- --- --- --- --- ---
DLA-88*01001    --- --- --- --- --- --- --- --- --- --- --- --- --- --- --- --- --- --- --- --- --- --- --- --G --- --- --- --- --- --- ---
DLA-88*01301    --- --- --- --- TT- --- TGG --- -TT --- --- --- --- --- --- --- --- --- --- --- --- --- --- --- --- -T- --- --- --- --- ---
DLA-88*01401    --- --- --- --- AT- --- TGG -CT -CG --- --C --- G-- --- --G -A- --- --- --- --- --- --- --T GAC --- TT- --- --- --- --- ---
DLA-88*01501    --- --- --- --- --- --- --- --- --- --- --- --- --- --- --- --- --- --- --- --- --- --- --- --G --- --G --- --- --- --- ---
DLA-88*01601    --- --- --- --- --- --- --- --- --- --- --- --- --- --- --- --- --- --- --- --- --- --- --- --G --- --G --- --- --- --- ---
DLA-88*01602    --- --- --- --- --- --- --- --- --- --- --- --- --- --- --- --- --- --- --- --- --- --- --- --G --- --G --- --- --- --- ---
DLA-88*01701    --- --- --- --- AT- --- --- --- --- --- --T --- --- --- --- --- --- --- --- --- --- --- --- --G --- -T- --- --- --- --- ---
DLA-88*01801    --- --- --- --- AT- --- --- --- --- --- --T --- --- --- --- --- --- --- --- --- --- --- --- --G --- -T- --- --- --- --- ---
DLA-88*01901    --- --- --- --- --- --- --- --- --- --- --- --- --- --- --- --- --- --- --- --- --- --- --- --- --- -T- --- --- --- --- ---
DLA-88*02001    --- --- --- --- --- --- --- --- -TT --- --- --- --- --- --- --- --- --- --- --- --- --- --- T-G --- --- --- --- --- --- ---
DLA-88*02201    --- --- --- --- --- --- --- --- --- --- --- --- --- --- --- --- --- --- --- --- --- --- --- --- --- --- --- --- --- --- ---
DLA-88*02301    --- --- --- --- AT- --- --- --- --- --- --- --- --- --- --- --- --- --- --- --- --- --- --- --G --- --- --- --- --- --- ---
DLA-88*02401    --- --- --- --- AT- --- --- --- --- --- --- --- --- --- --- --- --- --- --- --- --- --- --- --G --- --- --- --- --- --- ---
DLA-88*02501    --- --- --- --- --- --- --- --- --- --- --- --- --- --- --- --- --- --- --- --- --- --- --- --G --- --G --- --- --- --- ---
DLA-88*02601    --- --- --- --- AT- --- --- --- --- --- --- --- --- --- --- --- --- --- --- --- --- --- --- --G --- --- --- --- --- --- ---
DLA-88*02701    --- --- --- --- AT- --- --- --- --- --- --- --T --- --- --- --- --- --- --- --- --- --- --- --G --- --- --- --- --- --- ---
DLA-88*02801    --- --- --- --- AT- --- --- --- --- --- --- --- --- --- --- --- --- --- --- --- --- --- --- --- --- --- --- --- --- --- ---
DLA-88*02802    --- --- --- --- AT- --- --- --- --- --- --- --- --- --- --- --- --- --- --- --- --- --- --- --- --- --- --- --- --- --- ---
DLA-88*03001    --- --- --- --- AT- --- --- --- --- --- --- --- --- --- --- --- --- --- --- --- --- --- --- --G --- --- --- --- --- --- ---
DLA-88*03101    --- --- --- --- --- --- --- --- --- --- --- --- --- --- --- --- --- --- --- --- --- --- --- --- --- --- --- --- --- --- ---
DLA-88*03401    --- --- --- --- AT- --- --- --- --- --- --- --- --- --- --- --- --- --- --- --- --- --- --- --- --- --- --- --- --- --- ---
DLA-88*03501    --- --- --- --- --- --- --- --- --- --- --- --- --- --- --- --- --- --- --- --- --- --- --- --G --- --- --- --- --- --- ---
DLA-88*03701    --- --- --- --- --- --- --- -C- --- --- --- --- --- --- --- --- --- --- --- --- --- --- --- --- --- --- --- --- --- --- ---
DLA-88*03801    --- --- --- --- --- --- --- --- --- --- --- --- --- --- --- --- --- --- --- --- --- --- --- T-G --- -T- --- --- --- --- ---
DLA-88*03901    --- --- --- --- --- --- --- --- --- --- --- --- --- --- --- --- --- --- --- --- --- --- --- --G --- -T- --- --- --- --- ---
DLA-88*04101    --- --- --- --- AT- --- --- --- --- --- --- --- --- --- --- --- --- --- --- --- --- --- --- --G --- --- --- --- --- --- ---
DLA-88*04201    --- --- --- --- AT- --- --- --- --- --- --- --- --- --- --- --- --- --- --- --- --- --- --- --G --- --- --- --- --- --- ---
DLA-88*04301    --- --- --- --- --- --- --- --- --- --- --- --- --- --- --- --- --- --- --- --- --- --- --- T-G --- --- --- --- --- --- ---
DLA-88*04401    --- --- --- --- --- --- --- --- --- --- --- --- --- --- --- --- --- --- --- --- --- --- --- --G --- --- --- --- --- --- ---
DLA-88*04801    --- --- --- --- --- --- --- --- --- --- --- --- --- --- --- --- --- --- --- --- --- --- --- --G --- -T- --- --- --- --- ---
DLA-88*50101    --- --- --- --- --- --- --- --- --- --- --- --- --- --- --- --- --- --- --- --- --- --- --- --G --- --- --- --- --- --- ---
DLA-88*50201    --- --- --- --- --- --- --- --- --- --- --- --- --- --- --- --- --- --- --- --- --- --- --- --- --- --- --- --- --- --- ---
DLA-88*50301    --- --- --- --- --- --- --- --- --- --- --- --- --- --- --- --- --- --- --- --- --- --- --- --- --- --- --- --- --- --- ---
DLA-88*50401    --- --- --- --- AT- --- --- --- --- --- --- --- --- --- --- --- --- --- --- --- --- --- --- --- --- --- --- --- --- --- ---
DLA-88*50501    --- --- --- --- AT- --- --- --- --- --- --- --- --- --- --- --- --- --- --- --- --- --- --- --- --- -T- --- --- --- --- ---
DLA-88*50601    --- --- --- --- AT- --- --- --- --- --- --- --- --- --- --- --- --- --- --- --- --- --- --- --- --- -T- --- --- --- --- ---
DLA-88*50701    --- --- --- --- --- --- --- --- --- --- --- --- --- --- --- --- --- --- --- --- --- --- --- --G --- -T- --- --- --- --- ---
DLA-88*50801    --- --- --- --- AT- --- --- --- --- --- --- --- --- --- --- --- --- --- --- --- --- --- --- --G --- --- --- --- --- --- ---
DLA-88*00101    GGG TCT CAC ACC CGC CAG ACC ATG TAC GGC TGT GAC CTG GGG CCC GGC GGG CGC CTC CTC CGC GGG TAC AGT CAG GAC GCC TAC GAC GGC GCC
```

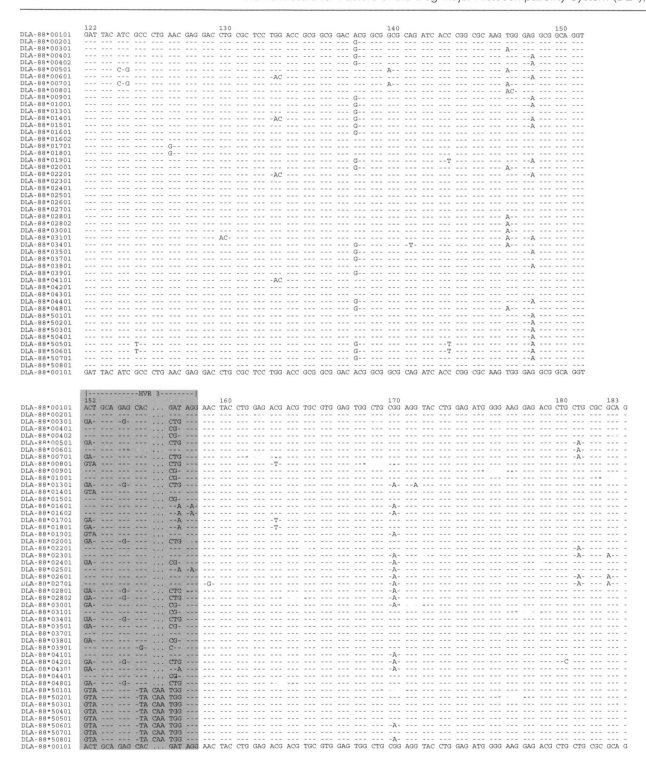

Figure 6.1   (*continued*)

```
                                          |------HVR 1----|
                1    10        20         30        40        50        60        70        80        90
DLA-88*00101    GSHSLRYFYT SVSRPGRGDP RFIAVGYVDD TQFVRFDSDA ATGRMEPRAP WMEQEGPEYW DRETRTAKET AQRYRVDLDT LRGYYNQSEA
DLA-88*00201    ---------- ---------- ---------- ---------- ---T----- -V-------- -Q--F---- --V------ ----------
DLA-88*00301    --------R- ---------- LY-S------ A--------- -SQKV----- -V-------- ----S--N --V------ ----------
DLA-88*00401    ---------- ---------- ---------- ---------- ---V----- -V-------- -PQ---I-- --RTF---- ----------
DLA-88*00402    ---------- ---------- ---------- ---------- ---------- -V-------- -PQ---I-- --RTF---- ----------
DLA-88*00501    ---------- ---------- ---------- ---------- ---T----- -V-------- -Q--NF--- -RNF----- ----------
DLA-88*00601    ---------- ---------- ---------- ---------- ---T----- -V-------- -PQ---I-- --L------ ----------
DLA-88*00701    ---------- ---------- ---------- ---------- ---T----- -V-------- ----NF--- -RNF----- ----------
DLA-88*00801    ---------- ---------- ---------- A--------- ---T----- ---------- ---------- ---------- ----------
DLA-88*00901    ---------- ---------- ---------- ---------- ---------- -V-------- -PQ--AI-- --RTF---- ----------
DLA-88*01001    ---------- ---------- ---------- ---------- ---------- -V-------- -Q----I-- --RTF---- ----------
DLA-88*01301    ---------- A--------- LY-S------ A--------- ---------- -V-------- -G---NV-- --RTF--N- ----------
DLA-88*01401    --------D- A--------- LY-T----- ---------- ---------- -V-------- -G---KV-- --RTF--S- ----------
DLA-88*01501    ---------- ---------- ---------- ---------- ---T----- -V-------- -Q---F--- --RTF---- ----------
DLA-88*01601    ---------- ---------- ---------- ---------- ---------- -V-------- -PQ---I-- ------S-- ----------
DLA-88*01602    ---------- ---------- ---------- ---------- ---------- -V-------- -PQ---I-- ------S-- ----------
DLA-88*01701    ---------- ---------- ---------- ---------- ---------- -V-------- -PQ---I-- ------S-- ----------
DLA-88*01801    ---------- ---------- ---------- ---------- ---------- -V-------- -PQ--IR-- ------S-- ----------
DLA-88*01901    --------H- ---------- LY-S------ ---------- ---------- -V-------- -GQ--IS-N --L------ ----------
DLA-88*02001    --------R- ---------- LY-S------ A--------- -SQKV----- -V-------- ----S--N --V------ --H-------
DLA-88*02201    ---------- ---------- ---------- ---------- ---T----- -V-------- -PQ---I-- --RTF---- ----------
DLA-88*02301    ---------- ---------- ---------- ---------- ---T----- -V-------- -GQ--KV-- --L-----P ----------
DLA-88*02401    ---------- ---------- ---------- ---------- ---T----- -V-------- -Q----I-- --RG--G-- ----------
DLA-88*02501    ---------- ---------- ---------- ---------- ---T----- -V-------- ----NF--- --RG------ ----------
DLA-88*02601    ---------- ---------- ---------- ---------- ---------- -V-------- ----NF--- --RG--G-- ----------
DLA-88*02701    ---------- ---------- ---------- ---------- ---T----- -V-------- -Q--NF--- --RG--G-- ----------
DLA-88*02801    ---------- ---------- ---------- ---------- ---------- -V-------- ----NF--- --V--G--- ----------
DLA-88*02802    ---------- ---------- ---------- A--------- ---L----- -V-------- ----NF--- --V--G--- ----------
DLA-88*03001    ---------- ---------- ---------- A--------- ---L----- -V-------- ----NL--- --V--G--- ----------
DLA-88*03101    ---------- ---------- ---------- ---------- ---L----- -V-------- PQ---F--- --L------ ----------
DLA-88*03401    ---------- ---------- ---------- ---------- ---------- -V-------- -G---KV-- --L------ ----------
DLA-88*03501    ---------- ---------- ---------- ---------- ---T----- -V-------- -Q----I-- --RTF---- ----------
DLA-88*03701    ---------- ---------- ---------- ---------- ---T----- -V-------- -Q---F--- --VF----- ----------
DLA-88*03801    -------D- ---------- ---S----- ---------- ---------- -V-------- -Q---S--N --VC----- ----------
DLA-88*03901    -------D- ---------- ---------- ---------- ---------- -V-------- -G---KV-- --L------ ----------
DLA-88*04101    ---------- ---------- ---------- ---------- ---------- -V-------- -G---NL-- --V------ ----------
DLA-88*04201    ---------- ---------- ---------- ---------- ---------- -V-------- -Q---F--- --RG--G-- ----------
DLA-88*04301    ---------- ---------- ---------- ---------- ---------- -V-------- ---------- --V------ ----------
DLA-88*04401    ---------- ---------- ---------- ---------- ---T----- -V-------- ---------- --L------ ----------
DLA-88*04801    ---------- ---------- ---------- ---------- ---------- -V-------- -PQ---I-- --RTF---- ----------
DLA-88*50101    ---------- ---------- ---------- A--------- ---------- -V-------- -PQ---I-- --RTF---- ----------
DLA-88*50201    ---------- ---------- ---------- A--------- ---------- -V-------- -PQ---I-- --RTF---- ----------
DLA-88*50301    ---------- ---------- ---------- A--------- ---------- -V-------- -PQ---I-- --RTF---- ----------
DLA-88*50401    ---------- ---------- ---------- ---------- ---------- -V-------- -Q----I-- --RTF---- ----------
DLA-88*50501    --------H- ---------- LY-S------ ---------- ---------- -V-------- -PQ---I-- --RTF---- ----------
DLA-88*50601    --------H- ---------- LY-S------ -------N-- ---------- -V-------- -Q--IS--N --V------ ----------
DLA-88*50701    ---------- ---------- ---------- ---------- ---------- -V-------- -PQ--IS--N --V--S-- ----------
DLA 88*50001    ---------- ---------- ---------- ---------- ----T----- -V-------- -G---KV-- V ------ ----------
DLA-88*00101    GSHSLRYFYT SVSRPGRGDP RFIAVGYVDD TQFVRFDSDA ATGRMEPRAP WMEQEGPEYW DRETRTAKET AQRYRVDLDT LRGYYNQSEA
```

Figure 6.2(a) DLA-88 exon 2 amino acid alignment

```
                |----------HVR 2----------|                                        HVR 3-
                100       110        120       130       140       150       160       170       180
DLA-88*00101    GSHTRQTMYG CDLGPGGRLL RGYSQDAYDG ADYIALNEDL RSWTAADTAA QITRRKWEAA GTAEH.DRNY LETTCVEWLR RYLEMGKETL LRA
DLA-88*00201    -------T-- ---------- ---------- ---- ---------- -------A-- ---------- ---------- ---------- ---
DLA-88*00301    ------TF-- ---------- ---W------ ---- ---------- -------A-- ------R--- -D-G-.L-- ---- ------------ ---
DLA 88*00401    ---------- ---------- ---P------ ---- ---------- -------A-- ---------- ------.R-- ---------- ---
DLA-88*00402    ---------- ---------- ---R------ ---- ---------- -------A-- ---------- ------.R-- ---------- ---
DLA-88*00501    A  L W--- ---------- ---W------ ---- ---L------ --------T ------R--- -D--.L-- ---------- Q--
DLA-88*00601    ---------- ---------- ---------- ---- ---------- --Y------ ---------- ------.R-- ---------- Q--
DLA-88*00701    -A-L-W--- ---------- ---W------ ---- ---L------ --------T ------R--- -D--.L-- ---------- Q--
DLA-88*00801    ---------- ---------- ---------- ---- ---------- -------T-- -V--.L-- --M------- --- ---
DLA-88*00901    ---------- ---------- ---R------ ---- ---------- -------A-- ---------- ------.R-- ---------- ---
DLA-88*01001    ---------- ---------- ---R------ ---- ---------- -------A-- ---------- ------.R-- ---------- ---
DLA-88*01301    ---F-W-F- ---------- ---V------ ---- ---------- -------A-- ---------- -D-G-.L-- ----------Q ---
DLA-88*01401    ---I-WTS- --V-D---- --D-F----- ---- --Y----A-- ---------- -V--.-- ---------- ---
DLA-88*01501    ---------- ---------- ---R-E---- ---- ---------- -------A-- ---------- ------.R-- ---------- ---
DLA-88*01601    ---------- ---------- ---R-E---- ---- ---------- -------A-- ---------- ------.EK-- ----------Q ---
DLA-88*01602    ---------- ---------- ---P-E---- ---- ---------- ---------- ---------- ------.EK-- ----------Q ---
DLA-88*01701    ---I----- ---------- ---R-V---- ------D--- ---------- ---------- -D--.E-- ---M------ ---
DLA-88*01801    ---I----- ---------- ---R-V---- ------D--- ---------- ---------- -D--.E-- ---M------ ---
DLA-88*01901    ---------- ---------- ----V----- ---- ---------- -------A-- ---------- -V--.-- ----------Q ---
DLA-88*02001    -----F-- ---------- ---W------ ---- ---------- -------A-- ---R--- -D-G-.L-- ---------- ---
DLA-88*02201    ---------- ---------- ---------- ---- ---------- --Y------ ---------- ---------- ---------- Q--
DLA-88*02301    ---I----- ---------- ---R------ ---- ---------- ---------- ---------- ----------Q Q-T
DLA-88*02401    ---I----- ---------- ---R------ ---- ---------- ---------- ---------- -D--.R-- ----------Q ---
DLA-88*02501    ---------- ---------- ---P-E---- ---- ---------- ---------- ---------- ------.EK-- ----------Q ---
DLA-88*02601    ---I----- ---------- ---R------ ---- ---------- ---------- ---------- ----------Q Q-T
DLA-88*02701    ---I----- ---------- ---R------ ---- ---------- ---------- ------S--- ----------Q Q-T
DLA-88*02801    ---I----- ---------- ---R------ ---- ---------- -------R-- -D-G-.L-- ----------Q ---
DLA-88*02802    ---I----- ---------- ---------- ---- ---------- -------R-- -D--.R-- ----------Q ---
DLA-88*03001    ---I----- ---------- ---R------ ---- ---------- -------R-- -D--.R-- ----------Q ---
DLA-88*03101    ---------- ---------- ---------- ---------T -------R-- ------.R-- ---------- ---
DLA-88*03401    ---I----- ---------- ---R------ ---- ---------- -------A-- L----R--- -D-G-.L-- ---------- ---
DLA-88*03501    ---------- ---------- ---R------ ---- ---------- -------A-- ---------- -D--.R-- ---------- ---
DLA-88*03701    ------T-- ---------- ---------- ---- ---------- -------A-- ---------- -D--.R-- ---------- ---
DLA-88*03801    ---------- ---------- ---W-V---- ---- ---------- ---------- ---------- -D--.R-- ---------- ---
DLA-88*03901    ---------- ---------- ---R-V---- ---- ---------- -------A-- ---------- ---R.H-- ---------- ---
DLA-88*04101    ---I----- ---------- ---R------ ---- ---------- --Y------ ---------- ----------Q ---
DLA-88*04201    ---I----- ---------- ---R------ ---- ---------- ---------- ---------- -D-G-.L-- ----------Q ---
DLA-88*04301    ---------- ---------- ---W------ ---- ---------- ---------- ---------- -D--.E-- ---------- ---
DLA-88*04401    ---------- ---------- ---R------ ---- ---------- -------A-- ---------- -D--.R-- ---------- ---
DLA-88*04801    ---------- ---------- ---P-V---- ---- ---------- ------R--- -D-G-.L-- ---------- ---
DLA-88*50101    ---------- ---------- ---R------ ---- ---------- ---------- -V--LQW-- ---------- ---
DLA-88*50201    ---------- ---------- ---------- ---- ---------- ---------- -V--LQW-- ---------- ---
DLA-88*50301    ---------- ---------- ---------- ---- ---------- ---------- -V--LQW-- ---------- ---
DLA-88*50401    ---I----- ---------- ---------- ---- ---------- ---------- -V--LQW-- ---------- ---
DLA-88*50501    ---I----- ---------- ----V----- ---S----- -------A-- ---------- -V--LQW-- ---------- ---
DLA-88*50601    ---I----- ---------- ----V----- ---S----- -------A-- ---------- -V--LQW-- ----------Q ---
DLA-88*50701    ---I----- ---------- ---R-V---- ---- ---------- -------A-- ---------- -V--LQW-- ---------- ---
DLA-88*50801    ---I----- ---------- ---------- ---- ---------- ---------- -V--LQW-- ---------- ---
DLA-88*00101    GSHTRQTMYG CDLGPGGRLL RGYSQDAYDG ADYIALNEDL RSWTAADTAA QITRRKWEAA GTAEH.DRNY LETTCVEWLR RYLEMGKETL LRA
```

Figure 6.2(b) DLA-88 exon 3 amino acid alignment

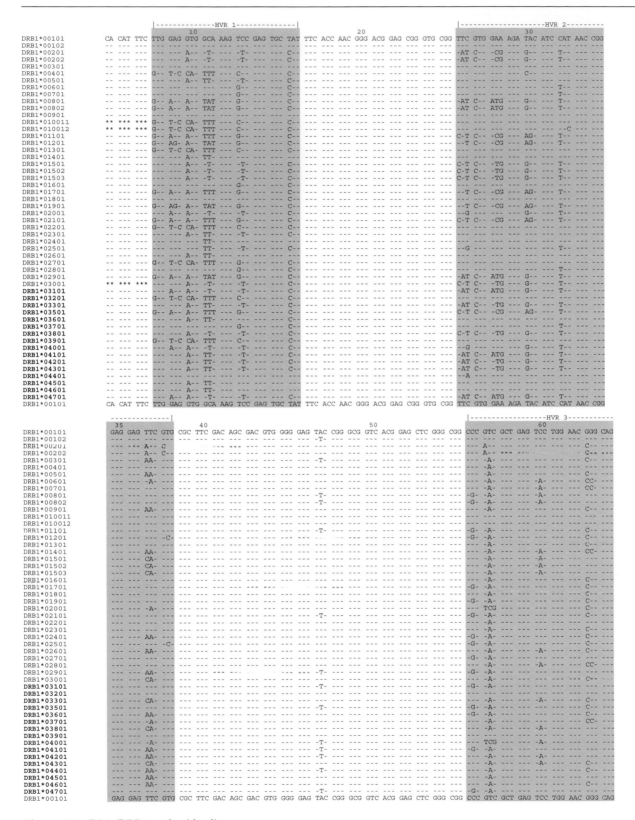

Figure 6.3   DLA-DRB1 nucleotide alignment

```
                    ---HVR-3-cont--------------------------|
                     65              70                                                        80                                  90              94
DRB1*00101          AAG GAG ATC TTG GAG CAG GAG CGG GCA ACG GTG GAC ACC TAC TGC AGA CAC AAC TAC GGG GTG ATT GAG AGC TTC ACG GTG CAG CGG CGA G
DRB1*00102          --- --- --- --- --- --- --- --- --- --- --- --- --- --- --- --- --- --- --- --- --- --- --- --- --- --- --- --- --- --- -
DRB1*00201          --- --- --- --- --- --- AG- --- --C G-- --- --- --- --- --- --- --- --- --- --- --- --- --- --- --- G-- --- --- --- --- -
DRB1*00202          --- --- --- --- --- --- AG- --- --C G-- --- --- --- --- --- --- --- --- --- --- --- --- --- --- --- --- --- --- --- --- -
DRB1*00301          --- --- C-- --- --- -G- A-- --- --C G-- --- --- --- --- --- --- --- --- --- --- --- --- --- --- --- --- --- --- --- --- -
DRB1*00401          --- --- C-- --- --- --- --- --- --- --- --- --- --- --- --- --- --- --- --- --- --- --- --- --- --- --- --- --- --- --- -
DRB1*00501          --- --- C-- --- --- --- --- --- --C G-- --- --- --- --- --- --- --- --- --- C-- --- GGC --- --- --- --- --- --- --- --- -
DRB1*00601          --- --- C-- --- --- -G- -C- --- --C G-- --- --- --- --- --- --- --- --- --- --- --- GGC --- --- --- --- --- --- --- --- -
DRB1*00701          --- --- C-- --- --- GG- -GC --- --C G-- --- --- --- --- --- --- --- --- --- --- --- GGC --- --- --- --- --- --- --- --- -
DRB1*00801          --- --- --- --- --- -G- A-- --- --C G-- --- --- --- --- --- --- --- --- --- C-- --- GGC --- --- --- --- --- --- --- --- -
DRB1*00802          --- --- --- --- --- -G- A-- --- --C G-- --- --- --- --- --- --- --- --- --- --- --- --- --- --- --- G-- --- --- --- --- -
DRB1*00901          --- --- T-- --- --- -G- A-- --- --C G-- --- --- --- --- --- --- --- --- --- --- --- --- --- --- --- --- --- --- --- --- -
DRB1*010011         --- --- C-- --- --- --- --- --- --- --- --- --- --- --- --- --- --- --- --- --- --- --- *** *** *** *** *** *** *** *** *
DRB1*010012         --- --- C-- --- --- --- --- --- --- --- --- --- --- --- --- --- --- --- --- --- --- --- *** *** *** *** *** *** *** *** *
DRB1*01101          --- --- C-- --- --- --- AG- --- --C G-- --- --- --- --- --- --- --- --- --- C-- --- GGC --- --- --- --- --- --- --- --- -
DRB1*01201          --- --- C-- --- --- -G- AG- --- --C GA- --- --- --- --- --- --- --- --- --- --- --- --- --- --- --- --- --- --- --- --- -
DRB1*01301          --- --- C-- --- --- --- --- --- --C G-- --- --- --- --- --- --- --- --- --- C-- --- GGC --- --- --- --- --- --- --- --- -
DRB1*01401          --- --- C-- --- --- -G- -C- --- --C G-- --- --- --- --- --- --- --- --- --- --- --- GGC --- --- --- --- --- --- --- --- -
DRB1*01501          --- --- C-- --- --- --- AG- --- --C GA- --- --- --G GTG --- --- --- --- --- --- --- --- --- --- --- --- --- --- --- --- -
DRB1*01502          --- --- C-- --- --- --- AG- --- --C GA- --- --- --G GTG --- --- --- --- --- --- --- --- --- --- --- G-- --- --- --- --- -
DRB1*01503          --- --- C-- --- --- --- AG- --- --C GA- --- --- --- --- --- --- --- --- --- --- --- --- --- --- --- --- --- --- --- --- -
DRB1*01601          --- --- C-- --- --- -G- A-- --- --C GA- --- --- --- --- --- --- --- --- --- --- --- --- --- --- --- --- --- --- --- --- -
DRB1*01701          --- --- C-- --- --- -G- -C- --- --C G-- --- --- --- --- --- --- --- --- --- --- --- --- --- --- --- --- --- --- --- --- -
DRB1*01801          --- --- C-- --- --- -G- A-- --- --C GA- --- --- --- --- --- --- --- --- --- --- --- --- --- --- --- --- --- --- --- --- -
DRB1*01901          --- --- C-- --- --- -G- AG- --- --C GA- --- --- --- --- --- --- --- --- --- --- --- --- --- --- --- --- --- --- --- --- -
DRB1*02001          --- --- T-- --- --- --- AG- --- --C GA- --- --- --G GTG --- --- --- --- --- --- --- GGC --- --- --- --- --- --- --- --- -
DRB1*02101          --- --- C-- --- --- --- AG- --- C-C G-- --- --- --- --- --- --- --- --- --- C-- --- GGC --- --- --- --- --- --- --- --- -
DRB1*02201          --- --- --- --- --- --- --- --- --C GA- --- --- --- --- --- --- --- --- --- --- --- --- --- --- --- --- --- --- --- --- -
DRB1*02301          --- --- C-- --- --- --- --- --- --C G-- --- --- --- --- --- --- --- --- --- C-- --- GGC --- --- --- --- --- --- --- --- -
DRB1*02401          --- --- C-- --- --- -G- A-- --- --C GA- --- --- --- --- --- --- --- --- --- C-- --- GGC --- --- --- --- --- --- --- --- -
DRB1*02501          --- --- C-- --- --- --- AG- --- --C G-- --- --- --- --- --- --- --- --- --- C-- --- GGC --- --- --- --- --- --- --- --- -
DRB1*02601          --- --- C-- --- --- -G- A-- --- --C GA- --- --- --- --- --- --- --- --- --- --- --- --- --- --- --- --- --- --- --- --- -
DRB1*02701          --- --- --- --- --- --- --- --- --- --- --- --- --- --- --- --- --- --- --- --- --- --- --- --- --- --- --- --- --- --- -
DRB1*02801          --- --- C-- --- --- -G- -C- --- --C G-- --- --- --- --- --- --- --- --- --- --- --- GGC --- --- --- --- --- --- --- --- -
DRB1*02901          --- --- --- --- --- -G- A-- --- --C G-- --- --- --- --- --- --- --- --- --- --- --- --- --- --- --- G-- --- --- --- --- -
DRB1*03001          --- --- C-- --- --- --- AG- --- --C GA- --- --- --G GTG --- --- --- --- --- --- --- --- *** *** *** *** *** *** *** *** *
DRB1*03101          --- --- C-- --- --- --- A-- --- --C G-- --- --- --- --- --- --- --- --- --- --- --- --- --- --- --- G-- --- --- --- --- -
DRB1*03201          --- --- --- --- --- --- --- --- --- --- --- --- --G GTG --- --- --- --- --- --- --- --- --- --- --- --- --- --- --- --- -
DRB1*03301          --- --- C-- --- --- -G- AG- --- --C GA- --- --- --- --- --- --- --- --- --- C-- --- GGC --- --- --- --- --- --- --- --- -
DRB1*03501          --- --- T-- --- --- --- AG- --- --C G-- --- --- --- --- --- --- --- --- --- C-- --- GGC --- --- --- --- --- --- --- --- -
DRB1*03601          --- --- C-- --- --- -G- A-- --- --C GA- --- --- --- --- --- --- --- --- --- --- --- --- --- --- --- --- --- --- --- --- -
DRB1*03701          --- --- C-- --- --- -G- -C- --- --C G-- --- --- --- --- --- --- --- --- --- --- --- GGC --- --- --- --- --- --- --- --- -
DRB1*03801          --- --- C-- --- --- -G- AG- --- --C GA- --- --- --G GTG --- --- --- --- --- C-- --- --- --- --- --- --- --- --- --- --- -
DRB1*03901          --- --- --- --- --- --- --- --- --- --- --- --- --- --- --- --- --- --- --- --- --- --- --- --- --- --- --- --- --- --- -
DRB1*04001          --- --- --- --- --- --- AG- --- --C GA- --- --- --G GTG --- --- --- --- --- --- --- --- --- --- --- G-- --- --- --- --- -
DRB1*04101          --- --- C-- --- --- --- A-- --- --C G-- --- --- --- --- --- --- --- --- --- --- --- --- --- --- --- G-- --- --- --- --- -
DRB1*04201          --- --- C-- --- --- --- A-- --- --C G-- --- --- --- --- --- --- --- --- --- --- --- --- --- --- --- G-- --- --- --- --- -
DRB1*04301          --- --- C-- --- --- -G- AG- --- --C GA- --- --- --G GTG --- --- --- --- --- C-- --- GGC --- --- --- --- --- --- --- --- -
DRB1*04401          --- --- C-- --- --- --- AG- --- --C G-- --- --- --- --- --- --- --- --- --- C-- --- GGC --- --- --- --- --- --- --- --- -
DRB1*04501          --- --- C-- --- --- -G- A-- --- --C GA- --- --- --- --- --- --- --- --- --- --- --- GGC --- --- --- --- --- --- --- --- -
DRB1*04601          --- --- C-- --- --- --- --- --- --C G-- --- --- --- --- --- --- --- --- --- --- --- --- --- --- --- --- --- --- --- --- -
DRB1*04701          --- --- C-- --- --- -G- AG- --- --C G-- --- --- --- --- --- --- --- --- --- --- --- --- --- --- --- --- --- --- --- --- -
DRB1*00101          AAG GAG ATC TTG GAG CAG GAG CGG GCA ACG GTG GAC ACC TAC TGC AGA CAC AAC TAC GGG GTG ATT GAG AGC TTC ACG GTG CAG CGG CGA G
```

Figure 6.3 DLA-DRB1 nucleotide alignment (*continued*)

```
                |--HVR 1-|          |---HVR 2----|                          |-----HVR 3---------|
                  10        20         30         40         50          60         70         80         90
DRB1*00101      HFLEV AKSECYFTNG TERVRFVERY IHNREEFVRF DSDVGEYRAV TELGRPVAES WNGQKEILEQ ERATVDTYCR HNYGVIESFT VQRR
DRB1*00102      ----- ---------- ---------- ---------- ------F--- ---------- ---------- ---------- ---------- ----
DRB1*00201      ----M V-F--H---- -----YLA-D -Y----IL-- ---------- ------I--- --R------- R--A------ ---------A ----
DRB1*00202      ----M V-F--H---- -----YLA-D -Y----IL-- ---------- ------I--- --R------- R--A------ ---------- ----
DRB1*00301      ----- ---------- ---------- ------N--- ------F--- ------D--- --R---L--R K--A------ ---------- ----
DRB1*00401      --VYQ F-P--H---- --------H- ---------- ---------- ------D--- ------L--- ---------- ---------- ----
DRB1*00501      ----M L-F--H---- ---------- ------N--- ---------- ------D--- --R---L--- ---A------ ---R-G---- ----
DRB1*00601      ----- --A--H---- ---------- -Y----Y--- ---------- ------D--Y --P---L--R A--A------ -----G---- ----
DRB1*00701      ----- --A--H---- ---------- -Y-------- ---------- ------D--Y --P---L--G G--A------ -----G---- ----
DRB1*00801      --VKM Y-A--H---- -----YLM-D -Y-------- ------F--- -----RD--Y ---------R K--A------ ---R-G---- ----
DRB1*00802      --VKM Y-A--H---- -----YLM-D -Y-------- ------F--- -----RD--Y ---------R K--A------ ---------A ----
DRB1*00901      ----- ---------- ---------- ------N--- ---------- ------D--- --R---F--R K--A------ ---------- ----
DRB1*010011     **VYQ F-P--H---- ---------- ---------- ---------- ---------- ------L--- ---------- ------**** ****
DRB1*010012     **VYQ F-P--H---- ---------- ---------- ---------- ---------- ------L--- ---------- ------**** ****
DRB1*01101      --VKM F-A--H---- -----LLA-S -Y-------- ------F--- -----RD--- --R---L--- R--A------ ---R-G---- ----
DRB1*01201      --VRM Y-A--H---- ------LA-S -Y-----A-- ---------- -----RD--- --R---L--R R--F------ ---------- ----
DRB1*01301      --VYQ F-P--H---- ---------- ---------- ---------- ------D--- --R---L--- ---A------ ---R-G---- ----
DRB1*01401      ----M L--------- ---------- ------N--- ---------- ------D--Y --P---L--R A--A------ -----G---- ----
DRB1*01501      ----M V-F--H---- -----LLV-D -Y----H--- ---------- ------D--Y ------L--- R--E---V-- ---------- ----
DRB1*01502      ----M V-F--H---- -----LLV-D -Y----H--- ---------- ------D--Y ------L--- R--E---V-- ---------A ----
DRB1*01503      ----M V-F--H---- -----LLV-D -Y----H--- ---------- ------D--Y ------L--- R--E------ ---------- ----
DRB1*01601      ----- --A--H---- ---------- ---------- ---------- ------D--- --R---L--R K--E------ ---------- ----
DRB1*01701      --VKM F-A--H---- ------LA-S -Y-------- ---------- -----RD--- --R---L--R A--A------ ---------- ----
DRB1*01801      ----- -----H---- ---------- ---------- ---------- ------D--- --R---L--R K--E------ ---------- ----
DRB1*01901      --VRM Y-A--H---- ------LA-S -Y-------- ---------- -----RD--- --R---L--R R--E------ ---------- ----
DRB1*02001      ---KM V-F--H---- -----L---D -Y----Y--- ---------- ------S--- --R---F--- R--E---V-- -----G---- ----
DRB1*02101      --VKM F-A--H---- -----LLA-S -Y-------- ------F--- -----RD--- --R---L--- R-PA------ ---R-G---- ----
DRB1*02201      --VYQ F-P--H---- ---------- ---------- ---------- ------D--- ---------- ---E------ ---------- ----
DRB1*02301      ----M L-F--H---- ---------- ---------- ---------- ------D--- --R---L--- ---A------ ---R-G---- ----
DRB1*02401      ----- L--------- ---------- ------N--- ---------- -----RD--- --R---L--R K--E------ ---R-G---- ----
DRB1*02501      ----- L-F--H---- -----L---- -Y-----A-- ---------- -----RD--- --R---L--- R--A------ ---R-G---- ----
DRB1*02601      ----M L--------- ---------- ------N--- ---------- ------D--Y --R---L--R K--E------ ---------- ----
DRB1*02701      --VYQ F-A--H---- ---------- ---------- ---------- -----RD--- ---------- ---------- ---------- ----
DRB1*02801      ----- --A--H---- ---------- -Y-------- ---------- ------D--Y --P---L--R A--A------ -----G---- ----
DRB1*02901      --VKM Y-A--H---- -----YLM-D -Y----N--- ------F--- -----RD--- ---------R K--A------ ---------A ----
DRB1*03001      **--M V-F--H---- -----LLV-D -Y----H--- ---------- ------D--- --R---L--- R--E---V-- ------**** ****
DRB1*03101      ---KM V-F--H---- -----YLM-D -Y-------- ------F--- -----RD--- ------L--- K--A------ ---------A ----
DRB1*03201      --VYQ F-P--H---- ---------- ---------- ---------- ---------- ---------- -------V-- ---------- ----
DRB1*03202      --VYQ F-P--H---- ---------- ---------- ---------- ---------- ---------- ---------- ---------- ----
DRB1*03301      ----M L-F--H---- -----YLV-D -Y----H--- ---------- ------D--Y --R---L--R R--E------ ---R-G---- ----
DRB1*03501      --VKM F-A--H---- -----LLA-S -Y-------- ------F--- -----RD--- --R---F--- R--A------ ---R-G---- ----
DRB1*03601      ----M L----H---- ---------- ------N--- ---------- -----RD--- --R---L--R K--E------ ---------- ----
DRB1*03701      ----- --A--H---- ---------- -Y----Y--- ---------- ------D--- --P---L--R A--A------ -----G---- ----
DRB1*03801      ----M V-F--H---- -----LLV-D -Y----H--- ---------- ------D--Y ------L--R R--E---V-- ---R------ ----
DRB1*04001      ---KM V-F--H---- -----L---D -Y----Y--- ------F--- ------S--Y ---------- R--E---V-- ---------A ----
DRB1*04101      ----M L-F--H---- -----YLM-D -Y----N--- ------F--- -----RD--- ------L--- K--A------ ---------A ----
DRB1*04201      ----M L-F--H---- -----YLV-D -Y----N--- ------F--- ------D--Y ------L--- K--A------ ---------A ----
DRB1*04301      ----M L-F--H---- -----YLV-D -Y----H--- ---------- ------D--Y --R---L--R R--E---V-- ---R-G---- ----
DRB1*04401      ----- ---------- -----L---- ------N--- ------F--- ------D--- --R---L--- R--A------ ---R-G---- ----
DRB1*04501      ----M L--------- ---------- ------N--- ---------- ------D--- ------L--R K--E------ -----G---- ----
DRB1*04601      ----M L--------- ---------- ------N--- ---------- ------D--- --R---L--- ---A------ ---------- ----
DRB1*04701      ---KM V-F--H---- -----YLM-D -Y-------- ------F--- -----RD--- ------L--R R--A------ ---------- ----
DRB1*00101      HFLEV AKSECYFTNG TERVRFVERY IHNREEFVRF DSDVGEYRAV TELGRPVAES WNGQKEILEQ ERATVDTYCR HNYGVIESFT VQRR
```

Figure 6.4 DLA-DRB1 amino acid alignment

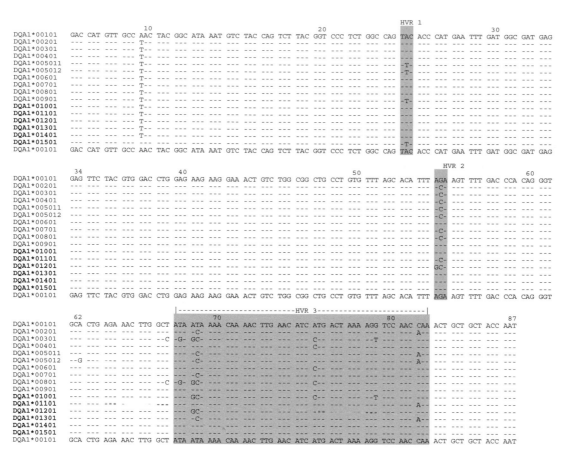

Figure 6.5 DLA-DQA1 nucleotide alignment

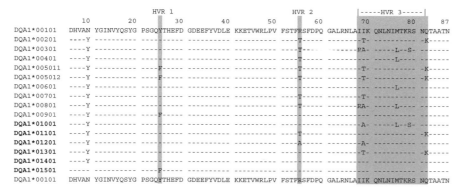

Figure 6.6 DLA-DQA1 amino acid alignment

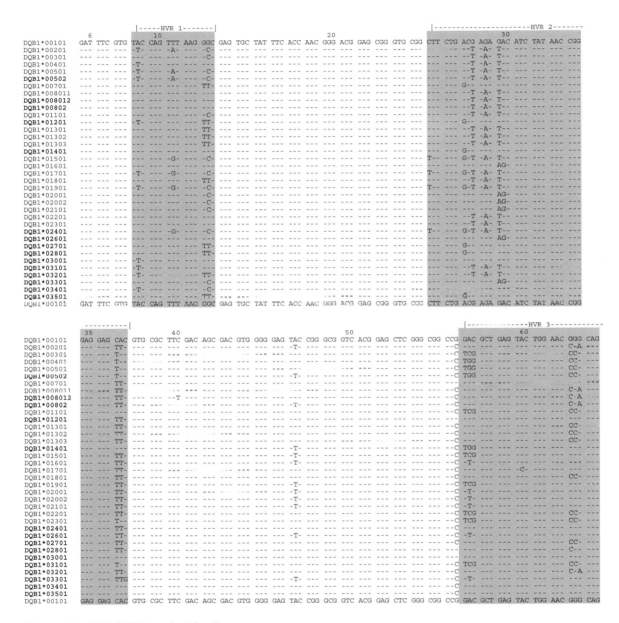

Figure 6.7 DLA-DQB1 nucleotide alignment

```
                ---------------HVR 3 cont-----------------|
              65                70                  80                          90              94
DQB1*00101    AAG GAG CTC TTG GAG CGG AGG CGG GCC GAG GTG GAC ACG GTG TGC AGA CAC AAC TAC GGG AGG GAA GAG CTC ACC ACG TTG CAG CGG CGA
DQB1*00201    --- --C GAG A-- --C --- GTA --- --- --- C-- --- --- --- --- --- --- --- --- --- --- --- --- --- --- --- --- --- --- ---
DQB1*00301    --- --C GAG A-- --C --- GTA --- --- --- C-- --- --- --- --- --- --- --- --- TT- --- --- --- --- --- --- --- --- --- ---
DQB1*00401    --- --C GAG A-- --C --- GTA --- --- --- C-- --- --- --- --- --- --- --- --- TT- --- --- TA- --- --- --- --- --- --- ---
DQB1*00501    --- --C GAG A-- --C --- GTA --- --- --- C-- --- --- --- --- --- --- --- --- TT- --- --- --- --- --- --- --- --- --- ---
DQB1*00502    --- --C GAG A-- --C --- GTA --- --- --- C-- --- --- --- --- --- --- --- --- TT- --- --- --- --- --- --- --- --- --- ---
DQB1*00701    --- --- --- --- --- -A- --- --- --- --- C-- --- --- --- --- --- --- --- --- TT- --- --- TA- --- --- --- --- --- --- ---
DQB1*008011   --- --C GAG A-- --C --- GTA --- --- --- C-- --- --- --- --- --- --- --- --- --- --- --- --- --- --- --- --- --- --- ---
DQB1*008012   --- --C GAG A-- --C --- GTA --- --- --- C-- --- --- --- --- --- --- --- --- --- --- --- --- --- --- --- --- --- --- ---
DQB1*00802    --- --C GAG A-- --C --- GTA --- --- --- C-- --- --- --- --- --- --- --- --- --- --- --- --- --- --- --- --- --- --- ---
DQB1*01101    --- --C GAG A-- --C --- GTA --- --- --- C-- --- --- --- --- --- --- --- --- TT- --- --- --- --- --- --- --- --- --- ---
DQB1*01201    --- --- --- --- --- -A- --- --- --- --- C-- --- --- --- --- --- --- --- --- TT- --- --- TA- --- --- --- --- --- --- ---
DQB1*01301    --- --C GAG A-- --C --- GTA --- --- --- C-- --- --- --- --- --- --- --- --- --- --- --- --- --- --- --- --- --- --- ---
DQB1*01302    --- --C GAG A-- --C --- GTA --- --- --- C-- --- --- --- --- --- --- --- --- GT- --- --- --- --- --- --- --- --- --- ---
DQB1*01303    --- --C GAG A-- --C --- GTA --- --- --- C-- --- --- --- --- --- --- --- --- GT- --- --- TA- --- --- --- --- --- --- ---
DQB1*01401    --- --- A-- --- --- -A- --- --- --- --- C-- --- --- --- --- --- --- --- --- GT- --- --- TA- --- --- --- --- --- --- ---
DQB1*01501    --- --- A-- --- --- -A- GA- --- --A AC- --- --- --- --- --- --- --- --- --- --- --- --- --- --- --- --- --- --- --- ---
DQB1*01601    --- --- A-- --- --- -A- --- --- --- --- --- --G --- --- --- --- --- --- --- --- --- --- --- --- --- --- --- --- --- ---
DQB1*01701    --- --- T-- --- --- -A- GA- --- --A AC- --- --- --- --- --- --- --- --- --- GT- --- --- TA- --- --- --- --- --- --- ---
DQB1*01801    --- --C GAG A-- --C --- GT- --- --- --- C-- --- --- --- --- --- --- --- --- GT- --- --- T-- --- --- --- --- --- --- ---
DQB1*01901    --- --- --- --- --- -AA GA- --- --A AC- --- --- --- --- --- --- --- --- --- --- --- --- --- --- --- --- --- --- --- ---
DQB1*02001    --- --- A-- --- --- -A- --- --- --- -C- --- --- --- --- --- --- --- --- --- --- --- --- --- --- --- --- --- --- --- ---
DQB1*02002    --- --- A-- --- --- -A- --- --- --- -C- --- --G --- --- --- --- --- --- --- --- --- --- --- --- --- --- --- --- --- ---
DQB1*02101    --- --- A-- --- --- -A- --- --- --- --- --- --- --- --- --- --- --- --- --- --- --- --- --- --- --- --- --- --- --- ---
DQB1*02201    --- --C GAG A-- --C --- GTA --- --- --- C-- --- --- --- --- --- --- --- --- TT- --- --- --- --- --- --- --- --- --- ---
DQB1*02301    --- --C GAG A-- --C --- GTA --- --- --- C-- --- --- --- --- --- --- --- --- TT- --- --- --- --- --- --- --- --- --- ---
DQB1*02401    --- --- --- --- --- -A- GA- --- --A AC- C-- --- --- --- --- --- --- --- --- GT- --- --- TA- --- --- --- --- --- --- ---
DQB1*02601    --- --- A-- --- --- -A- --- --- --- -C- --- --G --- --- --- --- --- --- --- --- --- --- --- --- --- --- --- --- --- ---
DQB1*02701    --- --C GAG A-- --C --- GTA --- --- --- C-- --- --- --- --- --- --- --- --- GT- --- --- TA- --- --- --- --- --- --- ---
DQB1*02801    --- --- --- --- --- -A- --- --- --- -C- --- --- --- --- --- --- --- --- --- TT- --- --- TA- --- --- --- --- --- --- ---
DQB1*03001    --- --- --- --- --- --- --- --- --- --- --- --- --- --- --- --- --- --- --- --- --- --- --- --- --- --- --- --- --- ---
DQB1*03101    --- --C GAG A-- --C --- GTA --- --- --- C-- --- --- --- --- --- --- --- --- TT- --- --- --- --- --- --- --- --- --- ---
DQB1*03201    --- --C GAG A-- --C --- GTA --- --- --- C-- --- --- --- --- --- --- --- --- --- --- --- --- --- --- --- --- --- --- ---
DQB1*03301    --- --- --- --- --- -A- --- --- --- -C- --- --G --- --- --- --- --- --- --- --- --- --- --- --- --- --- --- --- --- ---
DQB1*03401    --- --- --- --- --- --- --- --- --- --- --- --- --- --- --- --- --- --- --- --- --- --- --- --- --- --- --- --- --- ---
DQB1*03501    --- --- --- --- --- -A- --- --- --- --- C-- --- --- --- --- --- --- --- --- TT- --- --- TA- --- --- --- --- --- --- ---
DQB1*00101    AAG GAG CTC TTG GAG CGG AGG CGG GCC GAG GTG GAC ACG GTG TGC AGA CAC AAC TAC GGG AGG GAA GAG CTC ACC ACG TTG CAG CGG CGA
```

Figure 6.7 (*continued*)

```
                HVR 1              |---HVR 2----|              |------HVR 3---------|
              10        20      30        40          50      60        70        80        90
DQB1*00101    DFVYQ FKGECYFTNG TERVRLLTRD IYNREEHVRF DSDVGEYRAV TELGRPDAEY WNGQKELLER RRAEVDTVCR HNYGREELTT LQRR
DQB1*00201    ---F- Y-A------- --------KY ------F--- ------F--- ---------- --R--DEMD- V---L----- ----------- ----
DQB1*00301    ----- --A------- --------KY ------Y--- ---------- ------S--- --P--DEMD- V---L----- ----L----- ----
DQB1*00401    ---F- --------- --------KY ------Y--- ---------- ------W--- --P--DEMD- V---L----- ----L---Y- ----
DQB1*00501    ---F- Y-A------- --------KY ------Y--- ---------- ------W--- --P--DEMD- V---L----- ----L----- ----
DQB1*00502    ---F- Y-A------- --------KY ------Y--- ------F--- ------W--- --P--DEMD- V---L----- ----L----- ----
DQB1*00701    ----- --F------- -------A-- ------F--- ---------- ---------- ---------Q ----L----- ----L---Y- ----
DQB1*008011   ----- --------- --------KY ------F--- ---------- ---------- --R--DEMD- V---L----- ---------- ----
DQB1*008012   ----- --------- --------KY ------F--- ---------- ---------- --R--DEMD- V---L----- ---------- ----
DQB1*00802    ----- --------- --------KY ------F--- ------F--- ---------- --R--DEMD- V---L----- ---------- ----
DQB1*01101    ----- --A------- --------KY ------F--- ---------- ------S--- --P--DEMD- V---L----- ----L----- ----
DQB1*01201    ---F- --F------- -------A-- ------F--- ---------- ---------- ---------Q ----L----- ----L---Y- ----
DQB1*01301    ----- --F------- --------KY ------F--- ---------- ---------- --P--DEMD- V---L----- ---------- ----
DQB1*01302    ----- --F------- --------KY ------F--- ---------- ---------- --P--DEMD- V---L----- ----V----- ----
DQB1*01303    ----- --F------- --------KY ------F--- ---------- ---------- --P--DEMD- V---L----- ----V---Y- ----
DQB1*01401    ----- --------- -------A-- ------F--- ------F--- ------W--- ------F--- V---L----- ----V---Y- ----
DQB1*01501    ----- C-A------- ----F-AKY ------F--- ---------- ------S--- ------I-Q E--T---- ----V---Y- ----
DQB1*01601    ----- --------- --------S ------F--- ------F--- ------V--- ------I-- K-----R--- ---------- ----
DQB1*01701    ---F- C-A------- ----F-AKY ------F--- ---------- ------S------ ------F-Q E--T---- ----V---Y- ----
DQB1*01801    ----- --F------- --------KY ------F--- ---------- ---------- --P--DEMD- V--------- ----V---S- ----
DQB1*01901    ---F- C-A------- ----F-AKY ------F--- ------F--- ------S--- ---------Q E--T------ ---------- ----
DQB1*02001    ----- --A------- --------S ------F--- ------F--- ------V--- ------I-- K--A------ ---------- ----
DQB1*02002    ----- --A------- --------S ------F--- ------F--- ------V--- ------I-- K--A--R--- ---------- ----
DQB1*02101    ----- --A------- --------S ------F--- ------F--- ------V--- ------I K--------- ---------- ----
DQB1*02201    ----- --------- --------KY ------F--- ---------- ------S--- --P--DEMD- V---L----- ----L----- ----
DQB1*02301    ----- --------- --------KY ------Y--- ---------- ------S--- --P--DEMD- V---L----- ----L----- ----
DQB1*02401    ----- C-A------- ----F-AKY ---------- ---------- ---------- ---------Q E--TL----- ----V---Y- ----
DQB1*02601    ----- --------- --------S ------F--- ------F--- ------V--- ------I-- K--A--R--- ---------- ----
DQB1*02701    ----- --F------- -------A-- ------F--- ---------- ---------- --P--DEMD- V---L----- ----V---Y- ----
DQB1*02801    ----- --F------- -------A-- ------F--- ---------- ---------- --R-----Q ---A------ ----L---Y- ----
DQB1*03001    ----- F-------- ---------- ---------- ---------- ---------- ---------- ---------- ---------- ----
DQB1*03101    ---F- --------- --------KY ------Y--- ---------- ------S--- --P--DEMD- V---L----- ----L----- ----
DQB1*03201    ---F- --F------- --------KY ------F--- ---------- ---------- --R--DEMD- V---L----- ---------- ----
DQB1*03301    ----- --A------- --------S ------L--- ------F--- ------V--- ------I-- K--A--R--- ---------- ----
DQB1*03401    ---F- --A------- ---------- ---------- ---------- ---------- ---------- ---------- ---------- ----
DQB1*03501    ----- --F------- -------A-- ---------- ---------- ---------- ---------Q ----L----- ----L---Y- ----
DQB1*00101    DFVYQ FKGECYFTNG TERVRLLTRD IYNREEHVRF DSDVGEYRAV TELGRPDAEY WNGQKELLER RRAEVDTVCR HNYGREELTT LQRR
```

Figure 6.8   DLA-DQB1 amino acid alignment

KEY REFERENCES

1 Kennedy LJ, Altet L, Angles JM *et al.* (1999) Nomenclature for factors of the Dog Major Histocompatibility System (DLA), 1998: First report of the ISAG DLA Nomenclature Committee. *Tissue Antigens* 54: 312–321

2 Kennedy LJ, Altet L, Angles JM *et al.* (2000) Nomenclature for factors of the Dog Major Histocompatibility System (DLA), 1998: First report of the ISAG DLA Nomenclature Committee. *Animal Genetics* 31: 52–61

3 Graumann MB, DeRose SA, Ostrander EA and Storb R (1998) Polymorphism analysis of four canine MHC class I genes. *Tissue Antigens* 51: 374–381

4 Mellersh CS, Hitte C, Richman M *et al.* (2000) An Integrated Linkage-Radiation Hybrid Map of the Canine Genome. *Mammalian Genome* 11: 120–130

5 Burnett RC, Geraghty DE (1995) Structure and expression of a divergent canine class I gene. *Journal of Immunology* 155: 4278–4285

6 Burnett RC, DeRose SA, Wagner JL, Storb R (1997) Molecular analysis of six dog leukocyte antigen class I sequences including three complete genes, two truncated genes and one full–length processed gene. *Tissue Antigens* 49: 484–495

7 Sarmiento UM, Storb RF (1998) Characterization of class II alpha genes and DLA-D region allelic associations in the dog. *Tissue Antigens* 32: 224–234

8 Wagner JL, DeRose SA, Burnett RC, Storb R (1995) Nucleotide sequence and polymorphism analysis of Canine DRA cDNA clones. *Tissue Antigens* 45: 284–287

9 Sarmiento UM, Storb R (1990) Nucleotide sequence of a dog class I cDNA clone. *Immunogenetics* 31: 400–404

10 Wagner JL, Burnett RC, Storb R (1996) Molecular analysis of the DLA DR region. *Tissue Antigens* 48: 549–553

11 Wagner JL, Burnett RC, Works JD, Storb R (1996) Molecular analysis of DLA–DRBB1 polymorphism. *Tissue Antigens* 48: 554–561

12 Sarmiento UM, Storb R (1990) Nucleotide sequence of a dog DRB cDNA clone. *Immunogenetics* 31: 396–399

13 Wagner JL, Hayes-Lattin B, Works JD, Storb R (1998) Molecular analysis and polymorphism of the DLA-DQB genes. *Tissue Antigens* 52: 242–250

14 Wagner JL, Creer SA, Storb R (2000) Dog class I gene DLA-88 histocompatibility typing by PCR-SSCP and sequencing. *Tissue Antigens* 55: 564–567

15 Sarmiento UM, Sarmiento JI, Storb R (1990) Allelic variation in the DR subregion of the canine major histocompatibility complex. *Immunogenetics* 32: 13–19

16 Francino O, Amills M, Sanchez A (1997) Canine *Mhc* DRB1 genotyping by PCR–RFLP analysis. *Animal Genetics* 28: 41–45

17 Kennedy LJ, Carter SD, Barnes A *et al.* (1998) Nine new dog DLA-DRB1 alleles identified by sequence based typing. *Immunogenetics* 48: 296–301

18 Kennedy LJ, Hall LS, Carter SD *et al.* (2000) Identification of further DLA-DRB1 and DQA1 alleles in the dog. *European Journal of Immunogenetics* 27: 25–28

19 He YW, Ferencik S and Grosse-Wilde H (1994) A research on DLA-DRB1 genotyping by PCR-RFLP. I. To select a appropriate oligonucleotide primer pair. *J Tongji Med Univ* 14: 24–28

20 Wagner JL, Works JD, Storb R (1998) DLA-DRB1 and DLA-DQB1 histocompatibility typing by PSR-SSCP and sequencing. *Tissue Antigens* 52: 397–401

21 Sarmiento UM, DeRose S, Sarmiento JI, Storb R (1992) Allelic variation in the DQ subregion of the canine major histocompatibility complex: I. DQA. *Immunogenetics* 35: 416–420

22 Polvi A, Garden OA, Elwood CM *et al.* (1997) Canine major histocompatibility complex genes DQA and DQB in Irish setter dogs. *Tissue Antigens* 49: 236–243

23 Wagner JL, Burnett RC, DeRose SA, Storb R (1996) Molecular analysis and polymorphism of the DLA-DQA gene. *Tissue Antigens* 48: 199–204

24 Sarmiento UM, DeRose S, Sarmiento JI, Storb R (1993) Allelic variation in the DQ subregion of the canine major histocompatibility complex: II. DQB. *Immunogenetics* 37: 148–152

25 Wagner JL, Storb R, Storer B, Mignot E (2000) DLA-DQB1 alleles and bone marrow transplantation experiments in narcoleptic dogs. *Tissue Antigens* 56: 223–231

---

i

SEQUENCE DATABASE

- The DLA allele sequences have been integrated into the web-based IPD MHC database (see http://www.ebi.ac.uk/ipd/mhc/dla/index.html). The most up-to-date alignments are available from this site

ANTIGEN PRESENTATION

Antigen presentation is the expression of antigen molecules on the surface of a macrophage or other antigen-presenting cell in association with MHC class II molecules when the antigen is being presented to a CD4$^+$ helper T cell or in association with MHC class I molecules when presentation is to CD8$^+$ cytotoxic T cells. For appropriate presentation, it is essential that peptides bind securely to the MHC class II molecules, since those that do not bind or are bound only weakly are not presented and fail to elicit an immune response. Following interaction of the presented antigen and MHC class II molecules with the CD4$^+$ helper T cell receptor, the CD4$^+$ lymphocyte is activated, IL-2 is released, and IL-2 receptors are expressed on the CD4$^+$ lymphocyte surface. The IL-2 produced by the activated cell stimulates its own receptors, as well as those of mononuclear phagocytes, increasing their microbicidal activity. IL-2 also stimulates B cells to synthesize antibody. Whereas B cells may recognize a protein antigen in its native state, T cells only recognize the peptides, that result from antigen processing, in the context of major histocompatibility complex molecules.

ANTIGEN-PRESENTING CELLS

Antigen-presenting cells (APC) are cells that can process a protein antigen, break it into peptides, and present it in conjunction with class II MHC molecules on the cell surface where it may interact with appropriate T cell receptors. Professional APCs include dendritic cells, macrophages, and B cells, whereas nonprofessional APCs that function in antigen presentation for only brief periods include thymic epithelial cells and vascular endothelial cells. Dendritic cells, macrophages, and B cells are the principal antigen-presenting cells for T cells, whereas follicular dendritic cells are the main antigen-presenting cells for B cells. The immune system contains three types of antigen-presenting cells, i.e., macrophages, dendritic cells, and B cells. **Table 7.1** shows properties and functions of these three types of antigen-presenting cells.

ANTIGEN-PRESENTING PATHWAYS

Dendritic cells, macrophages, and B cells process and present antigen to immunoreactive lymphocytes such as CD4$^+$helper/inducer T cells. A MHC transporter gene-encoded peptide supply factor may mediate peptide antigen presentation. Other antigen-presenting cells that serve mainly as passive antigen transporters include B cells, endothelial cells, keratinocytes, and Kupffer cells. This group of APCs present exogenous antigen processed in their endosomal compartment and presented together with class II MHC molecules. Other APCs present antigen that has been endogenously produced by the body's own cells with processing in an intracellular compartment and presentation together with class I MHC molecules. A third group of APCs present exogenous antigen that is taken into the cell and processed, followed by presentation together with class I MHC molecules. In addition to processing and presenting antigenic peptides in association with Class II MHC molecules, an antigen-presenting cell must also deliver a co-stimulatory signal that is necessary for T cell activation. The outcome of an appropriate immune response is dependent on the bidirectional communication between T cells and APCs. **Figure 7.1** depicts the 'cross-talk' between APCs and T cells.

Optimal activation of a naïve lymphocyte requires two signals, an antigen-specific signal initiated by engagement of TCR or BCR and a costimulatory signal independent of the antigen receptor complex. C28 is the most studied T costimulatory pathway. Engagement of CD28 augments proliferation and promotes survival of T cells. The other major T

Table 7.1 Characteristics of antigen presenting cells

	Dendritic cells	Macrophages	B cells
Immune response	Innate immunity	Innate immunity	Adaptive immunity
Specific antigen receptors	No	No	Surface immunoglobulins
Location	Skin and mucosal epithelium (Langerhans cells), lymphoid tissue, connective tissue	Lymphoid tissue, connective tissue, body cavities	Blood, lymphoid tissue
Antigen type	Intracellular antigens and extracellular antigens	Extracellular antigens	Extracellular antigens
MHC molecule associated with antigen presentation	Class I MHC and class II MHC	Class II MHC	Class II MHC
Co-stimulation	High level B7 expression	Low level B7 expression, induced by bacteria/cytokines	No B7 expression unless induced upon activation by Th cells

costimulatory molecule is ICOS, Inducible Costimulator. It is suggested that CD28 acts on resting/naive T cells while ICOS acts on activated/effector T cells. Other T cell costimulatory molecules include members of the TNF/TNFR family. **Tables 7.2–7.4** depict the T costimulatory molecules.

Among the three major antigen-presenting cells, dendritic cells are the only ones that continuously express high levels of costimulatory B7 and can present antigen via both class I MHC molecules and class II MHC molecules. Thus they can activate both CD8 and CD4 T cells, directly. Dendritic cells are derived from hematopoietic stem cells in the bone marrow and are widely distributed as immature cells within all tissues, especially those that interface with the environment (e.g. skin and mucosal epithelium, where they are referred to as Langerhans cells) and in lymphoid organs. Upon pathogen invasion, immature dendritic cells are recruited to sites of inflammation. Internalization of foreign antigens subsequently triggers their maturation and migration from peripheral tissues to lymphoid organs. This antigen capture or uptake by immature dendritic cells is accomplished by phagocytosis, macropinocytosis or via interaction with a variety of cell surface receptors and endocytosis. A number of these cell surface receptors are downregulated upon dendritic cell maturation, indicating their specific roles in antigen uptake. **Table 7.5** summarizes the prevalent antigen receptors expressed by dendritic cells in antigen uptake.

Chemokine responsiveness and chemokine receptor expression are essential components of the dendritic cell recruitment and migration process. During dendritic cell maturation, chemokine and chemokine receptor expression is modulated. A change in expression levels of CCR6 and CCR7 contributes to the functional shifts observed during dendritic cell maturation. The chemokine and chemokine receptor expression profiles of immature and mature dendritic cells are summarized in **Table 7.6** and **Table 7.7**, respectively.

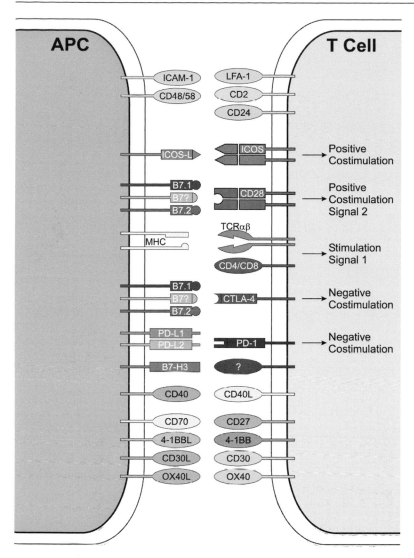

Figure 7.1 Cellular interactions between APCs and T cells

Table 7.2 T cell costimulatory molecules, CD 28 family

Receptor: CD28 family	Other names	Function	Expression		Ligand: B7 family	Other names	Expression
CD28	T44	Costimulation	T and NK cells		B7.1	CD80	Activated APC
					B7?		
CTLA-4	CD152	Inhibition	Activated T cells		B7.2	CD86	APC (upregulated) activated T cells
ICOS	H4, CRP-1, AILIM	Costimulation	Activated T cells		B7RP-1	ICOSL, GL50, B7-H2, B7h	APC
?					B7-H3		

Table 7.3   T cell costimulatory molecules, TNFR family

TNFR family	Other names	Function	Expression	Ligand: TNF family	Other names	Expression
CD27	T14	Costimulation	T cells, B subset, NK	CD70		Activated B cells
CD30	Ki-1	Costimulation, apoptosis	Activated T, NK and B cells	CD153	CD30L	Neutrophils activated B and T cells
CD40L	CD154, gp39, TRAP	Activation	Activated T cells	CD40		APC, T subset, endothelium, cardiac myocytes, fibroblasts
4-1BB	CD137	Costimulation	Activated T cells	4-1BBL		Activated B, DC, peritoneal cells
OX-40	CD134	Activation, differentiation, apoptosis	Activated T cells	OX40L		Activated B cells, cardiac myocytes
Fas	CD95, Apo-1	Activation, apoptosis	Leukocytes	FasL	CD95L, CD178	Activated T cells

Table 7.4   T cell costimulatory molecules, inhibitory factors

Inhibitory molecules	Other names	Function	Expression		Ligand: B7 family	Other names	Expression
PD-1		Inhibition	Activated T and B cells		PD-L1	B7-H1	Leukocytes
?					PD-L2	B7-DC	Monocytes, macrophages, DC
					B7.1	CD80	Activated APC
CTLA-4	CD152	Inhibition	Activated T cells		B7?		
					B7.2	CD86	APC (upregulated) activated T cells

Table 7.5 Antigen receptors of antigen uptake in immature dendritic cells

Molecule	Type of protein	Function	Additional notes
DEC-205	Type I C-type lectin	Antigen uptake	Expression is downregulated upon DC maturation
Dectin-1	Type II C-type lectin	Antigen presentation	Contains an immunoreceptor tyrosine-based activation motif (ITAM)
Dectin-2	Type II C-type lectin	Antigen update	Ligand has not been identified as of yet
CLEC-1	Type II C-type lectin	Antigen uptake	Ligand has not been identified as of yet
DCIR	Type II C-type lectin	Antigen uptake	Contains an immunoreceptor tyrosine-based inhibitory motif (ITIM); binds glycosylated ligands and is downregulated upon maturation
DC-SIGN	Type II C-type lectin	T cell interaction	ICAM-3 ligand; binds gp120 and promotes HIV-1 infection of T cells
DORA	Ig superfamily member	Antigen uptake	CD40L activation of DCs downregulates DORA
CD83	Ig superfamily member	Antigen presentation	Maturation marker
ILT3	Ig superfamily member	Antigen presentation	Contains an ITIM motif; could negatively regulate DC activation and maturation
DC-LAMP	Lysosomal glycoprotein	Antigen presentation	Specifically expressed in the lysosomal MHC class II compartment; upregulated upon CD40 ligation
DC-STAMP	Multi-membrane spanner	Unknown	Downregulated upon CD40 ligation
TLR2	Toll-like receptor family member	Antigen uptake	Binding of bacterial lipopeptides can induce DC maturation
TLR3	Toll-like receptor family member	Antigen uptake	Expressed by immature DCs/ downregulated upon maturation
MR	Mannose receptor	Antigen uptake	Mediates endocytosis and targets the endosomal/lysosomal compartment; expression is downregulated upon DC maturation
$Fc\gamma R$	Fc receptor	Antigen uptake	Expression is downregulated upon DC maturation
$Fc\varepsilon R$	FC receptor	Antigen uptake	Expression is downregulated upon DC maturation
CD36	Multifunctional receptor for collagen, thrombospondin, oxidized LDL, and long-chain fatty acids	Antigen uptake (apoptotic cells)	Expression is downregulated upon DC maturation
$\alpha_v\beta_5$	lintegrin	Antigen uptake (apoptotic cells)	Expression is downregulated upon DC maturation
$\alpha_v\beta_3$	Integrin	Antigen uptake (apoptotic cells)	

Table 7.5 Antigen receptors of antigen uptake in immature dendritic cells (*continued*)

Molecule	Type of protein	Function	Additional notes
LFA-1	Integrin	Antigen presentation	Adhesion receptor
ICAM-1	Ig superfamily member	Antigen presentation	Adhesion receptor
LFA-3	Integrin	Antigen presentation	Adhesion receptor
CD44	Cell adhesion receptor	Antigen presentation	Adhesion receptor
CD47	Thrombospondin receptor		Ligation inhibits cytokine production and maturation of DCs
CD91	Hsp receptor (*i.e.* gp96)	Antigen uptake	All hsp utilize CD91; complexes of gp96+antigen are processed and presented via the MHC1 pathway; gp96 can induce DC maturation; gp96R(s) are downregulated upon DC maturation

Table 7.6 Chemokine and chemokine receptor expression profile of immature dendritic cells

Molecule	Tyoe of protein	Function	Reference
MIP-3α/CCL20	CC chemokine family member	Attracts immature DCs	28,29
CCR1, CCR2, CCR3, CCR5 and CCR6	CC chemokine receptor family members	Used in response to chemokines, *e.g.* MIP-3α, RANTES, MIP-1α	18,26,28,58,78
CXCR1	CXC chemokine receptor family member	Used in response to IL-8	78
DEC-205	Type I C-type lectin; multilectin receptor	Antigen capture	49
BDCA-2, -3, -4	Type II C-type lectin	Antigen capture	31,32
DCIR	DC immunoreceptor; type II C-type lectin	Antigen capture	10
Dectin-2	DC-associated C-type lectin-2; type II C-type lectin	Antigen capture	4
CLEC-I	C-type lectin receptor 1; type II C-type lectin	Antigen capture	23
DORA	Downregulated by activation; Ig superfamily member	Antigen capture	11
DC-ASGPR	DC-asialoglycoprotein receptor; type II C-type lectin	Antigen capture	100
CD68	Macrosialin	Antigen capture	14,73,94
MR	Mannose receptor; multilectin receptor	Antigen capture	50,79
TLR2	Toll-like receptor family member	Antigen capture (binds bacterial lipopeptides)	45,97,98
TLR3	Toll-like receptor family member	Antigen capture (binds bacterial lipopeptides)	67
FcγR	FC receptor	Antigen capture (immune complexes)	74

Table 7.6 Chemokine and chemokine receptor expression profile of immature dendritic cells (*continued*)

Molecule	Tyoe of protein	Function	Reference
FcεR	Fc receptor	Antigen capture (immune complexes)	62,80
CD36	Receptor for collagen, thrombospondin, oxidized LDL, and long-chain fatty acids	Antigen capture (apoptotic cells)	3,80
gp96	hsp (heat shock protein)	Antigen capture (receptor-targeted cross-priming carrier)	9,87,89
CD91	hsp receptor (*i.e.* gp96)	Antigen capture	8,87,88
$\alpha_v\beta_5$	Integrin	Antigen capture (apoptotic cells)	3
$\alpha_v\beta_3$	Integrin	Antigen capture (apoptotic cells)	77
TAP-1 and -2	Transporter associated with antigen-processing protein	Antigen uptake/processing	47,86
Cdc42 and Rac1	Rho GTPase family members	Antigen uptake/processing	39,69,106
di-ubiquitin	Ubiquitin protein family member	Antigen uptake/processing	12
Dectin-1	DC-associated C-type lectin-1; type II C-type lectin	Antigen presentation	5,109
ILT3	Immunoglobulin-like transcript 3; Ig superfamily member	Antigen presentation	20
PI-11	Serpin (serine protease inhibitor)	Antigen presentation	65
Decysin	ADAM family member	Antigen presentation	66
Cathepsin S	Cathepsin protease family member	Antigen presentation	30,62
p55/fascin	Actin-bundling protein	Migration	64
uPAR	Urokinase plasminogen activator receptor	Migration	34
CD47	Thrombospondin receptor	Ligation inhibits cytokine production and maturation of DCs	27
Fas/TNFRSF6	TNFR superfamily member	Ligation can induce DC maturation	75
LIGHT/TNFSF14	TNF superfamily member	Costimulates induction of DC maturation (with CD40L)	63,96
DC-STAMP	DC-specific transmembrane protein; multi-membrane spanner	Unknown	44
S100b	Cytosolic Ca^{2+}-binding protein	Unknown	7,93
MADDAM/ADAM19	ADAM family member	Unknown	36,103
Langerin	Type II C-type lectin	Formation of Birbeck granules (Langerhans cells)	101

Table 7.7   Chemokine and chemokine receptor expression profile of mature dendritic cells

Molecule	Type of protein	Function	Reference
CCR7	CCR family member	Induces directional migration of mature DCs	28,29,58,78,90,108
CXCR4	CXCR family member	Induces directional migration of mature DCs	26,58
MIP-3β/CCL19	CC family member	Attracts DCs	19
IL-16	Interleukin 16	Potent chemoattractant from DCs toward themselves and T cells	54
6Ckine/CCL21	CC family member	Co-localizes DC and naïve T cells	22,29,43
PARC/CCL18	CC family member	Attracts naïve T cells	1
TARC/CCL17	CC family member	Attracts activated and memory T cells	57
MDC/CCL22	CC family member	Attracts activated and memory T cells	41,53,102
Fractalkine/CX3CL1	CX3C chemokine	T cell/DC interaction	53,71
CXCL16	CXC family member	T cell/DC interaction	61
DC-SIGN	DC-specific ICAM-3 grabbing non-integrin; type II C-type lectin	T cell/DC interaction	40,92
LFA-1, LFA-3, ICAM-1, CD11	Adhesion molecules	T cell/DC interaction	84
CD1a,b,c,d	Membrane glycoproteins (structurally related to MHC1 proteins)	Antigen presentation	15,41,83,91,95
CD83	Ig superfamily member	Antigen presentation	55,111
DC-LAMP	DC-specific lysosome-associated membrane glycoprotein	Antigen presentation	24
MHCI	MHC family member	Antigen presentation	16,76
MHCII	MHC family member	Antigen presentation	21,48,81,99
HLA-DM	MHCII accessory molecule/chaperone	Antigen presentation	6
B7-1/CD80,B7-2/CD86	Amplification of TCR signaling and activation of T cells	Antigen presentation	104,110
TNF-α/TNFSF2	TNF superfamily member	Costimulatory molecule involved in DC-mediated immune responses	96
4-1BBL/TNFSF9	TNF superfamily member	Costimulates CD8+ T cells	25,85
OX40L/TNFSF4	TNF superfamily member	Costimulates both T cells and DC activation and differentiation	70
CD40/TNFRSF5	TNFR superfamily member	Costimulates both T cell and DC activation and differentiation	93
IFN-γ	Type II interferon	Costimulates both T cell and DC activation and differentiation	37,72,75

Table 7.7 Chemokine and chemokine receptor expression profile of mature dendritic cells (*continued*)

Molecule	Type of protein	Function	Reference
ICOS L	CD28/CD152 receptor family member	Costimulatory molecule for regulating T cell activation	2
IL-1β	Interleukin 1 beta	Promotes DC production of IL-12	38,75,105
IL-12	Interleukin 12	DC production induces Th1 differentiation	33,107
IL-18	Interleukin 18	Enhances IL-12-induced IFN-γ production	37
Rel A/p65, Rel B, Rel C, p50 and p52	NF-κB transcriptional control protein family members	Regulate expression of genes encoding immune and inflammatory proteins	46,68
IFN-α	Type I interferon	Promotes DC maturation	13,59
IL-6	Interleukin 6	Accessory cytokine for DC development	52,82
TRANCE R/ TNFRSF11	TNFR superfamily member	Can enhance DC viability	51,60
IL-10	Interleukin 10	Downregulates DC maturation	17,35
cFLIP	Cellular FLICE-Inhibitory protein	Influences DC apoptosis	56

KEY REFERENCES (*see* Tables 7.6 and 7.7)

1 Adema GJ et al. (1997) Nature 387: 713
2 Alcher A et al. (2000) J Immunol 164: 4689
3 Albert ML et al. (1998) J Exp Med 188: 1359
4 Ariizumi K et al. (2000) J Biol Chem 275: 11957
5 Ariizumi K et al. (2000) J Biol Chem 275: 20157
6 Arndt SO et al. (2000) EMBO J 19: 1241
7 Banchereau J, Steinman RM (1998) Nature 392: 245
8 Basu S et al. (2001) Immunity 14: 303
9 Basu S et al. (2001) Int Immunity 12: 1539
10 Bates EE et al. (1997) Eur J Immunol 27: 2471
11 Bates EE et al. (1998) Mol Immunol 35: 513
12 Bates EE et al. (1999) J Immunol 163: 1973
13 Bendriss-Vermare N et al. (2001) J Clin Invest 107: 835
14 Betjes MG et al. (1991) Immunobiol 183: 79
15 Blumberg RS et al. (1995) Immunol.Rev 147: 5
16 Brossart P, Bevan MJ. (1997) Blood 90: 1594
17 Buelens C et al. (1997) Eur J Immunol 27: 1848
18 Carramolino L et al. (1999) J Leukoc Biol 66: 837
19 Caux C et al. (2000) Springer Semin Immunopathol 22: 345
20 Cella M et al. (1997) J Exp Med 185: 1743
21 Cella M et al. (1997) Nature 388: 782
22 Chan VW et al. (1999) Blood 93: 3610
23 Colonna M et al. (2000) Eur J Immunol 30: 697
24 de Saint-Vis B et al. (1998) Immunity 9: 325
25 DeBenedette MA et al. (1999) J Immunol 163: 4833
26 Delgado E et al. (1998) Immunobiol 198: 490
27 Demeure CE et al. (2000) J Immunol 164: 2193
28 Dieu MC et al. (1998) J Exp Med 188: 373
29 Dieu-Nosjean MC et al. (1999) J Leukoc Biol 66: 252
30 Driessen C et al. (1999) J Cell Biol 147: 775
31 Dzionek A et al. (2000) J Immunol 765: 6037.
32 Dzionek A et al. (2001) J Exp Med 194: 1823.
33 Ebner S et al. (2001) J Immunol 166: 633
34 Ferrero E et al. (2000) J Immunol 164: 712.
35 Fiebiger E et al. (2001) J Exp Med 193: 881
36 Fritsche J et al. (2000) Blood 96: 732
37 Fukao T et al. (2000) J Immunol 164: 64
38 Gardella S et al. (2000) Blood 95: 3809
39 Garrett WS et al. (2000) Cell 102: 325
40 Geijtenbeek TB et al. (2000) Cell 100: 575
41 Gerlini G et al. (2001) J Invest Dermatol 117: 576
42 Godiska R et al. (1997) J Exp Med 185: 1595
43 Gunn MD et al. (1998) J Exp Med189: 451
44 Hartgers FC et al. (2000) Eur J Immunol 30: 3585
45 Hertz C et al. (2001) J Immunol 166: 2444
46 Hofer S et al. (2001) Microbes Infect 3: 259
47 Huang AY et al. (1996) Immunity 4: 349
48 Inaba K et al. (2000) J Exp Med 191: 927
49 Jiang W et al. (1995) Nature 375: 151
50 Jordens R et al. (1999) Int Immunol 11: 1775
51 Josien R et al. (1999) J Immunol 162: 2562.
52 Kalinski P et al. (1999) J Immunol 162: 3232
53 Kanazawa N et al. (1999) Eur J Immunol 29: 1925
54 Kaser A et al. (1999) J Immunol 163: 3232
55 Lechmann M et al. (2001) J Exp Med 194: 1813
56 Leverkus M et al. (2000) Blood 96: 2628
57 Lieberam I, Forster I. (1999) Eur J Immunol 29: 2684
58 Lin CL et al. (1998) Eur J Immunol 28: 4114
59 Luft T, et al (1998) J Immunol 161: 1947
60 Lum L et al. (1999) J Biol Chem 274: 13613

61   Matloubian M *et al.* (2000) *Nat Immunol* 1: 298
62   Maurer D *et al.* (1998) *J Immunol* 161: 2731
63   Morel Y *et al.* (2001) *J Immunol* 167: 2479
64   Mosialos G *et al.* (1996) *Am J Pathol* 148: 593
65   Mueller CG *et al.* (1997) *Eur J Immunol* 27: 3130
66   Mueller CG *et al.* (1997) *J Exp Med* 186: 655
67   Muzio M *et al.* (2000) *J Immunol* 164: 5998
68   Neumann M *et al.* (2000) *Blood* 95: 277
69   Nobes C, Marsh M. (2000) *Curr Biol* 10: R739
70   Ohshima Y *et al.* (1997) *J Immunol* 159: 3838
71   Papadopoulos EJ *et al.* (1999) *Eur J Immunol* 29: 2551
72   Pashenkov M *et al.* (2000) *Eur Cytokine Netw* 11: 456
73   Petzelbauer P *et al.* (1993) *J Invest Dermatol* 101: 256
74   Regnault A *et al.* (1999) *J Exp Med* 189: 371
75   Rescigno M *et al.* (2000) *J Exp Med* 192: 1661
76   Rodriguez A *et al.* (1999) *Nat Cell Biol* 1: 362
77   Rubartelli A *et al.* (1997) *Eur J Immunol* 27: 1893
78   Sallusto F, Lanzavecchia A. (1995) *J Exp Med* 182: 389
79   Sallusto F, Lanzavecchia A. (1995) *J Exp Med* 179: 1109
80   Sallusto F *et al.* (1998) *Eur J Immunol* 28: 2760
81   Santambrogio L *et al.* (1999) *Proc Natl Acad Sci USA* 96: 15050
82   Santiago-Schwarz F *et al.* (1996) *Stem Cells* 14: 225
83   Schaible UE *et al.* (2000) *J Immunol* 164: 4843
84   Scheeren RA *et al.* (1991) *Eur J Immunol* 21: 1101
85   Shuford WW *et al.* (1997) *J Exp Med* 186: 47
86   Sigal LJ *et al.* (1999) *Nature* 398: 77

87   Singh-Jasuja H *et al.* (2000) *Cell Stress Chaperons* 5: 462
88   Singh-Jasuja H *et al.* (2000) *Eur J Immunol* 30: 2211
89   Singh-Jasuja H *et al.* (2000) *J Exp Med* 191: 1965
90   Sozzani S *et al.* (1998) *J Immunol* 161: 1083
91   Spada FM *et al.* (2000) *Eur J Immunol* 30: 3468
92   Steinman RM. (2000) *Cell* 100: 491
93   Steinman RM *et al.* (1997) *Immunol Rev* 156: 25
94   Strobl H *et al.* (1998) *J Immunol* 161: 740
95   Sugita M *et al.* (2000) *Proc Natl Acad USA* 97: 8445
96   Tamada K *et al.* (2000) *J Immunol* 164: 4105
97   Thoma-Uszynski S *et al.* (2000) *J Immunol* 165: 3804
98   Tsuji S *et al.* (2000) *Infect Immun* 68: 6883
99   Turley SJ *et al.* (2000) *Science* 288: 522
100   Valladeau J *et al.* (2001) *J Immunol* 167: 5767
101   Valladeau J *et al.* (2000) *Immunity* 12: 71
102   Vulcano M *et al.* (2001) *Eur J Immunol* 31: 812
103   Wei P *et al.* (2001) *Biochem Biophys Res Commun* 280: 744
104   Weissman D *et al.* (1995) *J Immunol* 155: 4111
105   Wesa AK, Galy A. (2001) *Int Immunol* 13: 1053
106   West MA *et al.* (2000) *Curr Biol* 10: 839
107   Winzler C *et al.* (1997) *J Exp Med* 185: 317
108   Yanagihara S *et al.* (1998) *J Immunol* 161: 3096
109   Yokota K *et al.* (2001) *Gene* 272: 51
110   Yoshimura S *et al.* (2001) *Int Immunol* 13: 675
111   Zhou LJ, Tedder TF. (1995) *J Immunol* 154: 3821

B Cells, Immunoglobulin Genes, and Immunoglobulin Structure

8

- **B CELLS**
- **IMMUNOGLOBULIN GENES**
- **IMMUNOGLOBULIN STRUCTURE**

B CELLS

B cells are cells of the B cell lineage that mature under the influence of the bursa of Fabricius in birds and the bursa equivalent (bone marrow) in mammals. B cells occupy follicular areas in lymphoid tissues and account for 5–25% of all human blood cells, which number 1000–2000 cells per mm³. They comprise most of the bone marrow cells, one-third to one-half of lymph node and spleen cells but less than 1% of those in the thymus. Non-activated B cells circulate through lymph nodes and spleen. They are concentrated in follicles and marginal zones around the follicles. Circulating B cells may interact and be activated by T cells at extrafollicular sites where the T cells are present in association with antigen-presenting dendritic cells. Activated B cells enter the follicles, proliferate, and displace resting cells. They form germinal centers and differentiate into both plasma cells that form antibody and long-lived memory B cells. Those B cells synthesizing antibodies provide defense against microorganisms, including bacteria and viruses. Surface and cytoplasmic markers reveal the stage of development and function of cells in the B cell lineage. Pre-B cells contain cytoplasmic immunoglobulins whereas mature B cells express surface immunoglobulin and complement receptors. B cell markers include CD9, CD19, CD20, CD24, Fc receptors, B1, BA-1, B4 and Ia. **Figure 8.1** shows the ontogeny of B cells and the surface markers expressed on B cells at different stages.

IMMUNOGLOBULIN GENES

The human immunoglobulins (Ig) are the products of three unlinked sets of genes – the immunoglobulin heavy (*IGH*), the immunoglobulin κ (*IGK*), and the immunoglobulin λ (*IGL*) genes – localized on chromosome 14 (14q32.33), 2 (2p12), and 22 (22q 11.2), respectively.

IGH locus

The human *IGH* locus at 14q32.33 spans 1250 kb. It consists of 123–129 *IGHV* genes depending on the haplotypes, 27 *IGHD* genes belonging to 7 subgroups, 9 *IGHJ* genes, and, in the most frequent haplotype, 11 *IGHC* genes. 82–88 *IGHV* genes belong to 7 subgroups, whereas 41 pseudogenes, which are too divergent to be assigned to subgroups, have been assigned to the 4 clans. Seven non-mapped *IGHV* genes have been described as corresponding to insertion/deletion polymorphisms but have not yet been precisely located. **Table 8.1** presents tabulated lists of the human immunoglobulin heavy genes named in accordance with the International ImMunoGeneTics database (IMGT) and approved by the Human Genome Organization (HUGO) Nomenclature Committee.

IGK locus

The human *IGK* locus at 2p12 spans 1820 kb. It consists of 76 *IGKV* genes belonging to 7 subgroups, 5 *IGKJ* genes, and a unique *IGKC* gene. In the most frequent haplotypes, the 76 *IGKV* genes are organized in two clusters separated by 800 kb, 40 genes comprising the proximal cluster, and the remaining 36 comprising the distal cluster. Twenty-eight *IGKV* orphon have been identified and sequenced: one on chromosome 1, three on the short arm of chromosome 2 but outside of the main *IGK* locus, thirteen on the long arm of chromosome 2, one on chromosome 15, six on chromosome 22, and four outside of chromosome 2. **Table 8.2** lists the human immunoglobulin kappa genes named in accordance with IMGT and approved by the HUGO Nomenclature Committee.

IGL locus

The human *IGL* locus at 22q11.2 spans 1050 kb. It consists of 70–71 *IGLV* genes, 7–11 *IGLJ* and 7–11 *IGLC* genes depend-

277

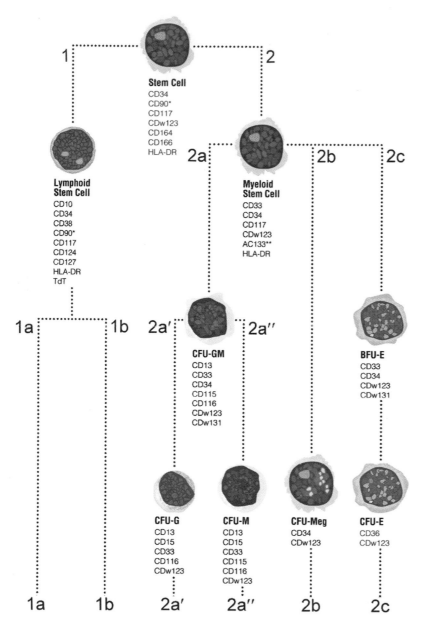

Figure 8.1 B cell ontogeny and surface markers

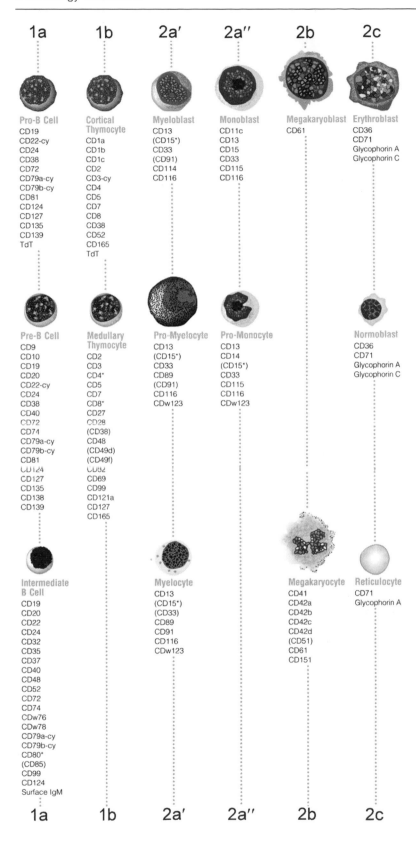

1a

Pro-B Cell
CD19
CD22-cy
CD24
CD38
CD72
CD79a-cy
CD79b-cy
CD81
CD124
CD127
CD135
CD139
TdT

1b

Cortical Thymocyte
CD1a
CD1b
CD1c
CD2
CD3-cy
CD4
CD5
CD7
CD8
CD38
CD52
CD165
TdT

2a'

Myeloblast
CD13
(CD15*)
CD33
(CD91)
CD114
CD116

2a''

Monoblast
CD11c
CD13
CD15
CD33
CD115
CD116

2b

Megakaryoblast
CD61

2c

Erythroblast
CD36
CD71
Glycophorin A
Glycophorin C

Pre-B Cell
CD9
CD10
CD19
CD20
CD22-cy
CD24
CD38
CD40
CD72
CD74
CD79a-cy
CD79b-cy
CD81
CD124
CD127
CD135
CD138
CD139

Medullary Thymocyte
CD2
CD3
CD4*
CD5
CD7
CD8*
CD27
CD28
(CD38)
CD48
(CD49d)
(CD49f)
CD52
CD69
CD99
CD121a
CD127
CD165

Pro-Myelocyte
CD13
(CD15*)
CD33
CD89
(CD91)
CD116
CDw123

Pro-Monocyte
CD13
CD14
(CD15*)
CD33
CD115
CD116
CDw123

Normoblast
CD36
CD71
Glycophorin A
Glycophorin C

Intermediate B Cell
CD19
CD20
CD22
CD24
CD32
CD35
CD37
CD40
CD48
CD52
CD72
CD74
CDw76
CDw78
CD79a-cy
CD79b-cy
CD80*
(CD85)
CD99
CD124
Surface IgM

Myelocyte
CD13
(CD15*)
(CD33)
CD89
CD91
CD116
CDw123

Megakaryocyte
CD41
CD42a
CD42b
CD42c
CD42d
(CD51)
CD61
CD151

Reticulocyte
CD71
Glycophorin A

1a   **1b**   **2a'**   **2a''**   **2b**   **2c**

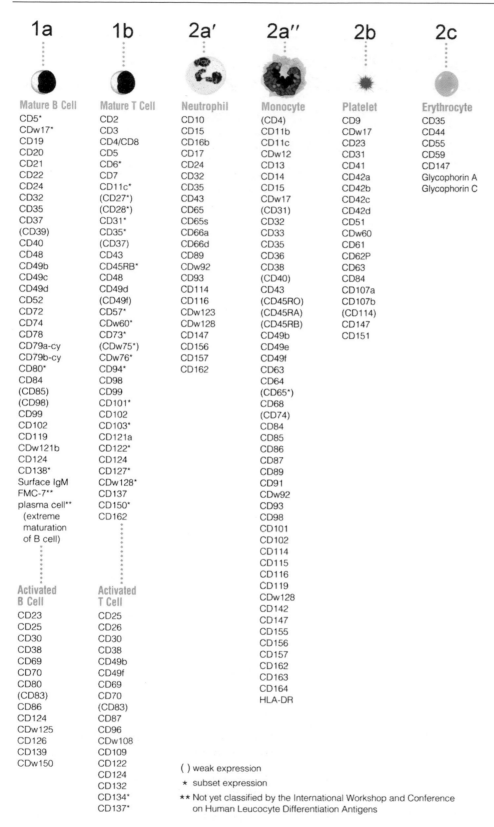

1a   **1b**   **2a′**   **2a″**   **2b**   **2c**

Mature B Cell	Mature T Cell	Neutrophil	Monocyte	Platelet	Erythrocyte
CD5*	CD2	CD10	(CD4)	CD9	CD35
CDw17*	CD3	CD15	CD11b	CDw17	CD44
CD19	CD4/CD8	CD16b	CD11c	CD23	CD55
CD20	CD5	CD17	CDw12	CD31	CD59
CD21	CD6*	CD24	CD13	CD41	CD147
CD22	CD7	CD32	CD14	CD42a	Glycophorin A
CD24	CD11c*	CD35	CD15	CD42b	Glycophorin C
CD32	(CD27*)	CD43	CDw17	CD42c	
CD35	(CD28*)	CD65	(CD31)	CD42d	
CD37	CD31*	CD65s	CD32	CD51	
(CD39)	CD35*	CD66a	CD33	CDw60	
CD40	(CD37)	CD66d	CD35	CD61	
CD48	CD43	CD89	CD36	CD62P	
CD49b	CD45RB*	CDw92	CD38	CD63	
CD49c	CD48	CD93	(CD40)	CD84	
CD49d	CD49d	CD114	CD43	CD107a	
CD52	(CD49f)	CD116	(CD45RO)	CD107b	
CD72	CD57*	CDw123	(CD45RA)	(CD114)	
CD74	CDw60*	CDw128	(CD45RB)	CD147	
CD78	CD73*	CD147	CD49b	CD151	
CD79a-cy	(CDw75*)	CD156	CD49e		
CD79b-cy	CDw76*	CD157	CD49f		
CD80*	CD94*	CD162	CD63		
CD84	CD98		CD64		
(CD85)	CD99		(CD65*)		
(CD98)	CD101*		CD68		
CD99	CD102		(CD74)		
CD102	CD103*		CD84		
CD119	CD121a		CD85		
CDw121b	CD122*		CD86		
CD124	CD124		CD87		
CD138*	CD127*		CD89		
Surface IgM	CDw128*		CD91		
FMC-7**	CD137		CDw92		
plasma cell**	CD150*		CD93		
(extreme	CD162		CD98		
maturation			CD101		
of B cell)			CD102		
			CD114		
			CD115		
			CD116		
			CD119		
			CDw128		
			CD142		
			CD147		
			CD155		
			CD156		
			CD157		
			CD162		
			CD163		
			CD164		
			HLA-DR		

Activated B Cell	Activated T Cell
CD23	CD25
CD25	CD26
CD30	CD30
CD38	CD38
CD69	CD49b
CD70	CD49f
CD80	CD69
(CD83)	CD70
CD86	(CD83)
CD124	CD87
CDw125	CD96
CD126	CDw108
CD139	CD109
CDw150	CD122
	CD124
	CD132
	CD134*
	CD137*

( ) weak expression

* subset expression

** Not yet classified by the International Workshop and Conference
on Human Leucocyte Differentiation Antigens

Figure 8.1   B cell ontogeny and surface markers (*continued*)

Table 8.1 Immunoglobulin heavy (*IGH*) genes[a]

IMGT gene group	IMGT gene name[b]	Functionality	IMGT reference sequence accession number	Number of alleles	IMGT gene definition[c]	GDB accession ID	Locus-link number
IGH locus on chromosome 14 at 14q32.33							
IGHC	*IGHA1*	F	J00220	1	Immunoglobulin heavy constant α 1	GDB:119332	3493
	IGHA2	F	J00221	3	Immunoglobulin heavy constant α 2 (A2m marker)	GDB:119333	3494
	IGHD	F	K02875-K02882	2	Immunoglobulin heavy constant δ	GDB:120084	3495
	IGHE	F	J00222	3	Immunoglobulin heavy constant ε	GDB:119335	3497
	IGHEP1	P	J00223	3	Immunoglobulin heavy constant ε P1	GDB:119336	3498
	IGHG1	F	J00228	2	Immunoglobulin heavy constant γ1 (G1m marker)	GDB:120085	3500
	IGHG2	F	J00230	2	Immunoglobulin heavy constant γ2 (G2m marker)	GDB:119338	3501
	IGHG3	F	M12958	4	Immunoglobulin heavy constant γ3 (G3m marker)	GDB:119339	3502
	IGHG4	F	K01316	3	Immunoglobulin heavy constant γ4 (G4m marker)	GDB:119340	3503
	IGHGP	ORF	X06766	1	Immunoglobulin heavy constant γP	GDB:120689	3505
	IGHM	F	X14940, X14939	3	Immunoglobulin heavy constant μ	GDB:120086	3507
IGHD[d]	*IGHD1-1*	F	X97051	1	Immunoglobulin heavy diversity 1-1	GDB:9953175	28510
	IGHD1-7	F	X13972	1	Immunoglobulin heavy diversity 1-7	GDB:9953261	28509
	IGHD1-14	ORF	X13972	1	Immunoglobulin heavy diversity 1-14	GDB:9953263	28508
	IGHD1-20	F	X97051	1	Immunoglobulin heavy diversity 1-20	GDB:9953265	28507
	IGHD1-26	F	X97051	1	Immunoglobulin heavy diversity 1-26	GDB:9953266	28506

Table 8.1 Immunoglobulin heavy (*IGH*) genes[a] (*continued*)

IMGT gene group	IMGT gene name[b]	Functionality	IMGT reference sequence accession number	Number of alleles	IMGT gene definition[c]	GDB accession ID	Locus-link number
	IGHD2-2	F	J00232	3	Immunoglobulin heavy diversity 2-2	GDB:9953230	28505
	IGHD2-8	F	X13972	2	Immunoglobulin heavy diversity 2-8	GDB:9953278	28504
	IGHD2-15	F	J00234	1	Immunoglobulin heavy diversity 2-15	GDB:9953292	28503
	IGHD2-21	F	J00235	2	Immunoglobulin heavy diversity 2-21	GDB:9953294	28502
	IGHD3-3	F	X13972	2	Immunoglobulin heavy diversity 3-3	GDB:9953296	28501
	IGHD3-9	F	X13972	1	Immunoglobulin heavy diversity 3-9	GDB:9953298	28500
	IGHD3-10	F	X13972	2	Immunoglobulin heavy diversity 3-10	GDB:9953300	28499
	IGHD3-16	F	X93614	1	Immunoglobulin heavy diversity 3-16	GDB:9953302	28498
	IGHD3-22	F	X93616	1	Immunoglobulin heavy diversity 3-22	GDB:9953304	28497
	IGHD4-4	F	X13972	1	Immunoglobulin heavy diversity 4-4	GDB:9953306	28496
	IGHD4-11	ORF	X13972	1	Immunoglobulin heavy diversity 4-11	GDB:9953308	28495
	IGHD4-17	F	X97051	1	Immunoglobulin heavy diversity 4-17	GDB:9953310	28494
	IGHD4-23	ORF	X97051	I	Immunoglobulin heavy diversity 4-23	GDB:9953312	28493
	IGHD5-5	F	X13972	1	Immunoglobulin heavy diversity 5-5	GDB:9953314	28492
	IGHD5-12	F	X13972	1	Immunoglobulin heavy diversity 5-12	GDB:9953316	28491
	IGHD5-18	F	X97051	1	Immunoglobulin heavy diversity 5-18	GDB:9953318	28490
	IGHD5-24	ORF	X97051	1	Immunoglobulin heavy diversity 5-24	GDB:9953320	28489
	IGHD6-6	F	X13972	1	Immunoglobulin heavy diversity 6-6	GDB:9953322	28488
	IGHD6-13	F	X13972	1	Immunoglobulin heavy diversity 6-13	GDB:9953324	28487

Table 8.1 Immunoglobulin heavy (*IGH*) genes[a] (*continued*)

IMGT gene group	IMGT gene name[b]	Functionality	IMGT reference sequence accession number	Number of alleles	IMGT gene definition[c]	GDB accession ID	Locus-link number
	IGHD6-19	F	X97051	1	Immunoglobulin heavy diversity 6-19	GDB:9953326	28486
	IGHD6-25	F	X97051	1	Immunoglobulin heavy diversity 6-25	GDB:9953328	28485
	IGHD7-27	F	J00256	1	Immunoglobulin heavy diversity 7-27	GDB:9953330	28484
IGHJ	*IGHJ1*	F	J00256	1	Immunoglobulin heavy joining 1	GDB:9953332	28483
	IGHJ1P	P	J00256	—	Immunoglobulin heavy joining 1P	GDB:9953334	28482
	IGHJ2	F	J00256	1	Immunoglobulin heavy joining 2	GDB:9953336	28481
	IGHJ2P	P	J00256	—	Immunoglobulin heavy joining 2P	GDB:9953338	28480
	IGHJ3	F	J00256	2	Immunoglobulin heavy joining 3	GDB:9953340	28479
	IGHJ3P	P	J00256	—	Immunoglobulin heavy joining 3P	GDB:9953342	28478
	IGHJ4	F	J00256	3	Immunoglobulin heavy joining 4	GDB:9953344	28477
	IGHJ5	F	J00256	2	Immunoglobulin heavy joining 5	GDB:9953346	28476
	IGHJ6	F	J00256	3	Immunoglobulin heavy joining 6	GDB:9953348	28475
IGHV[e]	*IGHV1-2*	F	X07448	4	Immunoglobulin heavy variable 1-2	GDB:9931660	28474
	IGHV1-3	F	X62109	2	Immunoglobulin heavy variable 1-3	GDB:9931661	28473
	IGHV1-8	F	M99637	1	Immunoglobulin heavy variable 1-8	GDB:9931662	28472
	IGHV1-12	P	X92210	—	Immunoglobulin heavy variable 1-12	GDB:9931663	28471
	IGHV1-14	P	M99639	—	Immunoglobulin heavy variable 1-14	GDB:9931664	28470
	IGHV1-17	P	M99640	—	Immunoglobulin heavy variable 1-17	GDB:9931665	28469
	IGHV1-18	F	M99641	2	Immunoglobulin heavy variable 1-18	GDB:9931666	28468

Table 8.1   Immunoglobulin heavy (*IGH*) genes[a] (*continued*)

IMGT gene group	IMGT gene name[b]	Functionality	IMGT reference sequence accession number	Number of alleles	IMGT gene definition[c]	GDB accession ID	Locus-link number
	IGHVI-24	F	M99642	1	Immunoglobulin heavy variable 1-24	GDB:9931667	28467
	IGHVI-45	F	X92209	3	Immunoglobulin heavy variable 1-45	GDB:9931668	28466
	IGHVI-46	F	X92343	3	Immunoglobulin heavy variable 1-46	GDB:9931669	28465
	IGHVI-58	F	M29809	1	Immunoglobulin heavy variable 1-58	GDB:9931670	28464
	IGHVI-67	P	X92212	—	Immunoglobulin heavy variable 1-67	GDB:9931671	28463
	IGHVI-68	P	AB019437	—	Immunoglobulin heavy variable 1-68	GDB:9931672	28462
	IGHVI-69	F	L22582	7	Immunoglobulin heavy variable 1-69	GDB:9931673	28461
	IGHV1-c	ORF	Z18904	1	Immunoglobulin heavy variable 1-c (provisional)	GDB:9931674	28460
	IGHV1-f	F	Z12305	2	Immunoglobulin heavy variable 1-f (provisional)	GDB:9931676	28458
	IGHV2-5	F	X62111	9	Immunoglobulin heavy variable 2-5	GDB:9931677	28457
	IGHV2-10	P	M99647	—	Immunoglobulin heavy variable 2-10	GDB:9931678	28456
	IGHV2-26	F	M99648	1	Immunoglobulin heavy variable 2-26	GDB:9931679	28455
	IGHV2-70	F	L21969	12	Immunoglobulin heavy variable 2-70	GDB:9931680	28454
	IGHV3-6	P	M99650	—	Immunoglobulin heavy variable 3-6	GDB:9931681	28453
	IGHV3-7	F	M99649	2	Immunoglobulin heavy variable 3-7	GDB:9931682	28452
	IGHV3-9	F	M99651	1	Immunoglobulin heavy variable 3-9	GDB:9931683	28451
	IGHV3-11	F, P	M99652	3	Immunoglobulin heavy variable 3-11	GDB:9931684	28450
	IGHV3-13	F	X92217	2	Immunoglobulin heavy variable 3-13	GDB:9931685	28449

Table 8.1 Immunoglobulin heavy (*IGH*) genes[a] (*continued*)

IMGT gene group	IMGT gene name[b]	Functionality	IMGT reference sequence accession number	Number of alleles	IMGT gene definition[c]	GDB accession ID	Locus-link number
	IGHV3-15	F	X92216	8	Immunoglobulin heavy variable 3-15	GDB:9931686	28448
	IGHV3-16	ORF	M99655	1	Immunoglobulin heavy variable 3-16	GDB:9931687	28447
	IGHV3-19	P	M99656	—	Immunoglobulin heavy variable 3-19	GDB:9931688	28446
	IGHV3-20	F	M99657	1	Immunoglobulin heavy variable 3-20	GDB:9931689	28445
	IGHV3-21	F	Z14073	2	Immunoglobulin heavy variable 3-21	GDB:9931690	28444
	IGHV3-22	P	M99659	—	Immunoglobulin heavy variable 3-22	GDB:9931691	28443
	IGHV3-23	F	M99660	3	Immunoglobulin heavy variable 3-23	GDB:9931692	28442
	IGHV3-25	P	M99661	—	Immunoglobulin heavy variable 3-25	GDB:9931693	28441
	IGHV3-29	P	M99662	—	Immunoglobulin heavy variable 3-29	GDB:9931694	28440
	IGHV3-30	F	M83134	19[f]	Immunoglobulin heavy variable 3-30	GDB:9931735	28439
	IGHV3-30-2	P	AB019439	—	Immunoglobulin heavy variable 3-30-2	GDB:9931695	28438
	IGHV3-30-3	F	X92283	2	Immunoglobulin heavy variable 3-30-3	GDB:9931696	28437
	IGHV3-30-5	F	f	f	Immunoglobulin heavy variable 3-30-5	GDB:9931697	28436
	IGHV3-32	P	M99664	—	Immunoglobulin heavy variable 3-32	GDB:9931698	28435
	IGHV3-33	F	L06618	5	Immunoglobulin heavy variable 3-33	GDB:9931699	28434
	IGHV3-33-2	P	AB019439	—	Immunoglobulin heavy variable 3-33-2	GDB:9931700	28433
	IGHV3-35	ORF	M99666	1	Immunoglobulin heavy variable 3-35	GDB:9931701	28432
	IGHV3-36	P	M99667	—	Immunoglobulin heavy variable 3-36	GDB:9931702	28431
	IGHV3-37	P	M99668	—	Immunoglobulin heavy variable 3-37	GDB:9931703	28430

Table 8.1　　Immunoglobulin heavy (*IGH*) genes[a] (*continued*)

IMGT gene group	IMGT gene name[b]	Functionality	IMGT reference sequence accession number	Number of alleles	IMGT gene definition[c]	GDB accession ID	Locus-link number
	IGHV3-38	ORF	M99669	2	Immunoglobulin heavy variable 3-38	GDB:9931704	28429
	IGHV3-41	P	M99670	—	Immunoglobulin heavy variable 3-41	GDB:9931705	28428
	IGHV3-42	P	M99671	—	Immunoglobulin heavy variable 3-42	GDB:9931706	28427
	IGHV3-43	F	M99672	2	Immunoglobulin heavy variable 3-43	GDB:9931707	28426
	IGHV3-47	ORF, P	Z18900	3	Immunoglobulin heavy variable 3-47	GDB:9931708	28425
	IGHV3-48	F	M99675	3	Immunoglobulin heavy variable 3-48	GDB:9931709	28424
	IGHV3-49	F	M99676	3	Immunoglobulin heavy variable 3-49	GDB:9931710	28423
	IGHV3-50	P	M99677	—	Immunoglobulin heavy variable 3-50	GDB:9931711	28422
	IGHV3-52	P	M99678	—	Immunoglobulin heavy variable 3-52	GDB:9931712	28421
	IGHV3-53	F	M99679	2	Immunoglobulin heavy variable 3-53	GDB:9931713	28420
	IGHV3-54	P	M99680	—	Immunoglobulin heavy variable 3-54	GDB:9931714	28419
	IGHV3-57	P	M29815	—	Immunoglobulin heavy variable 3-57	GDB:9931715	28418
	IGHV3-60	P	M29813	—	Immunoglobulin heavy variable 3-60	GDB:9931716	28417
	IGHV3-62	P	M29814	—	Immunoglobulin heavy variable 3-62	GDB:9931717	28416
	IGHV3-63	P	M99681	—	Immunoglobulin heavy variable 3-63	GDB:9931718	28415
	IGHV3-64	F	M99682	5	Immunoglobulin heavy variable 3-64	GDB:9931719	28414
	IGHV3-65	P	Z27503	—	Immunoglobuli heavy variable 3-65	GDB:9931720	28413
	IGHV3-66	F	X92218	3	Immunoglobulin heavy variable 3-66	GDB:9931736	28412
	IGHV3-71	P	AB019437	—	Immunoglobulin heavy variable 3-71	GDB:9931721	28411

Table 8.1 Immunoglobulin heavy (*IGH*) genes[a] (*continued*)

IMGT gene group	IMGT gene name[b]	Functionality	IMGT reference sequence accession number	Number of alleles	IMGT gene definition[c]	GDB accession ID	Locus-link number
	IGHV3-72	F	X92206	2	Immunoglobulin heavy variable 3-72	GDB:9931722	28410
	IGHV3-73	F	Z27508	1	Immunoglobulin heavy variable 3-73	GDB:9931723	28409
	IGHV3-74	F	Z12353	3	Immunoglobulin heavy variable 3-74	GDB:9931724	28408
	IGHV3-75	P	Z27510	—	Immunoglobulin heavy variable 3-75	GDB:9931725	28407
	IGHV3-76	P	Z12341	—	Immunoglobulin heavy variable 3-76	GDB:9931726	28406
	IGHV3-79	P	AB019437	—	Immunoglobulin heavy variable 3-79	GDB:9931727	28405
	IGHV3-d	F	Z18898	1	Immunoglobulin heavy variable 3-d (provisional)	GDB:9931729	28404
	IGHV3-g	P	Z12336	—	Immunoglobulin heavy variable 3-g (provisional)	GDB:9931728	28403
	IGHV3-h	—	Z29981	1	Immunoglobulin heavy variable 3-h (provisional)	GDB:9931730	28402
	IGHV4-4	F	X05713	8	Immunoglobulin heavy variable 4-4	GDB:9931731	28401
	IGHV4-28	F	X05714	5	Immunoglobulin heavy variable 4-28	GDB:9931732	28400
	*IGHV4-30-1	F	g	g	Immunoglobulin heavy variable 4-30-1	GDB:9931734	28399
	*IGHV4-30-2	F	L10089	4	Immunoglobulin heavy variable 4-30-2	GDB:9953352	28398
	*IGHV4-30-4	F	Z14238	6	Immunoglobulin heavy variable 4-30-4	GDB:9953354	28397
	IGHV4-31	F	L10098	10[g]	Immunoglobulin heavy variable 4-31	GDB:9931737	28396
	IGHV4-34	F	X92278	13	Immunoglobulin heavy variable 4-34	GDB:9931738	28395
	IGHV4-39	F	L10094	6	Immunoglobulin heavy variable 4-39	GDB:9931739	28394

Table 8.1   Immunoglobulin heavy (*IGH*) genes[a] (*continued*)

IMGT gene group	IMGT gene name[b]	Functionality	IMGT reference sequence accession number	Number of alleles	IMGT gene definition[c]	GDB accession ID	Locus-link number
	IGHV4-55	P	M99685	—	Immunoglobulin heavy variable 4-55	GDB:9931740	28393
	IGHV4-59	F	L10088	9	Immunoglobulin heavy variable 4-59	GDB:9931741	28392
	IGHV4-61	F, ORF	M29811	7	Immunoglobulin heavy variable 4-61	GDB:9931742	28391
	IGHV4-80	P	AB019437	—	Immunoglobulin heavy variable 4-80	GDB:9931743	28390
	IGHV4-b	F	Z12367	2	Immunoglobulin heavy variable 4-b (provisional)	GDB:9931744	28389
	IGHV5-51	F	M99686	5	Immunoglobulin heavy variable 5-51	GDB:9931745	28388
	IGHV5-78	P	X92213	—	Immunoglobulin heavy variable 5-78	GDB:9931746	28387
	IGHV5-a	F, P	X92227	4	Immunoglobulin heavy variable 5-a (provisional)	GDB:9931747	28386
	IGHV6-1	F	X92224/J04097	2	Immunoglobulin heavy variable 6-1	GDB:9931748	28385
	*IGHV7-4-1	F	Z12323	3	Immunoglobulin heavy variable 7-4-1	GDB:9931749	28384
	IGHV7-27	P	M99643	—	Immunoglobulin heavy variable 7-27	GDB:9931750	28383
	IGHV7-34-1	P	AB019739	—	Immunoglobulin heavy variable 7-34-1	GDB:9931751	28382
	IGHV7-40	P	M99644	—	Immunoglobulin heavy variable 7-40	GDB:9953358	28381
	IGHV7 56	P	M29810	—	Immunoglobulin heavy variable 7-56	GDB:9931752	28380
	IGHV7-77[h]	—	—	—	Immunoglobulin heavy variable 7-77 (provisional)	GDB:9931753	28379
	IGHV7-81	ORF	Z27509	1	Immunoglobulin heavy variable 7-81	GDB:9931754	28378
	IGHV(II)-1-1	P	AB019441	—	Immunoglobulin heavy variable (II)-1-1	GDB:9931755	28377

Table 8.1 Immunoglobulin heavy (*IGH*) genes[a] (*continued*)

IMGT gene group	IMGT gene name[b]	Functionality	IMGT reference sequence accession number	Number of alleles	IMGT gene definition[c]	GDB accession ID	Locus-link number
	IGHV(II)-15-1	P	AB019440	—	Immunoglobulin heavy variable (II)-15-1	GDB:9931756	28376
	IGHV(II)-20-1	P	AB019440	—	Immunoglobulin heavy variable (II)-20-1	GDB:9931757	28375
	IGHV(II)-22-1	P	AB019439	—	Immunoglobulin heavy variable (II)-22-1	GDB:9931758	28374
	IGHV(II)-26-2	P	AB019439	—	Immunoglobulin heavy variable (II)-26-2	GDB:9933358	28373
	IGHV(II)-28-1	P	AB019439	—	Immunoglobulin heavy variable (II)-28-1	GDB:9931759	28372
	IGHV(II)-30-1	P	AB019439	—	Immunoglobulin heavy variable (II)-30-1	GDB:9931760	28371
	IGHV(II)-31-1	P	AB019439	—	Immunoglobulin heavy variable (II)-31-1	GDB:9931761	28370
	IGHV(II)-33-1	P	AB019439	—	Immunoglobulin heavy variable (II)-33-1	GDB:9931762	28369
	IGHV(II)-40-1	P	AB019438	—	Immunoglobulin heavy variable (II)-40-1	GDB:9953360	28368
	IGHV(II)-43-1	P	AB019438	—	Immunoglobulin heavy variable (II)-43-1	GDB:9953362	28367
	IGHV(II)-44-2	P	AB019438	—	Immunoglobulin heavy variable (II)-44-2	GDB:9931763	28366
	IGHV(II)-46-1	P	AB019438	—	Immunoglobulin heavy variable (II)-46-1	GDB:9931764	28365
	IGHV(II)-49-1	P	AB019438	—	Immunoglobulin heavy variable (II)-49-1	GDB:9931765	28364
	IGHV(II)-51-2	P	AB019438	—	Immunoglobulin heavy variable (II)-51-2	GDB:9931766	28363

Table 8.1 Immunoglobulin heavy (*IGH*) genes[a] (*continued*)

IMGT gene group	IMGT gene name[b]	Functionality	IMGT reference sequence accession number	Number of alleles	IMGT gene definition[c]	GDB accession ID	Locus-link number
	IGHV(II)-53-1	P	AB019438	—	Immunoglobulin heavy variable (II)-53-1	GDB:9931767	28362
	IGHV(II)-60-1	P	AB019437	—	Immunoglobulin heavy variable (II)-60-1	GDB:9931768	28361
	IGHV(II)-62-1	P	AB019437	—	Immunoglobulin heavy variable (II)-62-1	GDB:9931769	28360
	IGHV(II)-65-1	P	AB019437	—	Immunoglobulin heavy variable (II)-65-1	GDB:9931770	28359
	IGHV(II)-67-1	P	AB019437	—	Immunoglobulin heavy variable (II)-67-1	GDB:9953364	28358
	IGHV(II)-74-1	P	AB019437	—	Immunoglobulin heavy variable (II)-74-1	GDB:9931771	28357
	IGHV(II)-78-1	P	AB019437	—	Immunoglobulin heavy variable (II)-78-1	GDB:9931772	28356
	IGHV(III)-2-1	P	AB019441	—	Immunoglobulin heavy variable (III)-2-1	GDB:9931773	28355
	IGHV(III)-5-1	P	AB019440	—	Immunoglobulin heavy variable (III)-5-1	GDB:9931774	28354
	IGHV(III)-5-2	P	AB019440	—	Immunoglobulin heavy variable (III)-5-2	GDB:9931775	28353
	IGHV(III)-11-1	P	AB019440	—	Immunoglobulin heavy variable (III)-11-1	GDB:9931776	28352
	IGHV(III)-13-1	P	AB019440	—	Immunoglobulin heavy variable (III)-13-1	GDB:9931777	28351
	IGHV(III)-16-1	P	AB019440	—	Immunoglobulin heavy variable (III)-16-1	GDB:9931778	28350
	IGHV(III)-22-2	P	AB019439	—	Immunoglobulin heavy variable (III)-22-2	GDB:9931779	28349

Table 8.1 Immunoglobulin heavy (*IGH*) genes[a] (*continued*)

IMGT gene group	IMGT gene name[b]	Functionality	IMGT reference sequence accession number	Number of alleles	IMGT gene definition[c]	GDB accession ID	Locus-link number
	IGHV(III)-25-1	P	AB019439	—	Immunoglobulin heavy variable (III)-25-1	GDB:9931780	28348
	IGHV(III)-26-1	P	AB019439	—	Immunoglobulin heavy variable (III)-26-1	GDB:9953366	28347
	IGHV(III)-38-1	P	AB019439	—	Immunoglobulin heavy variable (III)-38-1	GDB:9953368	28346
	IGHV(III)-44	P	M99673	—	Immunoglobulin heavy variable (III)-44	GDB:9953370	28345
	IGHV(III)-47-1	P	AB019438	—	Immunoglobulin heavy variable (III)-47-1	GDB:9953372	28344
	IGHV(III)-51-1	P	AB019438	—	Immunoglobulin heavy variable (III)-51-1	GDB:9953381	28343
	IGHV(III)-67-2	P	AB019437	—	Immunoglobulin heavy variable (III)-67-2	GDB:9953382	28342
	IGHV(III)-67-3	P	AB019437	—	Immunoglobulin heavy variable (III)-67-3	GDB:9953383	28341
	IGHV(III)-67-4	P	AB019437	—	Immunoglobulin heavy variable (III)-67-4	GDB:9953385	28340
	IGHV(III)-76-1	P	AB019437	—	Immunoglobulin heavy variable (III)-76-1	GDB:9953386	28339
	IGHV(III)-82	P	AB019437	—	Immunoglobulin heavy variable (III)-82	GDB:9953388	28338
	IGHV(IV)-44-1	P	AB019438	—	Immunoglobulin heavy variable (IV)-44-1	GDB:9953374	28337

***IGH* orphons**
On chromosome 9 at 9p24.2-p24.1

IGHC	*IGHEP2*	P	K01241	—	Immunoglobulin heavy constant ε P2	GDB:119337	3499

On chromosome 15 at 15q11.2

IGHD	*IGHD1/OR15-1a*	ORF	X55575	—	Immunoglobulin heavy diversity 1/OR15-la	GDB:9953376	28335

Table 8.1 Immunoglobulin heavy (*IGH*) genes[a] (*continued*)

IMGT gene group	IMGT gene name[b]	Functionality	IMGT reference sequence accession number	Number of alleles	IMGT gene definition[c]	GDB accession ID	Locus-link number
	IGHD1/OR15-1b	ORF	X55576	—	Immunoglobulin heavy diversity 1/OR15-lb	GDB:9953378	28334
	IGHD2/OR15-2a	ORF	X55577	—	Immunoglobulin heavy diversity 2/OR15-2a	GDB:9953380	28333
	IGHD2/OR15-2b	ORF	X55578	—	Immunoglobulin heavy diversity 2/OR 15-2b	GDB:9953382	28332
	IGHD3/OR15-3a	ORF	X55579	—	Immunoglobulin heavy diversity 3/OR15-3a	GDB:9953384	28331
	IGHD3/ORl5-3b	ORF	X55580	—	Immunoglobulin heavy diversity 3/OR 15-3b	GDB:9953386	28330
	IGHD4/ORl5-4a	ORF	X55581	—	Immunoglobulin heavy diversity 4/OR15-4a	GDB:9953388	28329
	IGHD4/OR15-4b	ORF	X55582	—	Immunoglobulin heavy diversity 4/OR 15-4b	GDB:9953390	28328
	IGHD5/ORl5-5a	ORF	X55583	—	Immunoglobulin heavy diversity 5/OR15-5a	GDB:9953392	28327
	IGHD5/OR15-5b	ORF	X55584	—	Immunoglobulin heavy diversity 5/OR15-5b	GDB:9953394	28326
IGHV	*IGHV1/ORl5-1*	ORF	Z29631	—	Immunoglobulin heavy variable 1 /OR 15-1	GDB:9931784	28325
	IGHV1/ORl5-2	P	L25543	—	Immunoglobulin heavy variable 1 /OR 15-2	GDB:9931787	28324
	IGHV1/OR15-3	P, ORF	Z29595	—	Immunoglobulin heavy variable 1/OR15-3	GDB:9953396	28323
	IGHV1/OR15-4	P	Z29596	—	Immunoglobulin heavy variable 1/OR15-4	GDB:9931789	28322
	IGHV1/OR15-5	P, ORF	Z29633	—	Immunoglobulin heavy variable 1 /OR 15-5	GDB:9953398	28321

Table 8.1 Immunoglobulin heavy (*IGH*) genes[a] (*continued*)

IMGT gene group	IMGT gene name[b]	Functionality	IMGT reference sequence accession number	Number of alleles	IMGT gene definition[c]	GDB accession ID	Locus-link number
	IGHV1/OR15-6	P	Z29634	—	Immunoglobulin heavy variable 1/OR15-6	GDB:9931790	28320
	IGHV1/OR15-9	ORF	L25542	—	Immunoglobulin heavy variable 1/OR15-9	GDB:9931791	28319
	IGHV3/OR15-7	P, ORF	Z29597	—	Immunoglobulin heavy variable 3/OR15-7	GDB:9931792	28318
	IGHV4/OR15-8	ORF	Z29598	—	Immunoglobulin heavy variable 4/OR15-8	GDB:9953400	28317
On chromosome 16 at 16p11.2							
IGHV	*IGHV1/OR16-1*	P	Z29599	—	Immunoglobulin heavy variable 1/OR16-1	GDB:9953402	28315
	IGHV1/OR16-2	P	Z29600	—	Immunoglobulin heavy variable 1/OR16-2	GDB:9953404	28314
	IGHV1/OR16-3	P	Z29639	—	Immunoglobulin heavy variable 1/OR16-3	GDB:9931793	28313
	IGHV1/OR16-4	P	Z17397	—	Immunoglobulin heavy variable 1/OR16-4	GDB:9931794	28312
	IGHV2/OR16-5	ORF	L25544	—	Immunoglobulin heavy variable 2/OR16-5	GDB:9931795	28311
	IGHV3/OR16-6	P	L25545	—	Immunoglobulin heavy variable 3/OR16-6	GDB:9953406	28310
	IGHV3/OR16-7	P	Z29604	—	Immunoglobulin heavy variable 3/OR16-7	GDB:9931796	28309
	IGHV3/OR16-8	ORF	Z29605	—	Immunoglobulin heavy variable 3/OR16-8	GDB:9931797	28308
	IGHV3/OR16-9	ORF	Z29606	—	Immunoglobulin heavy variable 3/OR16-9	GDB:9953408	28307
	IGHV3/OR16-10	ORF	Z29607	—	Immunoglobulin heavy variable 3/OR16-10	GDB:9953410	28306

Table 8.1   Immunoglobulin heavy (*IGH*) genes[a] (*continued*)

IMGT gene group	IMGT gene name[b]	Functionality	IMGT reference sequence accession number	Number of alleles	IMGT gene definition[c]	GDB accession ID	Locus-link number
	IGHV3/OR16-11	P	Z29608	—	Immunoglobulin heavy variable 3/OR 16-11	GDB:9953412	28305
	IGHV3/OR16-12	ORF	Z29609	—	Immunoglobulin heavy variable 3/OR16-12	GDB:9953414	28304
	IGHV3/OR16-13	ORF	Z29610	—	Immunoglobulin heavy variable 3/OR16-13	GDB:9953416	28303
	IGHV3/OR16-14	P	Z29611	—	Immunoglobulin heavy variable 3/OR 16-14	GDB:9953418	28302
	IGHV3/OR16-15	P	L25546	—	Immunoglobulin heavy variable 3/OR16-15	GDB:9953420	28301
	IGHV3/OR16-16	P	Z29613	—	Immunoglobulin heavy variable 3/OR16-16	GDB:9953422	28300

Notes:

[a]Gene names are according to the IMGT gene name nomenclature for IG and TcR of all vertebrates. (IMGT Scientific chart at http://imgt.cines.fr:8104).

[b]IMGT *IGH* gene names have been approved by the HUGO Nomenclature Committee in 1999. Note that, in the HUGO symbols, parentheses of the truncated pseudogene names and slashes of the orphon names are omitted, and capital letters replace the lower case letters found in seven provisional IMGT gene names. Otherwise all the gene names (gene symbols) are identical in IMGT and HUGO nomenclatures.

[c]Gene definitions (full names) are identical (including parentheses and slashes) in IMGT and HUGO nomenclatures. Note that in the databases, the Greek letters are written in full (i.e., α = alpha, γ = gamma, δ = delta, ε = epsilon, and μ = mu).

[d]*IGHD* genes are designated by a number for the subgroup, followed by a hyphen and a number for the localization from 5′ to 3′ in the locus.

[e]*IGHV* genes are designated by a number for the subgroup, followed by a hyphen and a number for the localization from 3′ to 5′ in the locus. Pseudogenes which could not be assigned to subgroups with functional genes are designated by a roman number between parentheses corresponding to the clan (clan I: *IGHV1*, *IGHV5*, and *IGHV7* subgroup genes; clan II: *IGHV2*, *IGHV4*, and *IGHV6* subgroup genes, and pseudogenes *IGHV(II)*; clan III: *IGHV3* subgroup genes, and pseudogenes *IGHV(III)*; clan IV: pseudogene *IGHV(IV)-44*), followed by a hyphen, and a number for the localization from 3′ to 5′ in the locus. All of these pseudogenes have truncations. Seven genes which have been described as insertion polymorphisms but which have not been precisely located are designated by a number for the subgroup, followed by a hyphen and a small letter: *IGHV1-c*, *IGHV1-f*, *IGHV3-d*, *IGHV3-g*, *IGHV3-h*, *IGHV4-6*, and *IGHV5-a*. These genes are not counted in the potential repertoire and have a provisional designation. An asterisk (*) indicates allelic polymorphisms by insertion/deletion which concern (1) a 50 kb insertion of 5 genes (3-30-5, 4-30-4, 3-30-3, 4-30-2, 4-30-1) observed in 45% Caucasoids (2) the *IGHV7-4-1* gene.

[f]Sequences of the polymorphic *IGHV3-30-5* gene cannot be differentiated from those of the *IGHV3-30* gene. All sequences are described therefore as 'IGHV3-30 alleles' by comparison to the allele*01 of *IGHV3-30* (M83134); however, it is not excluded that some of these "alleles" belong exclusively to *IGHV3-30-5*.

[g]Sequences of the polymorphic *IGHV4-30-1* gene cannot be differentiated from those of the *IGHV4-31* gene. All sequences are described therefore as 'IGHV4-31' alleles by comparison to the allele*01 of *IGHV4-31* (L10098); however, it is not excluded that some of these 'alleles' belong exclusively to *IGHV4-30-1*.

[h]A putative gene *IGHV7-77* was not found by Matsuda *et al.* (1998). This gene is not counted in the potential repertoire and has a provisional designation.

Table 8.2 Immunoglobulin kappa (*IGK*) genes[a]

IMGT gene group	IMGT gene name[b]	Functionality	IMGT reference sequence accession number	Number of alleles	IMGT gene definition[c]	GDB accession ID	Locus-link number
*IGK* locus on chromosome 2 at 2p12							
Proximal cluster							
IGKC	*IGKC*	F	J00241/V 00557	4	Immunoglobulin κ constant	GDB:120088	3514
IGKJ	*IGKJ1*	F	J00242	1	Immunoglobulin κ joining 1	GDB:9953169	28950
	IGKJ2	F	J00242	1	Immunoglobulin κ joining 2	GDB:9953424	28949
	IGKJ3	F	J00242	1	Immunoglobulin κ joining 3	GDB:9953426	28948
	IGKJ4	F	J00242	1	Immunoglobulin κ joining 4	GDB:9953428 28947	
	IGKJ5	F	J00242	1	Immunoglobulin κ joining 5	GDB:9953430	28946
IGKV[d]	*IGKV1-5*	F	Z00001	3	Immunoglobulin κ variable 1-5	GDB:9953432	28944
	IGKV1-6	F	M64858	1	Immunoglobulin κ variable 1-6	GDB:9953434	28943
	IGKV1-8	ORF	Z00014	1	Immunoglobulin κ variable 1-8	GDB:9953436	28942
	IGKV1-9	F	Z00013	1	Immunoglobulin κ variable 1-9	GDB:9953438	28941
	IGKV1-12	F	V01577	$1(+1?)^e$	Immunoglobulin κ variable 1-12	GDB:9953440	28940
	IGKV1-13	P, (ORF?)	Z00010	$1(+1?)^e$	Immunoglobulin κ variable 1-13	GDB:9953442	28939
	IGKV1-16	F	J00248	1	Immunoglobulin κ variable 1-16	GDB:9953444	28938
	IGKV1-17	F	X72808	1	Immunoglobulin κ variable 1-17	GDB:9953446	28937
	IGKV1-22	P	X71885	—	Immunoglobulin κ variable 1-22	GDB:9953448	28936
	IGKV1-27	F	X63398	1	Immunoglobulin κ variable 1-27	GDB:9953450	28935
	IGKV1-32	P	X71883	—	Immunoglobulin κ variable 1-32	GDB:9953452	28934

Table 8.2 Immunoglobulin kappa (*IGK*) genes[a] (*continued*)

IMGT gene group	IMGT gene name[b]	Functionality	IMGT reference sequence accession number	Number of alleles	IMGT gene definition[c]	GDB accession ID	Locus-link number
	IGKV1-33	F	M64856	1	Immunoglobulin κ variable 1-33	GDB:9953454	28933
	IGKV1-35	P	X71890	—	Immunoglobulin κ variable 1-35	GDB:9953456	28932
	IGKV1-37	ORF	X59316	1	Immunoglobulin κ variable 1-37	GDB:9953458	28931
	IGKV1-39	F, P	X59315	2	Immunoglobulin κ variable 1-39	GDB:9953460	28930
	IGKV2-4	P	X72814	—	Immunoglobulin κ variable 2-4	GDB:9953462	28929
	IGKV2-10	P	Z00012	—	Immunoglobulin κ variable 2-10	GDB:9953464	28928
	IGKV2-14	P	X72810	—	Immunoglobulin κ variable 2-14	GDB:9953466	28927
	IGKV2-18	P	X63400	—	Immunoglobulin κ variable 2-18	GDB:9953468	28926
	IGKV2-19	P	X12692	—	Immunoglobulin κ variable 2-19	GDB:9953470	28925
	IGKV2-23	P	X71885	—	Immunoglobulin κ variable 2-23	GDB:9953472	28924
	IGKV2-24	F	X12684	1	Immunoglobulin κ variable 2-24	GDB:9953474	28923
	IGKV2-26	P	X71884	—	Immunoglobulin κ variable 2-26	GDB:9953476	28922
	IGKV2-28	F	X63397	1	Immunoglobulin κ variable 2-28	GDB:9953478	28921
	IGKV2-29	P, F	X63396	2	Immunoglobulin κ variable 2-29	GDB:9953480	28920
	IGKV2-30	F	X63403	1	Immunoglobulin κ variable 2-30	GDB:9953482	28919
	IGKV2-36	P	X71889	—	Immunoglobulin κ variable 2-36	GDB:9953484	28918
	IGKV2-38	P	X71888	—	Immunoglobulin κ variable 2-38	GDB:9953486	28917
	IGKV2-40	F	X59314	2	Immunoglobulin κ variable 2-40	GDB:9953488	28916
	IGKV3-7	ORF	X02725	3	Immunoglobulin κ variable 3-7	GDB:9953490	28915

Table 8.2 Immunoglobulin kappa (*IGK*) genes[a] (*continued*)

IMGT gene group	IMGT gene name[b]	Functionality	IMGT reference sequence accession number	Number of alleles	IMGT gene definition[c]	GDB accession ID	Locus-link number
	IGKV3-11	F	X01668	2	Immunoglobulin κ variable 3-11	GDB:9953492	28914
	IGKV3-15	F	M23090	1	Immunoglobulin κ variable 3-15	GDB:9953494	28913
	IGKV3-20	F	X12686	2	Immunoglobulin κ variable 3-20	GDB:9953496	28912
	IGKV3-25	P	X06583	—	Immunoglobulin κ variable 3-25	GDB:9953498	28911
	IGKV3-31	P	X71883	—	Immunoglobulin κ variable 3-31	GDB:9953500	28910
	IGKV3-34	P	X71891	—	Immunoglobulin κ variable 3-34	GDB:9953502	28909
	IGKV4-1	F	Z00023	1	Immunoglobulin κ variable 4-1	GDB:9953504	28908
	IGKV5-2	F	X02485	1	Immunoglobulin κ variable 5-2	GDB:9953506	28907
	IGKV6-21	ORF	X63399	1	Immunoglobulin κ variable 6-21	GDB:9953508	28906
	IGKV7-3	P	X12682	—	Immunoglobulin κ variable 7-3	GDB:9953510	28905
Distal cluster							
IGKV	*IGKV1D-8*	F	Z00008	1	Immunoglobulin κ variable 1D-8	GDB:9953512	28904
	IGKV1D-12	F	X17263	1(+1?)[e]	Immunoglobulin κ variable 1D-12	GDB:9953514	28903
	IGKV1D-13	ORF	X17262	1(+1?)[e]	Immunoglobulin κ variable 1D-13	GDB:9953516	28902
	IGKV1D-16	F	K01323	2	Immunoglobulin κ variable 1D-16	GDB:9953518	28901
	IGKV1D-17	F	X63392	1	Immunoglobulin κ variable 1D-17	GDB:9953520	28900
	IGKV1D-22	P	X71887	—	Immunoglobulin κ variable 1D-22	GDB:9953522	28899
	IGKV1D-27	P	Z00004	—	Immunoglobulin κ variable 1D-27	GDB:9953524	28898
	IGKV1D-32	P	X71896	—	Immunoglobulin κ variable 1D-32	GDB:9953526	28897
	IGKV1D-33	F	M64855	1	Immunoglobulin κ variable 1D-33	GDB:9953528	28896
	IGKV1D-35	P	X71894	—	Immunoglobulin κ variable 1D-35	GDB:9953530	28895

Table 8.2 Immunoglobulin kappa (*IGK*) genes[a] (*continued*)

IMGT gene group	IMGT gene name[b]	Functionality	IMGT reference sequence accession number	Number of alleles	IMGT gene definition[c]	GDB accession ID	Locus-link number
	IGKV1D-37	ORF	X71893	1	Immunoglobulin κ variable 1D-37	GDB:9953532	28894
	IGKV1D-39	F	X59312	1	Immunoglobulin κ variable 1D-39	GDB:9953534	28893
	IGKV1D-42	ORF	X72816	1	Immunoglobulin κ variable 1D-42	GDB:9953536	28892
	IGKV1D-43	F	X72817	1	Immunoglobulin κ variable 1D-43	GDB:9953538	28891
	IGKV2D-10	P	X17265	—	Immunoglobulin κ variable 2D-10	GDB:9953540	28890
	IGKV2D-14	P	X72811	—	Immunoglobulin κ variable 2D-14	GDB:9953542	28889
	IGKV2D-18	P	X63395	—	Immunoglobulin κ variable 2D-18	GDB:9953544	28888
	IGKV2D-19	P	X71882	—	Immunoglobulinκ variable 2D-19	GDB:9953546	28887
	IGKV2D-23	P	X71887	—	Immunoglobulin κ variable 2D-23	GDB:9953548	28886
	IGKV2D-24	F	X63401	1	Immunoglobulin κ variable 2D-24	GDB:9953550	28885
	IGKV2D-26	P	X12689	—	Immunoglobulin κ variable 2D-26	GDB:9953552	28884
	IGKV2D-28	F	X12691	1	Immunoglobulin κ variable 2D-28	GDB:9953554	28883
	IGKV2D-29	F, ORF	M31952	2	Immunoglobulin κ variable 2D-29	GDB:9953556	28882
	IGKV2D-30	F	X63402	1	Immunoglobulin κ variable 2D-30	GDB:9953558	28881
	IGKV2D-36	P	X71893	—	Immunoglobulin κ variable 2D-36	GDB:9953560	28880
	IGKV2D-38	P	X71892	—	Immunoglobulin κ variable 2D-38	GDB:9953562	28879
	IGKV2D-40	F	X59311	1	Immunoglobulin κ variable 2D-40	GDB:9953564	28878
	IGKV3D-7	F	X72820	1	Immunoglobulin κ variable 3D-7	GDB:9953566	28877
	IGKV3D-11	F	X17264	1	Immunoglobulin κ variable 3D-11	GDB:9953568	28876

Table 8.2 Immunoglobulin kappa (*IGK*) genes[a] (*continued*)

IMGT gene group	IMGT gene name[b]	Functionality	IMGT reference sequence accession number	Number of alleles	IMGT gene definition[c]	GDB accession ID	Locus-link number
	IGKV3D-15	F, P	X72815	2	Immunoglobulin κ variable 3D-15	GDB:9953570	28875
	IGKV3D-20	F	X12687	1	Immunoglobulin κ variable 3D-20	GDB:9953572	28874
	IGKV3D-25	P	X71886	—	Immunoglobulin κ variable 3D-25	GDB:9953574	28873
	IGKV3D-31	P	X71896	—	Immunoglobulin κ variable 3D-31	GDB:9953576	28872
	IGKV3D-34	P	X71895	—	Immunoglobulin κ variable 3D-34	GDB:9953578	28871
	IGKV6D-21	ORF	X12683	1	Immunoglobulin κ variable 6D-21	GDB:9953580	28870
	IGKV6D-41	ORF	X12688	1	Immunoglobulin κ variable 6D-41	GDB:9953582	28869
	***IGKV* orphons**
	**On chromosome 1 at 1 pter-1 qter**
IGKV	*IGKV1/OR1-1*	P	M20809	—	Immunoglobulin κ variable 1/OR1-1	GDB:9953584	3525
	**On chromosome 2 at 2p12**
IGKV	*IGKV1/OR2-0*	ORF	Y08392	—	Immunoglobulin κ variable 1/OR2-0	GDB:9953586	28867
	**On chromosome 2 at 2cen-2q11**
	IGKV1/OR2-3	P	X05102	—	Immunoglobulin κ variable 1/OR2-3	GDB:9953588	28866
	IGKV1/OR2-6	P	X05103	—	Immunoglobulin κ variable 1/OR2-6	GDB:9953590	28865
	IGKV1/OR2-9	P	X51879	—	Immunoglobulin κ variable 1/OR2-9	GDB:9953592	28864
	IGKV1/OR2-11	P	X51885	—	Immunoglobulin κ variable 1/OR2-11	GDB:9953594	28863
	**On chromosome 2 at 2q12-2q14**
	IGKV1/OR2-108	ORF	X51887	—	Immunoglobulin κ variable 1/OR2-108	GDB:9953596	28862
	**On chromosome 2 at 2cen-2q11**
	IGKV2/OR2-1	P	X05101	—	Immunoglobulin κ variable 2/OR2-1	GDB:9953600	28861

Table 8.2 Immunoglobulin kappa (*IGK*) genes[a] (*continued*)

IMGT gene group	IMGT gene name[b]	Functionality	IMGT reference sequence accession number	Number of alleles	IMGT gene definition[c]	GDB accession ID	Locus-link number
	IGKV2/ OR2-la	P	X76074	—	Immunoglobulin κ variable 2/OR2-la	GDB:9953602	28860
	IGKV2/ OR2-2	P	X51884	—	Immunoglobulin κ variable 2/OR2-2	GDB:9953604	28859
	IGKV2/ OR2-4	P	X51883	—	Immunoglobulin κ variable 2/OR2-4	GDB:9953606	28858
	IGKV2/ OR2-7	P	X51881	—	Immunoglobulin κ variable 2/OR2-7	GDB:9953608	28857
	IGKV2/ OR2-8	P	X51880	—	Immunoglobulin κ variable 2/OR2-8	GDB:9953610	28856
	IGKV2/ OR2-10	P	X51886	ND	Immunoglobulin κ variable 2/OR2-10	GDB:9953612	28855
	IGKV3/ OR2-5	P	X51882	—	Immunoglobulin κ variable 3/OR2-5	GDB:9953614	28854
On chromosome 2 at 2p12							
	IGKV3/ OR2-268	ORF	X74459	—	Immunoglobulin κ variable 3/OR2-268	GDB:9953616	3523
	IGKV3/ OR2-286a	ORF	X74460	—	Immunoglobulin κ variable 3/OR2-268a	GDB:9953618	28852
On chromosome 15 at 15pter-15qter							
IGKV	*IGKV1/ OR15-118*	P	M20812	—	Immunoglobulin κ variable 1/OR15-118	GDB:9953598	3526
On chromosome 22 at 22pter-22qter							
IGKV	*IGKV1/ OR22-1*	P	Z00040	—	Immunoglobulin κ variable 1/OR22-1	GDB:9953620	3530
	IGKV1/ OR22-5	P	Z00003	—	Immunoglobulin κ variable 1/OR22-5	GDB:9953622	28850
	IGKV1/ OR22-5a	P	Z00002	—	Immunoglobulin κ variable 1/OR22-5a	GDB:9953624	29973
	IGKV2/ OR22-3	P	Z00041	—	Immunoglobulin κ variable 2/OR22-3	GDB:9953626	3529
	IGKV2/ OR22-4	P	M20707	—	Immunoglobulin κ variable 2/OR22-4	GDB:9953628	28847
	IGKV3/ OR22-2	P	Z00042	—	Immunoglobulin κ variable 3/OR22-2	GDB:9953630	3527

Table 8.2 Immunoglobulin kappa (*IGK*) genes[a] (*continued*)

IMGT gene group	IMGT gene name[b]	Functionality	IMGT reference sequence accession number	Number of alleles	IMGT gene definition[c]	GDB accession ID	Locus-link number
	Outside chromosome 2, not localized (NL)

IMGT gene group	IMGT gene name[b]	Functionality	IMGT reference sequence accession number	Number of alleles	IMGT gene definition[c]	GDB accession ID	Locus-link number
IGKV	*IGKV1/ OR-1*	P	M23653	—	Immunoglobulin κ variable 1/OR-1	GDB:9953632	3531
	IGKV1/ OR-2	P	X64640	—	Immunoglobulin κ variable 1/OR-2	GDB:9953633	3532
	IGKV1/ OR-3	P	X64641	—	Immunoglobulin κ variable 1/OR-3	GDB:9953634	3533
	IGKVI/ OR-4	P	X64642	—	Immunoglobulin κ variable 1/OR-4	GDB:9953635	3534

Notes:

[a]Gene names are according to the IMGT gene name nomenclature for IG and TcR of all vertebrates. (IMGT Scientific chart at http://imgt.cines.fr:8104).

[b]IMGT *IGK* gene names have been approved by the HUGO Nomenclature committee in 1999. Note that, in the HUGO symbols, slashes of the orphon names are omitted. Otherwise all the gene names (gene symbols) are identical in IMGT and HUGO nomenclatures.

[c]Gene definitions (full names) are identical (including slashes) in IMGT and HUGO nomenclatures. Note that in the databases, the Greek letters are written in full (e.g., κ = kappa).

[d]*IGKV* genes are designated by a number for the subgroup, followed by a hyphen and a number for the localization from 3′ to 5′ in the locus. The *IGKV* genes of the distal duplicated V-CLUSTER are designated by the same number as the corresponding genes in the proximal V-CLUSTER, with the letter D added.

[e]Alleles that could not be assigned to the proximal or distal cluster gene are shown between parentheses followed by a number and a question mark.

Table 8.3 Immunoglobulin lambda (*IGL*) genes[a]

IMGT gene group	IMGT gene name[b]	Functionality	IMGT reference sequence accession number	Number of alleles	IMGT gene definition[c]	GDB accession ID	Locus-link number
	*IGL* locus on chromosome 22 at 22q11.2[d]

IMGT gene group	IMGT gene name[b]	Functionality	IMGT reference sequence accession number	Number of alleles	IMGT gene definition[c]	GDB accession ID	Locus-link number
IGLC[e]	*IGLC1*	F	J00252	3	Immunoglobulin λ constant 1	GDB:120690	3537
	IGLC2	F	J00253	2	Immunoglobulin λ constant 2	GDB:120691	3538
	IGLC3	F	J00254	4	Immunoglobulin λ constant 3	GDB:120692	3539
	IGLC4	P	J03009	2	Immunoglobulin λ constant 4	GDB:120693	3540
	IGLC5	P	J03010	2	Immunoglobulin λ constant 5	GDB:120694	3541
	IGLC6	F, P	J03011	5	Immunoglobulin λ constant 6	GDB:120524	3542
	IGLC7	F	X51755	2	Immunoglobulin λ constant 7	GDB:9953636	28834

Table 8.3 Immunoglobulin lambda (*IGL*) genes[a] (*continued*)

IMGT gene group	IMGT gene name[b]	Functionality	IMGT reference sequence accession number	Number of alleles	IMGT gene definition[c]	GDB accession ID	Locus-link number
IGLJ	*IGLJ1*	F	X04457	1	Immunoglobulin λ joining 1	GDB:9953638	28833
	IGLJ2	F	M15641	1	Immunoglobulin λ joining 2	GDB:9953640	28832
	IGLJ3	F	M15642	2	Immunoglobulin λ joining 3	GDB:9953642	28831
	IGLJ4	ORF	X51755	1	Immunoglobulin λ joining 4	GDB:9953644	28830
	IGLJ5	ORF	X51755	2	Immunoglobulin λ joining 5	GDB:9953646	28829
	IGLJ6	ORF	M18338	1	Immunoglobulin λ joining 6	GDB:9953648	28828
	IGLJ7	F	X51755	2	Immunoglobulin λ joining 7	GDB:9953650	28827
IGLV[f]	*IGLV1-36*	F	Z73653	1	Immunoglobulin λ variable 1-36	GDB:9953652	28826
	IGLV1-40	F	M94116	3	Immunoglobulin λ variable 1-40	GDB:9953654	28825
	IGLV1-41	ORF, P	M94118	2	Immunoglobulin λ variable 1-41	GDB:9953656	28824
	IGLV1-44	F	Z73654	1	Immunoglobulin λ variable 1-44	GDB:9953658	28823
	IGLV1-47	F	Z73663	2	Immunoglobulin λ variable 1-47	GDB:9953660	28822
	IGLV1-50	ORF	M94112	1	Immunoglobulin λ variable 1-50	GDB:9953662	28821
	IGLV1-51	F	Z73661	2	Immunoglobulin λ variable 1-51	GDB:9953664	28820
	IGLV1-62	P	D87022	—	Immunoglobulin λ variable 1-62	GDB:9953666	28819
	IGLV2-5	P	Z73641	—	Immunoglobulin λ variable 2-5	GDB:9953668	28818
	IGLV2-8	F	X97462	3	Immunoglobulin λ variable 2-8	GDB:9953670	28817
	IGLV2-11	F	Z73657	3	Immunoglobulin λ variable 2-11	GDB:9953674	28816
	IGLV2-14	F	Z73664	4	Immunoglobulin λ variable 2-14	GDB:9953676	28815

Table 8.3   Immunoglobulin lambda (*IGL*) genes[a] (*continued*)

IMGT gene group	IMGT gene name[b]	Functionality	IMGT reference sequence accession number	Number of alleles	IMGT gene definition[c]	GDB accession ID	Locus-link number
	IGLV2-18	F	Z73642	4	Immunoglobulin λ variable 2-18	GDB:9953679	28814
	IGLV2-23	F	X14616	3	Immunoglobulin λ variable 2-23	GDB:9953681	28813
	IGLV2-28	P	X97466	—	Immunoglobulin λ variable 2-28	GDB:9953683	28812
	IGLV2-33	ORF	Z273643	3	Immunoglobulin λ variable 2-33	GDB:9953685	28811
	IGLV2-34	P	D87013	—	Immunoglobulin λ variable 2-34	GDB:9953687	28810
	IGLV3-1	F	X57826	1	Immunoglobulin λ variable 3-1	GDB:9953689	28809
	IGLV3-2	P	X97468	—	Immunoglobulin λ variable 3-2	GDB:9953691	28808
	IGLV3-4	P	D87024	—	Immunoglobulin λ variable 3-4	GDB:9953693	28807
	IGLV3-6	P	X97465	—	Immunoglobulin λ variable 3-6	GDB:9953695	28806
	IGLV3-7	P	X97470	—	Immunoglobulin λ variable 3-7	GDB:9953697	28805
	IGLV3-9	F, P	X97473	3	Immunoglobulin λ variable 3-9	GDB:9953699	28804
	IGLV3-10	F	X97464	2	Immunoglobulin λ variable 3-10	GDB:9953701	28803
	IGLV3-12	F	Z73658	2	Immunoglobulin λ variable 3-12	GDB:9953703	28802
	IGLV3-13	P	X97463	—	Immunoglobulin λ variable 3-13	GDB:9953705	28801
	IGLV3-15	P	D87015	—	Immunoglobulin λ variable 3-15	GDB:9953707	28800
	IGLV3-16	F	X97471	1	Immunoglobulin λ variable 3-16	GDB:9953709	28799
	IGLV3-17	P	X97472	—	Immunoglobulin λ variable 3-17	GDB:9953711	28798
	IGLV3-19	F	X56178	1	Immunoglobulin λ variable 3-19	GDB:9953713	28797
	IGLV3-21	F	X71966	3	Immunoglobulin λ variable 3-21	GDB:9953715	28796
	IGLV3-22	F, P	Z73666	2	Immunoglobulin λ variable 3-22	GDB:9953717	28795

Table 8.3 Immunoglobulin lambda (*IGL*) genes[a] (*continued*)

IMGT gene group	IMGT gene name[b]	Functionality	IMGT reference sequence accession number	Number of alleles	IMGT gene definition[c]	GDB accession ID	Locus-link number
	IGLV3-24	P	X71968	—	Immunoglobulin λ variable 3-24	GDB:9953719	28794
	IGLV3-25	F	X97474	3	Immunoglobulin λ variable 3-25	GDB:9953721	28793
	IGLV3-26	P	X97467	—	Immunoglobulin λ variable 3-26	GDB:9953723	28792
	IGLV3-27	F	D86994	1	Immunoglobulin λ variable 3-27	GDB:9953725	28791
	IGLV3-29	P	Z73644	—	Immunoglobulin λ variable 3-29	GDB:9953727	28790
	IGLV3-30	P	Z73646	—	Immunoglobulin λ variable 3-30	GDB:9953729	28789
	IGLV3-31	P	X97469	—	Immunoglobulin λ variable 3-31	GDB:9953731	28788
	IGLV3-32	ORF	Z73645	1	Immunoglobulin λ variable 3-32	GDB:9953733	28787
	IGLV4-3	F	X57828	1	Immunoglobulin λ variable 4-3	GDB:9953735	28786
	IGLV4-60	F	Z73667	2	Immunoglobulin λ variable 4-60	GDB:9953737	28785
	IGLV4-69	F	Z73648	2	Immunoglobulin λ variable 4-69	GDB:9953739	28784
	IGLV5-37	F	Z73672	1	Immunoglobulin λ variable 5-37	GDB:9953741	28783
	*IGLV5-39	ORF	Z73668	1	Immunoglobulin λ variable 5-39	GDB:9953743	28782
	IGLV5-45	F	Z73670	3	Immunoglobulin λ variable 5-45	GDB:9953745	28781
	IGLV5-48	ORF	Z73649	1	Immunoglobulin λ variable 5-48	GDB:9953747	28780
	IGLV5-52	F	Z73669	1	Immunoglobulin λ variable 5-52	GDB:9953749	28779
	IGLV6-57	F	Z73673	1	Immunoglobulin λ variable 6-57	GDB:9953751	28778
	IGLV7-35	P	Z73660	—	Immunoglobulin λ variable 7-35	GDB:9953753	28777
	IGLV7-43	F	X14614	1	Immunoglobulin λ variable 7-43	GDB:9953755	28776

Table 8.3 Immunoglobulin lambda (*IGL*) genes[a] (*continued*)

IMGT gene group	IMGT gene name[b]	Functionality	IMGT reference sequence accession number	Number of alleles	IMGT gene definition[c]	GDB accession ID	Locus-link number
	IGLV7-46	F, P	Z73674	3	Immunoglobulin λ variable 7-46	GDB:9953757	28775
	IGLV8-61	F	Z73650	2	Immunoglobulin λ variable 8-61	GDB:9953759	28774
	IGLV9-49	F	Z73675	3	Immunoglobulin λ variable 9-49	GDB:9953761	28773
	IGLV10-54	F	Z73676	3	Immunoglobulin λ variable 10-54	GDB:9953763	28772
	IGLV10-67	ORF	Z73651	—	Immunoglobulin λ variable 10-67	GDB:9953765	28771
	IGLV11-55	ORF	D86996	1	Immunoglobulin λ variable 11-55	GDB:9953767	28770
	IGLV(I)-20	P	D87007	—	Immunoglobulin λ variable (I)-20	GDB:9953769	28769
	IGLV(I)-38	P	D87009	—	Immunoglobulin λ variable (I)-38	GDB:9953771	28768
	IGLV(I)-42	P	X14613	—	Immunoglobulin λ variable (I)-42	GDB:9953773	28767
	IGLV(I)-56	P	D86996	—	Immunoglobulin λ variable (I) 56	GDB:9953775	28766
	IGLV(I)-63	P	D87022	—	Immunoglobulin λ variable (I)-63	GDB:9953777	28765
	IGLV(I)-68	P	D86993	—	Immunoglobulin λ variable (I)-68	GDB:9953779	28764
	IGLV(I)-70	P	D86993	—	Immunoglobulin λ variable (I)-70	GDB:9953781	28763
	IGLV(IV)-53	P	D86996	—	Immunoglobulin λ variable (IV)-53	GDB:9953783	28762
	IGLV(IV)-59	P	D87000	—	Immunoglobulin λ variable (IV)-59	GDB:9953785	28761
	IGLV(IV)-64	P	D87022	—	Immunoglobulin λ variable (IV)-64	GDB:9953787	28760
	IGLV(IV)-65	P	D87022	—	Immunoglobulin λ variable (IV)-65	GDB:9953789	28759
	IGLV(IV)-66-1	P	D87004	—	Immunoglobulin λ variable (IV)-66-1	GDB:9991231	
	IGLV(V)-58	P	D87000	—	Immunoglobulin λ variable (V)-58	GDB:9953791	28758
	IGLV(V)-66	P	D87004	—	Immunoglobulin λ variable (V)-66	GDB:9953793	28757

Table 8.3 Immunoglobulin lambda (*IGL*) genes[a] (*continued*)

IMGT gene group	IMGT gene name[b]	Functionality	IMGT reference sequence accession number	Number of alleles	IMGT gene definition[c]	GDB accession ID	Locus-link number
IGL orphons							
On chromosome 8 at 8q11.2							
IGLV	*IGLV8/ OR8-1*	ORF, P	Y08831	—	Immunoglobulin λ variable 8/OR8-1	GDB:9953795	28756
	IGLV/ OR8-2			—	Immunoglobulin λ variable/OR8-2 (provisional)		
On chromosome 22 at 22q12.2-22q12.3[g]							
IGLC	*IGLC/ OR22-1*		AL008723	—	Immunoglobulin λ constant /OR22-1	GDB:9991233	
	IGLC/ OR22-2		AL021937	—	Immunoglobulin λ constant /OR22-2	GDB:9991234	
On chromosome 22 at 22q11.2-22q12.1[g]							
IGLV	*IGLV(IV)/ OR22-1*		AL008721	—	Immunoglobulin λ variable (IV)/OR22-1	GDB:9991235	
On chromosome 22 at 22q12.2-22q12.3g							
	IGLV(IV)/ OR2 2-2		AL021937	—	Immunogiobulin λ variable (IV)/OR22-2	GDB:9991236	
Not localized							
IGLV	*IGLV8/ OR-1*	—	—	—	Immunoglobulin λ variable 8/OR-1 (provisional)		

Notes:
[a]Gene names are according to the IMGT gene name nomenclature for IG and TcR of all vertebrates. (IMGT Scientific chart at http://imgt.cines.fr:8104).

[b]IMGT *IGL* gene names have been approved by the HUGO Nomenclature Committee in 1999. Note that, in the HUGO symbols, parentheses of the 'clan assigned pseudogene' names and slashes of the orphon names are omitted. Otherwise all the gene names (gene symbols) are identical in IMGT and HUGO nomenclatures.

[c]Gene definitions (full names) are identical (including parentheses and slashes) in IMGT and HUGO nomenclatures. Note that in the databases, the Greek letters are written in full (e.g., λ = lambda).

[d]Sequencing of the long arm of chromosome 22 showed that it encompasses-35 Mb of DNA and that the *IGL locus is* localized at 6 Mb from the centromere. Although the correlation between DNA sequence and chromosomal bands has not yet been made, the localization of the *IGL locus* can be refined to 22q11.2.

[e]One, two, three, or four additional *IGLC* genes, each one probably preceded by one *IGLJ*, have been shown to characterize *IGLC* haplotypes with 8, 9, 10, or 11 genes, but these genes have not yet been sequenced and are not shown in this table.

[f]*IGLV* genes are designated by a number for the subgroup followed by a hyphen and a number for the localization from 3′ to 5′ in the locus. In the *IGLV* gene name column, the *IGLV* genes are listed, for each subgroup, according to their position from 3′ to 5′ in the locus. Pseudogenes which could not be assigned to subgroups with functional genes are designated by a roman number between parentheses, corresponding to the clan (clan I: *IGLV1, IGLV2, IGLV6,* and *IGLV10* subgroup genes, and pseudogenes *IGLV(I)-20, -38, -42, -56, -63, -68,* and *-70*; clan II: *IGLV3* subgroup genes; clan III: *IGLV7* and *IGLV8* subgroup genes; clan IV: *IGLV5* and *IGLV11* subgroup genes, and pseudogenes *IGLV(IV)-53, -59, -64, -65,* and *-66-1*; clan V: *IGLV4* and *IGLV9* subgroup genes, and pseudogenes *IGLV(V)-58* and *-66*), followed by a hyphen and a number for the localization from 3′ to 5′ in the locus. An asterisk (*) indicates allelic polymorphism by insertion/deletion which concerns the *IGLV5-39* gene.

[g]Not sequenced.

Table 8.4 Constituent chain structures of immunoglobulins

	α chain	δ chain	ε chain	γ chain	μ chain	λ chain	κ chain
Molecular weight (kDa)	58	64	72	51	72	23	23
Amino acid residue	470	500	550	450	570	214	214
Constant domain	3: C_H1, C_H2, C_H3	3: C_H1, C_H2, C_H3	4: C_H1, C_H2, C_H3, C_H4	3: C_H1, C_H2, C_H3	4: C_H1, C_H2, C_H3, C_H4	1	1
Variable domain	1: V_H	1: V_H	1: V_H	1: V_H	1: V_H	1	1
Hinge region	Hinge region situated between C_H1 and C_H2 domains	58-amino acid hinge region encoded by 2 exons	No hinge region	Hinge region located between the Fab arms and the two carboxy-terminal domains C_H2 and C_H3	No hinge region	No hinge region	No hinge region

Table 8.5 Comparative description of immunoglobulin superfamily molecules

	IgA	IgD	IgE	IgG	IgM
Structure	Four-chain monomeric structure, or dimer, trimer, and multimer	Four-chain monomeric structure	Four-chain monomeric structure	Four-chain monomeric structure	Pentameric structure with five four-chain monomers
Heavy-chain class	α	δ	ε	γ	μ
Molecular weight (kDa)	160	185	190	154	900
Total Ig (%)	5–15	<1	<1	85	5–10
Half-life	6 days	2–3 days	2.5 days	23 days	5 days
Placental transfer	No	No	No	Yes	No
Complement fixation	No	No	No	Yes	Yes
Mast cell binding	No	No	Yes	No	No
Phagocyte binding	No	No	No	Yes	No
Principal biological effect	Resistance prevents movement across mucous membranes	Elusive	Anaphylaxis to hypersensitivity	Resistance-opsonin	Resistance-precipitin
Principal site of action	Secretions	Receptors for B cells	Mast cells	Serum	Serum
Antibacterial lysis	+	?	?	+	+++
Antiviral lysis	+++	?	?	+	+
Subclass	Ig A1, IgA2	N/A	N/A	IgG1, IgG2, IgG3, IgG4	N/A

Table 8.6 Properties of IgG subclasses

	IgG1	IgG2	IgG3	IgG4
% of total IgG	43–75	16–48	1.7–7.5	0.8–11.7
Molecular weight (kDa)	146	146	170	146
Heavy chain type	$\gamma 1$	$\gamma 2$	$\gamma 3$	$\gamma 4$
Number of disulfide bonds	2	4	11	2
Number of allotypes	4	1	13	0
Half life	21 days	21 days	7 days	21 days
Complement fixation: C1q binding	Moderate	Weak	Strong	No
Antibody response to proteins	Strong	Marginal	Strong	Marginal
Antibody response to polysaccharides	Weak	Strong	No	No
Antibody response to allergens	Weak	No	No	Strong
Susceptibility to protease	Moderate	Marginal	Strong	Weak
Ability to cross placenta	Yes	Yes	Yes	Yes

ing on the haplotypes, each *IGLC* gene being preceded by one *IGLJ* gene. Fifty-six to 57 genes belong to 11 subgroups, whereas 14 pseudogenes which are too divergent to be assigned to subgroups have been assigned to 3 clans. Two *IGLV* orphons have been identified on chromosome 8. Two *IGLC* orphons and two *IGLV* orphons have also been characterized on 22q outside of the major *IGL* locus. **Table 8.3** presents tabulated lists of the human immunoglobulin lambda genes named in accordance with the International ImMunoGeneTics database (IMGT) and approved by the Human Genome Organization (HUGO) Nomenclature Committee.

IMMUNOGLOBULIN STRUCTURE

Immunoglobulins are composed of four polypeptide chains: two 'light' chains (lambda or kappa), and two 'heavy' chains (alpha, delta, gamma, epsilon or mu). The four polypeptide chains are held together by covalent disulfide bonds. Light chains are composed of 220 amino acids while heavy chains are comprised of 440–550 amino acids. Each chain has 'constant' and 'variable' regions. Variable regions are contained within the NH_2 terminal end of the polypeptide chain and have distinct amino acid sequences when comparing one antibody to another. Constant regions, however, are rather uniform within the same isotype when comparing one

antibody to another. 'Hypervariable' regions, or 'Complementarity Determining Regions' (CDRs) are found within the variable regions of both the heavy and light chains. These regions serve to recognize and bind specifically to antigen. The constituent chain structures of immunoglobulins with features of each are listed in **Table 8.4.**

Immunoglobulin superfamily

The type of heavy chain determines the immunoglobulin isotype (IgA, IgD, IgG, IgE, IgM, respectively). More specifically, an isotype is determined by the primary sequence of amino acids in the constant region of the heavy chain, which in turn determines the three-dimensional structure of the molecule. **Table 8.5** gives a comparative description of immunoglobulin superfamily molecules.

Immunoglobulins are the key elements in the adaptive immune system. The glycoprotein immunoglobulin G (IgG), a major effector molecule of the humoral immune response in man, plays a pivotal role in the antibody response. There are four distinct subgroups of human IgG, which are designated IgG1, IgG2, IgG3, and IgG4, respectively. They have differences in their hinge regions and differ in the number and position of disulfide bonds that link the two γ chains in each IgG molecule. Different properties of IgG subclasses are summarized in **Table 8.6.**

KEY REFERENCES

1   Lefranc MP (2000) Nomenclature of the human immunoglobulin genes. *Curr Protocol Immunol* A.1 P.1–A.1 P.37

2   Pallarès N, Lefebvre S, Contet V, Matsuda F, Lefranc MP (1999) The human immunoglobulin heavy variable (*IGHV*) genes. *Exp Clin Immunogenet* 16: 36–60

3   Ruiz M, Pallarès N, Contet V, Barbiè V, Lefranc MP (1999) The human immunoglobulin heavy diversity *(IGHD)* and joining (IGHJ) segments. *Exp Clin Immunogenet* 16: 173–184

4   Scaviner D, Barbiè V, Ruiz M, Lefranc MP (1999) Protein displays of the human immunoglobulin heavy, kappa, and lambda variable and joining regions. *Exp Clin Immunogenet* 16: 234–240

5   Barbiè V, Lefranc MP (1998) The human immunoglobulin kappa variable (*IGKV*) gene and joining (*IGKJ*) segments. *Exp Clin Immunogenet* 15: 171–183

6   Pallarès N, Frippiat JP, Giudicelli V, Lefranc MP (1998) The human immunoglobulin lambda variable (*IGLV*) genes and joining (*IGLJ*) segments. *Exp Clin Immunogenet* 15: 8–18

7   Matsuda F, Ishii K, Bouvagnet P, Kuma KI, Hayashida H, Miyata T, Honjo T (1998) The complete nucleotide sequence of the human immunoglobulin heavy chain variable region locus. *J Exp Med* 188: 2151–2162

KEY DATABASES

- The Genome Database: http://www.gdb.org
- HUGO Gene Nomenclature Committee website: http://www.gene.ucl.ac.uk/nomenclature
- IMGT, the international ImMunoGeneTics database: http://imgt.cines.fr:8104
- NCBI Locuslink: http://www.ncbi.nlm.nih.gov/LocusLink

T Cells and the Thymus | 9

THE THYMUS

Anatomical structure

The thymus is a triangular bilobed structure situated above the heart in the anterior mediastinum. This organ is conspicuously large in early life and undergoes atrophy at the time of puberty. The thymus is responsible for the production and differentiation of T cells and is essential for immunity. Each lobe of the thymus is subdivided by prominent trabeculae into interconnecting lobules, and each lobule comprises two histologically and functionally distinct areas, cortex and medulla. The prothymocytes, which migrate from the bone marrow to the subcapsular regions of the cortex, are influenced by this microenvironment which directs their further development. The process of education is governed by hormonal substances produced by the thymic epithelial cells. The cortical cells proliferate extensively. Some of these cells are short-lived and die. The surviving cells acquire characteristics of thymocytes. The cortical cells migrate to the medulla and from there to the peripheral lymphoid organs, sites of their main residence. The blood supply to the cortex comes from capillaries that form anastomosing arcades. Drainage is mainly through veins; the thymus has no afferent lymphatic vessels. The blood–thymus barrier protects thymocytes from contact with antigen. Cells reaching the thymus are prevented from contact with antigen by a physical barrier. The first level is represented by the capillary wall with endothelial cells inside the pericytes outside of the lumen. Potential antigenic molecules which escape the first level of control are taken over by macrophages present in the pericapillary space. Further protection is provided by a third level, represented by the mesh of interconnecting epithelial cells which enclose the thymocyte population. The anatomical constituents of the thymus together with their features are shown in **Table 9.1**.

Role in T cell differentiation and maturation

The effects of thymus and thymic hormones on the differentiation of T cells are demonstrated in **Figure 9.1**. Differentiation is associated with surface markers whose presence or disappearance characterizes the different stages of cell differentiation. There is extensive proliferation of the subcapsular thymocytes. The largest proportion of these cells die, but the remaining cells continue to differentiate. The differentiating cells become smaller in size and move through interstices in the thymic medulla. The fully developed thymocytes pass through the walls of the postcapillary venules to reach the systemic circulation and seed the peripheral lymphoid organs. Part of them recirculate, but do not return to the thymus.

T CELLS

T cell ontogeny

T cells are cells that are derived from hematopoietic precursors that migrate to the thymus where they undergo differentiation which continues thereafter to completion in the various lymphoid tissues throughout the body or during their circulation to and from these sites. The T cells are primarily involved in the control of the immune responses by providing specific cells capable of helping or suppressing these responses. They also have a number of other functions related to cell-mediated immune phenomena. Thymus-derived cells confer cell-mediated immunity and cooperates with B cells enabling them to synthesize antibody specific for thymus-dependent antigens, including switching from IgM to IgG and/or IgA production. T cells exiting the thymus recirculate in the blood and lymph and in the peripheral lymphoid organs. They migrate to the deep cortex of lymph nodes. Those in the blood may attach to postcapillary venule endothelial cells of lymph nodes and to the

Table 9.1 Anatomic constituents of thymus

	Cortex	Medulla
Percent thymic tissue	80%	20%
Cell composition	Epithelial-reticular cells, immature thymocytes	Richer in epithelial cells, antigen-presenting cells (dendritic cells and macrophages), mature cells
Epithelial island	Epithelial-reticular network, not island	Cystic structure termed Hassal's corpuscles
Cell markers	'Double positive' $CD4^+8^+$	'Single positive' $CD4^+8^-$ and $CD4^-CD8^+$
T cell receptor	$\alpha\beta$ TCR, <1% $\gamma\delta$ TCR	$\alpha\beta$ TCR, <1% $\gamma\delta$ TCR
HE section	Cells densely packed, dark appearance	Cells loosely packed, pale appearance
Sensitivity to hormones	Highly sensitive to glucocorticoids and other steroids such as sex hormones	Sensitive to steroids
Thymic atrophy	Undergoes atrophy more progressively, preferentially affected by stress	Involutes first with pyknosis and beading of the nuclei of small cells, giving a false impression of an increased number of Hassal's corpuscles

marginal sinus in the spleen. After passing across the venules into the splenic white pulp or lymph node cortex, they reside there for 12–24 hours, exit by the efferent lymphatics, proceed to the thoracic duct, and from there proceed to the left subclavian vein where they enter the blood circulation. Mature T cells are classified on the basis of their surface markers, such as CD4 and CD8. $CD4^+$ T cells recognize antigens in the context of MHC class II histocompatibility molecules, whereas $CD8^+$ T cells recognize antigen in the context of class I MHC histocompatibility molecules. The $CD4^+$ T cells participate in the afferent limb of the immune response to exogenous antigen, which is presented to them

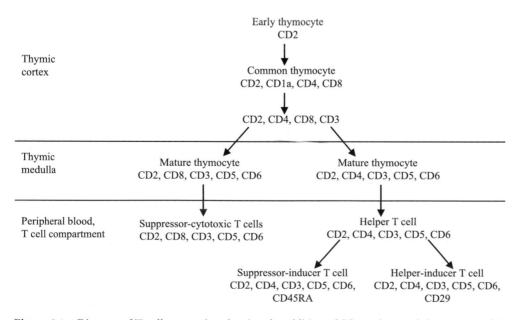

Figure 9.1 Diagram of T cell maturation showing the addition of CD markers and the position of the T cell when the marker is added

by antigen-presenting cells. This stimulates the synthesis of IL-2, which activates CD8$^+$ T cells, NK cells, and B cells, thereby orchestrating an immune response to the antigen. Thus, they are termed helper T cells. They also mediate delayed-type hypersensitivity reactions. CD8$^+$ T cells include cytotoxic and suppressor cell populations. They react to endogenous antigen and often express their effector function by a cytotoxic mechanism, e.g., against a virus-infected cell. Other molecules on mature T cells in humans include the E rosette receptor CD2 molecule, the T cell receptor, the pan-T cell marker termed CD3, and transferrin receptors. **Figure 9.2** depicts the ontogeny of T cells and the surface markers expressed on T cells of different stages.

T cell subpopulations

Most of the T cells in the body belong to one of two subsets. These are distinguished by the presence on their surface of one or the other of two glycoproteins designated CD4 or CD8. Which of these molecules is present determines what types of cells to which the T cell can bind.

The best understood CD8$^+$ cells are cytotoxic T cells (CTLs). They secrete molecules that destroy the cell to which they have bound. CTLs are a subset of antigen-specific effector T cells that have a principal role in protection and recovery from viral infection, mediate allograft rejection, participate in selected autoimmune diseases, participate in protection and recovery from selected bacterial and parasitic infections, and are active in tumor immunity. They are CD8$^+$, Class I major histocompatability complex (MHC)-restricted, nonproliferating endstage effector cells. However, this classification also includes T cells that evoke one or several mechanisms to produce cytolysis including perforin/granzyme, FasL/Fas, tumor necrosis factor α (TNF-α), synthesize various lymphokines by T$_H$1 and T$_H$2 cells and recognize foreign antigen in the context of either class I or class II MHC molecules.

The other subset of T cells are T helper cells which express CD4 as their surface marker. They potentiate immunoglobulin production and isotype switching by B cells, cooperate with cytotoxic T cells in the recognition of allografts and virally infected target cells, and release cytokines which are capable of activating macrophages and other cell types. **Table 9.2** summarizes the features of these two types of T cells.

T helper cell subsets

The subsets of T helper cells are listed in **Table 9.3** together with their origin, function, receptors, lymphokine secretion, and types of immunity in which they participate. Th0 cells are a CD4$^+$ T cell subset in both humans and mice based on cytokine production and effector functions. They synthesize multiple cytokines and are responsible for

effects intermediate between those of Th1 and Th2 cells, based on the cytokines synthesized and the responding cells. Th0 cells may be precursors of Th1 and Th2 cells. Th1 cells are a CD4$^+$ T cell subset in both humans and mice based on cytokine production and effector functions. Th1 cells synthesize interferon-gamma (IFN-γ), IL-2, and tumor necrosis factor (TNF)-β. They are mainly responsible for cellular immunity against intracellular microorganisms and for delayed-type hypersensitivity reactions. They affect IgG2a antibody synthesis and antibody-dependent cell-mediated cytotoxicity. Th1 cells activate host defense mediated by phagocytes. Intracellular microbial infections induce Th1 cell development which facilitates elimination of the microorganisms by phagocytosis. Th1 cells induce synthesis of antibody that activates complement and serves as an opsonin that facilitates phagocytosis. The IFN-γ they produce enhances macrophage activation. The cytokines released by Th1 cells activate NK cells, macrophages, and CD8$^+$ T cells. Their main function is to induce phagocyte-mediated defense against infections, particularly by intracellular microorganisms. Th2 cells are a CD4$^+$ T cell subset in both humans and mice based on cytokine production and effector functions. Th2 cells synthesize IL-4, IL-5, IL-6, IL-9, IL-10, and IL-13. They greatly facilitate IgE and IgG1 antibody responses, and mucosal immunity, by synthesis of mast cell and eosinophil growth and differentiation factors, and facilitation of IgA synthesis. IL-4 facilitates IgE antibody synthesis. IL-5 is an eosinophil-activating substance. IL-10, IL-13, and IL-4 suppress cell-mediated immunity. Th2 cells are principally responsible for host defense exclusive of phagocytes. They are crucial for the IgE and eosinophil response to helminths and for allergy attributable to activation of basophils and mast cells through IgE.

THYMUS-DEPENDENT ANTIGEN AND THYMUS-INDEPENDENT ANTIGEN

Thymus-dependent antigen

A thymus-dependent antigen is an immunogen that requires T cell cooperation with B cells to synthesize specific antibodies. Presentation of thymus-dependent antigen to T cells must be in the context of MHC class II molecules. Thymus-dependent antigens include proteins, polypeptides, hapten-carrier complexes, erythrocytes, and many other antigens that have diverse epitopes. T dependent antigens contain some epitopes that T cells recognize and others that B cells identify. T cells produce cytokines and cell surface molecules that induce B cell growth and differentiation into antibody-secreting cells. Humoral immune responses to T-dependent antigens are associated with isotype switching, affinity maturation, and memory. The response to thymus-dependent antigens shows only minor heavy chain

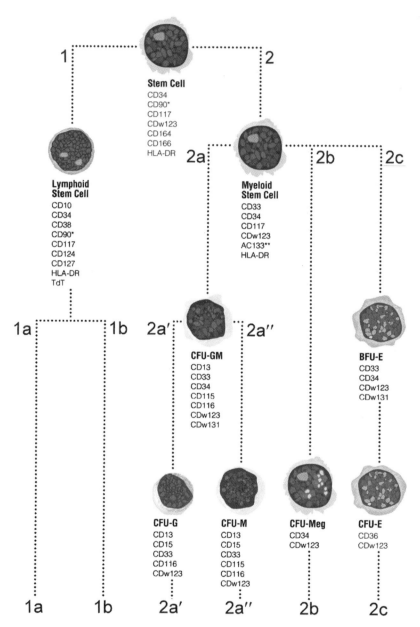

Figure 9.2   T cell ontogeny and surface markers

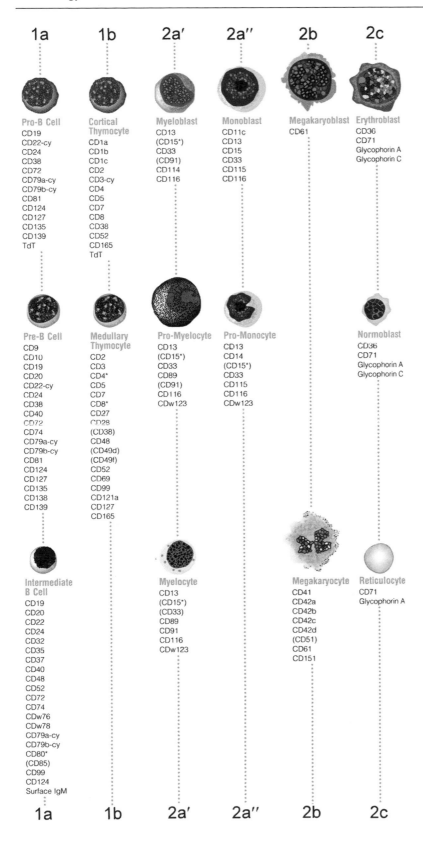

1a

Pro-B Cell
CD19
CD22-cy
CD24
CD38
CD72
CD79a-cy
CD79b-cy
CD81
CD124
CD127
CD135
CD139
TdT

Pre-B Cell
CD9
CD10
CD19
CD20
CD22-cy
CD24
CD38
CD40
CD72
CD74
CD79a-cy
CD79b-cy
CD81
CD124
CD127
CD135
CD138
CD139

Intermediate B Cell
CD19
CD20
CD22
CD24
CD32
CD35
CD37
CD40
CD48
CD52
CD72
CD74
CDw76
CDw78
CD79a-cy
CD79b-cy
CD80*
(CD85)
CD99
CD124
Surface IgM

1b

Cortical Thymocyte
CD1a
CD1b
CD1c
CD2
CD3-cy
CD4
CD5
CD7
CD8
CD38
CD52
CD165
TdT

Medullary Thymocyte
CD2
CD3
CD4*
CD5
CD7
CD8*
CD27
CD28
(CD38)
CD48
(CD49d)
(CD49f)
CD52
CD69
CD99
CD121a
CD127
CD165

2a′

Myeloblast
CD13
(CD15*)
CD33
(CD91)
CD114
CD116

Pro-Myelocyte
CD13
(CD15*)
CD33
CD89
(CD91)
CD116
CDw123

Myelocyte
CD13
(CD15*)
(CD33)
CD89
CD91
CD116
CDw123

2a″

Monoblast
CD11c
CD13
CD15
CD33
CD115
CD116

Pro-Monocyte
CD13
CD14
(CD15*)
CD33
CD115
CD116
CDw123

2b

Megakaryoblast
CD61

Megakaryocyte
CD41
CD42a
CD42b
CD42c
CD42d
(CD51)
CD61
CD151

2c

Erythroblast
CD36
CD71
Glycophorin A
Glycophorin C

Normoblast
CD36
CD71
Glycophorin A
Glycophorin C

Reticulocyte
CD71
Glycophorin A

1a   **1b**   **2a′**   **2a″**   **2b**   **2c**

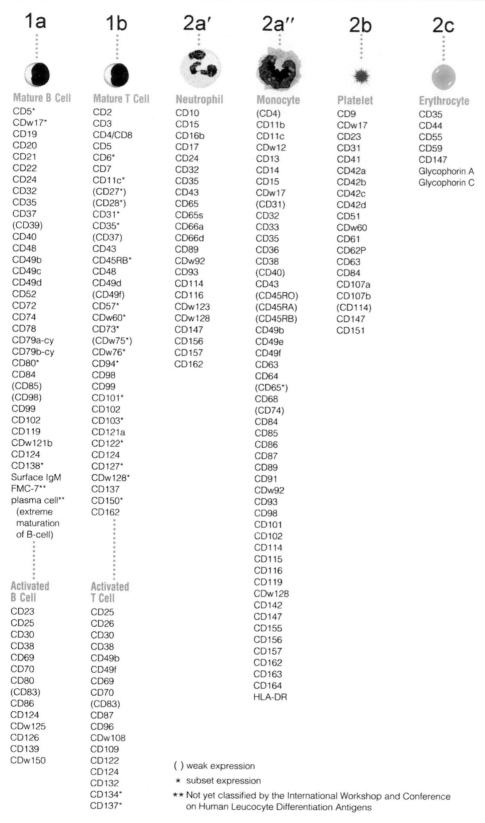

1a

Mature B Cell
CD5*
CDw17*
CD19
CD20
CD21
CD22
CD24
CD32
CD35
CD37
(CD39)
CD40
CD48
CD49b
CD49c
CD49d
CD52
CD72
CD74
CD78
CD79a-cy
CD79b-cy
CD80*
CD84
(CD85)
(CD98)
CD99
CD102
CD119
CDw121b
CD124
CD138*
Surface IgM
FMC-7**
plasma cell**
  (extreme
  maturation
  of B-cell)

Activated B Cell
CD23
CD25
CD30
CD38
CD69
CD70
CD80
(CD83)
CD86
CD124
CDw125
CD126
CD139
CDw150

1b

Mature T Cell
CD2
CD3
CD4/CD8
CD5
CD6*
CD7
CD11c*
(CD27*)
(CD28*)
CD31*
CD35*
(CD37)
CD43
CD45RB*
CD48
CD49d
(CD49f)
CD57*
CDw60*
CD73*
(CDw75*)
CDw76*
CD94*
CD98
CD99
CD101*
CD102
CD103*
CD121a
CD122*
CD124
CD127*
CDw128*
CD137
CD150*
CD162

Activated T Cell
CD25
CD26
CD30
CD38
CD49b
CD49f
CD69
CD70
(CD83)
CD87
CD96
CDw108
CD109
CD122
CD124
CD132
CD134*
CD137*

2a′

Neutrophil
CD10
CD15
CD16b
CD17
CD24
CD32
CD35
CD43
CD65
CD65s
CD66a
CD66d
CD89
CDw92
CD93
CD114
CD116
CDw123
CDw128
CD147
CD156
CD157
CD162

2a″

Monocyte
(CD4)
CD11b
CD11c
CDw12
CD13
CD14
CD15
CDw17
(CD31)
CD32
CD33
CD35
CD36
CD38
(CD40)
CD43
(CD45RO)
(CD45RA)
(CD45RB)
CD49b
CD49e
CD49f
CD63
CD64
(CD65*)
CD68
(CD74)
CD84
CD85
CD86
CD87
CD89
CD91
CDw92
CD93
CD98
CD101
CD102
CD114
CD115
CD116
CD119
CDw128
CD142
CD147
CD155
CD156
CD157
CD162
CD163
CD164
HLA-DR

2b

Platelet
CD9
CDw17
CD23
CD31
CD41
CD42a
CD42b
CD42c
CD42d
CD51
CDw60
CD61
CD62P
CD63
CD84
CD107a
CD107b
(CD114)
CD147
CD151

2c

Erythrocyte
CD35
CD44
CD55
CD59
CD147
Glycophorin A
Glycophorin C

( ) weak expression

* subset expression

** Not yet classified by the International Workshop and Conference
     on Human Leucocyte Differentiation Antigens

Figure 9.2 T cell ontogeny and surface markers (*continued*)

Table 9.2 T cell subpopulations

	Cytotoxic Tc cells	Helper Th cells
Antigen expression	CD8 antigen	CD4 antigen
Function	Kill cells that have pathogens in the cytosol and be responsible for the rejection of tissue and organ grafts	Eliminate pathogens residing intracellularly in vascular compartments (Th1) and facilitate antibody production by B cells (Th2)
Mechanism	By secreting molecules that destroy the cells to which they have bound	By binding to antigen presented by APCs and releasing lymphokines to induce inflammation and by binding to antigen presented by B cells to induce plasma cell development and antibody secretion
Branch of immune response	Cell-mediated immune response	Cell-mediated and antibody-mediated immune response
Antigen presentation and MHC	MHC-I	MHC-II

isotype switching or affinity maturation, both of which require helper T cell signals.

Thymus-independent antigen

A thymus-independent antigen is an immunogen that can stimulate B cells to synthesize antibodies without participation by T cells. These antigens are less complex than are thymus-dependent antigens. They are often polysaccharides that contain repeating epitopes or lipopolysaccharides derived from Gram-negative microorganisms. Thymus-independent antigens induce IgM synthesis by B cells without cooperation by T cells. They also do not stimulate immunological memory. Murine thymus independent antigens are classified as either TI-1 or TI-2 antigens. Lipopolysaccharide (LPS), which activates murine B cells without participation by T or other cells, is a typical TI-1

antigen. Low concentrations of LPS stimulate synthesis of specific antigen, whereas high concentrations activate essentially all B cells to grow and differentiate. TI-2 antigens include polysaccharides, glycolipids, and nucleic acids. When T cells and macrophages are depleted, no antibody response develops against them. A comparison of thymus-dependent and thymus-independent antigens is shown in **Table 9.4**.

T CELL RECEPTORS

The T cell receptor (TCR) is a T cell surface structure that is comprised of a disulfide-linked heterodimer of highly variable α and β chains expressed at the cell membrane as a complex with the invariant CD3 chains. Most T cells that bear this type of receptor are termed αβ T cells. A second receptor, the γδ TCR, is comprised of variable γ and δ

Table 9.3 T helper cell subpopulations

	Th 1	Th 2
Originate from	Dendritic cells descended from monocytes and secrete IL-12 (DC-1)	Dendritic cells derived from lymphocytes (DC-2)
Types of immunity participated in	Cell-mediated immunity	Antibody-mediated immunity
Function	Essential for controlling intracellular pathogens and delayed-type hypersensitivity reactions	Essential to control extracellular pathogens
Chemokine receptors	CCR5	CCR3
Lymphokine secretion	TNF-β, IFN-γ, IL-2	IL-4, IL-5, IL-6, IL-9, IL-10, IL-13

Table 9.4 Comparison of thymus-dependent (TD) antigens and thymus-independent (TI) antigens

	TD antigens	TI antigens
Activation of B cells	Can only activate B cells in the presence of Th cells	Can activate B cells in the absence of Th cells
Structural properties	Complex	Simple
Composition	Proteins, polypeptides, hapten-carrier complexes, erythrocytes, and many other antigens that have diverse epitopes	Polysaccharides that contain repeating epitopes or lipopolysaccharides derived from Gram-negative microorganisms
Presence in most pathogenic microbes	Yes	No
Antibody class induced	IgG, IgM, IgA, IgD, IgE	IgM
Immunological memory response	Yes	No
Examples	Microbial proteins, non-self or alter-self proteins	Pneumococcal polysaccharide, lipopolysaccharide, flagella

chains expressed with CD3 on a smaller subset of T cells that recognize different types of antigens. Both of these types of receptors are expressed with a disulfide-linked homodimer of ξ chains. The TCR is a receptor for antigen on CD4$^+$ and CD8$^+$ T cells that recognizes foreign peptide-self – MHC molecular complexes on the surface of antigen-presenting cells. In the predominant αβ TCR, the two disulfide-linked transmembrane α and β polypeptide chains each bear one N-terminal Ig-like variable (V) domain, one Ig-like constant (C) domain, a hydrophobic transmembrane region, and a short cytoplasmic region.

T cell receptor genes

The human T cell receptors (TCR) α-β and γ-δ are the products of four sets of genes on two chromosomes: T cell receptors α (*TRA*) and δ (*TRD*) on chromosome 14 at 14q11.2, T cell receptor β (*TRB*) on chromosome 7 at 7q35, and T cell receptor γ (*TRG*) on chromosome 7 at 7p15-p14.

T cell receptor α genes

The human *TRA* locus at 14q11.2 spans 1000 kb. It consists of 54 *TRAV* genes belonging to 41 subgroups, 61 *TRAJ* segments localized on 71 kb, and a unique *TRAC* gene. **Table 9.5** lists the human T cell receptor α genes named in accordance with the International ImMunoGeneTics database (IMGT) and approved by the Human Genome Organization (HUGO) Nomenclature Committee in 1999.

T cell receptor β genes

The human *TRB* locus at 7q35 spans 620 kb. It consists of 64 to 67 *TRBV* genes belonging to 32 subgroups. Except for *TRBV30*, localized downstream of the *TRBC2* gene, in inverted orientation of transcription, all the other *TRBV* genes are located upstream of a duplicated D-J-C-cluster, which comprises, for the first part, one *TRBD* gene, six TRBJ genes, and the *TRBCI* gene, and for the second part, one *TRBD* gene, eight *TRBJ* genes, and the *TRBC2* gene. **Table 9.6** is the tabulated list of the human T cell receptor β genes named in accordance with IMGT and approved by the HUGO Nomenclature Committee.

T cell receptor γ genes

The human TRG locus at 7p15-p14 spans 160 kb. It consists of 12 to 15 *TRGV* genes belonging to 6 subgroups, upstream of a duplicated J-C-cluster, which comprises, for the first part, three *TRGJ* genes and the *TRGCI* gene, and for the second part, two *TRGJ* genes and the *TRGC2* gene. **Table 9.7** lists the human T cell receptor γ genes named in accordance with the International ImMunoGeneTics database (IMGT) and approved by the Human Genome Organization (HUGO) Nomenclature Committee.

T cell receptor δ genes

The human *TRD* locus at 14q 11.2 spans 60 kb. It comprises a cluster of one *TRDV* gene (*TRDV2*), three *TRDD* genes, and four *TRDJ* genes, upstream of the unique *TRDC* gene. Another *TRDV* gene (*TRDV3*) is localized downstream of the *TRDC* gene, in inverted orientation of transcription.

This cluster is localized inside the *TRA* locus, between the *TRAV* genes and the *TRAJ* genes. The *TRD* locus also consists of one *TRDV* (*TRDV1*) localized at 360 kb upstream of the *TRDC* gene, among the *TRAV* genes, and the five genes described above as *TRAV/DV*. **Table 9.8** is the tabulated list of the human T cell receptor δ genes named in accordance with IMGT and approved by the HUGO Nomenclature Committee.

Two additional tables, **Table 9.9** and **Table 9.10**, list corresponding nomenclatures for these genes.

Table 9.5 T cell receptor α (*TRA*) genes[a]

IMGT gene group	IMGT gene name[b]	Functionality	IMGT reference sequence accession number	Number of alleles	IMGT gene definition[c]	GDB accession ID	Locus-link number
TRA locus on chromosome 14 at 14q11.2							
TRAC	*TRAC*	F	X02883	3	T cell receptor α constant	GDB:9953797	28755
TRAJ[d]	*TRAJ1*	ORF	X02884	1	T cell receptor α joining 1	GDB:9953799	28754
	TRAJ2	ORF	X02884	1	T cell receptor α joining 2	GDB:9953801	28753
	TRAJ3	F	X02884	1	T cell receptor α joining 3	GDB:9953803	28752
	TRAJ4	F	M94081	1	T cell receptor α joining 4	GDB:9953805	28751
	TRAJ5	F	M94081	1	T cell receptor α joining 5	GDB:9953807	28750
	TRAJ6	F	M16747	1	T cell receptor α joining 6	GDB:9953809	28749
	TRAJ7	F	M94081	1	T cell receptor α joining 7	GDB:9953811	28748
	TRAJ8	F	M94081	1	T cell receptor α joining 8	GDB:9953813	28747
	TRAJ9	F	M94081	1	T cell receptor α joining 9	GDB:9953815	28746
	TRAJ10	F	M94081	1	T cell receptor α joining 10	GDB:9953817	28745
	TRAJ11	F	M94081	1	T cell receptor α joining 11	GDB:9953819	28744
	TRAJ12	F	X02885	1	T cell receptor α joining 12	GDB:9953821	28743
	TRAJ13	F	M94081	1	T cell receptor α joining 13	GDB:9953823	28742
	TRAJ14	F	M94081	1	T cell receptor α joining 14	GDB:9953825	28741
	TRAJ15	F	X05775	2	T cell receptor α joining 15	GDB:9953827	28740

Table 9.5   T cell receptor α (*TRA*) genes[a] (*continued*)

IMGT gene group	IMGT gene name[b]	Functionality	IMGT reference sequence accession number	Number of alleles	IMGT gene definition[c]	GDB accession ID	Locus-link number
	TRAJ16	F	M94081	1	T cell receptor α joining 16	GDB:9953829	28739
	TRAJ17	F	X05773	1	T cell receptor α joining 17	GDB:9953831	28738
	TRAJ18	F	M94081	1	T cell receptor α joining 18	GDB:9953833	28737
	TRAJ19	ORF	M94081	1	T cell receptor α joining 19	GDB:9953835	28736
	TRAJ20	F	M94081	1	T cell receptor α joining 20	GDB:9953837	28735
	TRAJ21	F	M94081	1	T cell receptor α joining 21	GDB:9953839	28734
	TRAJ22	F	X02886	1	T cell receptor α joining 22	GDB:9953841	28733
	TRAJ23	F	M94081	1	T cell receptor α joining 23	GDB:9953843	28732
	TRAJ24	F	X02887	2	T cell receptor α joining 24	GDB:9953845	28731
	TRAJ25	ORF	X02888	1	T cell receptor α joining 25	GDB:9953847	28730
	TRAJ26	F	M94081	1	T cell receptor α joining 26	GDB:9953849	28729
	TRAJ27	F	M94081	1	T cell receptor α joining 27	GDB:9953851	28728
	TRAJ28	F	M94081	1	T cell receptor α joining 28	GDB:9953853	28727
	TRAJ29	F	X02889	1	T cell receptor α joining 29	GDB:9953855	28726
	TRAJ30	F	M94081	1	T cell receptor α joining 30	GDB:9953857	28725
	TRAJ31	F	M14905	1	T cell receptor α joining 31	GDB:9953859	28724
	TRAJ32	F	M94081	1	T cell receptor α joining 32	GDB:9953861	28723
	TRAJ33	F	M94081	1	T cell receptor α joining 33	GDB:9953863	28722
	TRAJ34	F	M35622	1	T cell receptor α joining 34	GDB:9953865	28721
	TRAJ35	ORF	M94081	1	T cell receptor α joining 35	GDB:9953867	28720

Table 9.5 T cell receptor α (*TRA*) genes[a] (*continued*)

IMGT gene group	IMGT gene name[b]	Functionality	IMGT reference sequence accession number	Number of alleles	IMGT gene definition[c]	GDB accession ID	Locus-link number
	TRAJ36	F	M94081	1	T cell receptor α joining 36	GDB:9953869	28719
	TRAJ37	F	M94081	1	T cell receptor α joining 37	GDB:9953871	28718
	TRAJ38	F	M94081	1	T cell receptor α joining 38	GDB:9953873	28717
	TRAJ39	F	M94081	1	T cell receptor α joining 39	GDB:9953875	28716
	TRAJ40	F	M35620	1	T cell receptor α joining 40	GDB:9953877	28715
	TRAJ41	F	M94081	1	T cell receptor α joining 41	GDB:9953879	28714
	TRAJ42	F	M94081	1	T cell receptor α joining 42	GDB:9953881	28713
	TRAJ43	F	M94081	1	T cell receptor α joining 43	GDB:9953883	28712
	TRAJ44	F	M35619	1	T cell receptor α joining 44	GDB:9953885	28711
	TRAJ45	F	M94081	1	T cell receptor α joining 45	GDB:9953887	28710
	TRAJ46	F	M94081	1	T cell receptor α joining 46	GDB:9953889	28709
	TRAJ47	F	M94081	1	T cell receptor α joining 47	GDB:9953891	28708
	TRAJ48	F	M94081	1	T cell receptor α joining 48	GDB:9953893	28707
	TRAJ49	F	M94081	1	T cell receptor α joining 49	GDB:9953895	28706
	TRAJ50	F	M94081	1	T cell receptor α joining 50	GDB:9953897	28705
	TRAJ51	P	M94081	—	T cell receptor α joining 51	GDB:9953899	28704
	TRAJ52	F	M94081	1	T cell receptor α joining 52	GDB:9953901	28703
	TRAJ53	F	M94081	1	T cell receptor α joining 53	GDB:9953903	28702
	TRAJ54	F	M94081	1	T cell receptor α joining 54	GDB:9953905	28701
	TRAJ55	P	M94081	—	T cell receptor α joining 55	GDB:9953907	28700

Table 9.5 T cell receptor α (*TRA*) genes[a] (*continued*)

IMGT gene group	IMGT gene name[b]	Functionality	IMGT reference sequence accession number	Number of alleles	IMGT gene definition[c]	GDB accession ID	Locus-link number
	TRAJ56	F	M94081	1	T cell receptor α joining 56	GDB:9953909	28699
	TRAJ57	F	M94081	1	T cell receptor α joining 57	GDB:9953911	28698
	TRAJ58	ORF	M94081	1	T cell receptor α joining 58	GDB:9953913	28697
	TRAJ59	ORF	M94081	1	T cell receptor α joining 59	GDB:9953915	28696
	TRAJ60	P	M94081	—	T cell receptor α joining 60	GDB:9953917	28695
	TRAJ61	ORF	M94081	1	T cell receptor α joining 61	GDB:9953919	28694
TRAV[e]	*TRAV1-1*	F	AE000658	2	T cell receptor α variable 1-1	GDB:9953921	28693
	TRAV1-2	F	AE000658	2	T cell receptor α variable 1-2	GDB:9953923	28692
	TRAV2	F	AE000658	2	T cell receptor α variable 2	GDB:9953925	28691
	TRAV3	F, (P)[f]	AE000658	2	T cell receptor α variable 3	GDB:9953927	28690
	TRAV4	F	AE000658	1	T cell receptor α variable 4	GDB:9953929	28689
	TRAV5	F	AE000659	1	T cell receptor α variable 5	GDB:9953931	28688
	TRAV6	F	AE000659	6	T cell receptor α variable 6	GDB:9953933	28687
	TRAV7	F	AE000659	1	T cell receptor α variable 7	GDB:9953935	28686
	TRAV8-1	F	ABO00659	2	T cell receptor α variable 8-1	GDB:9953937	28685
	TRAV8-2	F	AE000659	2	T cell receptor α variable 8-2	GDB:9953939	28684
	TRAV8-3	F	AE000659	3	T cell receptor α variable 8-3	GDB:9953941	28683
	TRAV8-4	F	AE000659	7	T cell receptor α variable 8-4	GDB:9953943	28682
	TRAV8-5	P	AE000659	—	T cell receptor α variable 8-5	GDB:9953945	28681

Table 9.5 T cell receptor α (*TRA*) genes[a] (*continued*)

IMGT gene group	IMGT gene name[b]	Functionality	IMGT reference sequence accession number	Number of alleles	IMGT gene definition[c]	GDB accession ID	Locus-link number
	TRAV8-6	F	X02850	2	T cell receptor α variable 8-6	GDB:9953947	28680
	TRAV8-7	F	AE000660	1	T cell receptor α variable 8-7	GDB:9953949	28679
	TRAV9-1	F	AE000659	1	T cell receptor α variable 9-1	GDB:9953951	28678
	TRAV9-2	F	AE000659	4	T cell receptor α variable 9-2	GDB:9953953	28677
	TRAV10	F	AE000659	1	T cell receptor α variable 10	GDB:9953955	28676
	TRAV11	F	AE000659	1	T cell receptor α variable 1	GDB:9953957	28675
	TRAV12-1	F	AE000659	2	T cell receptor α variable 12-1	GDB:9953959	28674
	TRAV12-2	F	AE000659	3	T cell receptor α variable 12-2	GDB:9953961	28673
	TRAV12-3	F	X06193	2	T cell receptor α variable 12-3	GDB:9953963	28672
	TRAV13-1	F	AE000659	3	T cell receptor α variable 13-1	GDB:9953965	28671
	TRAV13-2	F	AE000659	2	T cell receptor α variable 13-2	GDB:9953967	28670
	TRAV14/DV4[g]	F	M21626	4	T cell receptor α variable 14/δ variable 4	GDB:9953969	28669
	TRAV15	P	AE000659	—	T cell receptor α variable 15	GDB:9953971	28668
	TRAV16	F	AE000659	1	T cell receptor α variable 16	GDB:9953973	28667
	TRAV17	F	AE000660	1	T cell receptor α variable 17	GDB:9953975	28666
	TRAV18	F	AE000660	1	T cell receptor α variable 18	GDB:9953977	28665
	TRAV19	F	AE000660	I	T cell receptor α variable 19	GDB:9953979	28664
	TRAV20	F	AE000660	4	T cell receptor α variable 20	GDB:9953981	28663
	TRAV21	F	AE000660	2	T cell receptor α variable 21	GDB:9953983	28662

Table 9.5 T cell receptor α (*TRA*) genes[a] (*continued*)

IMGT gene group	IMGT gene name[b]	Functionality	IMGT reference sequence accession number	Number of alleles	IMGT gene definition[c]	GDB accession ID	Locus-link number
	TRAV22	F	AE000660	1	T cell receptor α variable 22	GDB:9953985	28661
	TRAV23/ DV6[g]	F	AE000660	4	T cell receptor α variable 23/δ variable 6	GDB:9953987	28660
	TRAV24	F	AE000660	2	T cell receptor α variable 24	GDB:9953989	28659
	TRAV25	F	AE000660	1	T cell receptor α variable 25	GDB:9953991	28658
	TRAV26-1	F	AE000660	3	T cell receptor α variable 26-1	GDB:9953993	28657
	TRAV26-2	F	AE000660	2	T cell receptor α variable 26-2	GDB:9953995	28656
	TRAV27	F	AE000660	3	T cell receptor α variable 27	GDB:9953997	28655
	TRAV28	P	AE000660	1	T cell receptor α variable 28	GDB:9953999	28654
	TRAV29/ DV5[g]	F, (P)[f]	AE000660	3	T cell receptor α variable 29/δ variable 5	GDB:9954001	28653
	TRAV30	F	AE000660	4	T cell receptor α variable 30	GDB:9954003	28652
	TRAV31	P	AE000660	—	T cell receptor α variable 31	GDB:9954005	28651
	TRAV32	P	AE000660	—	T cell receptor α variable 32	GDB:9954007	28650
	TRAV33	P	AE000660	—	T cell receptor α variable 33	GDB:9954009	28649
	TRAV34	F	AE000660	1	T cell receptor α variable 34	GDB:9954011	28648
	TRAV35	F	AE000660	2	T cell receptor α variable 35	GDB:9954013	28647
	TRAV36/ DV7[g]	F	AE000660	4	T cell receptor α variable 36/δ variable 7	GDB:9954015	28646
	TRAV37	P	AE000661	—	T cell receptor α variable 37	GDB:9954017	28645
	TRAV38-1	F	AE000661	4	T cell receptor α variable 38-1	GDB:9954019	28644

Table 9.5 T cell receptor α (*TRA*) genes[a] (*continued*)

IMGT gene group	IMGT gene name[b]	Functionality	IMGT reference sequence accession number	Number of alleles	IMGT gene definition[c]	GDB accession ID	Locus-link number
	TRAV38-2/DV8[g]	F	AE000661	1	T cell receptor α variable 38-2/δ variable 8	GDB:9954021	28643
	TRAV39	F	AE000661	1	T cell receptor α variable 39	GDB:9954023	28642
	TRAV40	F	X73521	1	T cell receptor α variable 40	GDB:9954025	28641
	TRAV41	F	AE000661	1	T cell receptor α variable 41	GDB:9954027	28640

Notes:

[a]Gene names are according to the IMGT gene name nomenclature for Ig and TcR of all vertebrates (IMGT Scientific chart; http://imgt.cines.fr:8104).

[b]IMGT *TRA* gene names have been approved by the HUGO Nomenclature Committee in 1999. Note that, in the HUGO symbols, slashes of the *TRAV/DV* gene names are omitted. Otherwise all the gene names (gene symbols) are identical in IMGT and HUGO nomenclatures.

[c]Gene definitions (full names) are identical (including slashes) in IMGT and HUGO nomenclatures. Note that in the databases, the Greek letters are written in full (e.g., α = alpha, δ = delta).

[d]*TRAJ* genes are designated by a number for the localization from 3′ to 5′ in the locus.

[e]*TRAV* genes are designated by a number for the subgroup followed, whenever there are several genes belonging to the same subgroup, by a hyphen and a number for their relative localization in the locus. Numbers increase from 5′ to 3′ in the locus.

[f]Functionality is shown between parentheses when the germline *TRAV* genes have not yet been isolated.

[g]The *TRAV14/DV4*, *TRAV23/DV6*, *TRAV29/DV5*, *TRAV36/DV7*, and *TRAV38-2/DV8* genes have been found rearranged to J genes of the *TRA* locus, and to D and J genes of the *TRD* locus.

Table 9.6 T cell receptor β (*TRB*) genes[a]

IMGT gene group	IMGT gene name[b]	Functionality	IMGT reference sequence accession number	Number of alleles	IMGT gene definition[c]	GDB accession ID	Locus-link number
			TRB locus on chromosome 7 at 7q35				
TRBC	TRBC1	F	M12887	2	T cell receptor β constant 1	GDB:9954029	28639
	TRBC2	F	M12888	2	T cell receptor β constant 2	GDB:9954031	28638
TRBD	TRBD1	F	X00936	1	T cell receptor β diversity 1	GDB:9954033	28637
	TRBD2	F	X02987	2	T cell receptor β diversity 2	GDB:9954035	28636
TRBJ[d]	TRBJ1-1	F	X00936	1	T cell receptor β joining 1-1	GDB:9954037	28635
	TRBJ1-2	F	X00936	1	T cell receptor β joining 1-2	GDB:9954039	28634

Table 9.6 T cell receptor β (*TRB*) genes[a] (*continued*)

IMGT gene group	IMGT gene name[b]	Functionality	IMGT reference sequence accession number	Number of alleles	IMGT gene definition[c]	GDB accession ID	Locus-link number
	TRBJ1-3	F	M14158	1	T cell receptor β joining 1-3	GDB:9954041	28633
	TRBJ1-4	F	M14158	1	T cell receptor β joining 1-4	GDB:9954043	28632
	TRBJ1-5	F	M14158	1	T cell receptor β joining 1-5	GDB:9954045	28631
	TRBJ1-6	F	M14158	1	T cell receptor β joining 1-6	GDB:9954047	28630
	TRBJ2-1	F	X02987	1	T cell receptor β joining 2-1	GDB:9954049	28629
	TRBJ2-2	F	X02987	1	T cell receptor β joining 2-2	GDB:9954051	28628
	TRBJ2-2P	ORF	X02987	1	T cell receptor β joining 2-2P	GDB:9954053	28627
	TRBJ2-3	F	X02987	1	T cell receptor β joining 2-3	GDB:9954055	28626
	TRBJ2-4	F	X02987	1	T cell receptor β joining 2-4	GDB:9954057	28625
	TRBJ2-5	F	X02987	1	T cell receptor β joining 2-5	GDB:9954059	28624
	TRBJ2-6	F	X02987	1	T cell receptor β joining 2-6	GDB:9954061	28623
	TRBJ2-7	F, ORF	M14159	2	T cell receptor β joining 2-7	GDB:9954063	28622
TRBV[e]	TRBV1	P	L36092	—	T cell receptor β variable 1	GDB:9954065	28621
	TRBV2	F	L36092	3	T cell receptor β variable 2	GDB:9954067	28620
	TRBV3-1	F	U07977	2	T cell receptor β variable 3-1	GDB:9954069	28619
	TRBV3-2	P	L36092	—	T cell receptor β variable 3-2	GDB:9954071	28618
	TRBV4-1	F	U07977	2	T cell receptor β variable 4-1	GDB:9954073	28617
	TRBV4-2	F	U07975	2	T cell receptor β variable 4-2	GDB:9954075	28616
	TRBV4-3	F	U07978	4	T cell receptor β variable 4-3	GDB:9954077	28615

Table 9.6 T cell receptor β (*TRB*) genes[a] (*continued*)

IMGT gene group	IMGT gene name[b]	Functionality	IMGT reference sequence accession number	Number of alleles	IMGT gene definition[c]	GDB accession ID	Locus-link number
	TRBV5-1	F	L36092	2	T cell receptor β variable 5-1	GDB:9954079	28614
	TRBV5-2	P	L36092	—	T cell receptor β variable 5-2	GDB:9954081	28613
	TRBV5-3	ORF	X61439	2	T cell receptor β variable 5-3	GDB:9954083	28612
	TRBV5-4	F	L36092	4	T cell receptor β variable 5-4	GDB:9954085	28611
	TRBV5-5	F	L36092	3	T cell receptor β variable 5-5	GDB:9954087	28610
	TRBV5-6	F	L36092	1	T cell receptor β variable 5-6	GDB:9954089	28609
	TRBV5-7	ORF	L36092	1	T cell receptor β variable 5-7	GDB:9954091	28608
	TRBV5-8	F	L36092	2	T cell receptor β variable 5-8	GDB:9954093	28607
	TRBV6-1	F	X61446	1	T cell receptor β variable 6-1	GDB:9954095	28606
	TRBV6-2	F, (P)[f]	X61445	3	T cell receptor β variable 6-2	GDB:9954097	28605
	TRBV6-3	F	U07978	1	T cell receptor β variable 6-3	GDB:9954099	28604
	TRBV6-4	F	X61653	2	T cell receptor β variable 6-4	GDB:9954101	28603
	TRBV6-5	F	L36092	1	T cell receptor β variable 6-5	GDB:9954103	28602
	TRBV6-6	F	L36092	5	T cell receptor β variable 6-6	GDB:9954105	28601
	TRBV6-7	ORF	L36092	1	T cell receptor β variable 6-7	GDB:9954107	28600
	TRBV6-8	F	L36092	1	T cell receptor β variable 6-8	GDB:9954109	28599
	TRBV6-9	F	X61447	1	T cell receptor β variable 6-9	GDB:9954111	28598
	TRBV7-1	ORF	X61444	1	T cell receptor β variable 7-1	GDB:9954113	28597
	TRBV7-2	F	X61442	4	T cell receptor β variable 7-2	GDB:9954115	28596

Table 9.6   T cell receptor β (*TRB*) genes[a] (*continued*)

IMGT gene group	IMGT gene name[b]	Functionality	IMGT reference sequence accession number	Number of alleles	IMGT gene definition[c]	GDB accession ID	Locus-link number
	TRBV7-3	F, ORF	X61440	5	T cell receptor β variable 7-3	GDB:9954117	28595
	TRBV7-4	F, (P)[f]	L36092	3	T cell receptor β variable 7-4	GDB:9954119	28594
	TRBV7-5	P	L36092	—	T cell receptor β variable 7-5	GDB:9954121	28593
	TRBV7-6	F	L36092	2	T cell receptor β variable 7-6	GDB:9954123	28592
	TRBV7-7	F	L36092	2	T cell receptor β variable 7-7	GDB:9954125	28591
	TRBV7-8	F	M11953	3	T cell receptor β variable 7-8	GDB:9954127	28590
	TRBV7-9	F	L36092	7	T cell receptor β variable 7-9	GDB:9954129	28589
	TRBV8-1	P	L36092	—	T cell receptor β variable 8-1	GDB:9954131	28588
	TRBV8-2	P	L36092	—	T cell receptor β variable 8-2	GDB:9954133	28587
	TRBV9	F	L36092	3	T cell receptor β variable 9	GDB:9954135	28586
	TRBV10-1	F, (P)[f]	U17050	3	T cell receptor β variable 10-1	GDB:9954137	28585
	TRBV10-2	F	U17049	2	T cell receptor β variable 10-2	GDB:9954139	28584
	TRBV10-3	F	U03115	4	T cell receptor β variable 10-3	GDB:9954141	28583
	TRBV11-1	F	M33233	1	T cell receptor β variable 11-1	GDB:9954143	28582
	TRBV11-2	F	L36092	3	T cell receptor β variable 11-2	GDB:9954145	28581
	TRBV11-3	F	M33234	4	T cell receptor β variable 11-3	GDB:9954147	28580
	TRBV12-1	P	X07224	—	T cell receptor β variable 12-1	GDB:9954149	28579
	TRBV12-2	P	X06936	—	T cell receptor β variable 12-2	GDB:9954151	28578
	TRBV12-3	F	X07192	1	T cell receptor β variable 12-3	GDB:9954153	28577

Table 9.6 T cell receptor β (*TRB*) genes[a] (*continued*)

IMGT gene group	IMGT gene name[b]	Functionality	IMGT reference sequence accession number	Number of alleles	IMGT gene definition[c]	GDB accession ID	Locus-link number
	TRBV12-4	F	K02546	2	T cell receptor β variable 12-4	GDB:9954155	28576
	TRBV12-5	F	X07223	1	T cell receptor β variable 12-5	GDB:9954157	28575
	TRBV13	F	U03115	2	T cell receptor β variable 13	GDB:9954159	28574
	TRBV14	F	X06154	2	T cell receptor β variable 14	GDB:9954161	28573
	TRBV15	F	U03115	3	T cell receptor β variable 15	GDB:9954163	28572
	TRBV16	F, P	L26231	3	T cell receptor β variable 16	GDB:9954165	28571
	TRBV17	ORF	U03115	1	T cell receptor β variable 17	GDB:9954167	28570
	TRBV18	F	L36092	1	T cell receptor β variable 18	GDB:9954169	28569
	TRBV19	F	U48260	3	T cell receptor β variable 19	GDB:9954171	28568
	*TRBV20-1*g	F	M11955	7	T cell receptor β variable 20-1	GDB:9954173	28567
	*TRBV21-1*g	P	L36092	—	T cell receptor β variable 21-1	GDB:9954175	28566
	*TRBV22*g	P	L36092	—	T cell receptor β variable 22	GDB:9954177	28565
	*TRBV23-1*g	ORF	L36092	1	T cell receptor β variable 23-1	GDB:9954179	28564
	*TRBV24-1*g	F	M11951	1	T cell receptor β variable 24-1	GDB:9954181	28563
	*TRBV25-1*g	F	L36092	1	T cell receptor β variable 25-1	GDB:9954183	28562
	TRBV26	P	L36092	—	T cell receptor β variable 26	GDB:9954185	28561
	TRBV27	F	L36092	1	T cell receptor β variable 27	GDB:9954187	28560
	TRBV28	F	U08314	1	T cell receptor β variable 28	GDB:9954189	28559
	*TRBV29-1*g	F	L36092	3	T cell receptor β variable 29-1	GDB:9954191	28558

Table 9.6   T cell receptor β (*TRB*) genes[a] (*continued*)

IMGT gene group	IMGT gene name[b]	Functionality	IMGT reference sequence accession number	Number of alleles	IMGT gene definition[c]	GDB accession ID	Locus-link number
	TRBV30	F, P	L36092	5	T cell receptor β variable 30	GDB:9954193	28557
	TRBVA	P	L36092	—	T cell receptor β variable A	GDB:9954195	28556
	TRBVB	P	L36092	—	T cell receptor β variable B	GDB:9954197	28555
TRBV orphons on chromosome 9 at 9p21							
TRBV	*TRBV20/ OR9-2*	ORF	L05149	2	T cell receptor β variable 20/OR9-2	GDB:9954199	6962
	TRBV21/ OR9-2	ORF	L05151	1	T cell receptor β variable 21/OR9-2	GDB:9954201	6959
	TRBV23/ OR9-2	ORF	L27615	1	T cell receptor β variable 23/OR9-2	GDB:9954203	28552
	TRBV24/ OR9-2	ORF, P	L05153	2	T cell receptor β variable 24/OR9-2	GDB:9954205	6961
	TRBV25/ OR9-2	P	L05152	2	T cell receptor β variable 25/OR9-2	GDB:9954207	6960
	TRBV29/ OR9-2	ORF	L05150	2	T cell receptor β variable 29/OR9-2	GDB:9954209	6958

Notes:

[a]Gene names are according to the IMGT gene name nomenclature for Ig and TcR of all vertebrates (IMGT Scientific chart; http:// imgt.cines.fr:8104).

[b]IMGT *TRB* gene names have been approved by the HUGO Nomenclature Committee in 1999. Note that, in the HUGO symbols, slashes of the orphon names are omitted. Otherwise all the gene names (gene symbols) are identical in IMGT and HUGO nomenclatures.

[c]Gene definitions (full names) are identical (including slashes) in IMGT and HUGO nomenclatures. Note that in the databases, the Greek letters are written in full (e.g., β = beta).

[d]*TRBJ* genes are designated by a number for the cluster followed by a hyphen and a number for their relative localization in the locus. Numbers increase from 5′ to 3′ in the locus.

[e]*TRBV* genes are designated by a number for the subgroup followed, whenever there are several genes belonging to the same subgroup, by a hyphen and a number for their relative localization in the locus. Numbers increase from 5′ to 3′ in the locus.

[f]Functionality is shown between parentheses when the accession number refers to a rearranged sequence and the corresponding germline gene has not yet been isolated; brackets when the accession number refers to a DNA genomic sequence, but not known as being germline or rearranged.

[g]Since orphons (OR) have been described for each of the following *TRBV* subgroups: 20, 21, 23, 24, 25, and 29 (see *TRBV* orphons), the single member gene in the main locus is designated by the subgroup number followed by a hyphen and the number 1. To date, no orphon has been reported which belongs to subgroup 22, therefore the IMGT designation of the single member gene is *TRBV22*.

Table 9.7 T cell receptor γ (*TRG*) genes[a]

IMGT gene group	IMGT gene name[b]	Functionality	IMGT reference sequence accession number	Number of alleles	IMGT gene definition[c]	GDB accession ID	Locus-link number
			TRG locus on chromosome 7 at 7p15 - p14				
TRGC	*TRGC1*	F	M14996, 97, 98	2	T cell receptor γ constant 1	GDB:120408	6966
	TRGC2 (2x)	F	M15002/M13231	3	T cell receptor γ constant 2 (2x)	GDB:120409	6967
	TRGC2 (3x)	F	M17323/M25318	1	T cell receptor γ constant 2 (3x)	GDB:120409	6967
TRGJ	*TRGJ1*	F	M12960	2	T cell receptor γ joining 1	GDB:120410	6968
	TRGJ2	F	M12961	1	T cell receptor γ joining 2	GDB:120411	6969
	TRGJP	F	M12950	1	T cell receptor γ joining P	GDB:120412	6970
	TRGJP1	F	X08084	1	T cell receptor γ joining P1	GDB:120413	6971
	TRGJP2	F	M16016	1	T cell receptor γ joining P2	GDB:120414	6972
TRGV[d,e]	*TRGV1*	ORF	M12949	1	T cell receptor γ variable 1	GDB:120415	6973
	TRGV2	F	M13429	1	T cell receptor γ variable 2	GDB:120418	6974
	TRGV3	F	M13430	1	T cell receptor γ variable 3	GDB:120419	6976
	TRGV4	F	X15272	2	T cell receptor γ variable 4	GDB:120420	6977
	TRGV5	F	X13555	1	T cell receptor γ variable 5	GDB:120421	6978
	TRGV5P	P	M13431	—	T cell receptor γ variable 5P	GDB:120422	6979
	TRGV6	P	M13432	—	T cell receptor γ variable 6	GDB:120423	6980
	TRGV7	P	M13433	—	T cell receptor γ variable 7	GDB:120424	6981
	TRGV8	F	M13434	1	T cell receptor γ variable 8	GDB:120425	6982
	TRGV9	F	X07205	2	T cell receptor γ variable 9	GDB:120426	6983

Table 9.7   T cell receptor γ (*TRG*) genes[a] (*continued*)

IMGT gene group	IMGT gene name[b]	Functionality	IMGT reference sequence accession number	Number of alleles	IMGT gene definition[c]	GDB accession ID	Locus-link number
	TRGV10	ORF	X07206	2	T cell receptor γ variable 10	GDB:120416	6984
	TRGV11	ORF	Y11227	1	T cell receptor γ variable 11	GDB:120417	6985
	TRGVA	P	X07208	—	T cell receptor γ variable A	GDB:9953127	6986
	TRGVB	P	X07209	—	T cell receptor γ variable B	GDB:9953128	6987

Notes:

[a]Gene names are according to the IMGT gene name nomenclature for Ig and TcR of all vertebrates (IMGT Scientific chart; http://imgt.cines.fr:8104).

[b]IMGT *TRG* gene names have been approved by the HUGO Nomenclature Committee in 1999. All the gene names (gene symbols) are identical in IMGT and HUGO nomenclatures.

[c]Gene definitions (full names) are identical (including slashes) in IMGT and HUGO nomenclatures. Note that in the databases, the Greek letters are written in full (e.g., γ = gamma).

[d]*TRGV* genes are designated by a number (or a letter, for pseudogenes that are single members of their subgroup) for their position from 5′ to 3′ in the locus.

[e]The *IGHV3P* gene, a polymorphic gene by insertion, has been identified by Southern hybridization in a rare haplotype but has not been sequenced.

Table 9.8   T cell receptor δ (*TRD*) genes[a]

IMGT gene group	IMGT gene name[b]	Functionality	IMGT reference sequence accession number	Number of alleles	IMGT gene definition[c]	GDB accession ID	Locus-link number
			*TRD* locus on chromosome 14 at 14q11.2				
TRDC	*TRDC*	F	M22148-M22151	1	T cell receptor δ constant	GDB:9954211	28526
TRDD	*TRDD1*	F	M23325	1	T cell receptor δ diversity 1	GDB:9954213	28525
	TRDD2	F	M22153	1	T cell receptor δ diversity 2	GDB:9954215	28524
	TRDD3	F	M22152	1	T cell receptor δ diversity 3	GDB:9954217	28523
TRDJ	*TRDJ1*	F	M20289	1	T cell receptor δ joining 1	GDB:9954219	28522
	TRDJ2	F	L36386	1	T cell receptor δ joining 2	GD13:9954221	28521
	TRDJ3	F	M21508	1	T cell receptor δ joining 3	GDB:9954223	28520
	TRDJ4	F	AJ249814	1	T cell receptor δ joining 4	GDB:9953677	28519

Table 9.8 T cell receptor δ (*TRD*) genes[a] (*continued*)

IMGT gene group	IMGT gene name[b]	Functionality	IMGT reference sequence accession number	Number of alleles	IMGT gene definition[c]	GDB accession ID	Locus-link number
TRDV[d]	*TRDV1*	F	M22198	1	T cell receptor δ variable 1	GDB:9953671	28518
	TRDV2	F	X15207	2	T cell receptor δ variable 2	GDB:9953287	28517
	TRDV3	F	M23326	2	T cell receptor δ variable 3	GDB:9953273	28516

Notes:

[a]Gene names are according to the IMGT gene name nomenclature for Ig and TcR of all vertebrates (IMGT Scientific chart; http://imgt.cines.fr:8104).

[b]IMGT *TRD* gene names have been approved by the HUGO Nomenclature Committee in 1999. All the gene names (gene symbols) are identical in IMGT and HUGO nomenclatures.

[c]Gene definitions (full names) are identical in IMGT and HUGO nomenclatures. Note that in the database, the Greek letters are written in full (e.g., δ = delta).

[d]*TRDV* genes are designated by a number for their position from 5′ to 3′ in the locus. The *TRAV14/DV4*, *TRAV23/DV6*, *TRAV29/DV5*, *TRAV36/DV7*, and *TRAV38-2/DV8* genes, which have been found rearranged to J genes of the *TRA* locus, and to D and J genes of the *TRD* locus, are displayed in the human *TRAV* table.

KEY REFERENCES

1 Lefranc MP (2000) Nomenclature of the human T cell receptor genes. *Curr Protocol Immunol* A.10.1–A.10.23

2 Folch G, Lefranc MP (2000) The human T cell receptor beta variable (TRBV) genes. *Exp Clin Immunogenet* 17: 42–54

3 Folch G, Lefranc MP (2000) The human T cell receptor beta diversity (TRBD) and beta joining (TRBJ) genes. *Exp Clin Immunogenet* 17: 107–114

4 Lefranc MP, Rabbitts TH (1989) The human T cell receptor gamma (TRG) genes. *Trends Biochem Sci* 14: 214–218

5 Lefranc MP, Rabbitts TH. A nomenclature to fit the organization of the human T cell receptor gamma and delta genes. *Res Immunol* 141: 615–618

6 Lefranc MP, Chuchana P, Dariavach P, Nguyen C, Huck S, Brockly F, Jordan B, Lefranc G (1989) Molecular mapping of the human T cell receptor gamma (TRG) genes and linkage of the variable and constant regions. *Eur J Immunol* 19: 989–994

7 Lefranc MP (1990) The human T-cell receptor delta genes. *Res Immunol* 141: 692–693

8 Lefranc MP, Rabbitts TH (1990) Genetic organization of the human T-cell receptor gamma and delta loci. *Res Immunol* 141: 565–577

9 Scaviner D, Lefranc MP (2000) The human T cell receptor alpha variable (TRAV) genes. *Exp Clin Immunogenet* 17: 83–96

10 Scaviner D, Lefranc MP (2000) The human T cell receptor alpha joining (TRAJ) genes. *Exp Clin Immunogenet* 17: 97–106

Table 9.9 Correspondence between *TRAV* nomenclatures[a,b]

IMGT *TRAV* gene name (Scaviner and Lefranc, 2000)	Boysen *et al.*[c]	Arden *et al.*
TRAV41	41S1	19S1
TRAV40	40S1	31S1
TRAV39	39S1	27S1
TRAV38-2/DV8	hADV38S2	14S1-ADV14S1
TRAV38-1	38S1	14S2
TRAV37	37S1	—
TRAV36/DV7	hADV36S 1	28S1-DV28S1
TRAV35	35S1	25S1
TRAV34	34S1	26S1
TRAV26-2	26S2	4S1
TRAV33	33S1	—
TRAV32	32S1	—
TRAV31	31S1	—
TRAV30	30S1	29S1
TRAV29/DV5	hADV29S1	21S1-ADV21S1
TRAV28	28S1	—
TRAV27	27S1	10S1
TRAV8-7	8S7	—
TRAV26-1	26S1	4S2
TRAV25	25S1	32S1
TRAV24	24S1	18S1
TRAV23/DV6	hADV23S1	17S1-ADV17S1
TRAV22	22S1	13S1
TRAV21	21S1	23S1
TRAV20	20S1	30S1
TRAV19	19S1	12S1
TRAV18	18S1	—
TRAV17	17S1	3S1
TRAV16	16S1	9S1
TRAV8-6	8S6	1S3
TRAV12-3	12S3	2S2

Table 9.9 Correspondence between *TRAV* nomenclatures[a,b] (*continued*)

IMGT *TRAV* gene name (Scaviner and Lefranc, 2000)	Boysen *et al.*[c]	Arden *et al.*
TRAV15	15S1	—
TRAV9-2	9S2	22S1
TRAV14/DV4	hADV14S1	6S1-ADV6S1
TRAV13-2	13S2	8S2
TRAV8-5	8S5	—
TRAV8-4	8S4	1S2
TRAV12-2	12S2	2S1
TRAV13-1	13S1	8S1
TRAV8-3	8S3	1S4
TRAV8-2	8S2	1S5
TRAV12-1	12S1	2S3
TRAV11	11S1	—
TRAV10	10S1	24S1
TRAV9-1	9S1	—
TRAV8-1	8S1	1S1
TRAV7	7S1	—
TRAV6	6S1	5S1
TRAV5	5S1	15S1
TRAV4	4S1	20S1
TRAV3	3S1	16S1
TRAV2	2S1	11S1
TRAV1-2	1S2	7S2
TRAV1-1	1S1	7S1

Notes:
[a]*TRAV* genes are listed from $3'$ (top of the table) to $5'$ (bottom of the table). Cells with dashes indicate that no name exists for the gene in that system of nomenclature.
[b]See *TRA* locus and Table 3.5 for more information.
[c]IMGT reference sequence accession numbers: AE000658-AE000661.

References:
Arden *et al.* (1995)
Boysen *et al.* (Unpublished)
Scaviner and Lefranc (2001)

Table 9.10 Comparison of *TRBV* gene nomenclatures[a,b]

IMGT TRBV gene name (Folch and Lefranc, 2000)	Wei *et al.* (1994)	Arden *et al.* (1995)	Rowen *et al.* (1996)
TRBV30	20S1	20S1	30
TRBV29-1	4S1	4S1	29-1
TRBV28	3S1	3S1	28
TRBV27	14S1	14S1	27
TRBVB	—	34S1	—
TRBV26	—	28SI	26
TRBVA	—	33S1	—
TRBV25-1	11S1	11S1	25-1
TRBV24-1	15S1	15S1	24-1
TRBV23-1	19S1	19S1	23-1
TRBV22	—	29S1	22-1
TRBV21-1	10S1	10S1	21-1
TRBV20-1	2S1	2S1	20-1
TRBV19	17S1	17S1	19
TRBV18	18S1	18S1	18
TRBV17	26S1[c]	26S1	17
TRBV16	25S1	25S1	16
TRBV15	24S1	24S1	15
TRBV14	16S1	16S1	14
TRBV12-5	8S3	8S3	12-5
TRBV12-4	8S2	8S2	12-4
TRBV12-3	8S1	8S1	12-3
TRBV11-3	21S4	21S2	11-3
TRBV10-3	12S2	12S1	10-3
TRBV13	23S1	23S1	13
TRBV7-9	6S5	6S4	7-9
TRBV5-8	5S8	5S4	5-8
TRBV7-8	6S3	6S2	7-8
TRBV6-9	13S4	13S4	6-9
TRBV5-7	5S7	5S7	5-7
TRBV7-7	6S14	6S6	7-7

Table 9.10 Comparison of *TRBV* gene nomenclatures[a,b] (*continued*)

IMGT TRBV gene name (Folch and Lefranc, 2000)	Wei *et al.* (1994)	Arden *et al.* (1995)	Rowen *et al.* (1996)
TRBV6-8	13S7	13S7	6-8
TRBV5-6	5S2	5S2	5-6
TRBV7-6	6S4	6S3	7-6
TRBV6-7	13S8	13S8	6-7
TRBV5-5	5S3	5S3	5-5
TRBV7-5	6S12	6S9	7-5
TRBV6-6	13S6	13S6	6-6
TRBV5-4	5S6	5S6	5-4
TRBV7-4	6Sl1	6S8	7-4
TRBV6-5	13SI	13S1	6-5
TRBV12-2	8S5	8S5	12-2
TRBVI1-2	21S3	21S3	11-2
TRBV10-2	1253	1253	10-2
TRBV12-1	8S4	8S4	12-1
TRBV11-1	21S1	21S1	11-1
TRBV10-1	12S4	12S2	10-1
TRBV9	1S1	1S1	9
TRBV5-3	5S5	5S5	5-3
TRBV8-2	—	32S1	8-2
TRBV7-3	6S1	6S1	7-3
TRBV6-4	13S5	13S5	6-4
TRBV5-2	—	31S1	5-2
TRBV8-1	—	30S1	8-1
TRBV7-2	6S7	6S5	7-2
TRBV6-3	13S2b	13S2b	6-3
TRBV4-3	7S2	7S2	4-3
TRBV3-2	9S2	9S2	3-2
TRBV6-2	13S2a	13S2a	6-2
TRBV4-2	7S3	7S3	4-2
TRBV7-1	6S10	6S7	7-1
TRBV6-1	13S3	13S3	6-1

Table 9.10   Comparison of *TRBV* gene nomenclatures[a,b] (*continued*)

IMGT *TRBV* gene name (Folch and Lefranc, 2000)	Wei *et al.* (1994)	Arden *et al.* (1995)	Rowen *et al.* (1996)
TRBV5-1	5S1	5S1	5-1
TRBV4-1	7SI	7S1	4-1
TRBV3-1	9S 1	9S 1	3-1
TRBV2	22S1	22S1	2
TRBVI	—	27S1	1

Notes:
[a] *TRBV* genes are listed from 3′ in the *TRB* locus (top of the table) to 5′ (bottom of the table). Blank cells indicate that no corresponding name exists.
[b] See *TRB* locus and Table 3.6 for more information.
[c] IMGT note: 26S1 was defined in Slightom *et al.* (1994).

References:
Arden *et al.* (1995)
Folch and Lefranc (2000)
Rowen *et al.* (1996)
Slightom *et al.* (1994)
Wei *et al.* (1994)

i   **KEY DATABASES**

- The Genome Database: http://www.gdb.org
- HUGO Gene Nomenclature Committee website: http://www.gene.ucl. ac.uk/nomenclature
- IMGT, the international ImMunoGeneTics database: http://imgt. cines.fr:8104
- NCBI Locuslink: http://www.ncbi.nlm.nih.gov/LocusLink

CYTOKINES AND CHEMOKINES

10

- **CYTOKINES**
- **CHEMOKINES**
- **OTHER CYTOKINE FUNCTIONS**

CYTOKINES

Cytokines are soluble proteins or glycoproteins produced by leukocytes or other types of cells. They serve as chemical communicators from one cell to another. Cytokines are usually secreted, although some may be expressed on a cell membrane or maintained in reservoirs in the extracellular matrix. Cytokine secretion by different types of cells is depicted in **Figures 10.1–10.6**.

Monokines

Cytokines include monokines synthesized by macrophages and lymphokines produced by activated T cells and natural killer cells. A monokine is a cytokine produced by monocytes and macrophages that has a regulatory effect on the function of other cells such as lymphocytes. Monokines include interleukin-1, tumor necrosis factor (TNF), α and β interferons and colony-stimulating factors.

Lymphokines

A lymphokine is a nonimmunoglobulin polypeptide substance synthesized mainly by T cells that affects the function of other cells. It may either enhance or suppress an immune response, facilitate cell proliferation, growth and differentiation, and act on gene transcription to regulate cell function. Lymphokines include interleukins-2–6, γ interferon, granulocyte-macrophage colony-stimulating factor, migration inhibitory factor, and lymphotoxin.

Cytokine superfamilies

Large superfamilies of cytokines include the TGF-β superfamily (comprising various TGF-β isoforms, activin A, inhibins, BMP ((bone morphogenetic proteins)), dpp (decapentaplegic) and some others), the PDGF superfamily (including VEGF), the EGF superfamily (including EGF, TGF-α, AR (amphiregulin), betacellulin, HB-EGF, and some others), the VEGF family, chemokines (with various subfamilies defined by their structures), FGF (fibroblast growth factors), and the family of neurotrophins. **Table 10.1** summarizes the fibroblast growth factors, including their receptors, amino acid composition, chromosome location, and percent homology between human fibroblast growth factors and mouse fibroblast growth factors, and homology between human fibroblast growth factors and recombinant fibroblast growth factors.

Pro-inflammatory cytokines

In many respects the biological activities of cytokines resemble those of classical hormones produced in specialized glandular tissues. Some cytokines also behave like classical hormones in that they act at a systemic level, affecting, for example, biological phenomena such as inflammation, systemic inflammatory response syndrome, and acute phase reaction, wound healing, and the neuroimmune network. Pro-inflammatory cytokines include IL-1α, IL-1β, IL-6, IFN-γ, TNF-α, LT-α3. Their expression, structure, and major functions are summarized in **Table 10.2. Table 10.3** lists the expression, structure, and functions of pro-inflammatory cytokine receptors. A diagrammatic illustration of the involvement of these pro-inflammatory cytokines in the inflammatory response is given in **Figure 10.7**. Many different cytokines have been shown to be present in wound fluid although their detection does not necessarily correlate with biologic activity. Moreover, individual cytokines can influence wound repair in different ways as they may have diverse effects in similar physiological situations and usually have more than one specific effect on cells. Of the myriad of cytokines that have been studied in terms of wound healing, TGF-β1 has been shown to have the broadest effects. Other cytokines include PDGF, EGF, VEGF, IGF-1, FGF, etc. **Table 10.4** lists the cytokines that are thought to play a role

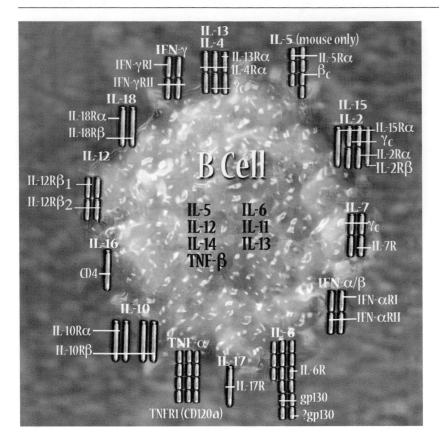

Figure 10.1 Cytokine secretion and cell surface cytokine receptors of B cells

in mediating the wound healing process and their cell sources.

Cytokine receptors

Cytokines include proteins synthesized by cells that affect the action of other cells. They combine with surface receptors on target cells that are linked to intracellular signal transduction and second messenger pathways. The cytokine receptors on the cell surface of B cells, T cells, monocytes, natural killer cells, eosinophils, and dendritic cells are also illustrated in Figures 10.1–10.6. The effects of cytokines may be autocrine, acting on cells that produce them, or paracrine, acting on neighboring cells.

CHEMOKINES

Chemokines are molecules that recruit and activate leukocytes and other cells at sites of inflammation. They comprise a family of pro-inflammatory activation-inducible cytokines previously referred to as members of SIS family of cytokines, SIG family of cytokines, SCY family of cytokines, platelet factor-4 superfamily or intercrines. These proteins are mainly chemotactic for different cell types (hence the

name, which is derived from chemotactic cytokines. Chemokines have molecular masses of 8–10 kDa and show approximately 20–50 percent sequence homology among each other at the protein level. They exhibit both chemoattractant and cytokine properties.

Chemokine families

There are two groups of chemokines. Those that mainly activate neutrophils are the α-chemokines (C-X-C chemokines). By contrast, those that activate monocytes, lymphocytes, basophils, and eosinophils are designated β-chemokines (C-C chemokines). Members of the α-chemokines are referred to also as the 4q chemokine family because the genes encoding members of this family map to human chromosome 4q12-21. The first two cysteine residues of members of this family are separated by single amino acids and these proteins, therefore, are called also CXC chemokines. Members of the β-chemokines or17q chemokine family map to human chromosome 17q11-32 (murine chromosome 11). The first two cysteine residues are adjacent and, therefore, these proteins are called also CC chemokines. The C chemokines or γ-chemokines differ from the other chemokines by the absence of a cysteine

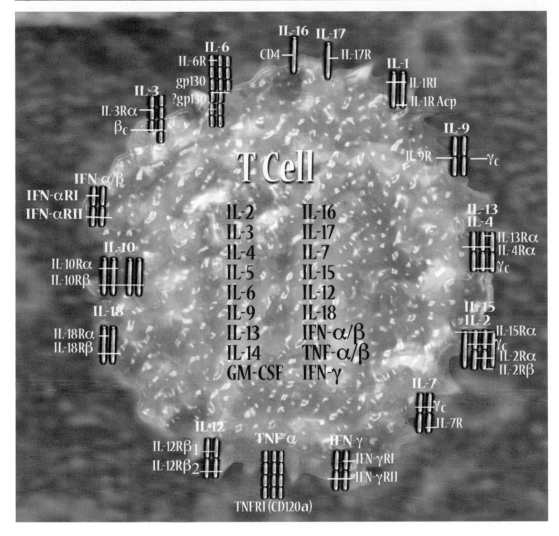

Figure 10.2 Cytokine secretion and cell surface cytokine receptors of T cells

residue. Members of the small group of chemokines with a CXXXC cysteine signature motif are referred to as δ-chemokines or CX3C chemokines or CXXXC chemokines. These four families of chemokines are depicted in **Table 10.5**.

Chemokine receptors

The biological activities of chemokines are mediated by specific receptors and also by receptors with overlapping ligand specificities that bind several of these proteins which always belong either to the CC chemokines or the group of CXC chemokines. The receptors that bind CXC chemokines are designated CXCR followed by a number while those binding CC chemokines are designated CCR followed by a number. Detailed descriptions of human chemokines and their receptors are shown in **Table 10.6** and **Table 10.7** respectively.

Cells require stimulation to become responsive to most known chemokines, and this process is linked closely to chemokine receptor expression. Chemokine activation of different types of leukocytes is demonstrated as simple diagrams in **Figures 10.8–10.16**. The chemokine signaling pathway is shown in **Figure 10.17**.

Table 10.8 and **Table 10.9** are intended to provide a quick reference guide for those contemplating the use of cell lines in chemokine research.

Table 10.10 lists the chemokine receptors on different cell types and can be considered as a tabular expression of Figures 10.8–10.16.

In **Table 10.11**, cytokines and chemokines are arranged in a tabular format and coded according to family membership.

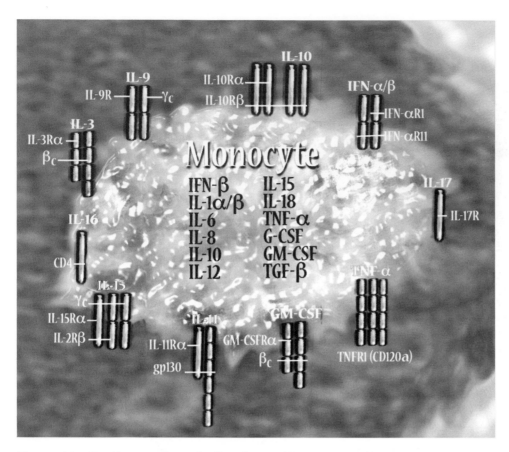

Figure 10.3   Cytokine secretion and cell surface cytokine receptors of monocytes

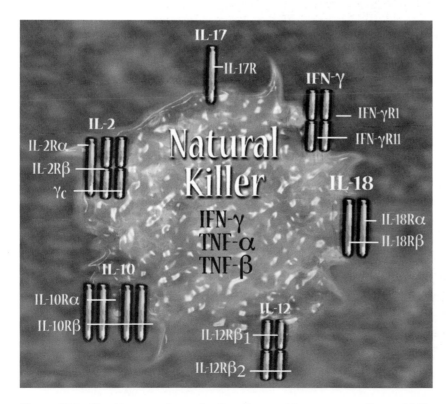

Figure 10.4   Cytokine secretion and cell surface cytokine receptors of natural killer cells

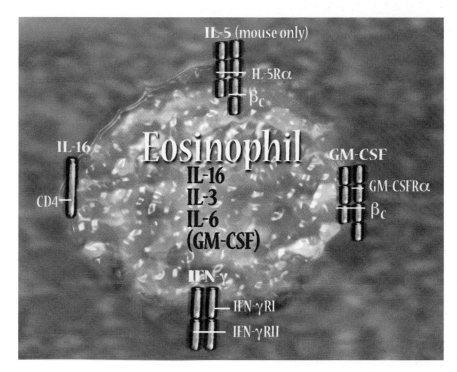

Figure 10.5   Cytokine secretion and cell surface cytokine receptors of eosinophils

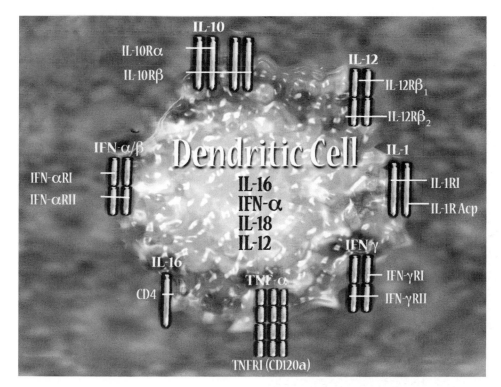

Figure 10.6   Cytokine secretion and cell surface cytokine receptors of dendritic cells

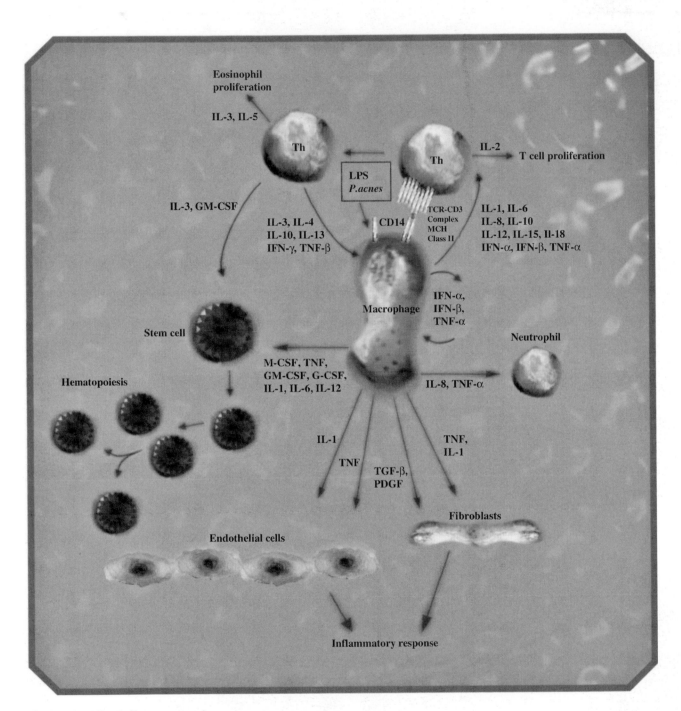

Figure 10.7   Pro-inflammatory pathway

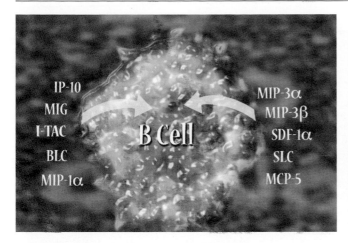

Figure 10.8   Chemokine activation of B cells

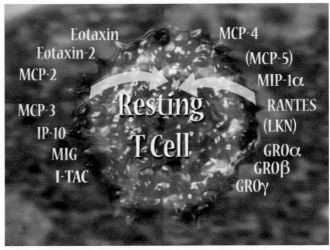

Figure 10.9   Chemokine activation of resting T cells

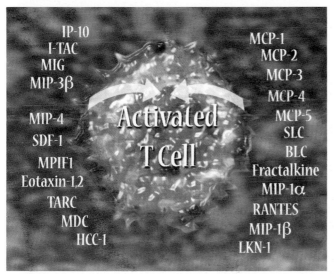

Figure 10.10   Chemokine activation of activated T cells

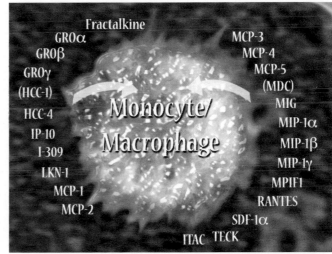

Figure 10.11   Chemokine activation of monocytes/macrophages

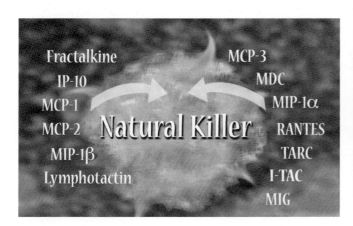

Figure 10.12   Chemokine activation of natural killer cells

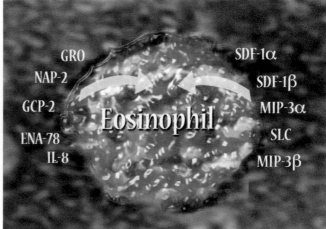

Figure 10.13   Chemokine activation of eosinophils

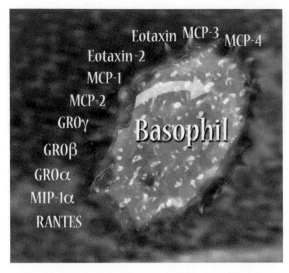

Figure 10.14   Chemokine activation of basophils

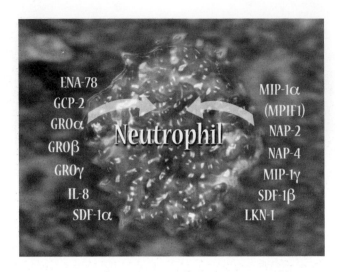

Figure 10.15   Chemokine activation of neutrophils

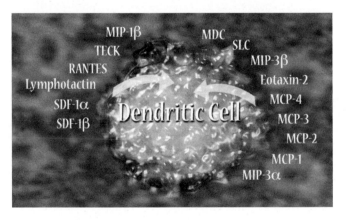

Figure 10.16   Chemokine activation of dendritic cells

OTHER CYTOKINE FUNCTIONS

Cytokines also regulate the expression of matrix metalloproteinases (MMP), which include active medically important enzymes such as angiotensin-converting enzyme, enkephalinase, and collagenase. **Table 10.12** lists the family of MMP, their cytokine substrates, amino acid sequence, and the relevant cytokines that induce or inhibit MMP expression.

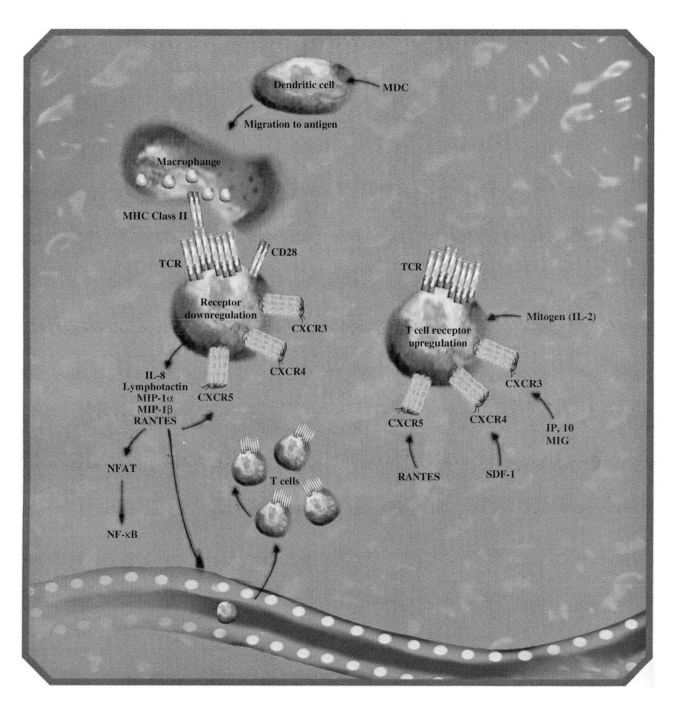

Figure 10.17   Chemokine signaling pathway

Table 10.1 Fibroblast growth factor mini-guide

	Receptors	Amino acids	Chromosomes	% homology (aa)
FGF-1 FGF acidic, aFGF, ECGF, HBGF-1	FGFR1IIIb and c, R2IIIb and c, R3IIIb and c and R4	155	5q31	hFGF-1 to mFGF-1 = 96%, hFGF-1 to rFGF-1 = 95%
FGF-2 FGF basic, bFGF, EDGF, HBGF-2	FGFR1IIIb and c, R2IIIc, R3IIIc and R4	155 (196, 201, 210, and 288 variants)	4q26-q27	hFGF-2 to mFGF-2 = 97%, hFGF-2 to rFGF-2 = 97%
FGF-3 *Int-2*	FGFR1IIIb andR2IIIb	222	11q13	hFGF-3 to mFGF-3 = 81%
FGF-4 K-FGF, KS-FGF, FGFK, HST	FGFR1IIIc, R2IIIc, 3RIIIc and R4	176	11q13.3	hFGF-4 to mFGF-4 = 87%
FGF-5 HBGF-5	FGFR1IIIc	251 (123 variant)	4q21	hFGF-5 to mFGF-5 = 88%, hFGF-5 to rFGF-5 = 84%
FGF-6 *hst-2*, HBGF-6	FGFR1IIIc, R2IIIc and R4	171 (136 and 149 variants)	12p13	hFGF-6 to mFGF-6 = 93%
FGF-7 KGF	KGFR	163	15q15-q21.1	hFGF-7 to mFGF-7 = 96%, hFGF-7 to rFGF -7 = 92%
FGF-8b AIGF	FGFR2IIIc, R3IIIc and R4	215 (8a = 204, 8e = 223 and 8f = 244)	10q24	hFGF-8 to mFGF-8 = 100%
FGF-9 HBGF-9, GAF	FGFR2IIIc, R3IIIb and c and R4	208 (precursor form)	13q11-q12	hFGF-9 to mFGF-9 = 98%, hFGF-9 to rFGF-9 = 98%
FGF-10 KGF-2	FGFR1IIIb and R2IIIb	174	5q12-p13	hFGF-10 to mFGF-10 = 94%, hFGF-10 to rFGF-10 = 100%
FGF-11 FHF-3	Not reported	225	17q12	hFGF-11 to mFGF-11 = 97%
FGF-12 FHF-1	Not reported	243 (181 variant)	3q28	hFGF-12 to mFGF-12 = 99%
FGF-13 FHF-2	Not reported	245 (192, 199, 226, and 255 variants)	Xq26	hFGF-13 to mFGF-13 = 99%
FGF-14 FHF-4	Not reported	247 (252 variant)	13q34	hFGF-14 to mFGF-14 = 98%
FGF-16	FGFR4	207	Unknown	hFGF-16 to mFGF-16 = 99%, hFGF-16 to rFGF-16 = 98%
FGF-17b	FGFR2IIIc, R3IIIc and R4	194	8q21	HFGF-17 to mFGF-17 = 93%, hFGF-17 to rFGF-17 = 93%
FGF-18	Not reported	181	14p11	hFGF-18 to mFGF-18 = 99%, hFGF-18 to rFGF-18 = 99%
FGF-19	FGFR4	194	11q13.1	Not reported
FGF-20	Not reported	211	8q21.3-p22	hFGF-20 to mFGF-20 = 94%, hFGF-20 to rFGF = 95%
FGF-21	Not reported	209 (precursor form)	19q13.1-qter	hFGF-21 to mFGF-21 = 75%

Table 10.2 Pro-inflammatory cytokines

Cytokine	Expression	Structure	Major functions
IL-1α	Monocytes, macrophages, NK cells, dendritic cells, B cells, T cells	17.4 kD protein, 159 amino acid residues	↑ fever and acute phase protein synthesis; ↑ thymocyte and T cell activation and B cell growth, differentiation, and immunoglobulin secretion. Binds and acts through the IL-1R Type I
IL-1β	Keratinocytes, monocytes, macrophages, NK cells, dendritic cells, B cells, T cells	17.3 kD protein, 153 amino acid residues	↑ fever and acute phase protein synthesis; ↑ thymocyte and T cell activation and B cell growth, differentiation, and immunoglobulin secretion. Binds and acts through the IL-1R Type I
IL-6	Many cell types including monocytes, macrophages, B cells, T cells, granulocytes, mast cells, chondrocytes, osteoblasts, vascular endothelial cells, fibroblasts, keratinocytes, astrocytes	26 kD glycoprotein, 185 amino acid residues	↑ acute phase protein synthesis; ↑ thymocyte and T cell activation; ↑ B cell growth, differentiation and Ig production. Acts through the IL-6Rα/gp 130 heterodimer
IFN-γ	T cell and NK cells	20–25 kD glycoprotein subunit, 146 amino acid residues, homodimer	Antiviral; ↑ Macrophage and NK cell function; ↑ MHC class I and II cell surface antigen expression. Binds and acts through the heterodimeric IFN-γRα,β
TNF-α	Many cell types including monocytes, macrophages, neutrophils, NK cells, B cells, T cells, astrocytes, microglial cells, fibroblasts	25 kD glycoprotein subunit, 171 amino acid residues. Expressed as cell surface homotrimer, shed in soluble form by enzymatic cleavage	↑ fever and septic shock; cytotoxic for many tumor cell types; activates macrophages and neutrophils; ↑ MHC class I and II cell surface antigen expression, acts through TNFRI and TNRFII
LT-α3	T cells, B cells, fibroblasts, astrocytes, endothelial cells	20–25 kD subunits, homotrimer	Homotrimeric form of LT-α acts through TNFRI and TNFII, cytotoxic for many tumor cell types, inhibits angiogenesis, activates macrophages and neutrophils

Table 10.3 Pro-inflammatory cytokine receptors

Cytokine receptor	Expression	Structure	Function
IL-1R Type I CD 121a	Low level expression on most cells	80 kD Type I transmembrane glycoprotein, 569 amino acid residues. A soluble form of the IL-1RI extracellular domain exists	Low affinity receptor for bioactive IL-1α, IL-1β and non-bioactive IL-1 receptor antagonist. Complexes with IL-1 Receptor Accessory Protein to form a high affinity receptor complex that mediates the cellular response to IL-1
IL-1R Type II CD121b	B-cells, some T-cells, myeloid cells, some epithelial tissues	60–68 kD Type 1 transmembrane glycoprotein, 398 amino acid residues. A soluble form of the IL-1RII extracellular domain exists	Binds IL-1α, IL-1β and IL-1 receptor antagonist. No known signaling function, IL-1 decoy receptor
IL-6Rα CD126	Monocytes, T cells, and activated B cells. Low on resting B cells. Little or no expression on NK cells, granulocytes, and erythrocytes. Hepatocytes	80 kD Type 1 transmembrane glycoprotein, 449 amino acid residues. A soluble form of the IL-6Ra extracellular domain exists	Low-affinity receptor for IL-6. Complex formed between IL-6 and IL-6Rα associates with gp130 (CD130) to form a signaling receptor for IL-6
GP130 CD130	T cells, B cells, NK cells, monocytes, granulocytes	130 kD Type I transmembrane glycoprotein, 896 amino acid residues	Associates with the complex of IL-6 and IL-6Rα (CD126) to form a signaling receptor for IL-6. The gp130 subunit does not bind IL-6 by itself. Common signaling subunit of receptors for IL-6, IL-11, Oncostatin M, LIF, CNTF, CT-1 receptors
IFN-γRα CD119	Moderate expression on B cells, T cells, NK cells, monocytes, granulocytes, platelets, epithelial cells, endothelium, many tumor cells. Not expressed by erythrocytes	52.6 kD Type I transmembrane glycoprotein, 489 amino acid residues. A soluble form of the IFN-γRα extracellular domain exists	Required for ligand binding and trafficking through the cell; it is necessary but not sufficient for signaling
IFN-γRβ	Low expression: B cells, T cells, NK cells, monocytes, granulocytes, platelets, epithelial cells, endothelium, many tumor cells. Not expressed by erythrocytes	35 kD type I transmembrane glycoprotein, 489 amino acid residues	Stabilization of the complex formed between the ligand and the IFN-γ receptor a subunit; required for signal transduction upon ligand ligation
TNFR Type I CD120a	Expressed by most nucleated cell types. Not expressed by erythrocytes	55 kD type I transmembrane glycoprotein, 435 amino acid residues. A soluble form of the human TNFRI extracellular domain exists	Receptor for both TNF-α and LT-α3 (TNF-β), An 80 AA residue 'death domain' triggers apototic pathway. Soluble TNFRI blocks TNF-α and LT-α$_3$ activities.
TNFR Type II CD120b	Expressed by most nucleated cell types. Upregulated expression by activated T and B cells. Not expressed by erythrocytes	75 kD Type I transmembrane glycoprotein, 435 amino acid residues. A soluble form if the human TNFRII extracellular domain exists	Receptor for both TNF-α and LT-α3 (TNF-β)

Table 10.4 Cytokines in wound healing

Cytokines	Platelets	Mast cells	PMN	Monocytes/ macrophages	Keratinocytes	Endothelium	Fibroblasts
TGF-βs	•	•	•	•	•	•	•
PDGF	•	•		•		•	•
EGF	•			•	•	•	•
TGF-α	•			•	•		•
VEGF	•	•	•	•	•	•	•
IGF-I	•			•			•
FGFs	•	•		•	•	•	•
Angiopoietin	•			•	•	•	•
FGF-7/KGF					•		•
Endothelin		•		•	•	•	
TNF-α		•	•	•	•	•	•
IL-1β	•	•	•	•	•	•	•
IL-6		•		•	•	•	•
IL-4		•	•				
IL-8		•	•	•	•	•	•
IL-10		•		•			
SLPI		•*	•r	•r	•		
MCP-1		•*	•	•			•
MIP-1α		•	•	•			•
MIP-2		•	•	•			•
IL-18		•	•	•	•		
IFN-α/β				•			•

r = rodent;   * = trace.

Table 10.5 Chemokines and their receptors

	Sources	Receptors	Proposed* Nomenclature
CC-chemokines			
I-309/TCA-3	T cells, MC	CCR8	CCL1
MCP-1	M, L, F, EC, EP	CCR2; CCR4	CCL2
MIP-1a	M, L, N, E, F, MC	CCR1; CCR4 − 5	CCL3
MIP-1b	M, L, N, F, MC	CCR5; CCR8	CCL4
RANTES	T cells, M, F, ME	CCR1; CCR3 − 5	CCL5
C10/MPR-1 (murine)	M	?	CCL6
MCP-3	Platelets, M, MC, F	CCR1 − 3	CCL7
MCP-2	PBMC, F	CCR2; CCR3	CCL8
MIP-1γ/MPR-2/CCF-18 (murine)	M, DC	?	CCL9
Eotaxin-1	EC, EP, E, lung	CCR3	CCL11
MCP-5 (murine)	M, LN, lung	CCR2	CCL12
MCP-4	Lung, colon, intestine	CCR2; CCR3	CCL13
HCC-1	BM, spleen, liver	CCR1	CCL14
HCC-2/MIP-1δ/LKN-1	Intestine, liver, lung	CCR1	CCL15
HCC-4/LEC	M	?	CCL16
TARC	Thymus	CCR4; CCR8	CCL17
PARC/DC-CK1	Lung, LN, thymus	?	CCL18
MIP-3β/ELC/Exodus-3	Thymus, LN	CCR7	CCL19
MIP-3α/LARC/Exodus-1	Liver, lung	CCR6	CCL20
SLC/TCA-4/6CKine/Exodus-2	LN, small intestine, spleen	CCR7	CCL21
MDC/STCP-1/ABCD-1	DC, M, T cells	CCR4; CCR8	CCL22
MPIF-1	DC	CCR1	CCL23
Eotaxin-2MPIF-2	M, T cells	CCR3	CCL24
TECK	DC, thymus, small intestine	CCR9	CCL25
Eotaxin-3	EC	CCR3	CCL26
CXC-chemokines			
GROα/MGSA	M, EC, tumor cells	CXCR2	CXCL1
GROβ/MIP-2α	MC, CM, ME	CXCR2	CXCL2
GROγ/MIP-2β	MC, CM, ME	CXCR2	CXCL3

Table 10.5 Chemokines and their receptors (*continued*)

	Sources	Receptors	Proposed* Nomenclature
CC-chemokines			
PF4	Platelets	?	CXCL4
ENA-78	EC, platelets	CXCR2	CXCL5
GCP-2	Osteosarcoma cells	CXCR1; CXCR2	CXCL6
NAP-2	Platelets	CXCR2	CXCL7
IL-8	M, T, EP, EC	CXCR1; CXCR2	CXCL8
MIG	M, N	CXCR3	CXCL9
IP-10/CRG-2	M, N, F, EC	CXCR3	CXCL10
I-TAC	Astrocytes, M, N	CXCR3	CXCL11
SDF-1	Stromal cells	CXCR4	CXCL12
BCA-1	Liver, spleen, LN	CXCR5	CXCL13
C-chemokine			
Lymphotactin/SCM-1	Activated T cells	XCR1	XCL1
CX3C-chemokine			
Fractalkine/neurotactin	EC, DC, T cells, brain	CX$_3$CR1	CX3CL1

*The nomenclature was proposed by Drs O Yoshie and A Zlotnik at the 1999 Keystone Symposium.

Abbreviations:
CM   cardiac myocytes
DC   dendritic cells
E   eosinophils
EC   endothelial cells
EP   epithelial cells
F   fibroblasts
L   lymphocytes
LN   lymph node
M   monocytes/macrophages
MC   mast cells
ME   mesangial cells
N   neutrophils

Table 10.6 Human chemokine mini-guide

	Full name	% homology (aa)	Family	Systematic name	Receptors	Chromosome	Amino acids (*chemokine domain)	Predicted MW (kDa) (*chemokine domain)
BCA-1/BLC	B cell activating chemokine-1/B cell chemoattractant	hBCA-1 to mBLC = 49% mouse equiv BLC	CXC chemokine (α chemokine)	CXCL13	CXCR5	4q21.1	87	10.3
BRAK Bolekine	Breast and kidney derived	hBRAK to mBRAK = 96%	CXC chemokine (α chemokine)	CXCL14	Unknown	5q31.1	77	8.8
CXCL16 SRPSOX		hCXCL16 to mCXCL16 = 70% (chemokine domain)	CXC chemokine (α chemokine)	CXCL16	CXCR6	17p13	90*	10*
ENA-78 SCYB5	Epithelial-cell derived neutrophil activating protein 78	hENA-78 to bENA-78 = 73% hENA -78 to hGCP-2 = 30% hENA-78 to hNAP-2 = 53% hENA-78 to hGROα = 52%	CXC chemokine ELR+ (α chemokine)	CXCL5	CXCR2	4q21.1	78	8
GCP-2	Granulocyte chemoattractant protein 2	hGCP to mGCP = 60% hGCP to hENA-78 = 79%	CXC chemokine ELR+ (α chemokine)	CXCL6	CXCR1, CXCR2	4q21.1	75	8
GROα MGSA-α, GRO-1, NAP-3, SCYB1	Growth-related oncogene alpha	hGROα to hENA-78 = 52% hGROα to mKC = 55% hGROα to rCINC = 55%	CXC chemokine ELR+ (α chemokine)	CXCL1	CXCR1, CXCR2	4q21.1	73	7.9
GROβ MGSA-β, MIP-2α, GRO-2, SCYB2	Growth-related oncogene beta	GROβ to GROγ = 85% hGROβ to mMIP-2 = 58%	CXC chemokine ELR+ (α chemokine)	CXCL2	CXCR2	4q21.1	73	8

Table 10.6 Human chemokine mini-guide (*continued*)

	Full name	% homology (aa)	Family	Systematic name	Receptors	Chromosome	Amino acids (*chemokine domain)	Predicted MW (kDa) (*chemokine domain)
GROγ MGSA-γ, MIP-2β, GRO-3, SCYB3	Growth-related oncogene gamma	hGROγ to hGROβ = 90%	CXC chemokine ELR+ (α chemokine)	CXCL3	CXCR2	4q21.2	73	8
IL-8 NCF, NAP-1, MDNCF, SCYB8	Interleukin-8	hIL-8 to pIL-8 = 76%	CXC chemokine ELR+ (α chemokine)	CXCL8	CXCR1, CXCR2	4q21.1	72, 77	8, 8.9
IP-10 SCYB10	Interferon-inducible protein-10 kD	hIP-10 to mCRG-2 = 67% hIP-10 to hMIG = 36% Mouse equiv CRG-2	CXC chemokine (α chemokine)	CXCL10	CXCR3	4q21.1	78	8.7
I-TAC b-R1, H174, SCYB9B	Interferon-inducible T cell alpha chemoattractant	hI-TAC to hMIG = 37% hI-TAC to hIP-10 = 33%	CXC chemokine (α chemokine)	CXCL11	CXCR3	4q21.1	73	8.3
MIG SCYB9	Monokine induced by interferon gamma	hMIG to mMIG = 69%	CXC chemokine (α chemokine)	CXCL9	CXCR3	4q21.1	103	11.7
NAP-2	Neutrophil activating peptide 2	hNAP-2 to hENA-78 = 36%	CXC chemokine ELR+ (α chemokine)	CXCL7	CXCR2	4q21.1	70	7.6
PF4	Platelet factor 4	hP4 to mP4 = 64%	CXC chemokine (α chemokine)	CXCL4	Unknown	4q21.1	70	7.8

Table 10.6 Human chemokine mini-guide (*continued*)

	Full name	% homology (aa)	Family	Systematic name	Receptors	Chromosome	Amino acids (*chemokine domain)	Predicted MW (kDa) (*chemokine domain)
SDF-1/PBSF SDF-1α (short form), SDF-1β (long form), SCYB12, hIRH	Stromal cell-derived factor 1/Pre B-cell stimulating factor	hSDF-1α to mSDF-1α = 89%	CXC chemokine (α chemokine)	CXCL12	CXCR4	10q11.21	68	7.8
MDC STCP-1	Macrophage-derived chemokine	MDC to other β chemokines = 35%	CC chemokine (β chemokine)	CCL22	CCR4	16q13	69	8
TARC	Thymus and activation-regulated chemokine	hTARC to hRANTES = 31% hTARC to hMIP-1α = 27% hTARC to hMIP-1β = 26%	CC chemokine (β chemokine)	CCL17	CCR4	16q13	71	8
TECK SCYA25	Thymus-expressed chemokine	hTECK to mTECK = 45%	CC chemokine (β chemokine)	CCL25	CCR9	19p13.3	127	14.3
Lymphotactin SCM-1α, ATAC		hLymphotactin to mLymphotactin = 60%	C chemokine (γ chemokine)	XCL1	XCR1	1q24.2	114	10
SCM-1β	Single C motif-1β		C chemokine (γ chemokine)	XCL2	XCR1	1q24.2	114 (precursor)	11 (precursor)
Fractalkine Neurotactin (NTN)		hFractalkine to mNTN = 64% hFractalkine to rFractalkine = 65% Mouse equiv. NTN	CX₃C chemokine (δ chemokine)	CX3CL1	CX3CR1	16q13	76, 324	8.5, 90
CTACK ILC, Eskine, Skinkine	Cutaneous T cell-attracting chemokine	hCTACK to mCTACK = 84%	CC chemokine (β chemokine)	CCL27	CCR10	9p13.3	88	10.1

Table 10.6 Human chemokine mini-guide (*continued*)

	Full name	% homology (aa)	Family	Systematic name	Receptors	Chromosome	Amino acids (*chemokine domain)	Predicted MW (kDa) (*chemokine domain)
Eotaxin (Eot)		hEot to mEot = 57%, hEot to hMCP-4 = 75%, hEot to hMCP-3 = 70%	CC chemokine (β chemokine)	CCL11	CCR3	17q11.2	74	8.4
Eotaxin-2 MPIF-2, Ckβ-6, SCYA24		hEot-2 to mEot-2 = 45%, hEot-2 to hMCP-3 = 40%, hEot-2 to hEot = 29%	CC chemokine (β chemokine)	CCL24	CCR3, CCR5	7q11.23	93	10.6
Eotaxin-3 SCYA26, TSC-1, MIP-4α			CC chemokine (β chemokine)	CCL26	CCR3	7q11.23	68, 71	8.2
HCC-1 MCIF, Ckβ-1, NCC-2	Hemofiltrate CC chemokine 1	hHCC-1 to hMIP-1α = 47%, HCC-1 to other β chemokines = 29-37%	CC chemokine (β chemokine)	CCL14	CCR1, CCR5	17q12	74	8.7
HCC-4 NCC-4, ILINCK, LEC, LMC, SCYA16, LCC-1	Hemofiltrate CC chemokine 4	hHCC-4 to other β chemokines <30%	CC chemokine (β chemokine)	CCL16	CCR1, CCR2	17q12	97	11
I-309		hI-309 to mTCA-3 = 41%, Mouse equiv. TCA-3	CC chemokine (β chemokine)	CCL1	CCR8	17q11.2	73	8.5
LD78β PAT 464.2		hLD78β to hMIP-1α = 94%	CC chemokine (β chemokine)	CCL3L1	CCR1, CCR5	17q12	93 (precursor)	7.8 (precusor)
MCP-1 MCAF, LDCF, GDCF, TDCF, SMC-CF, HC11, TSG8, SCYA-2	Monocyte chemoattractant protein 1	hMCP-1 to hMCP-2 = 69%, hMCP-1 to hMCP-3 = 74%, Mouse equiv JE	CC chemokine (β chemokine)	CCL2	CCR2	17q11.2	76	8.7

Table 10.6 Human chemokine mini-guide (*continued*)

	Full name	% homology (aa)	Family	Systematic name	Receptors	Chromosome	Amino acids (*chemokine domain)	Predicted MW (kDa) (*chemokine domain)
MCP-2 HC-14	Monocyte chemoattractant protein 2	hMCP-2 to hMCP-1 = 62% hMCP-3 to hMCP-2 = 58%	CC chemokine (β chemokine)	CCL8	CCR3, CCR5	17q11.2	76	9
MCP-3 SCYA7	Monocyte chemoattractant protein 3	hMCP-3 to hMCP-1 = 73% hMCP-3 to mMARC = 55% Mouse equiv. MARC	CC chemokine (β chemokine)	CCL7	CCR1, CCR2, CCR3	17q11.2	76	9
MCP-4 Ckβ-10, NCC-1, SCYA13	Monocyte chemoattractant protein 4	hMCP-4 to hMCP-1, hMCP-3 and hEot = 65-66%	CC chemokine (β chemokine)	CCL13	CCR2, CCR3	17q11.2	75	8.6
MEC	Mucosae-associated epithelial chemokine	hMEC to mMEC = 63%	CC chemokine (β chemokine)	CCL28	CCR3, CCR10	5p12	105	12.4
MIP-1α GOS19, LD78α, pAT464	Macrophage inflammatory protein 1 alpha	hMIP-1α to mMIP-1α = 75%	CC chemokine (β chemokine)	CCL3	CCR1, CCR5	17q12	66	7.5
MIP-1β pAT744 gene product, ACT-2, G-26, HC21, hH400, MAD-5, LAG-1	Macrophage inflammatory protein 1 beta	hMIP-1β to mMIP-1β = 78%	CC chemokine (β chemokine)	CCL4	CCR5	17q12	69	7.8
MIP-1δ Leukotactin-1 (LKN-1), MIP-5, HCC-2/NCC-3	Macrophage inflammatory protein 1 delta	hMIP-1δ to hMPIF-1 = 73% hMIP-1δ to mMIP-1γ = 42% hMIP-1δ to C10 = 45% hMIP-1δ to hHCC-1 = 30%	CC chemokine (β chemokine)	CCL15	CCR1, CCR3	17q12	92	10
MIP-3α LARC, Exodus-1, Mexikine	Macrophage inflammatory protein 3 alpha	hMIP-3α to mMIP-3α = 61% hMIP-3α to other β chemokines = 20-28%	CC chemokine (β chemokine)	CCL20	CCR6	2q36.3	70	8

Table 10.6 Human chemokine mini-guide (*continued*)

	Full name	% homology (aa)	Family	Systematic name	Receptors	Chromosome	Amino acids (*chemokine domain)	Predicted MW (kDa) (*chemokine domain)
MIP-3β ELC (EBI1-ligand chemokine), SCYA19, Exodus-3	Macrophage inflammatory protein 3 beta	hMIP-3β to other β chemokines = 20-30%	CC chemokine (β chemokine)	CCL19	CCR7	9p13.3	77	8.8
MPIF-1 CKβ-8, MIP-3, SCYA23	Myeloid progenitor inhibitory factor 1	hMPIF-1 to hMIP-1δ = 68% hMPIF-1 to hMIP-1α = 51%	CC chemokine (β chemokine)	CCL23	CCR1	17q12	99	11.5
PARC DC-CK1, MIP-4, AMAC-1, Dctactin	Pulmonary and activation-regulated chemokine	hPARC to hMIP-1α = 63%	CC chemokine (β chemokine)	CCL18	Unknown	17q12	69	7.8
RANTES	Regulated on activation of normal T cell expressed and secreted	hRANTES to mRANTES = 81%	CC chemokine (β chemokine)	CCL5	CCR1, CCR3, CCR5	17q12	68	7.8
6Ckine Secondary lymphoid tissue chemokine (SLC), Exodus-2, TCA-4, SCYA2	Six-cysteine chemokine	h6Ckine to m6Ckine = 70%	CC chemokine (β chemokine)	CCL21	CCR7	9p13.3	111	12

Table 10.7   Human chemokine receptor mini-guide

	Alternative names	Ligands	Chromosome	Predicted mol. wt (kDa)	Amino acid (aa)
CCR1	CMKBR1, CC CKRI, HM145, LD78 receptor, MIP-1α R, RANTES R	MIP-1α, RANTES, MCP-3, MCP-1, HCC-1, HCC-4, MIP-1δ, MPIF-1	3p21	41	355
CCR2A/ CCR2B	CMKBR2, CC CKR2, MCP-1R	MCP-1, MCP-2, MCP-3, MCP-4, MCP-5	3p21	42/41	374/360
CCR3	CMKBR3, CC CKR3, Eotaxin (Eot) R	Eot, Eot-2, MCP-4, Eot-3, RANTES, MCP-3, MCP-2, MIP-1δ	3p21.3	41	355
CCR4	CMKBR4, CC CKR4, K5-5	MDC, TARC	3p21	41	360
CCR5	CMKBR5, CC CKR5, ChemR13	MIP-1α, MIP-1β, RANTES	3p21	40	352
CCR6	CMKBR6, CC CKR6, GPR29, CKR-L3, CRPCY4, STRL22, DRY6, LARC receptor	MIP-3α	6p27	43	374
CCR7	CMKBR7, CC CKR7, EBI-1, BLR2	6Ckine, MIP-3β	17q12-q21.2	43	378
CCR8	CKMBR8, CC CKR8, TER1, CY6, ChemR1, CKR-L1	I-309, vMIP-I	3p22-p23	41	355
CCR9A/ CCR9B	CC CKR9, GPR-9-6	TECK	3pter-qter	42/41	369/357
CCR10		CTACK	17q21	39	362
CCR11			3q22	40	350
CX3CR1	CMKBRL1, V28	Fractalkine, vMIP-II	3p21	40	355
CXCR1	IL-8RA, IL-8 R1, IL-8Rα	IL-8, GCP-2 NAP-2	2p33-q36	40	350
CXCR2	IL-8RB, IL-8 R2, IL-8Rβ	IL-8, CGP-2, ENA-78, NAP-2, GROα, GROβ, GROγ	2q33-q35	41	360
CXCR3	IP-10/MIG R, GPR9	IP-10, MIG, I-TAC	8p12-p11.2	41	368
CXCR4	LESTR, HUMSTR, Fusin, HM89	SDF-1α/SDF-1β	2q21	40	352
CXCR5	Burkitt lymphoma receptor 1 (BLR-1)	BCA-1	11	42	372
XCR1	GPR5, SCM-1R, lymphotactin R	SCM-1β, SCM-1α, vMIP-II	3p21.3- p21.1	39	333

Table 10.8 Cells in chemokine research: ligands

Systemic name	Chemokine	Species	Type of expression	Cell line	Inducer/enhancer	Reference
CCL1	I-309	Human	mRNA	HMC-1	PMA	Selvan, RS *et al.* (1994) *J Biol Chem* 269: 13893
	I-309	Human	mRNA	IDP2	IL-2	Miller, MD *et al.* (1989) *J Immunol* 143: 2907
	I-309	Human	Protein	Mo	PMA, IL-2	Van Snick, J *et al.* (1996) *J Immunol* 157: 2570
	TCA-3	Mouse	mRNA	MC-9	None	Burd, PR *et al.* (1989) *J Exp Med* 170: 245
CCL2	MCP-1	Human	mRNA	Caco-2	None/IL-1, TNF-α	Warhurst, AC *et al.* (1998) *Gut* 42: 208
	MCP-1	Human	mRNA	HMC-1	PMA	Selvan, RS *et al.* (1994) *J Biol Chem* 269: 13893
	MCP-1	Human	mRNA	HT29-19A	TNF-α and IFN-γ	Warhurst, AC *et al.* (1998) *Gut* 42: 208
	MCP-1	Human	Protein	HEp-2	IL-1β	Van Damme, J *et al.* (1994) *J Immunol* 152: 5495
	MCP-1	Human	Protein	MG-63	IL-1β	Van Damme, J *et al.* (1994) *J Immunol* 152: 5495
	MCP-1	Human	Protein	U373MG	β-amyloid protein	Prat, E *et al.* (2000) *Neurosci Lett* 283: 177
	JE	Mouse	mRNA	BALB/c-3T3	None/TNF-α	Ohmori, Y *et al.* (1993) *Am J Pathol* 142: 861
	JE	Mouse	mRNA	NIH-3T3	LPS, TNF-α	Ohmori, Y *et al.* (1994) *J Immunol* 153: 2204
	JE	Mouse	mRNA	Swiss 3T3	LPS and TGF-β1	Smith, JB *et al.* (1995) *J Biol Chem* 270: 16756
	JE	Mouse	Protein	MO	LPA	Wuyts, A *et al.* (1996) *J Immunol* 157: 1736
	JE	Mouse	Protein	MTEC1	PMA	Wuyts, A *et al.* (1996) *J Immunol* 157: 1736
CCL3	MIP-1α	Human	mRNA	HMC-1	PMA	Selvan, RS *et al.* (1994) *J Biol Chem* 269: 13893
	MIP-1α	Human	mRNA	HUH7	None	Rowell, DL *et al.* (1997) *Am J Physiol* 273: G322
	MIP-1α	Human	mRNA	U937	IL-1α, TNF-α	Yoshida, T *et al.* (1995) *FEBS Letters* 360: 155
	MIP-1α	Mouse	mRNA	CTLL-R8	None	Youn, BS *et al.* (1995) *J Immunol* 155: 2661
	MIP-1α	Mouse	mRNA	DETC7-17	Con A	Boismenu, R *et al.* (1996) *J Immunol* 157: 985
	MIP-1α	Mouse	mRNA	p388D1	None	Youn, BS *et al.* (1995) *J Immunol* 155: 2661
	MIP-1α	Mouse	mRNA	RAW264.7	None	Youn, BS *et al.* (1995) *J Immunol* 155: 2661
	MIP-1α	Mouse	mRNA	WEH13	None	Youn, BS *et al.* (1995) *J Immunol* 155: 2661
CCL4	MIP-1β	Human	mRNA	HMC-1	PMA	Selvan, RS *et al.* (1994) *J Biol Chem* 269: 13893
	MIP-1β	Human	mRNA	HUH7	IL-1α	Rowell, DL *et al.* (1997) *Am J Physiol* 273: G322
	MIP-1β	Mouse	mRNA	DETC7-17	Con A	Boismenu, R *et al.* (1996) *J Immunol* 157: 895
CCL5	RANTES	Human	mRNA	BEAS-2B	TNF-α	Stellato, C (1999) *J Immunol* 163: 5624
	RANTES	Human	mRNA	HEP G2	None	Rowell, DL *et al.* (1997) *Am J Physiol* 273: G322
	RANTES	Human	mRNA	HH25	None	Rowell, DL *et al.* (1997) *Am J Physiol* 273: G322
	RANTES	Human	mRNA	HMC-1	PMA	Selvan, RS *et al.* (1994) *J Biol Chem* 269: 13893
	RANTES	Human	mRNA	HT29-19A	TNF-α and IFN-γ	Warhurst, AC *et al.* (1998) *Gut* 42: 208
	RANTES	Human	mRNA	HUH 7	None	Rowell, DL *et al.* (1997) *Am J Physiol* 273: G322
	RANTES	Mouse	mRNA	DETC7-17	Con A	Boismenu, R *et al.* (1996) *J Immunol* 157: 985

Table 10.8 Cells in chemokine research: ligands (continued)

Systemic name	Chemokine	Species	Type of expression	Cell line	Inducer/enhancer	Reference
CCL6	MRP-1	Mouse	mRNA	WEHI 3	None	Youn, BS et al. (1995) J Immunol 155: 2661
	C10	Mouse	Protein	32D cl3	G-CSF	Orlofsky, A et al. (1991) Cell Regul 2: 403
	C10	Mouse	Protein	DA3	G-CSF	Orlofsky, A et al. (1991) Cell Regul 2: 403
	C10	Mouse	Protein	P388D	GM-CSF	Orlofsky, A et al. (1991) Cell Regul 2: 403
CCL7	MCP-3	Human	mRNA	U937	PMA	Minty, A (1993) Eur Cytokine Net 4: 99
	MCP-3	Human	Protein	MG-63	IFN-γ	Menten, P et al. (1999) Eur J Immunol 29: 678
	MARC	Mouse	mRNA	Swiss 3T3	LPS and TGF-β1	Smith, JB et al. (1995) J Biol Chem 270: 16756
	MARC	Mouse	mRNA	WEHI-3	LPS	Thirion, S (1994) Biochem Biophys Res Commun 201: 493
CCL8	MCP-2	Human	Protein	HEp-2	IFN-β	Van Damme, J et al. (1994) J Immunol 152: 5495
	MCP-2	Human	Protein	MG-63	Measles virus	Van Damme, J et al. (1994) J Immunol 152: 5495
CCL9/10	CCF18	Mouse	mRNA	32D	None	Hara, T et al. (1995) J Immunol 155: 5352
	CCF18	Mouse	mRNA	MC-9	None	Hara, T et al. (1995) J Immunol 155: 5352
	CCF18	Mouse	mRNA	NIH-3T3	None	Hara, T et al. (1995) J Immunol 155: 5352
	MRP-2	Mouse	mRNA	p388D1	None	Youn, BS et al. (1995) J Immunol 155: 2661
	MRP-2	Mouse	mRNA	RAW 264.7	None	Youn, BS et al. (1995) J Immunol 155: 2661
	MRP-2	Mouse	mRNA	WEHI 3	None	Youn, BS et al. (1995) J Immunol 155: 2661
	CCF18	Mouse	mRNA	Y16	None	Hara, T et al. (1995) J Immunol 155: 5352
CCL11	Eotaxin	Human	mRNA	AS49	TNF-α	Miyamasu, M et al. (1999) Cytokine 11: 751
	Eotaxin	Human	mRNA	BEAS-28	TNF-α	Stellato, C (1999) J Immunol 163: 5624
	Eotaxin	Human	mRNA	U937	None	Garcia-Zepeda, E et al. (1996) Nature Med 2: 449
	Eotaxin	Mouse	mRNA	End-2	None	Gonzalo, JA et al. (1996) Immunity 4: 1
	Eotaxin	Mouse	mRNA	NIH-3T3	None	Gonzalo, JA et al. (1996) Immunity 4: 1
CCL12	MCP-5	Mouse	mRNA	RAW 264.7	IFN-γ and LPS	Saraf, MN (1997) J Exp Med 185: 99
CCL13	MCP-4	Human	mRNA	A549	TNF-α or IL-1β	Garcia-Zepeda, E et al. (1996) J Immunol 157: 5613
	MCP-4	Human	mRNA	BEAS-28	TNF-α	Garcia-Zepeda, E et al. (1996) J Immunol 157: 5613
	MCP-4	Human	mRNA	IB3-1	TNF-α	Stellato, C (1997) J Clin Invest 99: 926
	MCP-4	Human	mRNA	U937	None	Garcia-Zepeda, E et al. (1996) J Immunol 157: 5613
CCL14	HCC-1	Human	mRNA	HUH7	None	Schulz-Knappe, P et al. (1996) J Exp Med 183: 295
CCL15	Leukotactin-1	Human	mRNA	THP-1	IL-4	Youn, BS et al. (1995) J Immunol 155: 2661
CCL16	HCC-4	Human	mRNA	HEP G2	Induced	Yang, J-Y et al. (2000) Cytokine 12: 101

Table 10.8 Cells in chemokine research: ligands (*continued*)

Systemic name	Chemokine	Species	Type of expression	Cell line	Inducer/enhancer	Reference
CCL17	TARC	Human	mRNA	A549	None	Sekiya, T *et al.* (2000) *J Immunol* 165: 2205
	TARC	Human	mRNA	BEAS-28	None	Sekiya, T *et al.* (2000) *J Immunol* 165: 2205
	TARC	Mouse	mRNA	B220$^+$493$^-$	Anti-CD40, IL-4	Schaniel, C *et al.* (1999) *Eur J Immunol* 29: 2934
	TARC	Mouse	mRNA	bcl-2-5-8	Anti-CD40, IL-4	Schaniel, C *et al.* (1999) *Eur J Immunol* 29: 2934
	TARC	Mouse	mRNA	R2BFL	Anti-CD40, IL-4	Schaniel, C *et al.* (1999) *Eur J Immunol* 29: 2934
	TARC	Mouse	Protein	PAM212	TNF-α, IFN-γ, IL-1β	Vestergaard, C *et al.* (1999) *J Clin Invest* 104: 1097
CCL18	PARC	Human	mRNA	Bowes	PMA	Hieshima, K *et al.* (1997) *J Immunol* 159: 1140
	AMAC-1	Human	mRNA	THP-1	Phorbol ester and IL-4	Kodelja, V *et al.* (1998) *J Immunol* 160: 1411
CCL19	MIP-3β	Human	mRNA	Blood-derived IFN-γ activated dendritic cells		Hashimoto, S *et al.* (2000) *Blood* 96: 2206
	MIP-3β	Human	mRNA	Peripheral blood-derived dendritic cells		Vissers, J *et al.* (2001) *J Leukoc Biol* 69: 785
	MIP-3β	Human	mRNA	Primary BM stroma		Kim, CH *et al.* (1998) *J Immunol* 161: 2580
	MIP-3β	Mouse	mRNA	CD8* spleen dendritic cells		Luther, SA *et al.* (2000) *Proc Natl Acad Sci USA* 97: 12694
	MIP-3β	Mouse	mRNA	CD8* spleen dendritic cells		Luther, SA *et al.* (2000) *Proc Natl Acad Sci USA* 97: 12694
CCL20	MIP-3α	Human	mRNA	A549	None	Rossi, DL *et al.* (1997) *J Immunol* 158: 1033
	LARC	Human	mRNA	Bowes	PMA	Hieshima, K *et al.* (1997) *J Biol Chem* 272: 5846
	MIP-3α	Human	mRNA	G361	None	Rossi, DL *et al.* (1997) *J Immunol* 158: 1033
	MIP-3α	Human	mRNA	HL-60	None	Rossi, DL *et al.* (1997) *J Immunol* 158: 1033
	MIP-3α	Human	mRNA	K562	None	Rossi, DL *et al.* (1997) *J Immunol* 158: 1033
	MIP-3α	Human	mRNA	MOLT-4	None	Rossi, DL *et al.* (1997) *J Immunol* 158: 1033
	MIP-3α	Human	mRNA	Raji	None	Rossi, DL *et al.* (1997) *J Immunol* 158: 1033
	MIP-3α	Human	mRNA	SW480	None	Rossi, DL *et al.* (1997) *J Immunol* 158: 1033
	MIP-3α	Human	mRNA	THP-1	None	Rossi, DL *et al.* (1997) *J Immunol* 158: 1033
	Exodus-1	Human	mRNA	U937	PMA	Hromas, R *et al.* (1997) *Blood* 89: 3315
	LARC	Human	mRNA	THP-1	PMA	Hieshima, K *et al.* (1997) *J Biol Chem* 272: 5846
	LARC	Mouse	mRNA	J774	LPS	Yoshikazu, T *et al.* (1999) *Eur J Immunol* 29: 633
CCL21	Exodus-2	Human	mRNA	HEL	None	Hromas, R *et al.* (1997) *J Immunol* 159: 2554
CCL22	MDC	Human	mRNA	HL60	PMA	Godiska, R *et al.* (1997) *J Exp Med* 185: 1595
	ABCD-1 (MDC)	Mouse	mRNA	A20	Anti-CD40, IL-4	Schaniel, C *et al.* (1998) *J Exp Med* 188: 451
	ABCD-1 (MDC)	Mouse	mRNA	B220$^+$	Anti-CD40, IL-4	Schaniel, C *et al.* (1998) *J Exp Med* 188: 451
	ABCD-1 (MDC)	Mouse	mRNA	bcl-2-5-8	Anti-CD40, IL-4	Schaniel, C *et al.* (1998) *Eur J Immunol* 29: 2934
	ABCD-1 (MDC)	Mouse	mRNA	J558	None	Schaniel, C *et al.* (1998) *J Exp Med* 188: 451
	ABCD-1 (MDC)	Mouse	mRNA	R2BFL	Anti-CD40, IL-4	Schaniel, C *et al.* (1998) *Eur J Immunol* 29: 2934
	ABCD-1 (MDC)	Mouse	mRNA	WEHI-164	GM-CSF, IL-4, IFN-γ	Ross, R *et al.* (1999) *J Invest Dermatol* 113: 991
CCL23	MPIF-1	Human	mRNA	HL-60	PMA	Patel, VP *et al.* (1997) *J Exp Med* 185: 1163
	MPIF-1	Human	mRNA	THP-1	None	Patel, VP *et al.* (1997) *J Exp Med* 185: 1163

Table 10.8 Cells in chemokine research: ligands (*continued*)

Systemic name	Chemokine	Species	Type of expression	Cell line	Inducer/enhancer	Reference
CCL24	Eotaxin-2	Human	mRNA	Activated monocytes		Patel, VP et al. (1997) J Exp Med 185: 1163
CCL25	TECK	Human	mRNA	CD11b⁻ thymic dendritic cells		Vandenebeele, S et al. (2001) Blood 97: 1733
	TECK	Human	mRNA	CD11b⁺ thymic dendritic cells		Vandenebeele, S et al. (2001) Blood 97: 1733
	TECK	Human	mRNA	CD11c⁻ thymic epithelial cells		Wurbel, M-A et al. (2000) Eur J Immunol 30: 262
	TECK	Mouse	mRNA	CD11c⁺ thymic dendritic cells		Vicari, AP et al. (1997) Immunity 7: 291
CCL26	Eotaxin-3	Human	mRNA	HUVEC (Kurabo, Osaka Japan)		Shinkai, A et al. (1999) J Immunol 163: 1602
CCL27	CTACK	Human	mRNA	Keratinocytes (Clonetics)		Morales, J et al. (1999) Proc Natl Acad Sci USA 96: 14470
	CTACK	Mouse	mRNA	Keratinocytes (from mouse ear)		Morales, J et al. (1999) Proc Natl Acad Sci USA 96: 14470
CCL28		Human	mRNA	DLD-1	None	Wang, W et al. (2000) J Biol Chem 275: 22313
		Human	mRNA	HCT 116	None	Wang, W et al. (2000) J Biol Chem 275: 22313
		Mouse	mRNA	J774	None	Wang, W et al. (2000) J Biol Chem 275: 22313
CX3CL1	Fractalkine	Human	mRNA	T-84	IL-1β	Muehlhoefer, A et al. (2000) J Immunol 164: 3368
	Neurotactin	Mouse	mRNA	BMS-12	TPA, LPS	Pan, Y et al. (1997) Nature 387: 611
	Neurotactin	Mouse	mRNA	EOMA	None/TPA, LPS	Pan, Y et al. (1997) Nature 387: 611
	Neurotactin	Mouse	mRNA	STO	None/TPA, LPS	Pan, Y et al. (1997) Nature 387: 611
CXCL1	GROα	Human	mRNA	3229	IL-1β	Walz, A et al. (1997) J Leukoc Biol 62: 604
	GROα	Human	mRNA	Caco-2	None/IL-1β	Warhurst, AC et al. (1998) Gut 42: 208
	GROα	Human	mRNA	HEP G2	None	Rowell, DL et al. (1997) Am J Physiol 273: G322
	GROα	Human	mRNA	HH25	None	Rowell, DL et al. (1997) Am J Physiol 273: G322
	GROα	Human	mRNA	HT29-19A	None/TNF-α, IL-1β,	Warhurst, AC et al. (1998) Gut 42: 208
	GROα	Human	mRNA	HUH 7	LPS, LPA	Rowell, DL et al. (1997) Am J Physiol 273: G322
	GROα	Human	mRNA	VA13	None	Walz, A et al. (1997) J Leukoc Biol 62: 604
	GROα	Human	mRNA	W138	IL-1β	Walz, A et al. (1997) J Leukoc Biol 62: 604
	GROα	Human	Protein	A549	IL-1β	Walz, A et al. (1997) J Exp Med 174: 1355
	GROα	Human	Protein	MG-63	IL-1β	Proost, P et al. (1993) J Immunol 150: 1000
	KC	Mouse	mRNA	BALB/c-3T3	IL-1	Hall, DJ et al. (1989) J Cell Physiol 141: 154
	KC	Mouse	mRNA	BALB/c-3T3	PDGF, IL-1	Oquendo, P et al. (1989) J Biol Chem 264: 4133
	KC	Mouse	mRNA	NIH-3T3	PDGF	Ohmori, Y et al. (1994) J Immunol 153: 2204
	KC	Mouse	mRNA and protein	Mode-K	LPS, TNF-α, None/TNF-α and IFN-γ	Song, F et al. (1999) J Immunol 162: 2275
	KC	Mouse	Protein	L929	Virus	Wuyts, A et al. (1996) J Immunol 157: 1736

Table 10.8 Cells in chemokine research: ligands (*continued*)

Systemic name	Chemokine	Species	Type of expression	Cell line	Inducer/enhancer	Reference
CXCL2	GROβ	Human	mRNA	HEP G2	None	Rowell, DL et al. (1997) *Am J Physiol* 273: G322
	GROβ	Human	mRNA	HH25	None	Rowell, DL et al. (1997) *Am J Physiol* 273: G322
	GROβ	Human	mRNA	HUH7	None	Rowell, DL et al. (1997) *Am J Physiol* 273: G322
	MIP-2α	Human	mRNA	U937	LPS	Tekamp-Olson, P et al. (1990) *J Exp Med* 172: 911
CXCL3	GROγ	Human	mRNA	HUH 7	None	Rowell, DL et al. (1997) *Am J Physiol* 273: G322
	MIP-2β	Human	mRNA	U937	LPS	Tekamp-Olson, P et al. (1990) *J Exp Med* 172: 911
	GROγ	Human	Protein	U549	IL-1β	Walz, A et al. (1997) *J Exp Med* 174: 1355
	GROγ	Human	Protein	MG-63	IL-1	Proost, P et al. (1993) *J Immunol* 150: 1000
CXCL4	PF4	Mouse	mRNA	RAW 264.7	IFN-γ	Farber, JM (1990). *Proc Natl Acad Sci USA* 87: 5238
CXCL5	ENA-78	Human	mRNA	3229	IL-1β	Walz, A et al. (1997) *J Leukoc Biol* 62: 604
	ENA-78	Human	mRNA	A549	IL-1β and TNF-α	Walz, A et al. (1997) *J Leukoc Biol* 62: 604
	ENA-78	Human	mRNA	HEP G2	None	Rowell, DL et al. (1997) *Am J Physiol* 273: G322
	ENA-78	Human	mRNA	HH25	None	Rowell, DL et al. (1997) *Am J Physiol* 273: G322
	ENA-78	Human	mRNA	HUH 7	None	Rowell, DL et al. (1997) *Am J Physiol* 273: G322
	ENA-78	Human	mRNA	MG-63	IL-1β, TNF-α	Van Damme, J et al. (1997) *J Leukoc Biol* 62: 563
	ENA-78	Human	Protein	Caco-2	IL-1β or TNF-α	Keates, S et al. (1997) *Am J Physiol* 273: G75
	ENA-78	Human	Protein	T84	TNF-α	Keates, S et al. (1997) *Am J Physiol* 273: G75
	LIX	Mouse	mRNA	Swiss 3T3	LPS and TGF-β1	Smith, JB et al. (1995) *J Biol Chem* 270: 16756
	ENA-78	Mouse	Protein	MC-9	L-NAME	Lukacs, NW et al. (1998) *J Leukoc Biol* 63: 746
CXCL6	GCP-2	Bovine	Protein	MDBK	IFN-τ	Struyf, S et al. (2001) *Blood* 97: 2197
	GCP-2	Human	Protein	MG-63	IL-1	Proost, P et al. (1993) *J Immunol* 150: 1000
	GCP-2	Mouse	Protein	MO	LPS	Wuyts, A et al. (1996) *J Immunol* 157: 1736
	GCP-2	Mouse	Protein	MTEC1	PMA	Wuyts, A et al. (1996) *J Immunol* 157: 1736
CXCL7	NAP-2	Human	mRNA	LMVEC	None	Beck, C et al. (1999) *Clin Exp Immunol* 118: 298
CXCL8	IL-8	Bovine	Protein	MDBK	LPS	Struyf, S et al. (2001) *Blood* 97: 2197
	IL-8	Human	mRNA	3229	IL-1β	Walz, A et al. (1997) *J Leukoc Biol* 62: 604
	IL-8	Human	mRNA	Caco-2	None/IL-1β	Warhurst, AC et al. (1998) *Gut* 42: 208
	IL-8	Human	mRNA	HEP G2	None	Rowell, DL et al. (1997) *Am J Physiol* 273: G322
	IL-8	Human	mRNA	HH25	None	Rowell, DL et al. (1997) *Am J Physiol* 273: G322
	IL-8	Human	mRNA	HMC-1	PMA	Selvan, RS et al. (1994) *J Biol Chem* 269: 13893
	IL-8	Human	mRNA	HMEC-1	None	Hartmeyer, M et al. (1997) *J Immunol* 159: 1930
	IL-8	Human	mRNA	HT29-19A	None/TNF-α, IL-1β, LPS, LPA	Warhurst, AC et al. (1998) *Gut* 42: 208
	IL-8	Human	mRNA	HUH7	None	Rowell, DL et al. (1997) *Am J Physiol* 273: G322
	IL-8	Human	mRNA	VA13	IL-1β	Walz, A et al. (1997) *J Leukoc Biol* 62: 604
	IL-8	Human	mRNA	WI38	IL-1β	Walz, A et al. (1997) *J Leukoc Biol* 62: 604
	IL-8	Human	Protein	MG-63	IL-1	Proost, P et al. (1993) *J Immunol* 150: 1000

Table 10.8 Cells in chemokine research: ligands (*continued*)

Systemic name	Chemokine	Species	Type of expression	Cell line	Inducer/enhancer	Reference
CXCL9	MIG	Human	Protein	THP-1	IFN-γ	Farber, JM (1997) *J Leukoc Biol* 61: 246
	MIG	Mouse	mRNA	CA46	IL-12	Kanegane, C. et al. (1998) *J Leukoc Biol* 64: 384
	MIG	Mouse	mRNA	RAW 264.7	Con A	Farber, J.M. (1997) *J Leukoc Biol* 61: 246
	MIG	Mouse	Protein	RENCA	IFN-γ	Tannenbaum, C.S. et al. (1998) *J Immunol* 161: 927
CXCL10	IP-10	Human	mRNA	U937	IFN-γ	Luster, AD et al. (1985) *Nature* 315: 672
	IP-10	Human	Protein	MG-63	IL-1	Proost, P et al. (1993) *J Immunol* 150: 1000
	IP-10	Mouse	mRNA	BALB/c-3T3	IFN-γ, TNF-α	Ohmori, Y et al. (1993) *Am J Pathol* 142: 861
	IP-10	Mouse	mRNA	CA46	IL-12	Kanegane, C et al. (1998) *J Leukoc Biol* 64: 384
	IP-10	Mouse	mRNA	NIH-3T3	LPA, TNF-α	Ohmori, Y et al. (1994) *J Immunol* 153: 2204
	CRG-2 (IP-10)	Mouse	mRNA	RAW 264.7	IFN-γ	Vanguri, P et al. (1990) *J Biol Chem* 265: 15049
	CRG-2 (IP-10)	Mouse	mRNA	Swiss 3T3	LPA and TGF-β1	Smith, JB et al. (1995) *J Biol Chem* 270: 16756
	IP-10	Mouse	Protein	RENCA	IFN-γ	Tannenbaum, CS et al. (1998) *J Immunol* 161: 927
	Mob-1 (IP-10)	Rat	mRNA	Rat-1	Fetal bovine serum	Liang, P et al. (1994) *Proc Natl Acad Sci USA* 91: 12515
CXCL11	I-TAC	Human	mRNA	CRT	IFN-γ	Rani, MR et al. (1996) *J Biol Chem* 271: 22878
	I-TAC	Human	mRNA	SV-A3	IFN-γ	Cole, KE et al. (1998) *J Exp Med* 187: 2009
	I-TAC	Human	mRNA	THP-1	IFN-γ	Erdel, M et al. (1998) *Cytogenet Cell Genet* 81: 271
CXCL12	SDF-1α	Human	mRNA	498A	None	Begum, NA et al. (1999) *Int J Oncol* 14: 927
	SDF-1α	Human	mRNA	CRL 1496	None	Begum, NA et al. (1999) *Int J Oncol* 14: 927
	SDF-1α	Human	mRNA	MRC-5	None	Begum, NA et al. (1999) *Int J Oncol* 14: 927
	SDF-1α	Mouse	mRNA	BALB/3T3	None	Begum, NA et al. (1999) *Int J Oncol* 14: 927
	SDF-1α	Mouse	mRNA	MS-5	None	Bleul, CC et al. (1996) *J Exp Med* 184: 1101
	SDF-1α	Mouse	mRNA	PA6	None	Nagasawa, T et al. (1994) *Proc Natl Acad Sci USA* 91: 2305
	SDF-1α	Mouse	mRNA	ST-2	None	Tashiro, K et al. (1993) *Science* 261: 600
	SDF-1α	Mouse	mRNA	TC1	None	Begum, NA et al. (1999) *Int J Oncol* 14: 927
	SDF-1α	Mouse	mRNA	THP-1	None	Begum, NA et al. (1999) *Int J Oncol* 14: 927
CXCL13	BCA-1	Human	Protein* in IHC	CD21+ follicular dendritic cells	None	Shi, K et al. (2001) *J Immunol* 166: 650
	BCA-1	Human	Protein* in IHC	CD21+ follicular dendritic cells	None	Mazzucchelli, L et al. (1999) *J Clin Invest* 104: R49
CXCL14	BRAK	Human	mRNA	Activated B cells	LPS	Frederick, MJ et al. (2000) *Am J Pathol* 156: 1937
	BRAK	Human	mRNA	Activated monocytes	LPS	Frederick, MJ et al. (2000) *Am J Pathol* 156: 1937
	BRAK	Human	mRNA	Keratinocytes		Frederick, MJ et al. (2000) *Am J Pathol* 156: 1937
	BRAK	Human	mRNA	MDA-MB-435	None	Hromas, R et al. (1999) *Biochem Biophys Res Commun* 255: 703
	BRAK	Human	mRNA	SW485	None	Hromas, R et al. (1999) *Biochem Biophys Res Commun* 255: 703

Table 10.8 Cells in chemokine research: ligands (*continued*)

Systemic name	Chemokine	Species	Type of expression	Cell line	Inducer/enhancer	Reference
CXCL15	Lungkine	Mouse	mRNA	DAS 104-4	None	Ohneda, O et al. (2000) Immunity 12: 141
	Lungkine	Mouse	mRNA	DAS 104-8	None	Ohneda, O et al. (2000) Immunity 12: 141
CXCL16		Human	mRNA	CD14$^+$ monocytes		Wilbanks, A et al. (2001) J Immunol 166: 5145
		Human	mRNA	CD19$^+$ B cells		Wilbanks, A et al. (2001) J Immunol 166: 5145
		Mouse	mRNA	CD11c$^+$ dendritic cells		Matloubian, M et al. (2000) Nature Immunoi 1: 298
		Mouse	mRNA	CD8$^-$ dendritic cells		Matloubian, M et al. (2000) Nature Immunoi 1: 298
		Mouse	mRNA	CD8$^+$ dendritic cells		Matloubian, M et al. (2000) Nature Immunoi 1: 298
XCL1	Lymphotactin	Human	mRNA	HMC-1	IL-4, TGF-β	Rumsaeng, V et al. (1997) J Immunol 158: 1353
	SCM-1α	Human	mRNA	Jurkat	Mitogens	Yoshida, T et al. (1996) FEBS Letters 395: 82
	Lymphotactin	Human	mRNA	KU812	IL-4, TGF-β	Rumsaeng, V et al. (1997) J Immunol 158: 1353
	Lymphotactin	Mouse	mRNA	BMCMC	IL-4, TGF-β	Rumsaeng, V et al. (1997) J Immunol 158: 1353
	Lymphotactin	Mouse	mRNA	C1.MC/C57.1	IL-4, TGF-β	Rumsaeng, V et al. (1997) J Immunol 158: 1353
	Lymphotactin	Mouse	mRNA	DETC 7-17	Con A	Boismenu, R et al. (1996) J Immunol 157:985
XCL2	SCM-1β	Human	mRNA	Jurkat	Mitogens	Yoshida, T et al. (1996) FEBS Letters 395: 82
?	MIP-2	Mouse	mRNA	RAW 264.7	LPS	Tekamp-Olson, P et al. (1990) J Exp Med 172: 911
?	MIP-2	Mouse	Protein	L929	Virus	Wuyts, A et al. (1996) J Immunol 157: 1736
?	KC (CINC-1)	Rat	mRNA	NR8383	LPS	Huang, S et al. 1992 Biochem Biophys Res Commun 184: 922
?	KC (CINC-1)	Rat	mRNA	RFL-6	IL-1	Huang, S et al. 1992 Biochem Biophys Res Commun 184: 922
?	GRO (CINC-1)	Rat	Protein	NRK-52E	IL-1	Konishi, K et al. (1993) Gene 126: 285
?	CINC-2β	Rat	mRNA	RGM-1	IL-1α or TNF-α	Okada, A et al. (1998) J Lab Clin Med 131: 538
?	CINC-3	Rat	Protein	IEC-6	LPS and IL-1β	Ohno, Y et al. (1997) Proc Natl Acad Sci USA 94: 10279

*There are no reports in the current published literature of isolated cells expressing CXCL 13 (i.e., only immunohistochemical studies).

Table 10.9 Cells in chemokine research: receptors

Receptor	Species	Cell type	Reference
CCR1	Human	3T3.CD4	Deng, HK et al. (1997) Nature 388: 296
	Human	Basophils	Uguccioni, M et al. (1997) J Clin Invest 100: 137
	Human	CD14$^+$ monocytes	Weber, C (2000) J Leukoc Biol 67: 699
	Human	CD34$^+$ stem cells	Durig, J et al. (1998) Blood 92: 3073
	Human	CD4$^+$ Th1 cells (IFN-α activated)	Sallusto, F et al. (1998) J Exp Med 187: 875
	Human	CD4$^+$ Th2 cells (IFN-α activated)	Sallusto, F et al. (1998) J Exp Med 187: 875
	Human	CD4$^+$ tumor infiltrating lymphocytes (TIL) F9	Liao, F et al. (1997) J Exp Med 185: 2015
	Human	CD4$^+$CD62L$^-$ (naïve) T cells (activated and resting)	Rabin, RL et al. (1999) J Immunol 162: 3840
	Human	CD4$^+$CD45RO$^+$ (memory) T cells (activated and resting)	Rabin, RL et al. (1999) J Immunol 162: 3840
	Human	CD8$^+$ TIL R8	Liao, F et al. (1997) J Exp Med 185: 2015
	Human	CD8$^+$CD62L$^-$ (naïve) T cells (activated and resting)	Rabin, RL et al. (1999) J Immunol 162: 3840
	Human	CD8$^+$CD45RO$^+$ (memory) T cells (activated and resting)	Rabin, RL et al. (1999) J Immunol 162: 3840
	Human	Dendritic cells (immature)	Sallusto, F et al. (1998) Eur J Immunol 28: 2760
	Human	Eosinophils	Sabroe, I et al. (1999) J Immunol 162: 2946
	Human	Eosinophilic cell line (AML) 143D10	Zimmermann, N et al. (2000) J Immunol 164: 1055
	Human	Granulocytes	Liao, F et al. (1997) J Exp Med 185: 2015
	Human	Monocyte-derived dendritic cells	Sato, K et al. (1999). Blood 93: 34
	Human	Monocytes	Sozzani, S et al. (1998) J Exp Med 187: 439
	Human	PBLs	Liao, F et al. (1997) J Exp Med 185: 2015
	Human	PBMs	Liao, F et al. (1997) J Exp Med 185: 2015
	Human	PMNs (resting)	Bonecchi, R et al. (1999) J Immunol 162: 474
	Human	T cells	Alkhatib, G et al. (1997) Nature 388: 238
	Human	TIL B10	Liao, F et al. (1997) J Exp Med 185: 2015
	Human	TIL F9	Liao, F et al. (1997) J Exp Med 185: 2015
	Human	TIL R4	Liao, F et al. (1997) J Exp Med 185: 2015
	Human	TIL R8	Liao, F et al. (1997) J Exp Med 185: 2015
	Human	U937	Liao, F et al. (1997) J Exp Med 185: 2015
	Human	Vd2$^+$ T cells	Cipriani, B et al. (2000) Blood 95: 39
	Mouse	Astrocytes	Tanabe, S et al. (1997) J Neuroscience 17: 6522
	Mouse	Macrophage cell line 264.7	Gao, J-L et al. (1996) Biochem Biophys Res Comm 223: 679
	Mouse	Macrophages	Gao, J-L et al. (1996) Biochem Biophys Res Comm 223: 679
	Mouse	Mast cells (activated)	Gao, J-L et al. (1996) Biochem Biophys Res Comm 223: 679
	Mouse	PMNs	Gao, J-L et al. (1996) Biochem Biophys Res Comm 223: 679
	Mouse	XS52	Nibbs, RJB et al. (1997) J Biol Chem 272: 12495
	Rat	Hippocampal neurons	Meucci, O et al. (1998) Proc Natl Acad Sci USA 95: 14500

Table 10.9 Cells in chemokine research: receptors (*continued*)

Receptor	Species	Cell type	Reference
CCR2	Human	Basophils	Uguccioni, M et al. (1997) J Clin Invest 100: 1137
	Human	CD14+ monocytes	Weber, C (2000) J Leukoc Biol 67: 699
	Human	CD4+ Th1 cells (resting)	Sallusto, F et al. (1999) Eur J Immunol 29: 2037
	Human	CD4+ Th2 cells (resting)	Sallusto, F et al. (1999) Eur J Immunol 29: 2037
	Human	Mono Mac 6 (monocytic cell line)	Charo, IF et al. (1994) Proc Natl Acad Sci USA 91: 2752
	Human	Monocytes	Polentarutti, N et al. (1997) J Immunol 158: 2689
	Human	Monocytes	Sozzani, S et al. (1998) J Exp Med 187: 439
	Human	Mononuclear phagocytes	Polentarutti, N et al. (1997) J Immunol 158: 2689
	Human	NK cells	Polentarutti, N et al. (1997) J Immunol 158: 2689
	Human	NK cells (IL-2 activated)	Polentarutti, N et al. (1997) J Immunol 158: 2689
	Human	NK cells (activated)	Inngjerdingen, M et al. (2000) J Immunol 164: 4048
	Human	THP-1	Charo, IF et al. (1994) Proc Natl Acad Sci USA 91: 2752
	Human	THP-1	Polentarutti, N et al. (1997) J Immunol 158: 2689
	Human	Vd2+ T cells	Cipriani, B et al. (2000) Blood 95: 39
	Human	THP-1	Tangirala, RK et al. (1997) J Biol Chem 272: 8050
	Rat	Macrophage	Jiang, L et al. (1998) J Immunol 86:1
CCR2B	Human	Monocytes	Polentarutti, N et al. (1997) J Immunol 158: 2689
	Human	Mononuclear phagocytes	Polentarutti, N et al. (1997) J Immunol 158: 2689
	Human	NK cells	Polentarutti, N et al. (1997) J Immunol 158: 2689
	Human	NK cells (IL-2 activataed)	Polentarutti, N et al. (1997) J Immunol 158: 2689
	Human	T cells	Alkhatib, G et al. (1997) Nature 388: 238
	Human	THP-1	Polentarutti, N et al. (1997) J Immunol 158: 2689
	Human	TIL B10	Liao, F et al. (1997) J Exp Med 185: 2015
	Human	TIL F9	Liao, F et al. (1997) J Exp Med 185: 2015
	Human	TIL R4	Liao, F et al. (1997) J Exp Med 185: 2015
	Human	TIL R8	Liao, F et al. (1997) J Exp Med 185: 2015
	Human	Vd2+ T cells	Cipriani, B et al. (2000) Blood 95: 39
CCR3	Human	Basophils	Uguccioni, M et al. (1997) J Clin Invest 100: 1137
	Human	Basophils	Ochensberger, B et al. (1999) Eur J Immunol 29: 11
	Human	CD4+CD45RO+ (memory) T cells (activated)	Rabin, RL et al. (1999) J Immunol 162: 3840
	Human	CD4+ Th2 cells	Sallusto, F et al. (1999) J Exp Med 187: 875
	Human	CD4+ TIL F9	Liao, F et al. (1997) J Exp Med 185: 2015
	Human	CD4+ Th2 cells (resting)	Sallusto, F et al. (1999) Eur J Immunol 29: 2037
	Human	CD4+ Th2 cells (resting)	D'Ambrosio, D et al. (1998) J Immunol 161: 5111
	Human	CD8+CD62L- (naïve) T cells (activated)	Rabin, RL et al. (1999) J Immunol 162: 3840

Table 10.9 Cells in chemokine research: receptors (*continued*)

Receptor	Species	Cell type	Reference
CCR3 (cont.)	Human	CD8$^+$CD45RO$^+$ (memory) T cells (activated)	Rabin, RL *et al.* (1999) *J Immunol* 162: 3840
	Human	Eosinophils	Ponath, P *et al.* (1996) *J Exp Med* 183: 2437
	Human	Eosinophilic cell line (AML) 14.3D10	Zimmermann, N *et al.* (2000) *J Immunol* 164: 1055
	Human	Eosinophils	Zimmermann, N *et al.* (1999) *J Biol Chem* 274: 12611
	Human	Granulocytes	Liao, F *et al.* (1997) *J Exp Med* 185: 2015
	Human	HMMEC	Hoki, G *et al.* (1997) *Biochem Biophys Res Commun* 241: 136
	Human	HUVEC	Shahabuddin, S *et al.* (2000) *J Immunol* 164: 3847
	Human	Monocyte-derived dendritic cells	Sato, K *et al.* (1999). *Blood* 93: 34
	Human	Neurons (fetal)	Klein, RS *et al.* (1999) *J Immunol* 163: 1636
	Human	PMNs (resting)	Bonecchi, R *et al.* (1999) *J Immunol* 162: 474
	Human	T cells	Alkhatib, G *et al.* (1997) *Nature* 388: 238
	Human	Th2-polarized cells	Sallusto, F *et al.* (1997) *Science* 277: 2005
	Macaque	Lymphocytes	Westmoreland, S *et al.* (1998) *Am J Pathol* 152: 659
	Macaque	Pyramidal neurons	Westmoreland, S *et al.* (1998) *Am J Pathol* 152: 659
	Mouse	Eosinophils	Gao, J-L *et al.* (1996) *Biochem Biophys Res Commun* 223: 679
	Mouse	Macrophages	Gao, J-L *et al.* (1996) *Biochem Biophys Res Commun* 223: 679
	Mouse	PMNs	Gao, J-L *et al.* (1996) *Biochem Biophys Res Commun* 223: 679
CCR4	Human	Basophils	Power, CA *et al.* (1995) *J Biol Chem* 270: 19495
	Human	CD4$^+$ Th1 cells (activated)	D'Ambrosio, D *et al.* (1998) *J Immunol* 161: 5111
	Human	CD4$^+$ Th2 cells	Sallusto, F *et al.* (1999) *J Exp Med* 187: 875
	Human	CD4$^+$ TIL R4	Liao, F *et al.* (1997) *J Exp Med* 185: 2015
	Human	CD4$^+$ Th2 cells (activated)	D'Ambrosio, D *et al.* (1998) *J Immunol* 161: 5111
	Human	CD8$^+$ TIL R8	Liao, F *et al.* (1997) *J Exp Med* 185: 2015
	Human	CD8$^+$ Th2 cells (activated)	D'Ambrosio, D *et al.* (1998) *J Immunol* 161: 5111
	Human	CD34$^+$ stem cells	Durig, J *et al.* (1998) *Blood* 92: 3073
	Human	EBV414	Liao, F *et al.* (1997) *J Exp Med* 185: 2015
	Human	HUVEC	Shahabuddin, S *et al.* (2000) *J Immunol* 164: 3847
	Human	KU812 (basophilic cell line)	Power, CA *et al.* (1995) *J Biol Chem* 270: 19495
	Human	Monocyte-derived dendritic cells	Sato, K *et al.* (1999). *Blood* 93: 34
	Human	NK cells (activated)	Inngjerdingen, M *et al.* (2000) *J Immunol* 164: 4048
	Human	PBLs	Liao, F *et al.* (1997) *J Exp Med* 185: 2015
	Human	CD4$^+$ Th2 (resting)	D'Ambrosio, D *et al.* (1998) *J Immunol* 161: 5111
	Human	SUP-T1	Liao, F *et al.* (1997) *J Exp Med* 185: 2015
	Human	Th2 cells	Wilbanks, A *et al.* (2001) *J Immunol* 166: 5145
	Human	TIL B10	Liao, F *et al.* (1997) *J Exp Med* 185: 2015
	Human	TIL F9	Liao, F *et al.* (1997) *J Exp Med* 185: 2015
	Human	TIL R4	Liao, F *et al.* (1997) *J Exp Med* 185: 2015
	Human	TIL R8	Liao, F *et al.* (1997) *J Exp Med* 185: 2015
	Human	Vd2$^+$ T cells	Cipriani, B *et al.* (2000) *Blood* 95: 39
	Mouse	CTLL (cytotoxic T cell line)	Hoogewerf, AJ *et al.* (1996) *Biochem Biophys Res Commun* 218: 337
	Rat	Hippocampal neurons	Meucci, O *et al.* (1998) *Proc Natl Acad Sci USA* 95: 14500

Table 10.9 Cells in chemokine research: receptors (*continued*)

Receptor	Species	Cell type	Reference
CCR5	Human	3T3.CD4	Deng, HK *et al.* (1997) *Nature* 388: 296
	Human	Astrocytes (fetal)	Klein, RS *et al.* (1999) *J Immunol* 163: 1636
	Human	CD34$^+$ stem cells	Durig, J *et al.* (1998) *Blood* 92: 3073
	Human	CD4$^+$ B10	Liao, F *et al.* (1997) *J Exp Med* 185: 2015
	Human	CD4$^+$ T cells	Raport, CJ et al (1996) *J Biol Chem* 271: 17161
	Human	CD4$^+$ Th1 cells (resting)	Sallusto, F *et al.* (1999) *Eur J Immunol* 29: 2037
	Human	CD4$^+$ Th1 cells	Sallusto, F *et al.* (1998) *J Exp Med* 187: 875
	Human	CD4$^+$ Th2 cells (resting)	Sallusto, F *et al.* (1999) *Eur J Immunol* 29: 2037
	Human	CD4$^+$CD45RO$^+$ (memory) T cells	Lee, B *et al.* (1999) *Proc Natl Acad Sci USA* 96: 5215
	Human	CD4$^+$CD45RO$^+$ (memory) T cells (activated)	Rabin, RL et al (1999) *J Immunol* 162: 3840
	Human	CD8$^+$ T cells	Raport, CJ et al (1996) *J Biol Chem* 271: 17161
	Human	CD8$^+$ CD45RO$^+$ (memory) T cells (resting)	Rabin, RL et al (1999) *J Immunol* 162: 3840
	Human	CD8$^+$ Tc1 cells (activated)	D'Ambrosio, D *et al.* (1998) *J Immunol* 161: 5111
	Human	CD14$^+$ CD16 monocytes	Weber, C (2000) *J Leukoc Biol* 67: 699
	Human	CD56$^+$ NK cells	Lee, B *et al.* (1999) *Proc Natl Acad Sci USA* 96: 5215
	Human	Colon columnar epithelium (normal)	Dwinell, MB *et al.* (1999) *Gastroenterology* 117: 359
	Human	Eosinophilic cell line (AML) 14.3D10	Zimmermann, N *et al.* (2000) *J Immunol* 164: 1055
	Human	HT-29 (colon adenocarcinoma)	Dwinell, MB *et al.* (1999) *Gastroenterology* 117: 359
	Human	HUT78 (T cell line)	Lee, B *et al.* (1999) *Proc Natl Acad Sci USA* 96: 5215
	Human	HUT78 (T cell line)	Raport, CJ et al (1996) *J Biol Chem* 271: 17161
	Human	HUVEC	Shahabuddin, S *et al.* (2000) *J Immunol* 164: 3847
	Human	Immature dendritic cells	Sallusto, F *et al.* (1998) *Eur J Immunol* 28: 2760
	Human	Jijoye (B cell line)	Raport, CJ et al (1996) *J Biol Chem* 271: 17161
	Human	Jurkat-D	Lee, B *et al.* (1999) *Proc Natl Acad Sci USA* 96: 5215
	Human	MOLT4 (CD4+ T cell line)	Lee, B *et al.* (1999) *Proc Natl Acad Sci USA* 96: 5215
	Human	Monocyte-derived dendritic cells	Sato, K *et al.* (1999) *Blood* 93: 34
	Human	Monocytes	Sozzani, S *et al.* (1998) *J Exp Med* 187: 439
	Human	Neurons (fetal)	Klein, RS *et al.* (1999) *J Immunol* 163: 1636
	Human	T cells	Alkhatib, G *et al.* (1997) *Nature* 388: 238
	Human	T-84 (colon carcinoma)	Dwinell, MB *et al.* (1999) *Gastroenterology* 117: 359
	Human	Th0 cells (resting)	Annunziato, F *et al.* (1999) *J Leukoc Biol* 65: 691
	Human	Th1 cells (resting)	Annunziato, F *et al.* (1999) *J Leukoc Biol* 65: 691
	Human	THP-1	Schecter, AD *et al.* (2000) *J Biol Chem* 275: 5466
	Human	THP-1	Raport, CJ et al (1996) *J Biol Chem* 271: 17161
	Human	TIL B10	Liao, F *et al.* (1997) *J Exp Med* 185: 2015
	Human	TIL F9	Liao, F *et al.* (1997) *J Exp Med* 185: 2015
	Human	TIL R4	Liao, F *et al.* (1997) *J Exp Med* 185: 2015
	Human	TIL R8	Liao, F *et al.* (1997) *J Exp Med* 185: 2015
	Human	Vascular smooth muscle	Schecter, AD *et al.* (2000) *J Biol Chem* 275: 5466
	Human	Vd2$^+$ T cells	Cipriani, B *et al.* (2000) *Blood* 95: 39
	Macaque	Lymphocytes	Westmoreland, S *et al.* (1998) *Am J Pathol* 152: 659
	Macaque	Pyramidal neurons	Westmoreland, S *et al.* (1998) *Am J Pathol* 152: 659
	Rat	Hippocampal neurons	Meucci, O *et al.* (1998) *Proc Natl Acad Sci USA* 95: 14500
	Rat	Macrophages	Jiang, L *et al.* (1998) *J Immunol* 86: 1
	Rat	Microglial cells	Jiang, L *et al.* (1998) *J Immunol* 86: 1

Table 10.9   Cells in chemokine research: receptors (*continued*)

Receptor	Species	Cell type	Reference
CCR6	Human	B cells	Baba, M *et al.* (1997) *J Biol Chem* 272: 14893
	Human	Caco-2 (ileocecal colon carcinoma)	Dwinell, MB *et al.* (1999) *Gastroenterology* 117: 359
	Human	CD19$^+$ B cells	Zaballos, A *et al.* (1996) *Biochem Biophys Res Commun* 227: 846
	Human	CD4$^+$ B cells	Baba, M *et al.* (1997) *J Biol Chem* 272: 14893
	Human	CD4$^+$ T cells	Baba, M *et al.* (1997) *J Biol Chem* 272: 14893
	Human	CD4$^+$ T cells	Power, CA *et al.* (1997) *J Exp Med* 186: 825
	Human	CD4$^+$ Th1 cells (resting)	Sallusto, F *et al.* (1999) *Eur J Immunol* 29: 2037
	Human	CD8$^+$ B cells	Baba, M *et al.* (1997) *J Biol Chem* 272: 14893
	Human	CD8$^+$ T cells	Baba, M *et al.* (1997) *J Biol Chem* 272: 14893
	Human	CD8$^+$ T cells	Zaballos, A *et al.* (1996) *Biochem Biophys Res Commun* 227: 846
	Human	CD8$^+$ T cells	Power, CA *et al.* (1997) *J Exp Med* 186: 825
	Human	Dendritic cells (CD34$^+$ stem cell-derived)	Power, CA *et al.* (1997) *J Exp Med* 186: 825
	Human	Eosinophils	Sullivan, S *et al.* (1999) *J Leukoc Biol* 66: 674
	Human	HCA-7 (colon adenocarcinoma)	Dwinell, MB *et al.* (1999) *Gastroenterology* 117: 359
	Human	HT-29 (colon carcinoma)	Dwinell, MB *et al.* (1999) *Gastroenterology* 117: 359
	Human	Hut102 T cells	Baba, M *et al.* (1997) *J Biol Chem* 272: 14893
	Human	Immature dendritic cells	Dieu, M-C *et al.* (1998) *J Exp Med* 188: 373
	Human	Mature dendritic cells	Dieu, M-C *et al.* (1998) *J Exp Med* 188: 373
	Human	PBLs	Liao, F *et al.* (1997) *J Exp Med* 185: 2015
	Human	T-84 (colon carcinoma)	Dwinell, MB *et al.* (1999) *Gastroenterology* 117: 359
	Human	Th1 cells	Wilbanks, A *et al.* (2001) *J Immunol* 166: 5145
	Human	Th2 cells	Wilbanks, A *et al.* (2001) *J Immunol* 166: 5145
	Human	Tr1 cells	Wilbanks, A *et al.* (2001) *J Immunol* 166: 5145
	Human	TIL F9	Liao, F *et al.* (1997) *J Exp Med* 185: 2015
	Human	TIL R4	Liao, F *et al.* (1997) *J Exp Med* 185: 2015
	Human	TIL R8	Liao, F *et al.* (1997) *J Exp Med* 185: 2015
CCR7	Human	ATL cells	Hasegawa, H *et al.* (2000) *Blood* 95: 30
	Human	Caco-2 (ileocecal colon carcinoma)	Dwinell, MB *et al.* (1999) *Gastroenterology* 117: 359
	Human	CD34$^+$ HPC	Kim, C *et al.* (1998) *J Immunol* 161: 2580
	Human	CD4$^+$ Th1 cells (activated)	Sallusto, F *et al.* (1999) *Eur J Immunol* 29: 2037
	Human	CD4$^+$ Th2 cells (activated)	Sallusto, F *et al.* (1999) *Eur J Immunol* 29: 2037
	Human	CD4$^+$ Th2 lymphocytes (TGF-β activated)	Sallusto, F *et al.* (1998) *J Exp Med* 187: 875
	Human	CD4$^+$CD45RO$^+$ T cells	Hasegawa, H *et al.* (2000) *Blood* 95: 30
	Human	H9 (T cell line)	Schweickart, VI *et al.* (1994) *Genomics* 23: 643
	Human	HCA-7 (colon adenocarcinoma)	Dwinell, MB *et al.* (1999) *Gastroenterology* 117: 359
	Human	HS602 (B cell line)	Schweickart, VI *et al.* (1994) *Genomics* 23: 643
	Human	HUT78 (T cell line)	Yoshida, R *et al.* (1997) *J Biol Chem* 272: 13803

Table 10.9 Cells in chemokine research: receptors (*continued*)

Receptor	Species	Cell type	Reference
CCR7 (cont.)	Human	HUT78 (T cell line)	Schweickart, VI *et al.* (1994) *Genomics* 23: 643
	Human	Jijoye (B cell line)	Schweickart, VI *et al.* (1994) *Genomics* 23: 643
	Human	Mature dendritic cells	Sallusto, F *et al.* (1998) *Eur J Immunol* 28: 2760
	Human	Th1 cells	Wilbanks, A *et al.* (2001) *J Immunol* 166: 5145
	Human	Th2 cells	Wilbanks, A *et al.* (2001) *J Immunol* 166: 5145
	Human	Tr1 cells	Wilbanks, A *et al.* (2001) *J Immunol* 166: 5145
CCR8	Human	Caco-2 (ileocecal colon carcinoma)	Dwinell, MB *et al.* (1999) *Gastroenterology* 117: 359
	Human	CD4+ T cells	Zaballos, A *et al.* (1996) *Biochem Biophys Res Commun* 227: 846
	Human	CD4+ T cells (resting)	Samson, M (1996) *Eur J Immunol* 26: 3021
	Human	CD4+ Th2 cells (activated)	D'Ambrosio, D *et al.* (1998) *J Immunol* 161: 5111
	Human	CD8+ T cells	Zaballos, A *et al.* (1996) *Biochem Biophys Res Commun* 227: 846
	Human	CD8+ T cells (resting)	Samson, M (1996) *Eur J Immunol* 26: 3021
	Human	HCA-7 (colon adenocarcinoma)	Dwinell, MB *et al.* (1999) *Gastroenterology* 117: 359
	Human	Jurkat	Samson, M (1996) *Eur J Immunol* 26: 3021
	Human	Macrophages	Zaballos, A *et al.* (1996) *Biochem Biophys Res Commun* 227: 846
	Human	Molt-4	Samson, M (1996) *Eur J Immunol* 26: 3021
	Human	NK cells (activated)	Inngjerdingen, M *et al.* (2000) *J Immunol* 164: 4048
	Human	T-84 (colon carcinoma)	Dwinell, MB *et al.* (1999) *Gastroenterology* 117: 359
	Human	Vd2+ T cells	Cipriani, B *et al.* (2000) *Blood* 95: 39
	Mouse	BW5147 thymic lymphoma cells	Goya, I (1998) *J Immunol* 160: 1975
CCR9	Human	CD4+CD8+ thymocytes	Yu, C-R *et al.* (2000) *J Immunol* 164: 1293
	Human	MOLTA (CD4+ T cell line)	Yu, C-R *et al.* (2000) *J Immunol* 164: 1293
	Human	Sup T1 (CD4+ cell line)	Yu, C-R *et al.* (2000) *J Immunol* 164: 1293
	Mouse	AKR1 (Dp thymoma line)	Norment, A *et al.* (2000) *J Immunol* 164: 639
	Mouse	CD25+ DN thymocytes	Wurbel, M-A *et al.* (2000) *Eur J Immunol* 30: 262
	Mouse	CD25+ DP thymocytes	Wurbel, M-A *et al.* (2000) *Eur J Immunol* 30: 262
	Mouse	CD4+ SP thymocytes	Wurbel, M-A *et al.* (2000) *Eur J Immunol* 30: 262
	Mouse	CD8+ SP thymocytes	Wurbel, M-A *et al.* (2000) *Eur J Immunol* 30: 262
	Mouse	Thymocytes (DP)	Norment, A *et al.* (2000) *J Immunol* 164: 639
CCR10	Human	B cells (JY) (activated)	Wang, W *et al.* (2000) *J Biol Chem* 275: 22313
	Human	Pre-monocytes (U937) (activated)	Wang, W *et al.* (2000) *J Biol Chem* 275: 22313
	Human	Pre-monocytes (U937) (resting)	Wang, W *et al.* (2000) *J Biol Chem* 275: 22313
	Human	T cells (clone Mot 72) (activated)	Wang, W *et al.* (2000) *J Biol Chem* 275: 22313
	Human	T cells (clone Mot 81) (activated)	Wang, W *et al.* (2000) *J Biol Chem* 275: 22313
	Mouse	B cells (CH12)	Wang, W *et al.* (2000) *J Biol Chem* 275: 22313
	Mouse	CD44+/CD25+ pre-T cells	Wang, W *et al.* (2000) *J Biol Chem* 275: 22313
	Mouse	Macrophages (J774)	Wang, W *et al.* (2000) *J Biol Chem* 275: 22313
	Mouse	Mature B cells (leukemia A20)	Wang, W *et al.* (2000) *J Biol Chem* 275: 22313
	Mouse	Th2 cells (polarized)	Wang, W *et al.* (2000) *J Biol Chem* 275: 22313

Table 10.9 Cells in chemokine research: receptors (*continued*)

Receptor	Species	Cell type	Reference
CX3CR1	Human	CD14$^+$ Monocytes	Imai, T et al. (1997) Cell 91: 521
	Human	CD16$^+$ NK cells	Imai, T et al. (1997) Cell 91: 521
	Human	CD4$^+$ T cells (IL-12 induced)	Imai, T et al. (1997) Cell 91: 521
	Human	CD8$^+$CD45RO$^+$ T cells (memory)	Foussat, A et al. (2000) Eur J Immunol 30: 87
	Human	CD4$^+$ T cells (IL-12 induced)	Imai, T et al. (1997) Cell 91: 521
	Human	CD8$^+$CD45RO$^+$ T cells (naïve)	Foussat, A et al. (2000) Eur J Immunol 30: 87
	Human	CD8$^+$CD45RO$^+$ T cells (memory)	Foussat, A et al. (2000) Eur J Immunol 30: 87
	Human	EE110 (IEL cell line)	Muehlhoefer, A et al. (2000) J Immunol 164: 3368
	Human	Intestinal intraepithelial lymphocytes	Muehlhoefer, A et al. (2000) J Immunol 164: 3368
	Human	THP-1	Raport, CJ et al. (1995) Gene 163: 295
	Human	U937	Raport, CJ et al. (1995) Gene 163: 295
	Mouse	Monocytes	Combadiere, C et al. (1998) Biochem, Biophys Res Commun 253: 728
	Mouse	Neutrophils/PMNs	Combadiere, C et al. (1998) Biochem, Biophys Res Commun 253: 728
	Rat	Astrocytes	Jiang, L et al. (1998) J Immunol 86: 1
	Rat	Astrocytes (activated)	Maciejewski-Lenoir, D et al. (1999) J Immunol 163: 1628
	Rat	Hippocampal neurons	Meucci, O et al. (1998) Proc Natl Acad Sci USA 95: 14500
	Rat	Macrophages	Jiang, L et al. (1998) J Immunol 86: 1
	Rat	Microglial cells	Jiang, L et al. (1998) J Immunol 86: 1
	Rat	Microglial cells	Maciejewski-Lenoir, D et al. (1999) J Immunol 163: 1628
CXCR1	Human	Basophils	Ochensberger, B et al. (1999) Eur J Immunol 29: 11
	Human	ECV304	Murdoch, C et al. (1999) Cytokine 11: 704
	Human	HUVEC	Murdoch, C et al. (1999) Cytokine 11: 704
	Human	Immature dendritic cells	Sallusto, F et al. (1998) Eur J Immunol 28: 2760
	Human	Megakaryocytes	Gewirtz, AM et al. (1995) Blood 86: 2559
	Human	Neutrophils/PMNs	Sabroe, I et al. (1997) J Immunol 158: 1361
	Human	PMN (resting)	Bonecchi, R et al. (1999) J Immunol 162: 474
CXCR2	Human	Basophils	Power, CA et al. (1995) J Biol Chem 270: 19495
	Human	Basophils	Ochensberger, B et al. (1999) Eur J Immunol 29: 11
	Human	Eosinophils	Sabroe, I et al. (1999) J Immunol 162: 2946
	Human	KU81 2 (basophilic cell line)	Power, CA et al. (1995) J Biol Chem 270: 19495
	Human	Megakaryocytes	Gewirtz, AM et al. (1995) Blood 86: 2559
	Human	Monocytes	Sozzani, S et al. (1998) J Exp Med 187: 439
	Human	PMNs (resting)	Bonecchi, R et al. (1999) J Immunol 162: 474
	Rat	Hippocampal neurons	Meucci, O et al. (1998) Proc Natl Acad Sci USA 95: 14500

Table 10.9 Cells in chemokine research: receptors (*continued*)

Receptor	Species	Cell type	Reference
CXCR3	Human	CD4$^+$ Th1 cells (resting)	D'Ambrosio, D et al. (1998) *J Immunol* 161: 5111
	Human	CD4$^+$ Th1 cells (resting)	Sallusto, F et al. (1999) *Eur J Immunol* 29: 2037
	Human	CD4$^+$ Th2 cells	Sallusto, F et al. (1998) *J Exp Med* 187: 875
	Human	CD4$^+$ Th2 cells (resting)	Sallusto, F et al. (1999) *Eur J Immunol* 29: 2037
	Human	CD4$^+$CD62L$^+$ (naïve) T cells (activated and resting)	Rabin, RL et al. (1999) *J Immunol* 162: 3840
	Human	CD8$^+$ Tc1 cells (activated)	D'Ambrosio, D et al. (1998) *J Immunol* 161: 5111
	Human	CD8$^+$ Th2 cells (activated)	D'Ambrosio, D et al. (1998) *J Immunol* 161: 5111
	Human	CD8$^+$CD62L$^+$ (naïve) T cells (activated and resting)	Rabin, RL et al. (1999) *J Immunol* 162: 3840
	Human	CD8$^+$CD45RO$^+$ T cells	Sallusto, F et al. (1998) *J Exp Med* 187: 875
	Human	Naïve T cells (activated)	Annunziato, F et al. (1999) *J Leukoc Biol* 65: 691
	Human	Th0 cells (activated and resting)	Annunziato, F et al. (1999) *J Leukoc Biol* 65: 691
	Human	Th1 cells (activated and resting)	Annunziato, F et al. (1999) *J Leukoc Biol* 65: 691
	Human	Th2 cells (activated and resting)	Annunziato, F et al. (1999) *J Leukoc Biol* 65: 691
	Macaque	Lymphocytes	Westmoreland, S et al. (1998) *Am J Pathol* 152: 659
CXCR4	Human	3T3.CD4	Deng, HK et al. (1997) *Nature* 388: 296
	Human	Astrocytes (fetal)	Klein, RS et al. (1999) *J Immunol* 163: 1636
	Human	BFU-E	Kowalska, MA et al. (1999) *Br J Haematol* 104: 220
	Human	Caco-2 (ileocecal colon carcinoma)	Dwinell, MB et al. (1999) *Gastroenterology* 117: 359
	Human	CD 19$^+$ B cells	Lee, B et al. (1999) *Proc Natl Acad Sci USA* 96: 5215
	Human	CD34$^+$ stem cells	Kowalska, MA et al. (1999) *Br J Haematol* 104: 220
	Human	CD4$^+$ Th2 cells (resting)	Sallusto, F et al. (1999) *Eur J Immunol* 29: 2037
	Human	CD4$^+$CD45RA$^+$CD62L$^+$ naïve T cells	Lee, B et al. (1999) *Proc Natl Acad Sci USA* 96: 5215
	Human	CD4$^+$CD45RO$^+$ (Memory) T cells (activated and resting)	Rabin, RL et al. (1999) *J Immunol* 162: 3840
	Human	CD8$^+$CD45RO$^+$ (Memory) T cells (activated and resting)	Rabin, RL et al. (1999) *J Immunol* 162: 3840
	Human	CFU-GM	Kowalska, MA et al. (1999) *Br J Haematol* 104: 220
	Human	Colon columnar epithelium (normal)	Dwinell, MB et al. (1999) *Gastroenterology* 117: 359
	Human	ECV304	Murdoch, C et al. (1999) *Cytokine* 11: 704
	Human	Granulocytes	Liao, F et al. (1997) *J Exp Med* 185: 2015
	Human	HAEC (human aortic endothelial cells)	Salcedo, R (1999) *Am J Pathol* 154: 1125
	Human	HCA-7 (colon adenocarcinoma)	Dwinell, MB et al. (1999) *Gastroenterology* 117: 359
	Human	HT-29 (colon carcinoma)	Dwinell, MB et al. (1999) *Gastroenterology* 117: 359
	Human	HT-29 (colon carcinoma)	Jordan, NJ et al. (1999) *J Clin Invest* 104: 1061
	Human	Human dermal microvascular EC	Feil, C et al. (1998) *Biochem Biophys Res Commun* 247: 38
	Human	HUT78 (T cell line)	Lee, B et al. (1999) *Proc Natl Acad Sci USA* 96: 5215
	Human	HUVEC	Volin, MV (1998) *Biochem Biophys Res Commun* 242: 46
	Human	HUVEC	Murdoch, C et al. (1999) *Cytokine* 11: 704
	Human	HUVEC	Salcedo, R (1999) *Am J Pathol* 154: 1125

Table 10.9 Cells in chemokine research: receptors (*continued*)

Receptor	Species	Cell type	Reference
CXCR4 (cont.)	Human	HUVEC	Feil, C *et al.* (1998) *Biochem Biophys Res Commun* 247: 38
	Human	IM9 (B cell line)	Vila-Coro, AJ *et al.* (1999) *FASEB J* 13: 1699
	Human	Jurkat-D	Lee, B *et al.* (1999) *Proc Natl Acad Sci USA* 96: 5215
	Human	MOLT4 CD4⁺ T cells	Lee, B *et al.* (1999) *Proc Natl Acad Sci USA* 96: 5215
	Human	MOLT4 CD4⁺ T cells	Vila-Coro, AJ *et al.* (1999) *FASEB J* 13: 1699
	Human	Mono mac 1 (monocytic cell line)	Vila-Coro, AJ *et al.* (1999) *FASEB J* 13: 1699
	Human	Monocytes	Sozzani, S *et al.* (1998) *J Exp Med* 187: 439
	Human	MT-2	Lee, B *et al.* (1999) *Proc Natl Acad Sci USA* 96: 5215
	Human	Naïve T cells (resting)	Annunziato, F *et al.* (1999) *J Leukoc Biol* 65: 691
	Human	Neurons (fetal)	Klein, RS *et al.* (1999) *J Immunol* 163: 1636
	Human	Normal human colonic epithelium	Jordan, NJ *et al.* (1999) *J Clin Invest* 104: 1061
	Human	PBLs	Liao, F *et al.* (1997) *J Exp Med* 185: 2015
	Human	PBMs	Liao, F *et al.* (1997) *J Exp Med* 185: 2015
	Human	Platelets	Kowalska, MA *et al.* (1999) *Br J Haematol* 104: 220
	Human	PMNs (resting)	Bonecchi, R *et al.* (1999) *J Immunol* 162: 474
	Human	SUP T1 (CD4⁺ cell line)	Lee, B *et al.* (1999) *Proc Natl Acad Sci USA* 96: 5215
	Human	SUP-T1	Liao, F *et al.* (1997) *J Exp Med* 185: 2015
	Human	T cells	Alkhatib, G *et al.* (1997) *Nature* 388: 238
	Human	TIL R4	Liao, F *et al.* (1997) *J Exp Med* 185: 2015
	Human	TIL R8	Liao, F *et al.* (1997) *J Exp Med* 185: 2015
	Macaque	Lymphocytes	Westmoreland, S *et al.* (1998) *Am J Pathol* 152: 659
	Macaque	Pyramidal neurons	Westmoreland, S *et al.* (1998) *Am J Pathol* 152: 659
	Mouse	Astrocytes	Tanabe, S *et al.* (1997) *J Immunol* 159: 905
	Mouse	Microglial cells	Tanabe, S *et al.* (1997) *J Immunol* 159: 905
	Mouse	N9 (microglia)	Tanabe, S *et al.* (1997) *J Immunol* 159: 905
	Rat	Astrocytes	Jiang, L *et al.* (1998) *J Immunol* 86: 1
	Rat	Hippocampal neurons	Meucci, O *et al.* (1998) *Proc Natl Acad Sci USA* 95: 14500
	Rat	Microglial cells	Jiang, L *et al.* (1998) *J Immunol* 86: 1

Table 10.9 Cells in chemokine research: receptors (*continued*)

Receptor	Species	Cell type	Reference
CXCR5	Human	B cells	Legler, DF et al. (1998) *J Exp Med* 187: 655
	Human	BL106	Dobner, T et al. (1992) *Eur J Immunol* 22: 2795
	Human	BL21	Dobner, T et al. (1992) *Eur J Immunol* 22: 2795
	Human	BL64	Dobner, T et al. (1992) *Eur J Immunol* 22: 2795
	Human	CD19+ B cells	Forster, R et al. (1994) *Blood* 84: 830
	Human	CD4+CD45RO+CD44+ T cells	Forster, R et al. (1994) *Blood* 84: 830
	Human	Daudi (B cell line)	Barella, L et al. (1995) *Biochem J* 309: 773
	Human	Daudi (B cell line)	Dobner, T et al. (1992) *Eur J Immunol* 22: 2795
	Human	Es III	Dobner, T et al. (1992) *Eur J Immunol* 22: 2795
	Human	γδ T cells	Forster, R et al. (1994) *Blood* 90: 520
	Human	LY67	Dobner, T et al. (1992) *Eur J Immunol* 22: 2795
	Human	Monocytes	Barella, L et al. (1995) *Biochem J* 309: 773
	Human	Raji (B cell line)	Barella, L et al. (1995) *Biochem J* 309: 773
	Human	Raji (B cell line)	Dobner, T et al. (1992) *Eur J Immunol* 22: 2795
	Mouse	2PK3	Forster, R et al. (1994) *Cell Mol Biol* 40: 381
	Mouse	B1 B cells	Bowman, EP et al. (2000) *J Exp Med* 191: 1303
	Mouse	B2 B cells	Bowman, EP et al. (2000) *J Exp Med* 191: 1303
	Mouse	CD3−CD4+ mesenteric lymph node T cells	Mebius, RE et al. (1997) *Immunity* 7: 493
	Mouse	CD4+CD62L+ T cells	Walker, LSK et al. (1999) *J Exp Med* 190: 1115
	Mouse	Pro-B cells	Bowman, EP et al. (2000) *J Exp Med* 191: 1303
	Mouse	WEHI 231	Forster, R et al. (1994) *Cell Mol Biol* 40: 381
	Mouse	WEHI 279	Forster, R et al. (1994) *Cell Mol Biol* 40: 381
CXCR6	Human	3T3.CD4	Deng, HK ete al. (1998) *Nature* 388: 296
	Human	CD4+ TIL B10	Liao, F et al. (1997) *J Exp Med* 185: 2015
	Human	CD4+ TIL F9	Liao, F et al. (1997) *J Exp Med* 185: 2015
	Human	CD4+ TIL R4	Liao, F et al. (1997) *J Exp Med* 185: 2015
	Human	CD8+ TIL R8	Liao, F et al. (1997) *J Exp Med* 185: 2015
	Human	T cells	Alkhatib, G et al. (1997) *Nature* 388: 238
	Human	TIL B10	Liao, F et al. (1997) *J Exp Med* 185: 2015
	Human	TIL F9	Liao, F et al. (1997) *J Exp Med* 185: 2015
	Human	TIL R4	Liao, F et al. (1997) *J Exp Med* 185: 2015
	Human	TIL R8	Liao, F et al. (1997) *J Exp Med* 185: 2015
XCR1	Human	NKB1	Shan, L et al. (2000) *Biochem Biophy Res Commun* 268: 938
	Human	Th1 / HY06 (anergic)	Shan, L et al. (2000) *Biochem Biophy Res Commun* 268: 938
	Mouse	B cells	Huang, H et al. (2001) *Biochem Biophys Res Commun* 281: 378
	Mouse	CD8+ T cells (splenic)	Yoshida, T et al. (1999) *FEBS Lett* 458: 37
	Mouse	Neutrophils	Cairns, CM et al. (2001) *J Immunol* 167: 57
	Mouse	Neutrophils	Huang, H et al. (2001) *Biochem Biophys Res Commun* 281: 378

Table 10.10 Receptors on different cell types

Cell type	Receptors	Chemokines
Neutrophils	CXCR1	IL-8; GCP-2
	CXCR2	IL-8; GCP-2; GRO; NAP-2; ENA-78
Eosinophils	CCR1	MIP-1α; RANTES; MCP-3; MIP-1δ, HCC-1; MRIF-1
	CCR3	Eotaxin-1, -2, -3; RANTES; MCP-2, -3, -4
Basophils	CCR2	MCP-1, -2, -3, -4
	CCR3	Eotaxin-1, -2, -3; RANTES; MCP-2, -3, -4
Monocytes	CCR1	MIP-1α; RANTES; MCP-3; MIP-1δ, HCC-1; MRIF-1
	CCR2	MCP-1, -2, -3, -4
	CCR5	MIP-1α; MIP-1β; RANTES
	CCR8	1-309/TCA-3; TARC; MIP-1b
	CCR9	TECK
	CXCR4	SDF-1
	CX3CR1	Fractalkine
Dendritic cells	CCR1	MIP-1α; RANTES; MCP-3; MIP-1δ, HCC-1; MRIF-1
	CCR5	MIP-1α; MIP-1β; RANTES
	CCR6	MIP-3α/LARC
	CCR7	MIP-3β/ELC; SLC/TCA-4/6CKine
	CCR2	MCP-1, -2, -3, -4
	CCR4	MDC; TARC; MIP-1α; RANTES; MCP-1
	CXCR1	IL-8; GCP-2
	CXCR4	SDF-1
B cells	CXCR5	BCA-1
	CXCR4	SDF-1
Resting T cells	CCR7	MIP-3β/ELC; SLC/TCA-4/6CKine
	?	DC-CK1/PARC
	CXCR4	SDF-1
	XCR1	Lymphotactin
Activated T cells	CCR5 (Th1)	MIP-1α; MIP-1β; RANTES
	CXCR3 (Th1)	MIG; IP-10; I-TAC
	CCR4 (Th2)	MDC; TARC; MIP-1α; RANTES; MCP-1
	CCR8 (Th2)	I-309/TCA-3; TARC; MIP-1β
	CCR3 (Th2)	Eotaxin-1, -2, -3; RANTES; MCP-2, -3, -4
	CRTH2 (Th2)	?
	CCR1	MIP-1α; RANTES; MCP-3; MIP-1δ, HCC-1; MRIF-1
	CCR2	MCP-1, -2, -3, -4
	CCR6	MIP-3α/LARC
	CCR7	MIP-3β/ELC; SLC/TCA-4/6CKine
	?	DC-CK1/PARC
	CX3CR1	Fractalkine

Table 10.10 Receptors on different cell types (*continued*)

Cell type	Receptors	Chemokines
NK cells	CCR2	MCP-1, -2, -3, -4
	CCR4	MDC; TARC; MIP-1α; RANTES; MCP-1
	CCR5	MIP-1α; MIP-1β; RANTES
	CCR7	MIP-3β/ELC; SLC/TCA-4/6CKine
	CXCR3	MIG; IP-10; I-TAC
	CX3CR1	Fractalkine
Thymocytes	CCR4	MDC; TARC; MIP-1α; RANTES; MCP-1
	CCR7	MIP-3β/ELC; SLC/TCA-4/6CKine
	CCR8	1-309/TCA-3; TARC; MIP-1β
	CCR9	TECK
	?	DC-CK1/PARC
	CXCR3	MIG; IP-10; I-TAC

Table 10.11. Selected cytokines and chemokines

The human cytokines and chemokines listed in **Table 10.11** include the majority known and well characterized to date. The key below defines the components of each unit in the table. Additional information on some entries is given in the **Cytokine notes** listed by **note number**. A reference list to primary papers may also be found at http://www.rndsystems.com/cyt cat/references1.html

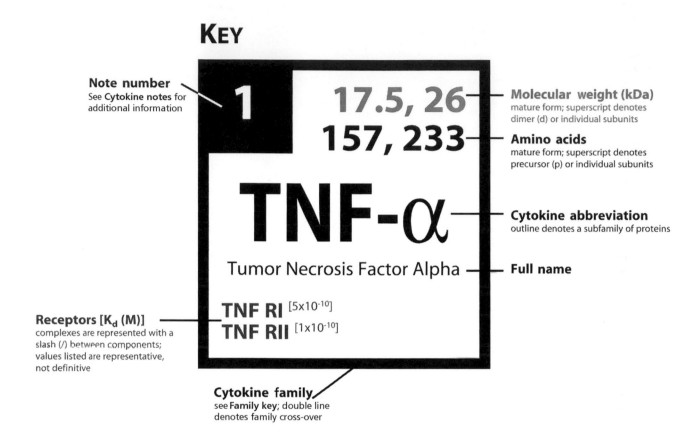

KEY

Note number
See **Cytokine notes** for additional information

1

17.5, 26

Molecular weight (kDa)
mature form; superscript denotes dimer (d) or individual subunits

157, 233

Amino acids
mature form; superscript denotes precursor (p) or individual subunits

TNF-α

Cytokine abbreviation
outline denotes a subfamily of proteins

Tumor Necrosis Factor Alpha

Full name

TNF RI $[5 \times 10^{-10}]$
TNF RII $[1 \times 10^{-10}]$

Receptors [K$_d$ (M)]
complexes are represented with a slash (/) between components; values listed are representative, not definitive

Cytokine family
see **Family key**; double line denotes family cross-over

CYTOKINE NOTES

1. Homotrimer; 17.5 kDa form is secreted, 26 kDa form is membrane-anchored; VEGI (**V**ascular **E**ndothelial **G**rowth **I**nhibitor, not listed here) is a novel member of the TNF family expressed predominantly in endothelial cells

2. Also known as CD70

3. Three isoforms (TGF-β1, 2, 3); biologically active forms are dimers; TGF-β RI and TGF-β RII are high affinity receptors; TGF-β RIII is a low affinity receptor; MIC-1 (**M**acrophage **I**nhibitory **C**ytokine 1, not listed here) bears structural characteristics of the TGF-β superfamily, yet has no strong homology to existing subfamilies

4. Homotrimer; also known as LT-α (lymphotoxin α); can form trimeric complex on membrane surface with LT-β and can bind LT-β R; HVEM (**H**erpes **V**irus **E**ntry **M**ediator)

7. Homodimer; 165 aa isoform shown here; 121 and 165 aa forms are secreted; 145, 183, 189, and 206 aa forms are cell- and matrix-associated

10. Splice variant of IL-6 exists that does not bind gp 130 (antagonist of IL-6)

11. Homodimer; 165 aa (soluble); 220, 248 aa (transmembrane); both soluble and membrane-bound forms are active

12. Approximately 20 total (homodimers and heterodimers); GDFs (growth differentiation factors, not listed here) are related to the BMPs

13. See note 4; LT-β is not secreted

14. Also known as CD154, gp39 and TRAP (**TNF-R**elated **A**ctivation **P**rotein); proteolytic cleavage may produce 15–18 kDa soluble forms; can form homotrimers

16. Homodimer or heterodimer (with VEGF-A); 167 and 186 aa forms are alternatively spliced variants

17. Not related to TGF-β

18. Also known as neuregulin

20. Native ligand-homodimer (65 kDa); sFlt-3L (30 kDa, 245 aa), transmembrane Flt-3L (36 kDa, 235 aa); multiple isoforms exist

21. Homodimer; GDNF Rα subunit (ligand-binding), Ret subunit (tyrosine kinase signaling); GDNF, neurturin, persephin, and artemin are members of the GDNF family

22. Homotrimer (soluble form); membrane-bound form is 40 kDa; DcR3 (**D**ecoy **R**eceptor 3)

23. Also known as TALL-2

26. Pro-domains are cleaved by caspase-1 and calpain to form active IL-1β and IL-1α, respectively; IL-1 ra competes with both IL-1α and IL-1β for binding to IL-1 RI or IL-1 RII; IL-1 RI accessory protein interacts with IL-1/IL-1 RI to create a functional signaling complex

27. Dimers, multimers, IFN-γ is a type II IFN; IFN-α, IFN-β, and IFN-Ω (not listed here) are type I IFNs; see also note 45

28. Also known as somatomedin C; may bind hybrids of IGF RI/ Insulin R

29. Heterodimer; pro-HGF is activated by proteolytic cleavage

30. Also known as FGF-1, β-ECGF; can form dimers

31. Homodimer

32. Amino terminus shares sequence homology with Epo; a splice variant (328 aa and proteolytic cleavage forms of Tpo also exist

33. Homodimer; PDGF Rα binds A chains [1-5×10⁻¹⁰]

34. Precursors dimers of 29/31 kDa chains are processed to form 21 kDa monomers which form non-covalently linked dimers

35. 98 aa is the long form; native forms of AR are 78 and 84 aa

37. Receptor complex contains the gp190 subunit (LIF R), gp130 and a ligand-specific α subunit

38. Three forms of M-CSF are derived from the 544 aa precursor (522, 406, 224 aa); active homodimers are released from the cell membrane by proteolytic cleavage

39. Dimers of β subunits: Activin A (βA/βA), Activin AB (βA/βB), Activin B (βB/βB)

40. Homotrimer; OPG (Osteoprotegerin)

41. Homotrimer; also known as BAFF (**B**-cell **A**ctivating **F**actor belonging to TNF Family), BLyS (**B** **L**ymphocyte **S**timulator), THANK (**T**NF **H**omologue activates **A**potosis, **N**F-κB, and c-Jun NH2-terminal **K**inase); membrane-bound form (52 kDa); C-terminus shares sequence homology with APRIL

42. IL-15 R contains both β and γc chains of the IL-2 R

43. Homodimer

44. See note 26

45. Homodimer; IFN-γ and IL-10 display topological similarity

46. Also known as somatomedin A; 10-25 kDa forms produced by some tumors; IGF RII is identical to the mannose-6-phosphate R; may also activate IGF RI and Insulin R

47. Heterodimer; pro-MSP is activated by proteolytic cleavage

48. Also known as FGF-2; can form multimers

49. Homodimer

51. Heterodimer; PDGF Rα binds both A and B chains [1-5×10⁻¹⁰]; PDGF Rβ binds only B chains [0.5-5×10⁻⁹]

52. Non-covalently linked dimers; most closely related to VEGF-C

54. 174 and 177 aa forms are alternatively spliced variants; the 174 aa form is predominant; G-CSF R is homologous to gp130

55. Receptor complex contains the gp190 subunit (LIF R), gp130, and a ligand-specific α subunit

56. PTP-ζ (**P**rotein-**T**yrosine **P**hosphatase zeta), RPTP-β (**R**eceptor-type **P**rotein-**T**yrosine **P**hosphatase beta)

57. Dimers of an α subunit and β subunit: Inhibin A (α/βA), Inhibin B (α/βB); differential proteolytic processing creates MW variants

58. Homotrimer; 177 aa form is shed; RANK (**R**eceptor **A**ctivator of **N**F-κB)

60. IL-4 Rα subunit is also found in various IL-13 R complexes; receptor contains γc; IL-4 Rα/γc [1×10⁻¹⁰], IL-4 Rα/IL-13 Rα1, IL-4Rα/IL-13Rα1/γc; IL-4δ2 (133 aa splice variant)

62. Homodimer; also known as IGIF (IFN-γ Inducing Factor); caspase-1 converts pro IL-18 to active IL-18 by proteolytic cleavage of the pro-domain

63. Homodimer; unglycosylated (15 kDa), glycosylated (22 kDa)

64. May form homotetramers; there is a discrepancy in the literature whether 121 or 130 aa is correct

65. Heterodimer (p35 = 197 aa / p40 = 306 aa); p40 homodimers are potent IL-12 antagonists

67. Homodimers (aa, MW, Rs listed are for NT-3); NT-3 binds trkA and trkB with lower efficacy; other human NTs include NT-4 and NT-6

68. Ligands for endothelial cell-specific tyrosine kinase receptors (*e.g*, Tie-2); agonists and antagonists

69. Homodimer; PDGF Rα binds both A and B chains [1-5×10⁻¹⁰]; PDGF Rβ binds only B chains [0.5-5×10⁻⁹]

70. Homodimer; 149, 170, and 219 aa forms are splice variants

71. HRG-α and HRG-β1 are splice variants

72. Ob R is homologous to gp130

74. PTP-ζ (**P**rotein-**T**yrosine **P**hosphatase zeta), RPTP-β (**R**eceptor-type **P**rotein-**T**yrosine **P**hosphatase beta)

75. 140 kDa homodimer can be processed into 115 kDa (N-terminal) and 25 kDa (TGF-β-like, C-terminal) segments

76. Homotrimer; homologous to **L**ymphotoxins, Inducible expression and competes with HSV **G**lycoprotein D for HVEM **T**-lymphocyte Receptor)

77. Homotrimer; also known as gp34; gp34 is expressed on HTLV-1 infected leukemic T cells and regulated by the *tax* gene

78. Four possible receptor complexes: IL-13 Rα1/α2; IL-13Rα1/ IL-4 Rα [3×10⁻¹¹]; IL-13 Rα1/IL-4 Rα/γc; IL-13 Rα1 (low levels)/IL-4 Rα/γc; 111 aa splice variant

79. Homotrimer; also known as Apo3L; a weak inducer of apoptosis; Apo3 is also known as DR3, WSL-1, TRAMP, LARD

80. Calculated mass is 20 kDa

Table 10.11. *Cytokine family (continued)*

Key

I — **TNF** – contain highly conserved carboxy terminal domains; can induce receptor trimerization influencing signaling pathways

II — γc – receptors contain a common γ chain (γc)

III — **IL-4** and **IL-3** – bind to shared heteromultimetric receptor complexes

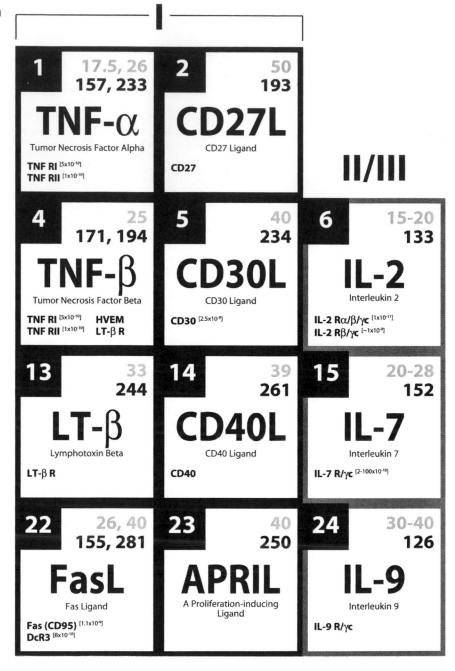

Table 10.11.   Cytokine family (*continued*)

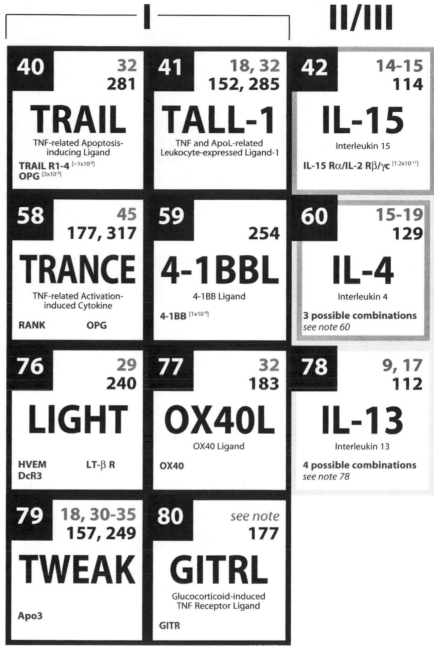

Table 10.11. Cytokine family (*continued*)

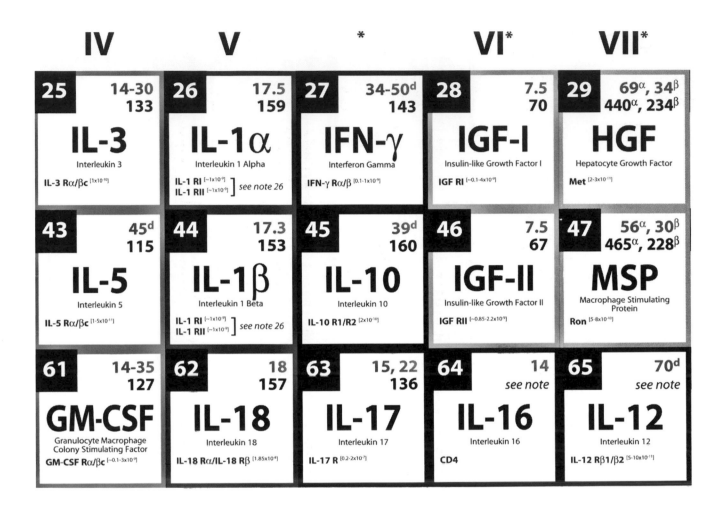

Key

| IV | βc – receptors contain a common β chain (βc) |

| V | **IL-1** – synthesized as glycosylated proforms lacking signal peptides |

| VI | **IGF** – share sequence homology with the insulin family of proteins |

| VII | **HGF** and **MSP** – contain a 4-kringle domain and a pseudo-serine protease domain that lacks enzymatic activity |

Table 10.11. Cytokine family (*continued*)

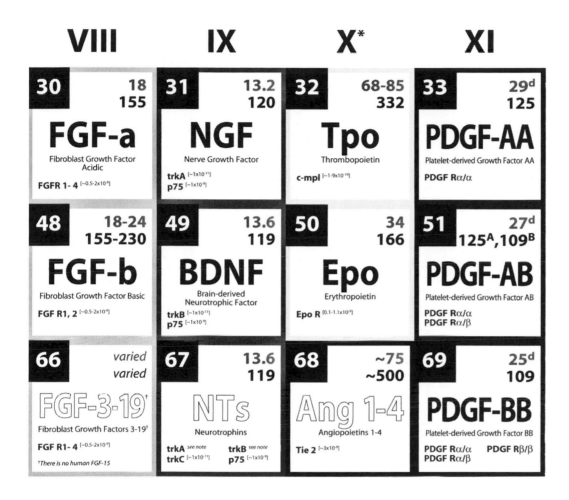

Key

VIII	**FGF** – heparin-binding polypeptides
IX	**Neurotrophic factors** – induce signal transduction through ligand-induced dimerization and activation of trk receptors
X	**Tpo** and **Epo** – share sequence homology
XI	**PDGF, VEGF, PlGF** – dimeric angiogenic factors containing an 8-cysteine motif

Note: Family XI is continued on the following page

Table 10.11. (*continued*)   Cytokine family

Key

| XII | **EGF** – contain at least one extracellular EGF structural unit (conserved 6-cysteine motif that forms 3 disulfide bonds) |

| XIII | **gp130** – receptors are homologous to or contain the gp130 subunit as the common signaling component |

| XIV | **SCF, Flt-3L, M-CSF** – contain a 4-helix bundle structure in the extracellular domain and 4 conserved cysteines; receptors are tyrosine kinases |

| XV | **MK** and **PTN** – products of retinoic acid-responsive genes; developmentally regulated molecules |

| XVI | **TGF-β Superfamily** – contain a highly conserved 7-cysteine domain that forms a characteristic cysteine knot |

XI

7 — 40-46^d / 165
VEGF-A
Vascular Endothelial Growth Factor A
VEGF R1 (flt-1)   Neuropilin-1
VEGF R2 (flk-1)   Neuropilin-2

16 — 46-60^d / 167, 186
VEGF-B
Vascular Endothelial Growth Factor B
VEGF R1 (flt-1)
Neuropilin-1

34 — 42^d / 125, 116
VEGF-C
Vascular Endothelial Growth Factor C
VEGF R2 (flk-1) $^{[4.1×10^{-10}]}$
VEGF R3 (flt-4) $^{[5×10^{-10}]}$

52 — 114
VEGF-D
Vascular Endothelial Growth Factor D
VEGF R2 (flk-1)
VEGF R3 (flt-4)

70 — 46-50^d / see note
PlGF
Placenta Growth Factor
VEGF R1 (flt-1) $^{[2×10^{-10}]}$
Neuropilin-1

XII

8 — 6 / 53
EGF
Epidermal Growth Factor
EGF R (erbB1) $^{[3-10×10^{-9}]}$

17 — 6 / 50
TGF-α
Transforming Growth Factor Alpha
EGF R (erbB1) $^{[0.2-5.3×10^{-9}]}$

35 — 11 / 98
AR
Amphiregulin
EGF R (erbB1)

53 — 32 / 178
BTC
Betacellulin
EGF R (erbB1) $^{[4.6×10^{-11}]}$
erbB3   erbB4

71 — ~45 $^{α, β1}$ / 230 $^{α, β1}$
HRGs
Heregulins
erbB4
erbB3

XII/XIII

9 — 19-23 / 86
HB-EGF
Heparin-binding Epidermal Growth Factor
EGF R (erbB1) $^{[1.4×10^{-10}]}$
erbB4 $^{[1.7×10^{-10}]}$

18 — 25-45 / 296
SMDF
Sensory and Motor Neuron-derived Factor
erbB3

36 — 23 / 178
IL-11
Interleukin 11
IL-11 Rα/gp130

54 — 21 / 174, 177
G-CSF
Granulocyte-colony Stimulating Factor
G-CSF R $^{[1-5×10^{-10}]}$

72 — 16 / 146
OB
Leptin
Ob R $^{[5.1×10^{-9}]}$

Table 10.11. Cytokine family (*continued*)

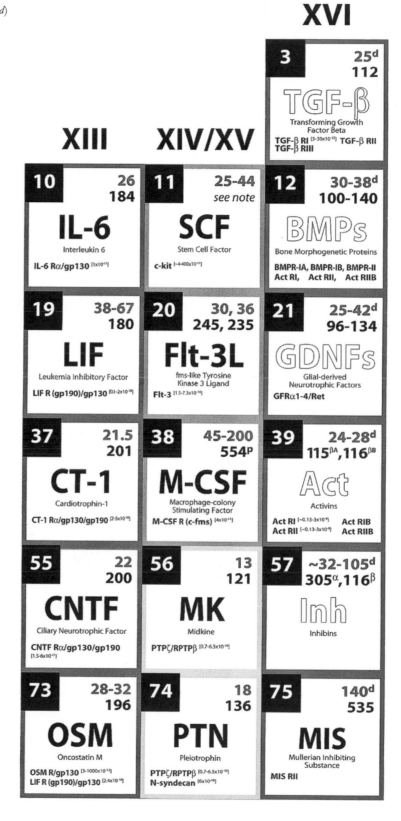

Table 10.11.   Chemokine family (*continued*)

Key

C   **Lymphotactin** – target populations include: lymphoid cells (T and NK cells)

CX₃C   **Fractalkine** – target populations include: lymphoid cells (T and NK cells), monocytes, and PMNs

CC   **β Subfamily** – target populations include: multiple leukocyte subsets (monocytes, basophils, eosinophils, T cells, dendritic cells, NK cells); generally inactive on PMNs

CXC   **α Subfamily** – target populations include: PMNs, T, and B cells

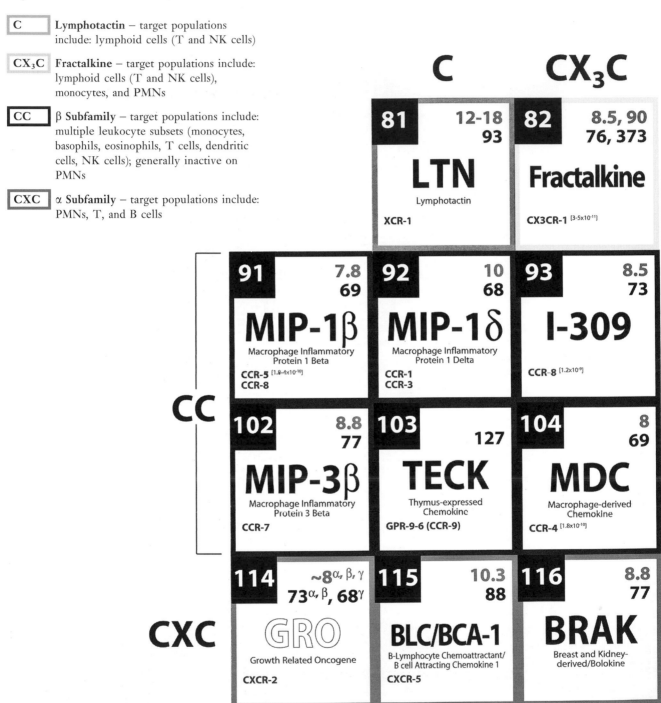

Table 10.11. Cytokine family (*continued*)

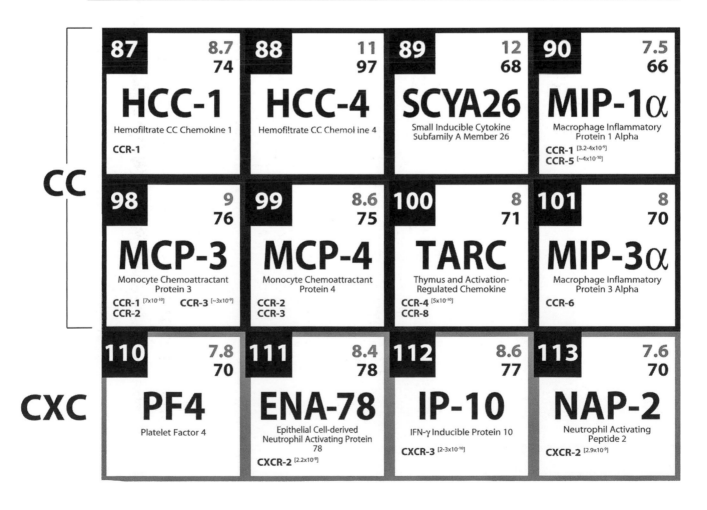

Table 10.12 The matrix metalloproteinases and cytokines

	Cytokine substrates	Amino acid sequence cleaved	Cytokine inducers of MMP expression	Cytokine inhibitors of MMP expression
MMP-1 Collagenase 1 Interstitial collagenase Mr (latent/active) 52/43 kDa	IGFBP-2[1] IGFBP-3[2] IL-1β[3,4] TNF-α[5]	Not defined L-R-A-Y L-L-P-A Not defined L-A-Q-A V-R-S-S	BTC, CD40L, EGF, FGF-a, FGF-b, FGF-7, FGF-9, GM-CSF, HGF, HRG-β1, IFN-β, IFN-γ, IL-1α, IL-1β, IL-4, IL-5, IL-6, sIL-6 Rα, IL-8, IL-10, MIF, NGF, oncostatin M, PD-ECGF, PDGF, PDGF-AA, PDGF-BB, PF4, TGF-α, TGF-β1, TNF-α, TNF-β, VEGF	BMP-2, CD40L, FGF-b, FGF-9, IFN-γ, IL-1ra, IL-4, IL-11, IL-13, TGF-β1, TGF-β2, TGF-β3
MMP-2 Gelatinase A 72 kDa gelatinase Type IV collagenase Mr (latent/active) 72/62 kDa	FGFR1[1] IGFBP-3[2] IGFBP-5[3] IL-1β[4,5] MCP-3[6] TGF-β1[7] THF-α[8]	R-P-A-V M-T-S-P L-R-A-Y L-L-P-A Not defined, G-P-T-E L-K-A-L Q-P-V-G I-N-T-S-T-T Not defined L-A-Q-A V-R-S-S	Activin A, CD40L, EGF, endothelin-1, FGF-b, FGF-3, G-CSF, CM-CSF, HGF, IFN-α, IFN-γ, IGF-1, IL-1α, IL-1β, IL-3, IL-6, sIL-6 Rα, IL-8, IL-13, LIF, M-CSF, MIF, oncostatin M, PDGF, SCF, TGF-α, TGF-β1, TNF-α, VEGF	IFN-β, IFN-γ, IL-4, IL-10
MMP-3 Stromelysin 1 Transin Proteoglycanase CAP Mr (latent/active) 52/43 kDa	HB-EGF[1] IGFBP-3[2] IGFBP-3[2] IGFBP-3[2] IL-1β[3,4] TNF-α[5-7]	L-P-V-E N-R-L-Y L-R-A-Y L-L-P-A A-P-G-N A-S-E-S F-S-S-E S-K-R-E Not defined L-A-Q-A V-R-S-S	BTC, CD40L, EGF, FGF-b, HGF, HRG-β1, IFN-β, IFN-γ, IGF-I, IL-1α, IL-1β, IL-6, IL-8, IL-10, IL-17, IL-18, MIF, NGF, oncostatin M, PD-ECGF, PDGF, TGF-β1, TNF-α, VEGF	CD40L, IFN-γ, IL-4, IL-11, IL-13, TGF-β1
MMP-7 Matrilysin Pump-1 Mr (latent/active) 28/19 kDa	TNF-α[1]	L-A-Q-A V-R-S-S	BTC, EGF, FGF-a, FGF-b, FGF-9, FGF-10, HRG-β1, IL-1α, IL-1β, IFN-γ, IL-4, IL-10, TGF-β1, TNF-α	
MMP-8 Collagenase 2 Neutrophil collagenase Mr (latent/active) 75/55 kDa	None identified to date		IL-1β, TNF-α	TGF-β1

Table 10.12 The matrix metalloproteinases and cytokines (*continued*)

	Cytokine substrates	Amino acid sequence cleaved	Cytokine inducers of MMP expression	Cytokine inhibitors of MMP expression
MMP-9 Gelatinase B 92 kDa gelatinase Type V collagenase Mr (latent/active) 92/82/65 kDa	CTAP-III/NAP-2[1] GROα[1] IL-1β[2] IL-8[1] PF-4[1] TGF-β1[3] TNF-α[4]	(7 different cleavage sites determined) Not defined Not defined A-V-L-P-R-S A-K-E-L-R Not defined L-A-Q-A V-R-S-S	AR, BTC, CD40L, EGF, FGF-b, FGF-3, fractalkine, GCP-2, G-CSF, GM-CSF, GROα, HB-EGF, HGF, HRG-β1, IFN-α, IFN-γ, IGF-I, IL-1α, IL-1β, IL-3, IL-6, IL-8, IL-13, IL-17, MCP-1, M-CSF, MIP-1α, MIP-1β, oncostatin M, PDGF, RANTES, SCF, TGF-α, TGF-β1, TNF-α, TNF-β, VEGF	IL-1ra, IL-4, IL-10, IFN-β, IFN-β1b, IFN-γ, TGF-β1, TGF-β2, TNF-α
MMP-10 Stromelysin 2 Transin 2 Mr (latent/active) 52/44 kDa	None identified to date		EGF, FGF-7, TGF-α, TGF-β1, TNF-α	
MMP-11 Stromelysin 3 Mr (latent/active) 51/46 kDa	IGFBP-1[1]	K-S-L-H V-T-N-I	EGF, FGF-b, IGF-II, IL-6, PDGF-BB	
MMP-12 Macrophage elastase Metalloelastase Mr (latent/active) 52/20 kDa	TNF-α[1]	Not defined	CD40L, GM-CSF, IL-1β, IL-13, MCP-1, M-CSF, PDGF-BB, TNF-α, VEGF	IFN-γ, M-CSF, TGF-β1
MMP-13 Collagenase 3 Mr (latent/active) 52/42 kDa	None identified to date		EGF, FGF-b, FGF-7, IGF-I, IL-1β, IL-6, IL-13, LIF, oncostatin M, PDGF, PDGF-BB, TGF-α, TGF-β1, TGF-β2, TNF-α	BMP-2, BMP-4, BMP-6, IFN-γ, IGF-I, IGF-II, IL-4, IL-13, TGF-β1

Table 10.12 The matrix metalloproteinases and cytokines (*continued*)

	Cytokine substrates	Amino acid sequence cleaved	Cytokine inducers of MMP expression	Cytokine inhibitors of MMP expression
MMP-14 MT1-MMP Mr (latent/active) 64/54 kDa	TNF-α[1]	Not defined	GM-CSF, HGF, IL-1α, IL-1β, IL-13, TNF-α	
MMP-15 MT2-MMP Mr (latent/active) 71/61 kDa	TNF-α[1]	Not defined		
MMP-17 MT4-MMP Mr (latent/active) 62/51 kDa	TNF-α[1,2]	Not defined		

Substrate references to MMP-1:
1 Rajah R *et al. Am J Physiol* 1996; 271: L1014
2 Fowlkes JL *et al. J Biol Chem* 1994; 269:25742
3 Hazuda DJ *et al. J Biol Chem* 1990; 265: 6318
4 Ito A *et al. J Biol Chem* 1996; 271:14657
5 Gearing AJ *et al. Nature* 1994; 370:555
Substrate references to MMP-2:
1 Levi E *et al. Proc Natl Acad Sci USA* 1996; 93:7069
2 Fowlkes JL *et al. J Biol Chem* 1994; 269:25742
3 Thrailkill KM *et al. Endocrinology* 1995; 136:3527
4 Ito A *et al. J Biol Chem* 1996; 271:14657
5 Schonbeck U *et al. J Immunol* 1998; 161:3340
6 McQuibban GA *et al. Science.* 2000; 289: 1202
7 Yu Q Stamenkovic I *Genes Dev* 2000; 14:163
8 Gearing AJ *et al. Nature* 1994; 370:555
Substrate references to MMP-3:
1 Suzuki M *et al. J Biol Chem* 1997; 272:31730
2 Fowlkes JL *et al. J Biol Chem* 1994; 269:25742
3 Ito A *et al. J Biol Chem* 1996; 271:14657
4 Schonbeck U *et al. J Immunol* 1998; 161:3340
5 Black RA *et al. Biochem Biophys Res Commun* 1996; 225:400
6 Gearing AJ *et al. Nature* 1994; 370:555
7 Gearing AJ *et al. J Leukocyt Biol* 1995; 57:774

Substrate references to MMP-7:
1 Gearing AJ *et al. Nature* 1994; 370:555
Substrate references to MMP-8:
1 Van den Steen PE *et al. Blood* 2000; 96:2673
2 Schonbeck U *et al. J Immunol* 1998; 161:3340
3 Yu Q and Stamenkovic I *Genes Dev* 2000; 14:163
4 Gearing AJ *et al. Nature* 1994; 370:555
Substrate references to MMP-11:
1 Manes S *et al. J Biol Chem* 1997; 272:25706
Substrate references to MMP-12:
1 Chandler S *et al. Biochem Biophys Res Commun* 1996; 228:421
Substrate references to MMP-14:
1 d'Ortho MP *et al. Eur J Biochem* 1997; 250:751
Substrate references to MMP-15:
1 d'Ortho MP *et al. Eur J Biochem* 1997; 250:751
Substrate references to MMP-17:
1 English WR *et al. J Biol Chem* 2000; 275:14046
2 Wang Y *et al. J Biol Chem* 1999; 274:33043

THE COMPLEMENT SYSTEM 11

- COMPLEMENT ACTIVATION
- COMPLEMENT CONSTITUENTS
- REGULATION OF THE COMPLEMENT CASCADE

- COMPLEMENT RECEPTORS
- COMPLEMENT GENETICS
- COMPLEMENT DEFICIENCY AND DISEASES

COMPLEMENT ACTIVATION

The complement system is comprised of multiple soluble plasma and other body fluid proteins, that function either as enzymes or as binding proteins, together with cellular receptors for many of them and regulatory membrane proteins found on blood and other tissue cells. These proteins play a critical role in facilitating phagocytosis of immune complexes. There are numerous biological activities associated with complement besides immune lysis. These include the formation of anaphylatoxin, chemotaxis, opsonization, phagocytosis, bacteriolysis, hemolysis, and other amplification mechanisms.

Complement proteins circulate in the blood in an inactivate form. These molecules and their fragments, resulting from the activation process, are significant in the regulation of immune responsiveness. The complement system can be activated by three different pathways: the classical, alternative, and lectin pathways. Each of the three pathways has distinct early events including binding of receptors and a unique set of components of the complement system that initiate an enzymatic cascade. **Table 11.1** compares the three complement activation pathways. The pathways differ in the manner in which they are activated and ultimately produce a key enzyme called C3 convertase. The end results and defense benefits of each pathway are the same, including triggering inflammation, attracting phagocytes to the infection site, promoting antigen attachment to phagocytes, causing lysis of Gram-negative bacteria, and removing harmful immune complexes from the body.

COMPLEMENT CONSTITUENTS

Complement components

The components of the classical pathway are numbered from 1 to 9 and prefixed by the letter C, i.e. C1 through C9. Characteristics of these complement components are shown in **Table 11.2**.

Complement factors

The early components of the alternative pathway are known as factors. Each molecule is named by a letter, for example factor B, factor D, factor P. **Table 11.3** lists the complement factors of the alternative pathway.

Membrane attack complex

All three complement pathways use the same terminal components C5–C9 in the later stages of activation that form the membrane attack complex (MAC) – $C5b678(9)_n$.

Initiation of the MAC assembly begins with C5 cleavage into C5a and C5b fragments. A $(C5b678)_1(C9)_n$ complex then forms either on natural membranes or, in their absence, in combination with such plasma inhibitors as lipoproteins, antithrombin III, and S protein. Mechanisms proposed for complement-mediated cytolysis include extrinsic protein channel incorporation into the plasma membrane or membrane deformation and destruction. Central regions of C6, C7, C8α, C8β, and C9 have been postulated to contain amphiphilic structures which may be membrane anchors. MAC assembly and insertion into the outer membrane is

393

Table 11.1   Three pathways of complement system activation

	Classical pathway	Alternative pathway	Lectin pathway
Activated by	Binding of antibody molecules (specifically IgM and IgG1, IgG2, IgG3) to a foreign particle	Invading microorganisms	Binding of MBP to the mannose groups of carbohydrates on microorganisms
Activation mechanism	Antibody-dependent	Antibody-independent	Antibody-independent
Limb of immunity	Adaptive immune response	Innate immune response	Innate immune response
Components	C1 (C1q, C1r, C1s) to C9	Factors B, D, P, H, I	C1 (C1r, C1s) to C9
Components that initiate enzyme cascade	C1 (q, r, s), C4, C2	C3, B, D	Lectin, MASP1, MASP2, C4, C2
C3 convertase	C4bC2a, C2b	C3bBb	C2b, C4bC2a
C5 convertase	C4bC2aC3b	C3bBbC3b	C4bC2aC3b
Terminal components	C5–C9, MAC (C5b678(9)$_n$)	C5–C9, MAC (C5b678(9)$_n$)	C5–C9, MAC (C5b678(9)$_n$)

Abbreviations:

B	Plasma factor B	I	Factor I
D	Plasma protease factor D	MBP	Mannan-binding protein
P	Properdin	MASP	MBP-associated serine protease
H	Protein H	MAC	Membrane attack complex

requisite for lysis of bacteria. A single C9 molecule per C5b678 leads to erythrocyte lysis. Nucleated cells may rid their surfaces of MAC through endocytosis or exocytosis. Control proteins acting at different levels may inhibit killing of homologous cells mediated by the MAC.

Complement fragments

Activation of each of the components results from a proteolytic cleavage event in a cascade mechanism which splits the native molecule into two fragments. The fragment which participates further in the complement cascade is designated the b fragment and is usually larger than the a fragment which possesses other biological activities. **Table 11.4** lists the active complement fragments, their complement sources and their functions.

Complement components in inflammation

The complement system is a potent mechanism for initiating and amplifying inflammation. This is mediated through fragments of complement components. Some important complement components in acute inflammatory reactions are summarized in **Table 11.5**.

REGULATION OF THE COMPLEMENT CASCADE

Activation of the complement cascade is finely tuned and under rigid control by complement regulatory proteins. Protein inhibitors that occur naturally and block the action of complement components include factor H, factor I, C1 inhibitor, and C4-binding protein (C4BP). Also included among complement inhibitors are heating to 56 °C to inactivate C1 and C2; combining with hydrazine and ammonia to block the action of C3 and C4; and the addition of zymosan or cobra venom factor to induce alternate pathway activation of C3, which consumes C3 in the plasma. **Table 11.6** lists some of these regulatory proteins and their action mechanisms.

COMPLEMENT RECEPTORS

Adherent receptors

Different fragments, released from individual components during complement activation, operate by a non-cytolytic mechanism through specific receptors present on various cell types. The direction and intensity of the biological

Table 11.2 Complement components of the classical pathway

	Molecular weight (kDa)	Structure	Genomic location	Function
C1	750	Multimeric with subcomponents C1q, C1r×2, and C1s×2,	Chromosome 1	Initiates classical pathway by binding with IgG or IgM, causes progression of complement cascade via serine esterase function of C1s to cleave C4 and C2
C2	110	Single polypeptide chain	Chromosome 6, short arm	C2a, associated with C4b, constitutes classical pathway C3 convertase C4b2a; C2b catalyzes enzymatic cleavage
C3	195	Glycoprotein heterodimer linked by disulfide bonds	Chromosome 19	Binds with C4b2a to form classical pathway C5 convertase C4b2a3b; functions as opsonin
C4	210	α, β and γ chains joined through connecting peptides	Chromosome 6, short arm	C4a is an anaphylatoxin; C4b enables complement activation to take place on cell surfaces where antibodies bind
C5	190	Dimeric molecule comprised of an α and β polypeptide chain linked by disulfide bonds	Chromosome 9	C5a has chemotactic action for neutrophils and anaphylatoxin activity; C5b has a binding site for C6 and complexes with it to begin generation of the membrane attack complex
C6	128	Single polypeptide chain	Chromosome 5	Participates in formation of the membrane attack complex
C7	110	Single polypeptide chain	Chromosome 5	Binds with C5b and C6 to form C5b67 that anchors to the cell membrane and provides a binding site for C8 and C9 in the formation of the membrane attack complex
C8	155	α, β, and γ chains with α and γ chains joined by disulfide bonds and then linked to β chain via non-covalent bonds	Chromosome 1 for α and β chains, chromosome 9 for γ chain	Forms C5b678 that inserts into the cell membrane's lipid bilayer and produces a transmembrane channel with limited cell lysis capacity
C9	69	Single polypeptide chain	Chromosome 5	Binds to C5b678 and polymerizes to complete formation of the membrane attack complex that produces 100 nm diameter hollow tubes in cell membrane to form transmembrane channels and cause cell lysis

response depend on the state of the receptors (affinity and density) and on the function of cells bearing receptors. From the functional standpoint, complement receptors can be divided into two types: the adherent and the other receptors. Adherent receptors mediate adherence of cells and other particles with bound C3b or C4b fragments and are known as CR1 to CR5. Proteolytic cleavage of human complement component C3 takes place following activation of either the classical or the alternative complement pathway. Following the generation of C3a and C3b, the C3b covalently binds to bacteria, immune complexes, or some other target and then unites with a high-affinity receptor termed the C3b/C4b receptor currently known as CR1. Subsequent proteolytic cleavage of the bound C3b is attributable to factor I and a cofactor. This action yields C3bi, C3dg, and C3c, which interact with specific receptors. CR2 is the C3dg receptor, and CR3 is the C3bi receptor.

Table 11.3 Factors of the alternative pathways

	Molecular weight (kDa)	Genomic location	Function
Factor B	93	Chromosome 6, short arm	Serves as a serine proteinase; Bb has both C3 convertase activity (as a C3bBb complex) and C5 convertase activity upon association with a second molecule of C3b
Factor D	25	Chromosome 10	Serves as a serine esterase and splits Factor B to produce Bb and Ba
Factor P (properdin)	220	Chromosome 10	Stabilizes the alternative pathway C3 convertase C3bBb complex to produce C3bBbP
Factor H	150	Chromosome 1	Blocks formation of C3 convertase in alternative pathway by uniting with C3b and facilitating dissociation of alternative complement pathway C3 convertase, designated C3bBb, into C3b and Bb
Factor I	90	Chromosome 4	Serves as an inhibitor of the alternative complement pathway by cleaving both C4b and C3b using several other inhibitors

Other receptors

The second group of receptors reacts with small complement fragments (C4a, C3a, C5a) as well as with C1q, Ba, Bb, and factor H. Stimulation of these receptors results in various biological effects, including chemotaxis, secretion of vasoactive amines, and release of mediators of the inflammatory and anaphylactic reaction.

The structure, specificity, and cellular distribution of complement receptors are summarized in **Table 11.7**.

COMPLEMENT GENETICS

Most genes encoding complement component-related proteins in humans have been sequenced and assigned to chromosomal loci. Structurally and/or functionally similar components are organized in major gene clusters in the human genome. For example, genes for complement regulatory proteins CR1, CR2, membrane cofactor protein (MCP), decay accelerating factor (DAF), C4 binding protein (C4BP), and factor H are clustered on chromosome 1. Genes for membrane attack complex components C6, C7, and C9 are clustered on chromosome 5. **Table 11.8** lists the chromosomal assignments of complement components and related proteins.

One important aspect of complement genetics is polymorphism. The evolution of polymorphism and structure–function relationships gives insight into deficiencies and genetically determined disease susceptibilities. **Table 11.9** summarizes polymorphisms and deficiencies of complement components and the methodology used to study polymorphism.

COMPLEMENT DEFICIENCY AND DISEASES

Deficiencies of the complement components, although rare, have been reported for most of the constituents. C2 deficiency is the most common complement deficiency disorder. These deficiencies can be inherited or acquired. Complement deficiency has been associated with autoimmune diseases or increased susceptibility to infections.

Deficiencies of complement components of the classical pathway are usually associated with immune complex diseases such as discoid or systemic lupus erythematosus. Deficiencies of complement components of the alternative pathway are associated with severe infections with a high mortality rate such as fulminant pyogenic *Neisseria*. Deficiencies of complement components of the lectin pathway are associated with recurrent infections and accelerated SLE and rheumatoid arthritis. **Table 11.10** gives the inheritance, related clinical conditions, and laboratory findings of inherited deficiencies of complement components and related proteins.

Table 11.4 Subcomponents and fragments of complement system

	Native component	Pathway	Function
C1q	C1	Classic	Actual recognition portion that binds to immunoglobulin Fc, activates C1r
C1r	C1	Classic, lectin	Protease that cleaves C1s
C1s	C1	Classic, lectin	Cleaves C2 and C4 to C2a/b and C4a/b
C2a	C2	Classic, lectin	Unknown
C2b	C2	Classic, lectin	Combines with C4b to produce C3 convertase that cleaves C3 to C3a/b
C3a	C3	Classic, alternative, lectin	Potent anaphylatoxin, mediates inflammation
C3b	C3	Classic, alternative, lectin	Combines with C4b2b to form C4b2b3b that cleaves C5, binds cell surfaces for opsonization and activation of alternative pathway
C4a	C4	Classic, lectin	Mediates inflammation
C4b	C4	Classic, lectin	Combines with C2b to produce C3 convertase that cleaves C3 to C3a/b, binds cell surfaces for opsonization
C5a	C5	Classic, alternative, lectin	Potent anaphylatoxin, chemotaxin, mediates inflammation
C5b	C5	Classic, alternative, lectin	Anchors on target cell surface and initiates MAC ($C5b678(9)_n$) assembly
C9n	C9	Classic, alternative, lectin	Polymerizes around C5b678 to form a hole in the cell membrane leading to lysis
B	Factor B	Alternative	Binds to cell surface bound C3b
Ba	Factor B	Alternative	Unknown
Bb	Factor B	Alternative	Contains active site for C3 convertase
D	Factor D	Alternative	Cleaves bound factor B to Ba/Bb
P	properdin	Alternative	Binds C3bBb to stabilize C3 convertase leading to cleavage of C3

Table 11.5 Complement fragments in acute inflammation

C2a	Opsonization of bacteria
C3a	Increase of vascular permeability; degranulation of mast cells and basophils, and release of histamine; chemotaxis for neutrophils
C3b	Opsonization of bacteria and immune complexes leading to phagocytosis; stimulation of respiratory burst of professional phagocytes; solubilization of circulating immune complexes
C4a	Increase of vascular permeability; degranulation of mast cells and basophils; smooth muscle contraction
C4b	Opsonization of bacteria and immune complexes leading to phagocytosis
C5a	Neutrophil activation and chemotaxis; increase of vascular permeability; releases of histamine from mast cells; stimulation of prostaglandin and leukotriene production; stimulation of respiratory burst of professional phagocytes
C567	Neutrophil chemotaxis
$C5b678(9)_n$	Cytolytic activity of bacteria and foreign cells
CR1	Solubilization of circulating immune complexes

Table 11.6 Complement regulatory proteins

Proteins	Functions
CR1, CD35	Binds C3b and C4b, processing of immune complexes and promotion of binding and phagocytosis of C3b-coated particles/cells
Membrane cofactor of proteolysis (MCP), CD46	Binds C3b and C4b, allowing their degradation by Factor I
Membrane attack complex inhibitory factor (MACIF), protective membrane inhibitor of reactive lysis (MIRL), homologous restriction factor 20 (HRF 20), CD59	Inhibits MAC formation by binding to sites on C8 and C9 which blocks the uptake and incorporation of multiple C9 molecules into the complex
C4 binding protein (C4-BP)	Accelerates decay of the classical pathway C3 convertase by binding to C4b and displacing C2a; functions as a cofactor for factor I-mediated cleavage of C4b
C1 inhibitor (C1 INH)	Blocks C1r activation, prevents C1r cleavage of C1s and inhibits C1s splitting of C4 and C2
Decay-accelerating factor (DAF)	Facilitates dissociation of classical complement pathway C3 convertase (C4b2a) into C4b and C2a; promotes the dissociation of alternative complement pathway C3 convertase (C3bBb) into C3b and Bb; inhibits C5 convertases (C4b2a3b and C3bBb3b) on the cell surface
Factor H	Unites with C3b and facilitates dissociation of alternative complement pathway C3 convertase, designated C3bBb, into C3b and Bb; facilitates Factor I to split C3b
Factor I	Splits the α chain of C3b to produce C3bi and the α chain of C4b to yield C4bi to prevent formation of convertase
Factor P (properdin)	Combines with C3b and stabilizes alternative pathway C3 convertase (C3bB) to produce C3bBbP

Table 11.7 Structure, specificity, and cellular distribution of complement receptors

Receptor type	Specificity	Structure	Cell type distribution
CR1 (CD35)	C3b> C4b> iC3b	Four allotypes that vary in size from 150 kDa to 250 kDa due to different numbers of repeating sequence motif units	Monocytes, neutrophils: high expression Tissue macrophages: low expression B cells, eosinophils: high expression Kidney podocytes: high expression T cells (~20%): low expression Peripheral nerves: low expression
CR2 (CD21)	iC3b = C3dg> C3d> C3b, Epstein–Barr virus, CD23	Made up of repeating sequence motif units similar to CR1; 140 kDa	B cells: high expression Lymph node follicular dendritic cells: very high expression Thymocytes, pharyngeal epithelial cells: low expression
CR3 (CD11b; Mac-1; $\alpha_m\beta_2$-integrin	iC3b, C3dg, C3d, β-glucan, ICAM-1, fibrinogen, factor X, collagen, heparan sulfate, CD14, CD16, CD87	Two non-covalently linked glycoprotein chains: 165 kDa α_m chain (CD11b), 95 kDa β_2 chain (CD18)	Neutrophils, monocytes: high expression Tissue macrophages: low expression Activated cytotoxic T cells, NK cells, eosinophils: high expression
CR4 (CD11c; p150,95; $\alpha_x\beta_2$-integrin	iC3b, C3dg, fibrinogen	Two non-covalently linked glycoprotein chains: 150 kDa α_x chain (CD11c), 95 kDa β_2 chain (CD18)	Neutrophils, monocytes: low expression Tissue macrophages: high expression Activated B cells: high expression Activated cytotoxic T cells, NK cells, eosinophils: low expression
CR5	C3d portion of fluid-phase iC3b, C3dg, C3d	Unknown; 95 kDa candidate molecule expressed by platelets	Neutrophils: low expression Platelets: low expression
gC1q-R	C1q globular 'head', as well as vitronectin and high molecular weight kininogen	33 kDa glycoprotein	All leukocyte types: variable expression Platelets: high expression Endothelial cells: high expression
cC1q-R	C1q collagen 'tail'	60 kDa glycoprotein; homologous to calreticulin	All leukocyte types: variable expression
C1q-R$_p$	C1q collagen-like fragment, mannose-binding protein, lung surfactant protein A	126 kDa glycoprotein	Neutrophils: high expression Monocytes: high expression
C3a-R	C3a, C4a	Guinea pig platelet: 95–105 kDa Human mast cell: 57 kDa and 97 kDa Human leukocyte: 482 amino acids, ~54 kDa, seven membrane-spanning domains	Mast cells: high expression Eosinophils, basophils: high expression Neutrophils, monocytes: low expression Guinea pig platelets: high expression
C5a-R (CD88)	C5a, C5a$_{des\ Arg}$	47 kDa glycoprotein binding unit with seven membrane-spanning domains expressed as an oligomer of 150–200 kDa	Mast cells, neutrophils: high expression Monocytes and tissue macrophages: high expression Eosinophils: high expression Hepatocytes, astrocytes, endothelial cells: high expression

Table 11.8 Chromosomal assignments of complement components and related proteins

Component (or subunit)	Gene symbol	Chromosomal location
C1q: α chain	C1QA	1p34.1 – p36.3
C1q: β chain	C1QB	1p34.1 – p36.3
C1q: γ chain	C1QG	1p34.1 – p36.3
C8: α chain	C8A	1p32
C8: β chain	C8B	1p32
C4-binding protein: α chain	C4BPA	1q32[a]
B4-binding protein: β chain	C4BPB	1q32[a]
Complement receptor 1 (CD35)	CR1	1q32[a]
Complement receptor 2 (CD21)	CR2	1q32[a]
Decay-accelerating factor (CD55)	DAF	1q32[a]
Membrane cofactor protein (CD46)	MCP	1q32[a]
Factor H	HF	1q32[a]
Factor I	IF	4q25
C6	C6	5p13[b]
C7	C7	5p13[b]
C9	C9	5p13[b]
C2	C2	6p21.3[c]
Factor B	BF	6p21.3[c]
C4A (isotype)	C4A	6p21.3[c]
C4B (isotype)	C4B	6p21.3[c]
C8: γ chain	C8G	9q22.3 – q32
C5	C5	9q33
Perforin	PRF1	10q22
Mannose-binding protein (lectin)	MBL	10q11.2 – q21
Surfactant protein A (SP-A)	SFTP1	10q22 – q23[d]
Surfactant protein D (SP-D)	SFTP4	10q22 – q23[d]
Membrane inhibitor of reactive lysis (MIRL, CD59)	CD59	11p13
C1 inhibitor	C1NH	11q12 – q13.1
C1r	C1R	12p13
C1s	C1S	12p13
Complement receptor 3: α chain	CR3A	16p11.2[e]

Table 11.8 Chromosomal assignments of complement components and related proteins

Component (or subunit)	Gene symbol	Chromosomal location
Vitronectin (S-protein)	VTN	17q11
C3	C3	19p13.3 – p13.2
C5a receptor 1	C5R1	19q13.3 – q13.4
Leukocyte adhesion molecule: β chain (CD18)	ITGB2	21q22.3[f]
Properdin	PFC	Xp11.4 – p11.2

[a]Regulators of complement activation (RCA) gene cluster.
[b]Membrane attack complex (MAC) gene cluster.
[c]MHC class III gene region.
[d]Surfactant protein (SP) gene cluster.
[e]Leukocyte adhesion α (LAA) gene cluster.
[f]Common β chain for the cell adhesion molecules CR3, LFA-1, and gp150,95.
No map assignment: C1q receptor (C1QR, collectin receptor), factor J (JF), C8-binding protein (C8BP, HRP).

Table 11.9 Summary of polymorphisms and deficiency of complement components in human and animals

Complement component	Typing technique	Total no. of known alleles	Deficiency	Polymophism in other species	Disease associations
C1q	Imm/Funct	—	Yes	—	Yes
C1r	IEF + WB	>10	Yes	—	Yes
C1s	IEF + WB	2	Yes	—	Yes
C2	IEF + WB RFLP/PCR	<10	Yes	Ch, Rh, Gp, Hm	Yes
C3	HVAGE + IFX RFLP/PCR	>30	Yes	Ch, Rh, Ba, Mc, Hl Ms, Rb, D	Yes
C4	HVAGE + IFX/WB HVAGE + HOV MAB SDS-PAGE (α/β chains) HAI (Rodgers/Chido) RFLP/PCR	>30	Yes	Ch, Mc, D, Ms, Gp, Xl	Yes
C5	IEF + WB	2	Yes	—	Yes
C6	IEF + WB	>20	Yes	Ch, Rh, Rb, Ms	Yes
C7	IEF + WB/HOV MAB	>10	Yes	—	Yes
C8	IEF + WB/HOV SDS-PAGE RFLP/PCR	<10	Yes	Ch	Yes

Table 11.9 Summary of polymorphisms and deficiency of complement components in human and animals (*continued*)

Complement component	Typing technique	Total no. of known alleles	Deficiency	Polymophism in other species	Disease associations
Factor B	HVAGE + IFX/HOV IEF RFLP/PCR	>20	Partial	Ch, Rh, Hl, Ba, Gp, Ms	No
Factor I	IEF + WB RFLP/PCR	<5	Yes	—	Yes
Factor H	IEF + WB	<5	Yes	—	Yes
P	Imm/Funct	–	Yes	—	Yes
C4BP	IEF + WB/IFX	<5	No	—	No
CR1	SDS-PAGE EXP RFLP/PCR	<5	Yes	—	Yes
CR2	SDS-PAGE RFLP/PCR	<2	No	—	No
CR3	SDS-PAGE	—	Yes	—	Yes
DAF	Imm (Cromer blood group antigen) RFLP/PCR	<2	Yes	—	Yes
MCP	EXP	<2	No	—	No
MBL	RFLP/PCR	3	No	—	Yes

Typing techniques:

EXP	Expression polymorphism (number of membrane-associated molecules per cell)
Funct	Functional assay
HAI	Hemagglutination inhibition with human alloantisera
HOV	Complement component-dependent hemolytic overlay
HVAGE	High-voltage agarose gel electrophoresis
IEF	Isoelectric focusing
IFX	Immunofixation
Imm	Immunologic detection with specific antisera
MAB	Monoclonal antibodies
RFLP/PCR	DNA restriction fragment length polymorphism detected by Southern blot or polymerase chain reaction analysis
SDS-Page	SDS polyacrylamide gel electrophoresis
WB	Western blot

Animal species:

Ba, baboon; Ch, chimpanzee; D, dog; Gp, guinea pig; Hm, hamster; Hl, *Hanuman langur* monkey; Mc, macaque; Ms, mouse; Rb, rabbit; Rh, Rhesus monkey; Xi, *Xenopus laevis*

Table 11.10 Inherited complement deficiencies in humans

Component	Inheritance	Remarks	Observed clinical condition	Laboratory findings
C1q	Autos. codom.	Three types: 1) partial; 2) complete; 3) inactive protein: combined with immunoglobulin deficiency	Infections, SLE	CH50 = 0; bactericidal activity = 0; C1r↑ C1s↑
C1r, C1s	Autos. codom. not MHC linked	Mostly combined C1r + C1s	Infections (meningitis); collagen diseases	Ch50 = 0; C1NH↑
C4	Autos. codom. (4 genes) MHC linked	Degree dependent on number of 'null' alleles	Collagen diseases; some combined with IgA disorders	CH50 = 0; bactericidal activity = 0; chemotaxis↓; opsonization↓; no IgM→IgG switch; MLC↓
C2	Autos. codom. MHC (A25, B18, DR2) linked	30–70% activity in heterozygotes	Often none; infections when combined with low factor B; SLE; juvenile rheumatoid arthritis	CH50 = 0; bactericidal activity = 0; phagocytosis↓
C3	Autos. codom. not MHC linked		Severe infections; collagen diseases; nephritis	C3<0.1%; CH50 = 0; bactericidal activity = 0; immune adherence = 0; opsonization↓; mobilization of PMN↓; Factor B cleavage = 0; chemotaxis↓
C5	Autos. codom. not MHC linked (regulatory gene?)	13–65% activity in heterozygotes	Collagen diseases; infections with meningococci, gonococci	CH50 = 0; bactericidal activity = 0; chemotaxis = 0; platelet aggregation = 0; opsonization normal (staphylococci)
C6	Autos. codom. not MHC linked; linked to C7 and C9 on chromosome 5		Gonococcal and meningococcal infections; streptococcal meningitis; chronic meningococcemia	CH50 = 0; bactericidal activity = 0; chemotaxis normal; opsonization and intracellular killing = normal; subtotal deficiency: truncated proteins hemolytically functional
C7	Autos. codom. not MHC linked		Some healthy; 50% *Neisseria* infections; 12% connective tissue disorders	CH50 = 0; reactive lysis positive; sera activatable up to C5,6
C8	Autos. codom. not MHC linked C8αγ and C8β controlled by two independent loci	C8αγ or C8β missing; can be reconstituted with missing part; dysfunctional C8β chain?	Meningococcal infections (often uncommon serogroups)	CH50 = 0
C9	Autos. codom. not MHC linked		None	Normal lysis and bactericidal activity at reduced rate (erythrocytes and *Escherichia coli*)

Table 11.10 Inherited complement deficiencies in humans (*continued*)

Component	Inheritance	Remarks	Observed clinical condition	Laboratory findings
C1 NH	Autos. codom. structural gene	Three types: I) synthesis deficiency; II) inactive protein; III) C1NH complexed to albumin (acquired deficiency sometimes by autoantibodies)	Angioedema following trauma, or stress; therapy: danazol, C1 NH infusion	C1 NH function$\downarrow$; C1 NH protein$\downarrow$/normal; C4$\downarrow$C2$\downarrow$; C1$\downarrow$ in acquired form
Factor I	?		Pyogenic infections, meningitis; therapy: plasma infusion	CH50$\downarrow$ C3$\downarrow$ C3b$\uparrow$; Factor B$\downarrow$; properdin$\downarrow$; Factor H$\downarrow$; bactericidal activity$\downarrow$; chemotaxis$\downarrow$; phagocytosis$\downarrow$
Factor H	Autos. recessive? codom. ?	Low levels present	Hemolytic uremic syndrome; dense deposit disease; lupus nephritis	CH50$\downarrow$; C3$\downarrow$; Factor B$\downarrow$; C3d present on erythrocytes; C5–C9 normal
Properdin	X-linked	1) Complete or 2) partial	Meningococcal meningitis; septicemia; none (?)	CH50 = normal; properdin = protein$\downarrow$/0 apH50$\downarrow$; opsonization$\downarrow$
Anaphylatoxin inactivator (carboxypeptidase N)	Autos. recessive	Low titers present	Chronic idiopathic urticaria/angioedema	CH50 normal during attacks; histamine $\uparrow$; C3a$\uparrow$; C4$\downarrow$; C3$\downarrow$; Factor B cleavage
CR3 (CD18 deficiency)	Autos. codom.	Three proteins missing; relative contribution unclear	Recurrent infections, LAD	Proteins missing on leukocytes; zymosan response of monocytes$\downarrow$
CR1	Acquired?		Immune complex disease	Diminished CR1 on erythrocytes; defective immune clearance
C8bp ('HRF')	Acquired	Reduced in PNH	Intravascular lysis	Acid lysis
C59 ('p18', 'MIRL' or 'HRF20')	Acquired	Reduced in PNH	Intravascular lysis	Acid lysis
DAF	Acquired	Reduced in PNH	Intravascular lysis	Acid lysis

Abbreviations:

ApH50	hemolytic test for alternative pathway activity
Autos. codom.	autosomal codominant
CH50	hemolytic test for classic pathway activity
DAF	decay-accelerating factor
HRF	homologous restriction factor
Ig	immunoglobulin
LAD	leukocyte adhesion deficiency syndrome
MHC	major histocompatibility complex
MIRL	membrane inhibitor of reactive lysis
MLC	mixed lymphocyte culture
PNH	paroxysmal nocturnal hemoglobinuria

TYPE I, II, III, AND IV HYPERSENSITIVITY

- **HYPERSENSITIVITY**
- **ALLERGEN**

- **ALLERGIC DISEASES**

HYPERSENSITIVITY

Hypersensitivity is the increased reactivity or increased sensitivity by the animal body to an antigen to which it has been previously exposed. The term is often used as a synonym for allergy, which describes a state of altered reactivity to an antigen. Hypersensitivity has been divided into categories based upon whether it can be passively transferred by antibodies or by specifically immune lymphoid cells. The most widely adopted current classification is that of Coombs and Gell, which designates immunoglobulin-mediated hypersensitivity reactions as types I, II, and III, and lymphoid cell-mediated (delayed-type) hypersensitivity/cell-mediated immunity as a type IV reaction. These four types of hypersensitivity have their own features. They are compared in **Table 12.1**, showing the immunoglobulins, effector cells, target cells, primary and secondary mediators, and the physiologic effects.

Type IV hypersensitivity is a form of hypersensitivity mediated by specifically sensitized cells. Whereas antibodies participate in type I, II, and III reactions, T cells mediate type IV hypersensitivity. Two types of reactions, mediated by separate T cell subsets, are observed. Delayed-type hypersensitivity (DTH) is mediated by CD4$^+$ T cells, and cellular cytotoxicity is mediated principally by CD8$^+$ T cells. A classic delayed hypersensitivity reaction is the tuberculin reaction. Following exposure to *Mycobacterium tuberculosis*, CD4$^+$ cells recognize the microbe's antigens complexed with class II MHC molecules on the surface of antigen-presenting cells that process the mycobacterial antigens. Memory T cells develop and remain in the circulation for prolonged periods. When tuberculin antigen is intradermally injected, sensitized T cells react with the antigen on the antigen-presenting cell's surface, undergo transformation, and secrete lymphokines that lead to the manifestations of hypersensitivity. In T cell-mediated cytotoxicity, CD8$^+$ T cells kill antigen-bearing target cells. The cytotoxic T

cells play a significant role in resistance to viral infections. Class I MHC molecules present viral antigens to CD8$^+$ T cells as a viral peptide-class I molecular complex, which is transported to the infected cell's surface. Cytotoxic CD8$^+$ cells recognize this and lyse the target before the virus can replicate, thereby stopping the infection. These two types of delayed hypersensitivity are compared and contrasted in **Table 12.2**.

ALLERGEN

An allergen is an antigen that induces an allergic or hypersensitivity response in contrast to a classic immune response produced by the recipient host in response to most immunogens. Allergens include such environmental substances as pollens, i.e., their globular proteins, from trees, grasses, and ragweed, as well as certain food substances, animal danders, and insect venom. Selected subjects are predisposed to synthesizing IgE antibodies in response to allergens and are said to be atopic. The crosslinking of IgE molecules anchored to the surfaces of mast cells or basophils through their Fc regions results in the release of histamine and other pharmacological mediators of immediate hypersensitivity from mast cells/basophils.

The criteria for an allergen to be included in the WHO/IUIS nomenclature are:

1 The molecular and structural properties should be clearly and unambiguously defined, including:
   - Purification of the allergen protein to homogeneity
   - Determination of molecular weight, pI, and carbohydrate composition
   - Determination of nucleotide and/or amino acid sequence
   - Production of monospecific or monoclonal antibodies to the allergen

Table 12.1 Comparison of four types of hypersensitivity

	Type I	Type II	Type III	Type IV
Induction	Antibody-mediated	Antibody-mediated	Antibody mediated	Cell-mediated
Onset	Immediate	Immediate	Immediate	Delayed
Antigen	Soluble antigen	Cell-associated antigen, matrix-associated antigen	Soluble antigen	Soluble antigen, cell-associated antigen
Immune reactant	IgE	IgG, IgM	IgG	Th1 cells, Th2 cells, cytotoxic T cells
Effector cells	Mast cells, basophils	Phagocytes, natural killer cells	Phagocytes, natural killer cells	CD4+ T cells, CD8+ T cells, macrophages, eosinophils
Primary mediators	Vasoactive amines	Complement, membrane attack complex	Antigen–antibody–complement complex	IFN-γ, IL-4, IL-5, IL-12, eotaxin
Secondary mediators	Leukotrienes, prostaglandin D$_2$, platelet-activating factor, cytokines	Lysosomal enzymes, perforin	Lysosomal enzymes	Lymphokines, chemokines, cytokines, cytotoxins
Physiologic effects	Increased vascular permeability, vasodilation, bronchial spasm, mucous secretion	Antibody/complement-mediated lysis, or antibody-dependent cell-mediated cytotoxicity	Immune-complex-mediated injury: immune complex deposition, PMN accumulation, lysosomal enzyme release and tissue injury	Edema, infiltration by lymphocytes and macrophages at local site, erythema, induration
Example of hypersensitivity reaction	Systemic anaphylaxis, asthma, allergic rhinitis	Drug allergy, Goodpasture's syndrome, Rh incompatibility	Serum sickness, Arthus reaction	Contact dermatitis, chronic asthma, chronic allergic rhinitis, tuberculin reaction

Table 12.2 Two groups of type IV hypersensitivity

	Delaycd-type hypersensitivity (DTH)	Cellular toxicity
Antigen	Soluble antigen	Cell-associated antigen
Responding cells	CD4+ T cells	CD8+ T cells
MHC and antigen presentation	MHC class II molecules	MHC class I molecules
Actions	Reaction of CD4+ T cells with antigen on APC surface ⟶ Release of lymphokines ⟶ Recruitment and activation of macrophages or eosinophils ⟶ Secretion of cytokines, chemokines, cytotoxins, and inflammatory mediators ⟶ Inflammatory response and manifestations of delayed hypersensitivity	Reaction of CD8+ T cells with peptide class I molecular complex on the infected cell's surface ⟶ Release of perforin ⟶ Cell lysis ⟶ Tissue damage
Examples	Tuberculin reaction, chronic asthma	Contact dermatitis

2   The importance of the allergen in causing IgE responses should be defined by:
   - Comparing the prevalence of serum IgE antibodies in large populations of allergic patients. Ideally, at least 50 or more patients should be tested
   - Demonstrating biologic activity, e.g. by skin testing or histamine release assay
   - Investigating whether depletion of the allergen from an allergic extract (e.g. by immunoabsorption) reduces IgE binding activity
   - Demonstrating, where possible, that recombinant allergens have comparable IgE antibody-binding activity to the natural allergen

Table 12.3 lists the allergens with known amino acid sequences. The molecular properties of common allergens are given in **Table 12.4**.

ALLERGIC DISEASES

Immunological responses involving specific T cells or antibodies can also cause adverse hypersensitivity reactions. *Hypersensitivity* generally represents the 'dark side', signifying the undesirable aspects of an immune reaction, whereas the term *immunity* implies a desirable effect. **Table 12.5** describes some common examples of allergic diseases showing their features, mechanism of action, and pathologic consequences.

Table 12.3   Some allergens with known amino acid sequences

Allergen source	Allergens; systematic and original names	MW (kDa)	Sequence data[a,b]
A: Weed pollens			
Asterales			
Ambrosia artemisiifolia (short ragweed)	Amb a 1; antigen E	38	C
	Amb a 2; antigen K	38	C
	Amb a 3; Ra3	11	C
	Amb a 5; Ra5	5	C
	Amb a 6; Ra6	10	C
	Amb a 7; Ra7	12	P
	Amb a ?	11	C
Ambrosia trifida (giant ragweed)	Amb t 5; Ra5G	4.4	C
Artemisia vulgaris (mugwort)	Art v 2	35	P
B: Grass pollens			
Poales			
Cynodon dactylon (Bermuda grass)	Cyn d 1	32	C
Dactylis glomerata (orchard grass)	Dec g 1; AgDg1	32	P
	Dec g 2	11	C
	Dec g 5		P
Lolium perenne (rye grass)	Lol p 1; group I	27	C
	Lol p 2; group II	11	C
	Lol p 3; group III	11	C
	Lol p 5	31	P
	Lol p 9; Lol p Ib	31/35	C
Phleum pratense (timothy)	Phl p 1	27	C
	Phl p 5; Ag25	32	C
Poa pratensis (Kentucky blue grass)	Poa p 1; group I	33	P
	Poa p 5	31	P
	Poa p 9	32/34	C
Sorghum halepense (Johnson grass) (Bermuda grass)	Sor h 1		C

Table 12.3   Some allergens with known amino acid sequences (*continued*)

Allergen source	Allergens; systematic and original names	MW (kDa)	Sequence data[a,b]
C: Tree pollens			
Fagales			
Alnus glutinosa (alder)	Aln g 1	17	C
Betula verrucosa (birch)	Bet v 1	17	C
	Bet v 2; profilin	15	C
Carpinus betulus (hornbeam)	Car b 1	17	C
Corylus avelana (hazel)	Cor a 1	17	C
Quercus alba (white oak)	Que a 1	17	P
Pinales			
Cryptomeria japonica (sugi)	Cry j 1	41–45	C
	Cry j 2		C
Juniper sabinoides	Jun s 1		C
Juniper virginiana	Jun v 1		C
Oleales			
Olea europea (olive)	Ole e 1		C
D: Mites			
Dermatophagoides pteronyssinus (mite)	Der p 1; antigen P1	25	C
	Der p 2	14	C
	Der p 3; trypsin	28/30	P
	Der p 4; amylase	60	P
	Der p 5	14	C
	Der p 6; chymotrypsin	25	P
	Der p 7	22–28	C
Dermatophagoides microceras (mite)	Der m 1	25	P
Dermatophagoides farinae (mite)	Der f 1	25	C
	Der f 2	14	C
	Der f 3	30	P
Lepidoglyphus destructor (storage mite)	Lep d?	15	P
E: Animals			
Canis domesticus[c]	Can d 1		C
	Can d 2		C
Felis domesticus (cat saliva)	Fel d 1; cat-1	38	C
Mus musculus (mouse urine)	Mus m 1; MUP	19	C
Rattus norvegius (rat urine)	Rat n 1	17	C

Table 12.3 Some allergens with known amino acid sequences (*continued*)

Allergen source	Allergens; systematic and original names	MW (kDa)	Sequence data[a,b]
F: Fungi			
Aspergillus fumigatus	Asp f 1	18	C
	Asp f ?	90	P
	Asp f ?	55	P
Candida albicans	Cand a?	40	C
Alternaria alternata	Alt a 1	28	P
Trichophyton tonsurans	Tri t 1	30	P
G: Insects			
Apis mellifera (honey bee)	Api m 1; phospholipase A2	16	C
	Api m 2; hyaluronidase	44	C
	Api m 4; melittin	3	C
Bombus pennsylvanicus (bumble bee)	Bom p 1; phospholipase	16	P
	Bom p 4; protease		
Blattaria germanica (cockroach)	Bla g 2	20	C
Chironomus thummi thummi (midges)	Chi t 1; hemoglobin	16	C
Dolichovespula maculata (white face hornet)	Dol m 1; phospholipase A1	35	C
	Dol m 2; hyaluronidase	44	C
	Dol m 5; antigen 5	23	C
Dolichovespula arenaria (yellow hornet)	Dol a 5; antigen 5	23	C
Polistes annularis (wasp)	Pol a 1; phospholipase A15	5	P
	Pol a 2; hyaluronidase	44	P
	Pol a 5; antigen 5	23	C
Polistes exclamans (wasp)	Pol a 1; phospholiase A1	34	P
	Pol a 5; antigen 5	23	C
Polistes fuscatus (wasp)	Pol f 5; antigen 5	23	C
Polistes metricus (wasp)	Pol m 5; antigen 5	23	P
Vespula flavopilosa (yellowjacket)	Ves f 5; antigen 5	23	C
Vespula germanica (yellowjacket)	Ves g 5; antigen 5	23	C
Vespula maculifrons (yellowjacket)	Ves m 1; phospholipase A1	33.5	C
	Ves m 2; hyaluronidase	44	P
	Ves m 5; antigen 5	23	C
Vespula pennsylvanica (yellowjacket)	Ves p 5; antigen 5	23	C
Vespula squamosa (yellowjacket)	Ves s 5; antigen 5	23	C
Vespula vidua (wasp)	Ves vi 5	23	C
Vespula vulgaris (yellowjacket)	Ves v 1; phopholipase A1	35	C
	Ves v 2; hyaluronidase	44	P
	Ves v 5; antigen 5	23	C

Table 12.3 Some allergens with known amino acid sequences (*continued*)

Allergen source	Allergens; systematic and original names	MW (kDa)	Sequence data[a,b]
Vespa crabo	Vesp c 1; phospholipase	34	P
	Vesp c 5.0101; antigen 5	23	C
	Vesp c 5.0102; antigen 5	23	C
Solenopsis invicta (fire ant)	Sol i 2	13	C
	Sol i 3	24	C
	Sol i 4	13	C
H: Foods			
Gadus callarias (cod)	Gad c 1; allergen M	12	C
Gallus domesticus (chicken)	Gal d 1; ovomucoid	28	C
	Gal d 2; ovalbumin	44	C
	Gal d 3; conalbumin (Ag22)	78	C
	Gal d 4; lysozyme	14	C
Penaeus aztecus (brown shrimp)	Pen a 1	36	P
	Pen a 2; tropomyosin	34	P
Brassica juncea (oriental mustard)	Bra j 1; 25 albumin	14	C
Hordeum vulgare (barley)	Hor v 1; BMAI-1	15	C
Sinapis albs (yellow mustard)	Sin a 1; 25 albumin	14	C
I: Others			
Ascaris suum	Asc s 1	10	P
Havea brasiliensis	Hev b 1; elongation factor 10	58	P

Notes:

[a]References are those where partial (P) or complete (C) amino acid sequence data are available. Original references describing the initial characterization studies are not given because of limited space.

[b]Sequence data for group 5 and 9 allergens from several grass pollens indicate that they are highly homologous proteins. Comparison of complete sequence data of group 5 and 9 allergens from a single grass species will clarify whether these two groups are the same protein.

[c]*Canis domesticus* is also designated as *Canis familiaris*.

Table 12.4 Molecular properties of common allergens

Source	Allergen	MW (kDa)	Homology/function
Inhalants:			
Indoor			
House dust mite (*Dermatophagoides pteronyssinus*)	Der p 1	25	Cysteine protease[b]
	Der p 2	14	Epididymal protein
	Der p 3	30	Serine protease
	Der p5	14	Unknown
Cat (*Felis domesticus*)	Fel d 1	36	(Uteroglobin)
Dog (*Canis familiaris*)	Can f 1	25	Calycin
Mouse (*Mus muscularis*)	Mus m 1	21	Pheromone binding protein (calycin)[b]
Rat (*Rattus norvegicus*)	Rat n 1	21	Pheromone binding protein (calycin)[b]
Cockroach (*Blattella germanica*)	Bla g 2	36	Aspartic protease
	Bla g 4	21	Calycin
Outdoor			
Pollens – grasses			
Rye (*Lolium perenne*)	Lol p 1	28	Unknown
Timothy (*Phleum pratense*)	Phl p 5	32	Unknown
Bermuda (*Cynodon dactylon*)	Cyn d 1	32	Unknown
Weeds			
Ragweed (*Artemisia artemisifolia*)	Amb a 1	38[a]	Pectate lyase
	Amb a 5	5	Neurophysins[b]
Trees			
Birch (*Betula verucosa*)	Bet v 1	17	Pathogenesis-related protein
	Bet v 2	14	Profilin
Foods:			
Milk	β-Lactoglobulin	36[a]	Retinol-binding protein (calycin)[b]
Egg	Ovomucoid	29	Trypsin inhibitor
Codfish (*Gadus callarias*)	Gad c 1	12	Ca 2^+-binding protein (muscle parvalbumin)
Peanut (*Arachis hypogea*)	Ara h 1	63	Vicilin (seed-storage protein)
Venoms:			
Bee (*Apis mellifera*)	Api m 1	19,5	Phospholipase A2
Wasp (*Polistes annularis*)	Pol a 5	23	Mammalian testis proteins
Hornet (*Vespa crabro*)	Ves c 5	23	Mammalian testis proteins
Fire ant (*Solenopsis invicta*)	Sol i 2	13	Unknown

Table 12.4   Molecular properties of common allergens (*continued*)

Source	Allergen	MW (kDa)	Homology/function
Fungi:			
Aspergillus fumigatus	Asp f 1	18	Cytotoxin (mitogillin)
Alternaria alternata	Alt a 1	29[a]	Unknown
Latex:			
Hevea brasiliensis	Hev b 1	58	Elongation factor
	Hev b 5	16	Unknown – homologous to kiwi fruit Protein of unknown function

Notes:
[a]Most allergens have a single polypeptide chain; dimers are indicated.
[b]Allergens of known three-dimensional structure are also indicated.

Table 12.5   Some common allergic diseases

	Anaphylaxis	Urticaria	Serum sickness	Contact dermatitis
Category of hypersensitivity	Type I	Type I	Type III	Type IV
Onset	Immediate	Immediate	Between the 5th and 14th day	1 day to 2 days
Cause of reaction	Injection of antigen or drug, bee sting	Immunologic sensitization, physical or chemical substances	Injection of a relatively large, single dose of serum (e.g., antitoxin)	Sensitization by topical drugs, cosmetics, other contact chemicals
Symptoms and signs	Embarrassed respiration due to laryngeal and bronchial constriction and shock associated with decreased blood pressure	Localized elevated, edematous, erythematous, and itching wheals with a pale center encircled by a red flare: wheal-and-flare reaction	Systemic vasculitis, glomerulonephritis, arthritis, fever, lymphadenopathy, urticaria	Rash, eczema, blistering skin lesions
Mechanism of action	Crosslinking of IgE with antigen or allergen; basophils or mast cells release of primary mediators (histamine, chemotactic factor, serotonin, heparin, etc.); formation of acute phase reactants; release of secondary mediators (slow reacting substance of anaphylaxis, platelet activating factor, bradykinin); increased vascular permeability, vasodilation, bronchial spasm	Binding of allergen with IgE antibodies; mast cell activation and release of histamine and other mediators and release of neurotransmitters from local nerve endings; increased vascular permeability, fluid extravasation and swelling, and vasodilation of surrounding cutaneous blood vessels	Escape of antigen into circulation from site of injection; formation of antigen–antibody complex; deposition of immune complex at microvasculature; fixation of complement; attraction of polymorphonuclear neutrophils through C5a; initiation of inflammation and tissue damage	Reaction of antigen with self protein in skin; formation of protein–hapten complex; conversion to hapten–peptide complex; binding with MHC molecules; recognition by T cells; release of T cell cytokines (INF-γ, IL-17); release of keratinocyte cytokines and chemokines; enhancement of inflammatory reaction at the site

- DEFENSE BARRIERS OF THE HUMAN BODY
- NONSPECIFIC DEFENSE MECHANISMS
- SPECIFIC DEFENSE MECHANISMS
- IMMUNOGLOBULIN

DEFENSE BARRIERS OF THE HUMAN BODY

Entry of a pathogenic microorganism into a susceptible host can be followed by invasion and colonization of tissues, circumvention of the host immune response, and injury to the host tissues. The human body, however, possesses natural barriers against infection.

External defense barriers

The skin and mucous membranes serve as barriers to the microorganisms. With the exception of a few organisms, most microorganisms cannot establish infections without penetrating the skin or mucous membranes. Cell shedding, mucus (motion of cilia), coughing, sneezing, flushing of microbes by tears, saliva, urine, perspiration, and other body fluids, and microbial elimination via emesis and diarrhea are the body's mechanical barriers against infection. **Table 13.1** lists the external defense barriers of the human body.

Phagocytosis

The immunological clearance of most pathogenic microbes requires phagocytic cells. Once the microorganism penetrates the body's physical barriers, inflammation is initiated to contain the infection and prevent its spread from the initial focus. At sites of typical local infection, the neutrophils dominate early (acute) responses (30 minutes); however, macrophages take over in longstanding (chronic) conditions. This takeover is generally observed within 48 hours (begins in 6 hours).

Neutrophils

The neutrophils constitute the first line of defense against infectious agents. Their targets include bacteria, fungi, protozoa, viruses, and virally infected cells. Neutrophils contain cytoplasmic granules (primary or azurophil and secondary or specific). These granules are of major importance for neutrophil function. They can be characterized morphologically or biochemically using enzyme markers or other substances. **Table 13.2** compares the two types of granules in neutrophils. Within the granules of neutrophils, there are numerous enzymes that can induce an oxygen-dependent as well as oxygen-independent response against invading organisms. **Table 13.3** depicts oxygen-dependent versus oxygen-independent killing 64 neutrophils.

Macrophages

Macrophages naturally phagocytose material in their surroundings without being activated. Certain microbial products, however, do activate these cells. It is only after this activation process that the macrophage starts an inflammatory process, i.e. effective antigen presentation and cytokine secretion. Secretory products from macrophages initiate the local tissue inflammation. These cytokines also activate T cells with the aid of effective antigen presentation. Therefore, macrophages constitute a bridge between innate resistance and specific immunity. **Table 13.4** lists the secreted products of macrophages that have a protective effect against infection. In addition to microbicidal activity, macrophages play a key role in the immune system. The roles of macrophages in the immune system are summarized in **Table 13.5**.

NONSPECIFIC DEFENSE MECHANISMS

Numerous enzymes, proteins, and other factors contribute to the host's nonspecific immunity. Some of these belong to humoral defenses, and others are cellular defenses.

Table 13.1 External defense barriers of the human body

Site	Functioning unit
Skin	Anatomic barrier; antimicrobial secretions
Eyes	Washing of tears; lysozyme
Respiratory tract	Mucus; ciliated epithelium; alveolar macrophages
Digestive tract	Stomach acidity; normal flora
Genitourinary tract	Washing of urine; urine acidity; vaginal lactic acid; lysozyme

Nonspecific humoral defense mechanisms

Table 13.6 lists the major factors of the nonspecific humoral defense mechanisms, together with their sources and functions.

Nonspecific cellular defense mechanisms

Table 13.7 lists the major factors of the nonspecific cellular defense mechanisms, together with the cell types that secrete them and their functions.

SPECIFIC DEFENSE MECHANISMS

Specific immune response to extracellular bacteria

Antibodies are the primary agents that protect the body against extracellular bacteria. **Table 13.8** lists the antimicrobial actions of antibodies.

Table 13.2 Comparison of azurophil granules and specific granules

	Azurophil	Specific
Synthesis	In the endoplasmic reticulum and concave side of golgi complex	In the endoplasmic reticulum and convex side of golgi complex
Bone marrow and blood stains	Azurophilic (purple-red)	Very small negative images ('white dots')
EM	More density	Less density
Size	Larger (0.8 m)	Smaller (0.5 m)
Proportion	1/3 of total granules	2/3 of total granules
Cytoplasmic membrane receptors	—	CR3, CR4, N-formylmethionyl-leucyl-phenylalanine receptors, laminin receptors
Neutral proteinases	Elastase, cathepsin G, proteinase 3	Collagenase, complement activator
Acid hydrolases	Cathepsin B, cathepsin D, β-D-glucuronidase, α-mannosidase, phospholipase A_2	Phospholipase A_2
Antimicrobial constituents	Myeloperoxidase, lysozyme, defensins, bactericidal permeability-increasing protein	Lysozyme, lactoferrin
Other constituents	Chondroitin-4-sulphate	Cytochrome b_{558}, monocyte-chemotactic factor, histaminase, vitamin B_{12} binding protein

Table 13.3 Intracellular killing of microorganisms by neutrophils

	Oxygen-independent mechanism	Oxygen-dependent mechanism
Enzymes	General lysosomal proteases and glycolases	Superoxide dismutase and myeloperoxidase
Action	Disrupt membrane functions of microorganisms,	Form toxic oxygen radicals (O_2-, OH, H_2O_2)
Other mechanisms	Defensins which insert into pathogen membrane and disrupt membrane permeability, lactoferrin which chelates iron required for bacterial growth, low pH inside phagocytic vacuoles, and catonic proteins	Lipid peroxidase which induces plasma membrane lipid oxidation
Target efficiency	Gram-negative bacteria, far less effective against Gram-positive bacteria	Gram-positive bacteria
Clinical correlate	Chronic skin infections or abscesses: cationic protein deficiency Chediak–Higashi syndrome: frequent infections (skin, oral, respiratory) caused by immature neutrophil granules that greatly diminish the bacterial killing ability of neutrophils	Chronic granulomatous disease: diminished cytochrome b and failure to form superoxide anions lead to impaired ability to oxidize NADPH and destroy bacteria through the oxidative pathway

Table 13.4 Secreted products of macrophages against infection

- Cell differentiation factors
- Colony stimulation factors
- Cytotoxic factors
- Tumor necrosis factor-α
- Cachectin
- Hydrolytic enzymes
  - collagenase
  - lipase
  - phosphatase
- Endogenous pyrogen
  - interleukin-1
- Complement components
  - C1 to C5
  - properdin
  - factors B, D, I, H
  - α-interferon
- Plasma proteins
- Coagulation factors
- Oxygen metabolites
  - H_2O_2
  - superoxide anion
- Arachidonic acid metabolites
  - prostaglandins
  - thromboxanes
  - leukotrienes

Specific immune response to intracellular bacteria and fungi

Cell-mediated immunity attributable to T cells is the principal mechanism whereby intracellular bacteria are eliminated by macrophages activated by γ-interferon derived from T cells.

Specific immune response to viruses

Antibodies specific for viral antigenic determinants may offer early protection following viral infection. However, antiviral immunity depends primarily on cytotoxic T cells.

Specific immune response to parasites

Parasites such as protozoa and helminths elicit a variety of immune responses. Helminths specifically stimulate CD4+ helper T cells that form IL-4 and IL-5. Antibody-dependent cell-mediated cytotoxicity (ADCC) involving eosinophils and IgE antibody is believed to be effective in immunity against helminths. Intracellular protozoa often activate specific cytotoxic T cells. They present a crucial mechanism to prevent dissemination of intracellular malarial parasites.

Table 13.5 Roles of macrophages in the immune response

Effect	Functioning mechanism
Anti-microbial activity	Natural mechanism: phagocytic killing via oxygen-dependent free radicals or oxygen-independent hydrolases Adaptive mechanism: inflammatory reaction following antigen presentation and cytokine (IL-1, IL-6, IL-8, IL-12, TNF-α) secretion
Lymphocyte activation	Antigen presenting cell function, cytokine secretion
Immune response modulation	Th-1 response: interleukin-12 secretion Th-2 response: interleukin-10 secretion
Tumor immunity	Tumor cell breakdown by toxic factors, free radicals, hydrolases, and TNF-α secretion
Tissue reorganization	Elastases, collagenases, fibroblast growth factors, and angiogenesis factors secretion

IMMUNOGLOBULIN

Antibodies induced in microbial invasion are immunoglobulins of defined specificity produced by plasma cells. Immunoglobulins are divided into five classes: three major classes (i.e., IgG, IgM, IgA) and two minor class (IgD and IgE). Secretory IgA is found in body secretions such as saliva, milk, and intestinal and bronchial secretions. IgD and IgM are present as membrane-bound immunoglobulins on B cells, where they interact with antigen to activate B cells. **Table 13.9** summarizes the five classes of immunoglobulins, including their physical and biological features. **Table 13.10** and **Table 13.11** depict the immunoglobulin serum levels of different age groups and the indications of changes in serum immunoglobulin levels and their clinical significance. Serum levels of IgG subclasses together with the application of quantitation are listed separately in **Table 13.12**.

Table 13.6 Nonspecific humoral defense mechanisms

Factor	Source	Function
Lysozyme	Tears, saliva, nasal secretions, body fluids, lysosomal granules	Catalyses hydrolysis of cell wall mucopeptide layer
Lactoferrin, transferrin	Specific granules of PMNs	Binds iron and competes with microorganisms for it
Lactoperoxidase	Milk and saliva	May be inhibitory to many microorganisms
Beta-lysin	Thrombocytes, normal serum	Effective mainly against Gram-positive bacteria
Chemotactic factors	Bacterial substances and products of cell injury and denatured proteins	Induce reorientation and directed migration of PMNs, monocytes, and other cells
Properdin	Normal plasma	Activates complement in the absence of antibody–antigen complex
Interferons	Leukocytes, fibroblasts, natural killer cells, T cells	Act as immunomodulators to increase the activities of macrophages
Defensins	Polymorphonuclear granules	Block cell transport activities

Table 13.7 Nonspecific cellular defense mechanisms

Factor	Source	Function
Monokine α-interferon	Leukocytes	Inhibits cell proliferation and tumor growth, enhances natural killer cell activity and phagocytosis
Interleukin-1	Dendritic cells, macrophages, B cells, PMNs, endothelial and smooth muscle cells, and others	Induces lymphokine production, enhances B cell proliferation and antibody production, increases phagocytosis, acts as chemoattractant, increases T cell activation and IL-2 receptor expression
Tumor necrosis factor α	Activated macrophage others	Many functions shared with IL-1
Colony-stimulating factors	Monocytes, fibroblasts, T cells, B cells, endothelial and epithelial cells, kidney cells	Specific factors stimulate the growth of specific cell lines such as neutrophils, monocytes, eosinophils, erythrocytes, megakaryocytes, and basophils
Lymphokines	Lymphocyte	T cell, B cell, and hematopoietic growth factors; multiple effector functions
γ-Interferon	Stimulated T cells, natural killer cells	Activates macrophages, maintains MHC class II expression on cell surfaces, inhibits cell proliferation, enhances accessory cell function of macrophages
Lymphotoxin (tumor necrosis factor β)	Lymphocyte	Target cell destruction
Interleukin-2	Activated CD4+ T cells T cells	Induces proliferation of activated T cells, B cells, and natural killer cells, stimulates lymphokine and immunoglobulin production
Interleukin-3	Activated T cells	Acts on pluripotent stem cells to stimulate growth of neutrophils, monocytes, erythrocytes, basophils, eosinophils, and megakaryocytes
Interleukin-4	T helper cells, mast cells	Stimulates B cells, promotes immunoglobulin subtype switching, stimulates mast cells and hemopoiesis, activates macrophages
Interleukin-5	T helper cells	Helps stimulate B cell proliferation and growth, stimulates eosinophils, promotes immunoglobulin subtype switching, enhances expression of IL-2 receptor
Interleukin-6	T and B cells, monocytes, fibroblasts, epithelial and endothelial cells	Increases immunoglobulin secretion, stimulates production of acute phase proteins, stimulates T cells and thymocytes, enhances differentiation of myelomonocytic cell lines
Interleukin-7	Bone marrow stromal cells	Stimulates pre-B cells and thymocytes, stimulates mature T cells, stimulates megakaryocytes and myeloid precursors
Interleukin-8	Monocytes, fibroblasts, epithelial and endothelial cells, synovial cells	Stimulates migration of monocytes and neutrophils, stimulates release of superoxide anions and lysosomal enzymes, chemotactic for basophils and T cells, stimulates release of histamine from basophils
Interleukin-9	T cells	Enhances mast cell growth and splenic CD4+ T cells
Interleukin-10	T cells	Regulates the class of immune response, modulates accessory cell (APC) function
Interleukin-11	Fibroblasts, stromal cells	Acts as megakaryocyte potentiator, stimulates IgG production

Table 13.8 Antimicrobial action of antibodies

- Opsonins – promote ingestion and killing by phagocytic cells (IgG)
- Block attachment (IgA)
- Neutralize toxins
- Agglutinate bacteria – may aid in cleaning
- Render motile organisms nonmotile
- Affect metabolism or growth of bacteria (*Mycoplasma*) – only rarely
- Antibodies, combining with antigens of the bacterial surface, activate the complement cascade, thus inducing an inflammatory response, and bring fresh phagocytes and serum antibodies into the site
- Antibodies, combining with antigens of the bacterial surface, activate the complement cascade, and through the final sequence the membrane attack complex (MAC) is formed involving C5b-C9

Table 13.9 Five classes of immunoglobulins

	IgA	IgD	IgE	IgG	IgM
Serum concentration (mg/dl)	140–420	0.3–0.4	<0.001	800–1700	50–190
Total Ig (%)	5–15	<1	<1	85	5–10
Molecular weight (kDa)	160	185	190	154	900
Heavy chain class	α	δ	ε	γ	μ
Half-life	6 days	2–3 days	2.5 days	23 days	5 days
Placental transfer	No	No	No	Yes	No
Complement fixation	–	–	–	+	++++
Mast cell binding	–	–	+	–	–
Phagocyte binding	–	–	–	+	–
Principal biological effect	Resistance prevents movement across mucous membranes	Elusive	Anaphylaxis to hypersensitivity	Resistance – opsonin; secondary response	Resistance – precipitin; primary response
Principal site of action	Secretions	Receptors for B cells	Mast cells	Serum	Serum
Antibacterial lysis	+	?	?	+	+++
Antiviral lysis	+++	?	?	+	+
Subclass	Ig A1, IgA2	–	–	IgG1, IgG2, IgG3, IgG4	–

Table 13.10 Serum immunoglobulin levels

Age (yr)	Serum immunoglobulin levels (range, mg/dl)		
	IgA	IgG	IgM
0–1	0–83	231–1411	0–145
1–3	20–100	453–916	19–146
4–6	27–195	504–1464	24–210
7–9	34–305	572–1474	31–208
10–11	53–204	698–1560	31–179
12–13	58–358	759–1549	35–239
14–15	47–249	716–1711	15–188
16–19	61–348	549–1584	23–259
>19	70–400	700–1600	40–230

Table 13.11 Indications for changes in immunoglobulin levels

Indications for changes in immunoglobulin levels			
	IgA	IgG	IgM
Increased level	Lymphoproliferative disorders, especially multiple myeloma and 'Mediterranean' lymphoma involving bowel, a wide range of conditions affecting mucosal surfaces. IgA monoclonal peak>2g/dl is a major criterion for myeloma	Responding to a wide variety of infectious or inflammatory insults. Oligoclonal IgG can be seen in multiple sclerosis and some chronic hepatitides. Increase in polyclonal IgG is seen in acquired immunodeficiency syndrome. Monoclonal IgG> 3 g/dl is a major diagnostic criterion for myeloma	Acute infection, congenital infection in a newborn, hyper-IgM immunodeficiency syndrome, Waldenstrom disease, primary biliary cirrhosis. IgM monoclonal peak>2g/dl is a major diagnostic criterion of myeloma
Decreased level	Chronic sinopulmonary disease, ataxia–telangiectasia, congenital IgA deficiency	Congenital or acquired IgG deficiency	Congenital or acquired hypogammaglobulinemia, increased and recurrent infection
Use of quantitation	Evaluate humoral immunity; monitor therapy in IgA myeloma	Evaluate humoral immunity; monitor therapy in IgG myeloma; evaluate patients, especially children and those with lymphoma, with propensity to infections	Evaluate humoral immunity; establish the diagnosis and monitor therapy in macroglobulinemia of Waldenstrom or plasma cell myeloma. IgM levels are used to evaluate likelihood of *in utero* infections or acuteness of infection

Table 13.12 Serum level of immunoglobulin G subclasses

Serum level of immunoglobulin G subclasses (mg/dl)			
Age	**Subclass 1**	**Subclass 2**	**Subclass 3**	**Subclass 4**
---	---	---	---	---
Cord	435–1084	143–453	27–146	1–47
0–2 mth	218–496	40–167	4–23	1–33
3–5 mth	143–394	23–147	4–100	1–14
6–8 mth	190–388	37–60	12–62	1–30
9–35 mth	286–680	30–327	13–82	1–65
3–4 yr	381–884	70–443	17–90	1–116
5–6 yr	292–816	83–513	8–111	1–121
7–8 yr	422–802	113–480	15–133	1–84
9–10 yr	456–938	163–513	26–113	1–121
11–12 yr	456–952	147–493	12–179	1–168
13–14 yr	347–993	140–440	23–117	1–83
Adults	422–1292	117–747	41–129	1–291
Use of quantitation	Evaluate T cell-dependent response in patients with poor response to viral or bacterial antigens, immunodeficiency syndromes	Evaluate antipolysaccharide antibodies, sinopulmonary infections, immunodeficiency syndrome, and patients who demonstrate a poor response to carbohydrate antigens (group A strep, pneumococcus)	Evaluate immunodeficiency syndromes; evaluate recurrent sinusitis and otitis media	Evaluate sinopulmonary infections, asthma; immunotherapy hyposensitization; evaluate allergies

IMMUNOREGULATION, TOLERANCE, AND THERAPEUTIC IMMUNOLOGY

14

- ● IMMUNOLOGIC TOLERANCE
- ● IMMUNOREGULATION

IMMUNOLOGIC TOLERANCE

Central and peripheral tolerance

Immunologic tolerance is an active, but carefully regulated response of lymphocytes to self antigens. Normal individuals are tolerant to their own antigens (self antigens). Even though many self antigens have free access to lymphocytes, lymphocytes do not normally mount an immune response against self antigens. This self tolerance is maintained by several mechanisms that prevent the maturation and activation of potentially self-reactive lymphocytes. Immunologic tolerance occurs in two forms: central and peripheral. **Table 14.1** compares central and peripheral immunologic tolerance with respect to B cell and T cell participation, positive and negative selection, clonal anergy, clonal deletion, etc.

T cell and B cell tolerance

When comparing the ease with which T and B cell tolerance may be induced, it was found that T cell tolerance is induced more rapidly and is longer lasting than B cell tolerance. For example, T cell tolerance may be induced in a single day, whereas B cells may require 10 days for induction. In addition, 100 times more tolerogen may be required for B cell tolerance than for T cell tolerance. The duration of tolerance is much greater in T cells, e.g., 150 days, compared to that in B cells, which is only 50–60 days. Maintenance of tolerance is considered to require the continued presence of specific antigens. Administration of antigen in a suboptimal dose can induce an antigen-specific immunosuppression termed **low-dose tolerance**. It is achieved easily in the neonatal period when the lymphoid cells of the animal are not sufficiently mature to mount an antibody or cell-mediated immune response. This renders helper T cells tolerant, thereby inhibiting them from signaling B cells to respond to immunogenic challenge. Although no precise inducing dose of antigen can be defined, in low-dose tolerance 10^{-8} mol antigen per kilogram of body weight is usually effective. In immunocompetent adult animals, repeated administration of large doses of protein antigen or a massive single dose administration of a polysaccharide antigen can induce a specific immunologic unresponsiveness termed **high-dose tolerance**. Although no precise inducing dose of antigen can be defined, in high-dose tolerance the antigen level usually exceeds 10^{-4} mol per kilogram of body weight. Low antigen doses may be effective in inducing tolerance in immature B cells leading to clonal abortion, whereas T cell tolerance does not depend upon the level of maturation. Another mechanism of B cell tolerance is clonal exhaustion, in which the immunogen activates all of the B cells specific for it, leading to maturation of cells and transient antibody synthesis and thereby exhausting and diluting the B cell response. Another mechanism of B cell tolerance is antibody-forming cell blockade. Antibody-expressing B cells are coated with excess antigen, rendering them unresponsive to the antigen. Characteristics of T cell and B cell tolerance are summarized and abbreviated in **Table 14.2**.

Variations of immunologic tolerance

Oral tolerance

Oral tolerance is antigen-induced specific suppression of humoral and cell-mediated immunity to an antigen following oral administration of that antigen as a consequence of anergy of antigen-specific T cells or the formation of immunosuppressive cytokines such as transforming growth factor-β. Oral tolerance may inhibit immune responses against food antigens and bacteria in the intestine. Based on the quantity of antigen fed, orally administered antigen may induce regulatory cells that suppress the antigen-specific response (low doses) or inhibit antigen-specific T cells by induction of clonal anergy (high doses).

Table 14.1   Central and peripheral immunologic tolerance

	Central tolerance	Peripheral tolerance
Features	Inactivation of cells required for initiation of an immune response	Inhibition of expression of the immune response
Site of tolerance induction	Generative lymphoid organs	Peripheral lymphoid tissues
Site of involvement	Afferent limb of the immune response, which is concerned with sensitization and cell proliferation	Efferent limb of immune response, which is concerned with the generation of effector cells
B cell participation	Immature B cells	Mature B cells
T cell participation	Immature thymocytes	Mature T cells
Mechanisms of tolerance	Clonal deletion (apoptotic cell death, negative selection)	Clonal deletion (apoptotic cell death); clonal anergy (functional inactivation without cell death); clonal ignorance (failure to recognize or recognition of antigens without costimulation); suppression of lymphocyte activation and effector functions by regulatory lymphocytes
Function	Eliminates potentially self-reactive lymphocytes	Maintains unresponsiveness to self antigens

Split tolerance

Split tolerance includes several mechanisms. It is seen either when specific immunological unresponsiveness affects the B cell (antibody) limb or the T cell (cell-mediated) limb of the immune response, leaving the unaffected limb intact to produce antibody or respond with cell-mediated immunity; or when immunologic tolerance is induced to some epitopes of allogeneic cells while the remaining epitopes are left capable of inducing an immune response characterized by antibody production and/or cell-mediated immunity.

Immune deviation

The selective suppression of certain phases of the immune response to an antigen without alteration of others is termed immune deviation. Immune deviation selectively suppresses delayed-type hypersensitivity and IgG_2 antibody production. Powerful cell-mediated responses (DTH) occur when T_H1 cells secrete IL-2 and IFN-γ. Immune deviation may involve conversion of a T cell response involving T_H1 cytokines that induce cell-mediated immunity to a T_H2 cytokine response that induces synthesis of selected antibody isotypes. This leads to deviation from the expected heightened delayed-type hypersensitivity and formation of IgG2 antibodies to result in little of either, i.e., negligible delayed-type hypersensitivity and suppression of IgG2 formation.

Immunological paralysis

Immunological paralysis is the immunologic unresponsiveness induced by the injection of large doses of pneumococcal polysaccharide into mice where it is metabolized slowly. Any antibody that is formed is consumed and not detectable. The pneumococcal polysaccharide antigen remains in tissues of the recipient for months, during which time the animal produces no immune response to the antigen. Immunologic paralysis is much easier to induce with polysaccharide than with protein antigens. It is highly specific for the antigen used for its induction.

Immunological ignorance

Immunological ignorance is a type of tolerance to self in which a target antigen and reactive lymphocytes capable of reacting with it are both present simultaneously in an individual without an autoimmune reaction occurring. The abrogation of immunologic ignorance may lead to autoimmune disease.

Immunological enhancement

Immunological enhancement refers to the prolonged survival, conversely the delayed rejection, of a tumor allograft in a host as a consequence of contact with specific antibody. Both the peripheral and central mechanisms have been pos-

Table 14.2   T cell tolerance and B cell tolerance

	T cell tolerance	B cell tolerance
Feature	Involves the processing and presentation of self proteins complexed with MHC molecules on antigen presenting cells of the thymus	Manifests a decreased number of antibody-secreting cells following antigenic stimulation
Site of tolerance induction	Thymus, peripheral lymphoid tissues	Bone marrow, peripheral lymphoid tissues
Tolerance-sensitive stage of maturation	$CD4^+CD8^+$ thymocytes	IgM^+IgD^- immature B lymphocytes
Stimulus for tolerance induction	High affinity recognition of antigen in thymus and/or activation of T cells by high antigen concentration (central tolerance); antigen presentation by antigen presenting cells in the absence of costimulation, or repeated stimulation by self antigens (peripheral tolerance)	Recognition of multivalent antigen in bone marrow (central tolerance); recognition of antigen without help of T cells (peripheral tolerance)
Factors that enhance tolerance induction	Interleukin 10 (IL-10), transforming growth factor β (TGF-β)	Prostaglandin E (PGE)
Factors that block tolerance induction	Interferon β (IFN-β), tumor necrosis factor β (TNF-β), leukotriene (LT)	Interleukin 1 (IL-1), lipopolysaccharide (LPS), 8-bromoguanosine
Mechanisms of tolerance	Central tolerance: clonal deletion (apoptotic cell death, negative selection) Peripheral tolerance: clonal anergy (functional inactivation without cell death); activation-induced cell death/supression	Central tolerance: clonal deletion (apoptotic cell death, negative selection); receptor editing Peripheral tolerance: clonal anergy (block in signal transduction); failure to enter lymphoid follicles
Ease of induction	Induced more rapidly (as soon as within a single day) and with less tolerogen	Induced less rapidly (requires 10 days) and with 100 times more tolerogen
Antigen dose	Low dose	High dose
Duration of tolerance	Longer lasting (6–8 months in T cells)	Shorter lasting (2 months in B cells of the bone marrow)

tulated. In the past, coating of tumor cells with antibody was presumed to interfere with the ability of specifically reactive lymphocytes to destroy them, but a central effect in suppressing cell-mediated immunity, perhaps through suppressor T cells, is more likely the main mechanism.

These variations of immunologic tolerance are compared and contrasted in **Table 14.3**.

IMMUNOREGULATION

Immunoregulation refers to control of the immune response usually by its own products such as the idiotypic network of antibody regulation described by Niels Jerne, feedback inhi-bition of antibody formation by antibody molecules, T cell receptor interaction with antibodies specific for them, the effect of immunosuppressive and immunoenhancing cytokines on the immune response in addition to other mechanisms. Results of these immunoregulatory interactions may lead to either suppression or potentiation of one or the other limb of the immune response.

Biological response modifier

A biological response modifier (BRM) is a substance that can alter the normal immune response and improve the body's natural response to infection and disease. BRMs cover a wide spectrum of molecules, such as cytokines,

Table 14.3   Variations of immunologic tolerance

	Stimulus for tolerance induction	Mechanisms of tolerance
Oral tolerance	Oral administration of an antigen	Anergy of antigen-specific T cells or the formation of immunosuppressive cytokines (TGF-β, IL-4, IL-10)
Split tolerance	Some epitopes of allogeneic cells	Specific immunological unresponsiveness affecting either the B cell (antibody) limb or the T cell (cell-mediated) limb of the immune response while the unaffected limb is left intact to produce antibody or respond with cell-mediated immunity
Immune deviation	An antigen capable of inducing formation of humoral antibody and development of delayed-type hypersensitivity	Conversion of a T cell response involving T_H1 cytokines that induce cell-mediated immunity to a T_H2 cytokine response that induces synthesis of selected antibody isotypes; selective suppression of delayed-type hypersensitivity and IgG_2 antibody production
Immunologic paralysis	Injection of large doses of pneumococcal polysaccharide	Slow metabolism of antigen that leads to antibody consumption and undetectable antibody to mount an immune response
Immunologic ignorance	Self antigen	Loss of lymphocytes known as regulatory or suppressor T cells
Immunologic enhancement	Tumor allograft	Peripheral mechanism: coating of tumor cells with antibody to interfere with the ability of specifically reactive lymphocytes to destroy them; Central mechanism: suppression of cell-mediated immunity through suppressor T cells

interleukins, interferons, hematopoietic colony-stimulating factors, tumor necrosis factor, B cell growth and differentiating factors, lymphotoxins, and macrophage-activating and chemotactic factors, as well as macrophage inhibitory factor, eosinophil chemotactic factor, osteoclast activating factor, etc. Research on pharmacological applications of BRMs has led to development of both immunosuppressive and immunostimulating drugs that are effective in preventing the rejection of transplanted organs, for the treatment of some autoimmune diseases, as cancer immunotherapy, or as adjuvants for vaccine construction. In addition to having potent immunomodulatory function, some BRMs may act directly on certain cancer cells to block their growth.

In recent years, research has focused on the mechanisms of action of these compounds as well as on the discovery of new ones. These biological response modifiers target specific chemicals in the immune system that contribute to disease processes and aim to reduce the signs and symptoms and to slow progression of the disease. The leading indications in development for available and emerging BRM drugs include anemia, bone marrow/stem cell transplantation, cancer, infectious diseases, inflammatory diseases, rheumatoid arthritis and multiple sclerosis.

Table 14.4 lists some of the biological response modifiers used in treatment of certain diseases.

Therapeutic monoclonal antibodies

Monoclonal antibody

Monoclonal antibody (mAb) is an antibody synthesized by a single clone of B cells or plasma cells. The identical copies of the antibody molecules produced contain only one class of heavy chain and one type of light chain. Köhler and Milestein in the mid-1970s developed B cell hybridomas by fusing an antibody-producing B cell with a mutant myeloma cell that was not secreting antibody. The B cell product provided the specificity, whereas the myeloma cell conferred immortality on the hybridoma clone. Today, monoclonal antibodies are produced in large quantities against a plethora of antigens for use in diagnosis and sometimes in treatment.

Monoclonal antibody therapy

Monoclonal antibody (mAb) therapy refers to treatment with monoclonal antibodies to suppress immune function, kill target cells or treat specific inflammatory diseases. MAbs demonstrate highly specific binding to precise cellular or molecular targets. Monoclonal antibodies with clinical implications can be divided into the following categories: murine monoclonal antibodies, chimeric antibodies, humanized antibodies, and human antibodies. Monoclonal anti-

Table 14.4 Biological response modifiers and clinical application

Category	Products	Clinical indications
Colony stimulating factors	Filgrastim (G-CSF), Neupogen, Granulokine	Prophylaxis of chemotherapy-associated neutropenia; after ablative chemotherapy and bone marrow transplantation in nonmyeloid cancers
	Pegfilgrastim (covalent conjugate of recombinant methionyl human G-CSF and monomethoxypolyethylene glycol), Neulasta	Decreasing the incidence of infection, as manifested by febrile neutropenia, in patients with nonmyeloid malignancies receiving myelosuppressive anti-cancer drugs
	Sargramostim (GM-CSF), Leukine, Prokine	Following chemotherapy-induced neutropenia and bone marrow or stem cell transplantation in non-Hodgkin's lymphoma, acute lymphocytic leukemia, or Hodgkin's disease; HIV-infected patients with neutropenia
Stem cell stimulation factors	Ancestim (recombinant-methionyl human SCF), Stemgen	Increasing the number of peripheral blood progenitor cells (PBPCs) in PBPC transplantation following chemotherapy; stem cell transplantation
Erythropoietin	Epoetin alfa, Epogen, Procrit	Anemia associated with chronic renal failure; anemia in zidovudine-treated HIV-infected patients; anemia in cancer patients on chemotherapy
	Darbepoitin alfa, Aranesp, Nespo	Kidney disease-related anemia; anemia in chemotherapy patients
	Human recombinant erythropoietin, Eprex	Kidney disease-related anemia; anemia in chemotherapy patients
	Epoetin beta, NeoRecormon	Anemia in patients with chronic kidney disease; anemia-related cancer
	Epoetin delta, Dynepo	Kidney disease-related anemia; anemia in chemotherapy patients
Thrombopoietin	Recombinant human thrombopoietin (rhTPO)	Patients with delayed platelet recovery after hematopoietic stem cell transplantation
Interferon α	Interferon alfa-n3, Alfernon N	Certain types of leukemia; certain AIDS-related illnesses; certain forms of hepatitis
	Interferon alfacon-1, Infergen	Chronic HCV infection
	Peginterferon alfa-2b, Peg-Intron	Chronic hepatitis C
	Interferon alfa-2b recombinant, Intron A	Chronic hepatitis B and C
	Natural human alpha interferon, Multiferon	Chronic hepatitis C
	Peginterferon alfa-2a, Pegasys	Chronic hepatitis C
	Interferon alfa-2a recombinant, Roferon-A	Chronic hepatitis C; hairy cell leukemia; AIDS-related Kaposi's sarcoma; chronic phase Philadelphia chromosome positive chronic myelogenous leukemia
Interferon β	Interferon beta-1a, Avonex, Rebif	Multiple sclerosis
	Interferon beta-1b, Betaseron, Betaferon	Multiple sclerosis

Table 14.4   Biological response modifiers and clinical application (*continued*)

Category	Products	Clinical indications
Interferon γ	Interferon gamma-1b, Actimmune	Prevention of excessive scarring; chronic granulomatous disease; severe, malignant osteopetrosis; idiopathic pulmonary fibrosis; liver fibrosis, ovarian cancer
Interleukins	Denileukin diftitox (IL-2), Ontak	Cutaneous T cell lymphoma
	Oprelvekin (IL-11), Neumega	Severe thrombocytopenia following myelosuppressive chemotherapy
	Aldesleukin (IL-2), Proleukin	Metastatic renal cell Carcinoma; metastatic melanoma; HIV-positive people
Cytokines	Multikine (mixture of naturally occurring cytokines including interleukins, interferons, chemokines, and colony-stimulating factors)	Head and neck cancer; HIV-infected women with cervical dysplasia
Tumor necrosis factor inhibitors	Etanercept, Enbrel	Moderate to severe rheumatoid arthritis

bodies have multiple uses in health care. Over two-thirds of mAb products are for transplant rejection, cancer, autoimmune diseases, infectious diseases, antiviral prophylaxis, and anti-thrombotic treatment. For example, Edrecolomab is used to treat solid tumors; Enlimomab is used to ameliorate organ transplant rejection; Infliximab is used as therapy for Crohn's disease and rheumatoid arthritis. OKT3 is used to treat organ transplant rejection; Palivizumab is used for respiratory syncytial virus; Rituximab is used for therapy of leukemias and lymphomas; Rhumabvegf is used to treat solid tumors; and Transtuzumab is used in subjects with metastatic breast cancer.

Table 14.5 lists the FDA approved mAbs or mAbs currently in clinical trials beyond phase II or phase III. Their potential clinical use is also listed.

Table 14.5   Therapeutic monoclonal antibodies

Antibody name	Target antigen	Conditions treated/prevented
Abciximab (ReoPro)	Glycoprotein II$_b$III$_a$ receptor	Complications of coronary angioplasty
ABX-CBL	CD147	GVHD
ABX-EGF	EGFr	EGF-dependent human tumor
ABX-IL8	IL-8	Rheumatoid arthritis, psoriasis
AcuTect		Diagnosis of acute venous thrombosis
Adalimumab (Humira)	TNF	Rheumatoid arthritis
AFP-Scan	AFP	Detection of liver and germ cell cancers
Alemtuzumab (Campath)	CD52	B cell chronic lymphocytic leukemia, multiple sclerosis, kidney transplant rejection
Apolizumab (Remitogen)	1D10 antigen	B cell non-Hodgkin's lymphoma, solid tumors
Arcitumomab (CEA-Scan) technetium-99m labeled	Carcinoembryonic antigen	Presence, location and detection of recurrent and metastatic colorectal cancer

Table 14.5 Therapeutic monoclonal antibodies (*continued*)

Antibody name	Target antigen	Conditions treated/prevented
Anti-CD11a hu1 124	CD11a	Psoriasis
Basiliximab (Simulect)	CD25 (IL-2 receptor)	Allograft rejection
Bectumomab		Non-Hodgkin's lymphoma
Bevacizumab (Avastin)	VEGF	Metastatic renal cell carcinoma
Capromab Pendetide (Prostascint) indium-111 labeled	Prostate membrane specific antigen (PMSA)	Radioimmunoscintigraphy for prostate cancer
Cetuximab	EGFr	Head and neck, breast, pancreatic, colorectal cancers
CEACide	Carcinoembryonic antigen	Colorectal cancer
Daclizumab (Zenapax)	CD25 (IL-2 receptor)	Allograft rejection
Edrecolomab (Panorex)	17-1A cell surface antigen	Colorectal cancer
Efalizumab (Xanelim)	CD11a	Rheumatoid arthritis
Enlimomab	CD54 (ICAM-1)	Organ transplant rejection
Epratuzumab (LymphoCide)	CD22	Non-Hodgkin's lymphoma
Gemtuzumab ozogamicin Mylotarg	CD33 calicheamicin	Acute myeloid leukemia
Hu23F2G (LeukArrest)	CD11/18 (leukointegrin)	Ischemic stroke
Hu1124	CD11a	Psoriasis
Ibritumomab tiuxetan (Zevalin)	CD20	B cell non-Hodgkin's lymphoma
Igovomab (Indimacis 125)	Tumor-associated antigen CA125	Detection of ovarian adenocarcinoma
Imciromab pentetate (Myoscint)	Human cardiac myosin	Myocardial infarction imaging
IMC-C225 (ERBITUX)	EGFR	EGF-dependent human tumor
Infliximab (Remicade)	TNF-α	Crohn's disease, rheumatoid arthritis
Inolimomab	IL-2 receptor	Organ transplant rejection
LDP-01	β2 integrin	Stroke, kidney transplant rejection
LDP-02	α4β7 integrin receptor	Crohn's disease, ulcerative colitis
LeuTech 99cTc-Anti-CD15 anti-granulocyte antibody	CD15	Imaging infection sites
Lerdelimumab	TGFb2	Glaucoma, cataract
Lym-1 yttrium-90 labeled	HLA-DR	Non-Hodgkin's lymphoma

Table 14.5 Therapeutic monoclonal antibodies (*continued*)

Antibody name	Target antigen	Conditions treated/prevented
LymphoScan	CD22	Detection of B cell non-Hodgkin's lymphoma
MAK-195F	TNF-α	Hyperinflammatory response in sepsis syndrome
MDX-33	CD64	Idiopathic thrombocytopenia purpura
MDX-H210	Bispecific HER2 × CD64	Breast, colorectal, kidney, ovarian, prostate cancers
MDX-447	Bispecific EGFR × CD64	Head, neck, renal cancers
Mitumomab (BEC2)	GD3-idiotypic	Small cell lung cancer, melanoma
Muromonab (Orthoclone OKT3)	CD3	Allograft rejection
Natalizumab (Antegren)	α-4 integrin (VLA-4)	Multiple sclerosis, Crohn's disease
Nebacumab (Centoxin)	Bacterial endotoxins	Gram-negative bacteria sepsis
Nofetumomab (Verluma)	Carcinoma-associated antigen	Detection of small cell lung cancer
OctreoScan indium-111 labeled	Somatostatin receptor	Immunoscintigraphic localization of primary and metastatic neuroendocrine tumors that contain somatostatin receptors
Olizuma, rhuMAb-E25	Ig-E	Allergic asthma, allergic rhinitis
Oncolym (131Lym-1) iodine-131 labeled	HLA-DA	B cell non-Hodgkin's lymphoma
Omalizumab (Xolair)	IgE	Allergic asthma, allergic rhinitis
Oregovomab (OvaRex)	Tumor-associated antigen CA125	Ovarian cancer
ORTHOCLONE OKT4A	CD4	CD4-mediated autoimmune diseases, allograft rejection
Palivizumab (Synagis)	Antigenic site of the F protein of respiratory syncytial virus (F gp)	Respiratory syncytial virus infection
Pexelizumab (5G1.1-SC)	Complement C5	AMI, UA, CPB, PTCA
Priliximab	CD4	Crohn's disease, multiple sclerosis
Regavirumab	Cytomegalovirus (CMV)	Acute CMV disease
Rituximab (Rituxan)	CD20	Non-Hodgkin's lymphoma
Satumomab pendetide (OncoScint CR/OV)	Tumor-associated glycoprotein-72	Detection of colorectal and ovarian cancers
Sevirumab (Protovir)	Cytomegalovirus (CMV)	Prevention of CMV infection in bone marrow transplant patients
Siplizumab (MEDI-507)	CD2	Acute GVHD, psoriatic arthritis
Smart M195	CD33	Acute myeloid leukemia, myelodysplastic syndrome
Sulesomab (LeukoScan) technetium-99m labeled	Surface granulocyte nonspecific cross-reacting antigen	Detection of osteomyelitis, acute atypical appendicitis

Table 14.5 Therapeutic monoclonal antibodies (*continued*)

Antibody name	Target antigen	Conditions treated/prevented
Tecnemab K1	High molecular weight melanoma-associated antigen	Diagnosis of cutaneous melanoma lesions
Tositumomab (Bexxar), iodine-131 attached	B cell surface protein	Non-Hodgkin's lymphoma
Trastuzumab (Herceptin)	Her2/neu	Her2 positive metastatic breast cancer
Visilizumab (Nuvion, Smart anti-CD3)	CD3	GVHD, ulcerative colitis
Vitaxin	αvβ3 integrin	Solid tumors
Votumumab (Humaspect)	Cytokeratin tumor-associated antigen	Detection of carcinoma of colon and rectum
YM-337	GPIIb/IIIa	Prevention of platelet aggregation
Zolimomab	CD5, ricin A-chain toxin	GVHD

Abbreviations:

GVHD	graft versus host disease	TGFb2	transforming growth factor b2
EGFr	epidermal growth factor receptor	HLA	human leukocyte antigen
IL-2	interleukin-2	VLA-4	very late antigen-4
IL-8	interleukin-8	AMI	acute myocardial infarction
AFP	α fetoprotein	UA	unstable angina
VEGF	vascular endothelial growth factor	CPB	cardiopulmonary bypass
ICAM-1	intercellular adhesion molecule-1	PTCA	percutaneous transluminal coronary angioplasty
TNF-α	tumor necrosis factor-α	GPIIb/IIIa	glycoprotein II_bIII_a

IMMUNOHEMATOLOGY AND TRANSFUSION MEDICINE

15

● **BLOOD GROUP**   ● **BLOOD TRANSFUSION**

BLOOD GROUP

Blood grouping

Blood grouping is the classification of erythrocytes based on their surface isoantigens. Among the well-known human blood groups are the ABO, Rh, and MNS systems. **Table 15.1** lists the blood group systems, as defined by the International Society of Blood Transfusion (ISBT) working party on blood group terminology, and their gene location.

ABO blood group system

The ABO blood group system is the first described of the human blood groups based upon carbohydrate alloantigens present on red cell membranes. Anti-A or anti-B isoagglutinins (alloantibodies) are present only in the blood sera of individuals not possessing that specificity. This serves as the basis for grouping humans into phenotypes designated A, B, AB, and O. Blood group methodology to determine the ABO blood type makes use of the agglutination reaction. **Table 15.2** shows ABO blood group antigens, antibodies, and the front and back typing.

The ABO system remains the most important in the transfusion of blood and is also critical in organ transplantation. **Table 15.3** gives the suggested ABO group selection order for transfusion of erythrocytes and plasma. Epitopes of the ABO system are found on oligosaccharide terminal sugars. The genes designated as *A/B*, *Se*, *H*, and *Le* govern the formation of these epitopes and of the Lewis (Le) antigens. The two precursor substances type I and type II differ only in that the terminal galactose is joined to the penultimate N-acetylglucosamine in the b 1–3 linkage in type I chains, but in the b 1–4 linkage in type II chains.

MNS blood group system

The MNS blood group system refers to human erythrocyte glycophorin epitopes. There are four distinct sialoglycoproteins (SGP) on red cell membranes. These include α-SGP

(glycophorin A, MN), β-SGP (glycophorin C), γ-SGP (glycophorin D), and δ-SGP (glycophorin B). MN antigens are present on α-SGP and δ-SGP. M and N antigens are present on α-SGP, with approximately 500 000 copies detectable on each erythrocyte. This transmembrane molecule has a carboxy terminus that stretches into the cytoplasm of the erythrocyte with a 23-amino acid hydrophobic segment embedded in the lipid bilayer. The amino terminal segment extends to the extracellular compartment. Blood group antigen activity is in the external segment. In α-SGP with M antigen activity, the first amino acid is serine and the fifth is glycine. When it carries N antigen activity, leucine and glutamic acid replace serine and glycine at positions 1 and 5, respectively. The Ss antigens are encoded by allelic genes at a locus closely linked to the MN locus. The U antigen is also considered a part of the MNSs system. Whereas anti-M and anti-N antibodies may occur without red cell stimulation, antibodies against Ss and U antigens generally follow erythrocyte stimulation. The MN and Ss alleles positioned on chromosome 4 are linked. **Table 15.4** shows the phenotypes and frequencies of the MNS blood group system.

Rhesus blood group system

The Rhesus blood group system is comprised of Rhesus monkey erythrocyte antigens such as the D antigen that are found on the red cells of most humans, who are said to be Rh+. This system is quite complex, and the rare Rh alloantigens are still not characterized biochemically. Three closely linked pairs of alleles designated Dd, Cc, and Ee are postulated to be at the Rh locus, which is located on chromosome 1. There are several alloantigenic determinants within the Rh system. More than 50 antigens of the Rh blood group system have been identified. They are listed in **Table 15.5**. Clinically, the D antigen is the one of greatest concern, since RhD$^-$ individuals who receive RhD$^+$ erythrocytes by transfusion can develop alloantibodies that may lead to severe reactions with further transfusions of RhD$^+$ blood. The D antigen also poses a problem in RhD$^-$ mothers

Table 15.1 Membrane component and chromosomal assignment of the human RBC blood group systems

ISBT number	Blood group	RBC membrane component	Chromosome location
001	ABO	Anion exchanger (AE-1), protein 4.5, lipids	9q34.1-q34.2
002	MNS	M,N: glycophorin A; S,s: glycophorin B	4q28-q31
003	P1	Glycolipid	22q11.2-qter
004	Rh	Rh proteins	1p36.2-p34
005	Lutheran	Lutheran glycoprotein	19q13.2
006	Kell	Kell proteins	7q33
007	Lewis	Type 2 oligosaccharides	19p13.3
008	Duffy	Chemokine receptor	1q22-q23
009	Kidd	Urea transporter	18q11-q12
010	Diego	AE-1	17q12-q21
011	Yt	Acetylcholinesterase	7q22
012	Xg	Xg glycoprotein	Xp22.32
013	Scianna	SC glycoprotein	1p36.2-p22.1
014	Dombrock	DO glycoprotein	12p13.2-p12.3
015	Colton	CHIP 28 (aquaporin)	7p14
016	LW	LW glycoprotein	19p13.3
017	Chido/Rodgers	C4A and C4B glycoproteins	6p21.3
018	Hh	AE-1, protein 4.5, lipids	19q13
019	Kx	Kx glycoprotein	Xp21.1
020	Gerbich	Glycophorins C and D	2q14-q21
021	Cromer	Decay accelerating factor (CD55)	1q32
022	Knops	CR1 (CD35)	1q32
023	Indian	CD44	11p13
024	Ok	CD147	19p13.3
025	Raph		11p15.5
026	John Milton Hagen		15q23-q24

Table 15.2 ABO blood group antigens, antibodies and grouping by front and back typing

Blood type	Erythrocyte surface antigen	Antibody in serum	Front typing		Back typing		
			Reaction of cells tested with		Reaction of serum tested against		
			Anti-A	Anti-B	A cells	B cells	O cells
A	A antigen	Anti-B	+	0	0	+	0
B	B antigen	Anti-A	0	+	+	0	0
AB	A, B antigens	No antibody	+	+	0	0	0
O	H antigen	Both anti-A and anti-B	0	0	+	+	0

Table 15.3 Suggested ABO group selection order for transfusion of erythrocytes and plasma

Recipient ABO group	Component ABO group							
	1st choice		2nd choice		3rd choice		4th choice	
	RBC	Plasma	RBC	Plasma	RBC	Plasma	RBC	Plasma
AB	AB	AB	A	(A)	B	(B)	O	(O)
A	A	A	O	AB		(B)		(O)
B	B	B	O	AB		(A)		(O)
O	O	O		A		B		AB

Table 15.4 MNSs blood group system

Phenotype	Reactions		Phenotype frequency (%)	
	Anti-M	Anti-N	Caucasians	African Americans
M+N−	+	0	28	26
M+N+	+	+	50	45
M−N+	0	+	22	30
	Anti-S	Anti-s		
S+s−	+	0	11	3
S+s+	+	+	43	28
S−s+	0	+	45	69
S−s−	0	0	0	<1

Table 15.5 Antigens of the Rh blood group system and their incidence

Numerical designation	Antigen name	Incidence (%) Cauc	Afr Am	Overall	Numerical designation	Antigen name	Incidence (%) Cauc	Afr Am	Overall
Rh1	D	85	92		Rh30	GO^a	0	<0.01	
Rh2	C	68	27		Rh31	hr^B			98
Rh3	E	29	22		Rh32	Rh32	<0.01	1	
Rh4	c	80	96		Rh33	Har			<0.01
Rh5	e			98	Rh34	Bastiaan			>99.9
Rh6	f	65	92		Rh35	Rh35			<0.01
Rh7	Ce	68	27		Rh36	Be^a			<0.1
Rh8	C^w	2	1		Rh37	Evans			<0.01
Rh9	C^x			<0.01	Rh39	C-like			>99.9
Rh10	V	1	30		Rh40	Tar			<0.01
Rh11	E^w			<0.01	Rh41	Ce-like	70		
Rh12	G	84	92		Rh42	CE^s	<0.1	2	
Rh17	Hr$_o$			>99.9	Rh43	Crawford			<0.01
Rh18	Hr			>99.9	Rh44	Nou			>99.9
Rh19	Hr^s			98	Rh45	Riv			<0.01
Rh20	VS	<0.01	32		Rh46	Rh46			>99.9
Rh21	C^G			68	Rh47	Dav			>99.9
Rh22	CE			<1	Rh48	JAL			<0.01
Rh23	D^w			<0.01	Rh49	STEM	<0.01	6	
Rh26		80	96		Rh50	FPTT			<0.01
Rh27	cE	28	22		Rh51	MAR			>99.9
Rh28				<0.01	Rh52	BARC			<0.01
Rh29	total Rh			>99.9	Rh53	JAHK			<0.01

who bear a child with RhD^+ red cells inherited from the father. The entrance of fetal erythrocytes into the maternal circulation at parturition or trauma during the pregnancy can lead to alloimmunization against the RhD antigen which may cause hemolytic disease of the newborn in subsequent pregnancies. Further confusion concerning this system has been caused by the use of separate designations by the Wiener and Fisher systems. Principal Rh genes and their frequencies of occurrence among Caucasians and African Americans are listed in **Table 15.6**.

Other blood group systems

The many antigens on erythrocytes are grouped into blood group systems. In addition to the above-mentioned blood groups, P, Kell, Duffy, Kidd, Lutheran, and Lewis blood groups are additional ones. The **P blood group system** consists of three ABH blood group-related antigens found on erythrocyte surfaces and is comprised of the three sugars galactose, N-isoacetyl-galactosamine, and n-acetyl-glucosamine. The P antigens are designated P_1, P_2, P^k, and P. P_2 subjects rarely produce anti-P_1 antibody which may lead to hemolysis in clinical situations. The **Kell blood group system** is named for an antibody that induces hemolytic disease of the newborn, which is specific for the K (KEL1) antigen. Anti-k (KEL2) antibodies react with the erythrocytes of more than 99% of the random population. Kell system antigens are present only in relatively low density on the erythrocyte membrane. The **Duffy blood group system** is comprised of human erythrocyte epitopes encoded by *Fya* and *Fyb* genes, located on chromosome 1. Mothers immunized through exposure to fetal red cells bearing the Duffy antigens, which she does not possess, may synthesize antibodies that cross the placenta and induce hemolytic disease of the newborn. The **Kidd blood group system** is named for the anti-Jk^a antibodies which were originally detected in the blood serum of a woman giving birth to a baby with hemolytic disease of the newborn. The anti-Jk^b antibodies were discovered in the serum of a patient following a transfusion reaction. Four phenotypes are revealed by the reactions of anti-Jk^a and anti-Jk^b antibodies. The **Lutheran blood group system** consists of human erythrocyte epitopes recognized by alloantibodies against Lu^a and Lu^b products. Antibodies developed against Lutheran antigens during pregnancy may induce hemolytic disease of the newborn. The **Lewis blood group system** is an erythrocyte antigen system that differs from other red cell groups in that the antigen is present in soluble form in the blood and saliva. Lewis antigens are adsorbed from the plasma onto the red cell membrane. The Lewis phenotype expressed is based on whether the individual is a secretor or a nonsecretor of the Lewis gene product. Expression of the Lewis phenotype is dependent also on the ABO phenotype. **Table 15.7** lists the phenotype,

antigen reaction, and frequencies of occurrence of these systems.

Other blood antigens

Chido (Ch) and Rodgers (Rg) antigens

Chido (Ch) and Rodgers (Rg) antigens are epitopes of C4d fragments of human complement component C4. They are not intrinsic to the erythrocyte membrane. The Chido epitope is found on C4d derived from C4B, whereas the Rodgers epitope is found on C4A derived from C4d. The Rodgers epitope is Val-Asp-Leu-Leu, and the Chido epitope is Ala-Asp-Leu-Arg. They are situated at residue positions 1188 and 1191 in the C4 α chain's C4d region. Antibodies against Ch and Rg antigenic determinants agglutinate saline suspensions of red blood cells coated with C4d. Since C4 is found in human serum, anti-Ch and anti-Rg are neutralized by sera of most individuals which contains the relevant antigens. Ficin and papain destroy these antigens. **Table 15.8** shows C4d component presence and frequency of occurrence of Ch and Rg antigen phenotypes.

Xg^a antigen

Anti-Xg^a, the sex-linked blood antigen, is an antibody more common in women than in men. It is specific for the Xg^a antigen, in recognition of its X-linked pattern of inheritance. **Table 15.9** gives phenotype frequencies in Caucasian males and females. The antibody is relatively uncommon and has not been implicated in hemolytic disease of the newborn or hemolytic transfusion reactions even though it can bind complement and may occasionally be an autoantibody. Anti-Xg^a antibodies might be of value in identifying genetic traits transmitted in association with the X chromosome.

Platelet antigens

The role of platelet antigens parallels in many ways that of erythrocyte antigens. Platelet antigens are surface epitopes on thrombocytes that may be immunogenic, leading to platelet antibody formation which causes such conditions as neonatal alloimmune thrombocytopenia and post-transfusion purpura. The Pl^{A1} antigen may induce platelet antibody formation in Pl^{A1} antigen-negative individuals. Additional platelet antigens associated with purpura include Pl^{A2}, Bak^a, and HLA-A2. Anti-Bak^a IgG antibody synthesized by a Bak^a negative pregnant woman may be passively transferred across the placenta to cause immune thrombocytopenia in the neonate. **Table 15.10** summarizes human platelet antigen systems.

Table 15.6   Rh blood group system

| | Wiener | Fisher–Race | Phenotype frequency (%) | |
			Cauc	Afr Am
R_1	Rh_1	CDe	42	17
R_2	Rh_2	cDE	14	11
R_0	Rh_0	cDe	4	44
R_z	Rh_z	CDE	Very rare	Very rare
r'	rh'	Cde	2	2
r''	rh''	cdE	1	<1
r	rh	cde	37	26
r^y	rh^y	CdE	Very rare	Very rare

Table 15.7   Other blood group systems

| Phenotype | Reactions with anti- | | | | | | Phenotype frequency (%) | |
							Cauc	Afr Am
P blood group								
	P_1	P	P^k	PP_1P^k				
P_1	+	+	0	+			79	94
P_2	0	+	0	+			21	6
P	0	0	0	0			Very rare	Very rare
P_1^k	+	0	+	+			Very rare	Very rare
P_2^k	0	0	+	+			Very rare	Very rare
Kell blood group								
	K	k	Kp^a	Kp^b	Js^a	Js^b		
K+k−	+	0					0.2	Rare
K+k+	+	+					8.8	2
K−k+	0	+					91	98
Kp (a+b−)			+	0			Rare	0
Kp (a+b+)			+	+			2.3	Rare
Kp (a−b+)			0	+			97.7	100
Js (a+b−)					+	0	0	1
Js (a+b+)					+	+	Rare	19

Phenotype	Reactions with anti-						Phenotype frequency (%)	
							Cauc	Afr Am
Js (a−b+)					0	+	100	80
K₀	0	0	0	0	0	0	Very rare	Very rare
Duffy blood group								
	Fyᵃ		Fyᵇ					
Fy (a+b−)	+		0				17	9
Fy (a+b+)	+		+				49	1
Fy (a−b+)	0		+				34	22
Fy (a−b−)	0		0				Very rare	68
Kidd blood group								
	Jkᵃ		Jkᵇ					
Jk (a+b−)	+		0				28	57
Jk (a+b+)	0		+				49	34
Jk (a−b+)	+		+				23	9
Jk (a−b−)	0		0				Very rare	Very rare
Lutheran blood group								
	Luᵃ		Luᵇ					
Lu (a+b−)	+		0				0.15	0.1
Lu (a+b+)	+		+				7.5	5.2
Lu (a−b+)	0		+				92.35	94.7
Lu (a−b−)	0		0				Very rare	Very rare
Lewis blood group								
	Leᵃ		Leᵇ					
Le (a+b−)	+		0				22	23
Le (a+b+)	+		+				Rare	Rare
Le (a−b+)	0		+				72	55
Le (a−b−)	0		0				6	22

Table 15.8   Chido (Ch) and Rodgers (Rg) antigens

Phenotype	C4d component	Frequency (%)
Ch (a+), Rg (a+)	C4dS, C4df	95
Ch (a−), Rg (a+)	C4df	2
Ch (a+), Rg (a−)	C4dS	3
Ch (a−), Rg (a−)	None	Very rare

Table 15.9 Xg^a, the sex-linked blood antigen

Phenotype	Reaction with anti-Xg^a	Phenotype frequency (%)	
		Males	Females
Xg (a+)	+	65.6	88.7
Xg (a−)	0	34.4	11.3

BLOOD TRANSFUSION

Blood transfusion covers a wide range of practice, including transfusion of red blood cells, platelets, granulocytes, special cellular blood components, and replacement of coagulation factors. Hematopoietic transplantation also falls into this category.

Complications of blood transfusion

In addition to transfer of infectious diseases, such as AIDS and hepatitis, transfusion reactions are the major complications of blood transfusion. **Table 15.11** demonstrates the potential problems with ABO- and Rh- incompatible hematopoietic progenitor cell transplantation.

Transfusion reactions include both immune and non-immune reactions that follow the administration of blood. Transfusion reactions with immune causes are considered serious and occur in 1 in 3000 transfusions. Patients may develop urticaria, itching, fever, chills, chest pains, cyanosis, and hemorrhage; some may even collapse. Immune, noninfectious transfusion reactions include allergic urticaria (immediate hypersensitivity); anaphylaxis, as in the administration of blood to IgA-deficient subjects, some of whom develop anti-IgA antibodies of the IgE class; and serum sickness, in which the serum proteins such as immunoglobulins induce the formation of precipitating antibodies that lead to immune complex formation. Transfusion reactions may cause intravascular lysis of red blood cells and when severe may lead to renal injury, fever, shock, and disseminated intravascular coagulation. Four broad categories of transfusion reactions are summarized in **Table 15.12**.

Table 15.10 Human platelet antigen systems

Alloantigen system	Other names	Alleles	Antigen frequency (%) in Caucasians	Amino acid substitution
HPA-1	P1^A, Zw	*HPA-1a (P1^{A1})*	98	GPIIIa
		HPA-1b (P1^{A2})	29	Leu $\leftrightarrow$ Pro$_{33}$
HPA-2	Ko, Sib	*HPA-2a (Ko^b)*	99	GPIb
		HPA-2b (Ko^a)	15	Thr $\leftrightarrow$ Met$_{145}$
HPA-3	Bak, Lek	*HPA-3a (Bak^a)*	81	GPIIb
		HPA-3b (Bak$_b$)	70	Ile $\leftrightarrow$ Ser$_{843}$
HPA-4	Pen, Yuk	*HPA-4a (Pen^a)*	>99	GPIIIa
		HPA-4b (Pen^b)	<1	Arg $\leftrightarrow$ Gln$_{143}$
HPA-5	Br, Hc, Zav	*HPA-5a (Br^b)*	99	GPIa
		HPA-5b (Br^a)	20	Glu $\leftrightarrow$ Lys$_{505}$
HPA-6w	Ca, Tu	*HPA-6a (Ca^b)*	>98	GPIIIa
		HPA-6b (Ca^a)	<2.4	Arg $\leftrightarrow$ Gln$_{489}$
HPA-7w	Mo	*HPA-7a (Mo^b)*	>99	GPIIIa
		HPA-7b (Mo^a)	<1	Pro $\leftrightarrow$ Ala$_{407}$
HPA-8w	Sra	*HPA-8a (Sr^b)*	>99	GPIIIa
		HPA-8b (Sr^a)	<1	Arg $\leftrightarrow$ Cys$_{636}$
		Gov^a	81	
		Gov^b	74	
		Vu	<1	
		Gro	<1	
		Iy^b	>99	GPIb
		Iy^a	<1	Gly $\leftrightarrow$ Glu$_{15}$
HPA-9w		*Max^b*	99	GPIIb
		Max^a	<1	Val $\leftrightarrow$ Met$_{337}$
HPA-10w		*La^a*	<1	GPIIIa
		La^b	>99	Arg $\leftrightarrow$ Gln$_{62}$

Table 15.11 Potential problems with ABO- and Rh- incompatible HPC transplantation

Incompatibility	Example		Potential problems
	Donor	Recipient	
ABO (major)	Group A	Group O	Hemolysis of infused donor RBCs, failure or delay of RBC engraftment, hemolysis at the time of donor RBC engraftment
ABO (minor)	Group O	Group A	Hemolysis of patient RBCs from infused donor plasma, hemolysis of patient RBCs 7–10 days after transplant due to passenger lymphocyte-derived isohemagglutinins
Rh	Negative	Positive	Hemolysis of patient RBCs by donor anti-D produced after engraftment
	Positive	Negative (with anti-D)	Hemolysis of donor RBCs from newly engrafted HPCs

Table 15.12 Categories and management of adverse transfusion reactions

Type	Incidence	Etiology	Presentation	Laboratory testing	Therapeutic/ prophylactic approach
Acute (<24 hours) transfusion reaction – immunologic					
Hemolytic	1:38,000 – 1:70,000	Red cell incompatibility	Chills, fever, hemoglobinuria, hypotension, renal failure with oliguria, DIC (oozing from IV sites), back pain, pain along infusion vein, anxiety	• Clerical check • DAT • Visual inspection (plasma-free Hb or methemalbumin) • Further tests as indicated to define possible incompatibility • Further tests as indicated to detect hemolysis (LDH, bilirubin, etc.)	• Keep urine output >100 ml/ hr with fluids and IV diuretic (furosemide) • Analgesics (may need morphine) • Pressors for hypotension (low-dose dopamine) • Hemostatic components (platelets, cryo, FFP) for bleeding
Fever/chill, nonhemolytic	RBCs: 1:200 – 1:17 (0.5–6%) Platelets: 1:100 – 1:3 (1-38%)	• Antibody to donor WBCs • Accumulated cytokines in platelet bag	Chills/rigors, rise in temperature, headache, vomiting	• Rule out hemolysis (DAT, inspect for Hb) • WBC antibody screen	• Antipyretic premedication (acetaminophen, no aspirin) • Leukocyte-reduced blood
Urticarial	1:100 – 1:33 (1–3%)	Antibody to donor plasma proteins	Urticaria, pruritis, flushing	• Rule out hemolysis (DAT, inspect for Hb)	• Antihistamine, treatment or premedication (PO or IV) • May restart unit slowly after antihistamine if symptoms resolve
Anaphylactic	1:20,000 – 1:50,000	Antibody to donor plasma proteins (includes IgA, C4)	Hypotension, urticaria, bronchospasm (respiratory distress, wheezing), local edema, anxiety	• Rule out hemolysis (DAT, inspect for Hb) • Anti-IgA • IgA, quantitative	• Trendelenberg (feet up) position • Fluids • Epinephrine (adult dose: 0.3– 0.5 ml of 1:000 solution SC or IM; in severe cases, 1:10,000 IV) • Antihistamine, corticosteroids, beta-2 agonists • IgA-deficient blood components

Table 15.12 Categories and management of adverse transfusion reactions (*continued*)

Type	Incidence	Etiology	Presentation	Laboratory testing	Therapeutic/ prophylactic approach
Acute (<24 hours) transfusion reaction – nonimmunologic					
Hypotension associated with ACE inhibition	Dependent on clinical setting	Inhibited metabolism of bradykinin with infusion of bradykinin (negatively charged filters) or activators of prekallikrein	Flushing, hypotension	• Rule out hemolysis (DAT, inspect for Hb)	• Withdraw ACE inhibition • Use of nonalbumin volume replacement for plasmapheresis • Avoid bedside leukocyte filtration
Transfusion-related acute lung injury	1:5000 – 1:190 000	Anti-WBC antibodies in donor (occas. in recipient), other WBC activating agents in components	Hypoxemia, respiratory failure, hypotension, fever	• WBC antibody screen in donor and recipient • WBC crossmatch	• Supportive care until recovery • Defer implicated donors
Circulatory overload	1%	Volume overload	Dyspnea, orthopnea, cough, tachycardia, hypertension, headache	None	• Upright posture • Oxygen • IV diuretic (furosemide) • Phlebotomy (250 ml increments)
Nonimmune hemolysis	Rare	Physical or chemical destruction of blood (heating, freezing, hemolytic drug or solution added to blood)	Hemoglobinuria	• Plasma-free Hb • DAT (should be negative) • Test unit for hemolysis	• Identify and eliminate cause
Air embolus	Rare	Air infusion via line	Sudden shortness of breath, acute cyanosis, pain, cough, hypotension, cardiac arrhythmia	None	• Lay patient on left side with legs elevated above chest and head
Hypocalcemia (ionized calcium)	Dependent on clinical setting	Rapid citrate infusion (massive transfusion of citrated blood, delayed metabolism of citrate, apheresis procedures)	Paresthesis, tetany, arrhythmia	• Ionized calcium • Prolonged Q-T interval on EKG	• Slow calcium infusion while monitoring ionized calcium levels in severe cases • PO calcium supplement for mild symptoms during apheresis procedures
Hypothermia	Dependent on clinical setting	Rapid infusion of cold blood	Cardiac arrhythmia	N/A	• Employ blood warmers

Table 15.12 Categories and management of adverse transfusion reactions (*continued*)

Type	Incidence	Etiology	Presentation	Laboratory testing	Therapeutic/ prophylactic approach
Delayed (>24 hours) transfusion reaction – immunologic					
Alloimmunization, RBC antigens HLA	1:100 (1%) 1:10 (10%)	Immune response to foreign antigens on RBCs, or WBCs and platelets (HLA)	Positive blood group antibody screening test, platelet refractoriness, delayed hemolytic reaction, hemolytic disease of the newborn	• Antibody screen • DAT • Platelet antibody screen • Lymphocytotoxicity test	• Avoid unnecessary transfusions • Leukocyte-reduced blood
Hemolytic	1:11 000– 1:5000	Anamnestic immune response to RBC antigens	Fever, decreasing hemoglobin, new positive antibody screening test, mild jaundice	• Antibody screen • DAT • Tests for hemolysis (visual inspection for hemoglobinemia, LDH, bilirubin, urinary hemosiderin as clinically indicated)	• Identify antibody • Transfuse compatible RBCs as needed
Graft-vs-host disease	Rare	Donor lymphocytes engraft in recipient and mount attack on host tissues	Erythroderma, maculopapular rash, anorexia, nausea, vomiting, diarrhea, hepatitis, pancytopenia, fever	• Skin biopsy • HLA typing	• Methotrexate, corticosteroids • Irradiation of blood components for patients at risk (including related donors and HLA-selected components)
Post-transfusion purpura	Rare	Recipient platelet antibodies (apparent alloantibody, usually anti-HPA1) destroy autologous platelets	Thrombocytopenic purpura, bleeding, 8–10 days following transfusion	Platelet antibody screen and identification	• IGIV • HPA1-negative platelets • Plasmapheresis
Immuno-modulation	Unknown	Incompletely understood interaction of donor WBC or plasma factors with recipient immune system	Increased renal graft survival, infection rate, post-resection tumor recurrence rate (controversial)	None specific	• Avoid unnecessary transfusions • Autologous transfusion • Leukocyte-reduced RBCs and platelets

Table 15.12 Categories and management of adverse transfusion reactions (*continued*)

Type	Incidence	Etiology	Presentation	Laboratory testing	Therapeutic/ prophylactic approach
Delayed (>24 hours) transfusion reactions – nonimmunologic					
Iron overload	Invariable after >100 units RBCs	Multiple transfusions with obligate iron load in transfusion-dependent patient	Diabetes, cirrhosis, cardiomyopathy	Iron studies	• Desferoxamine (iron chelator)

Abbreviations:

ACE	angiotensin-converting enzyme	IVIG	intravenous immunoglobulin
Antibody screen	blood group antibody screening test	LDH	lactate dehydrogenase
DIC	disseminated intravascular coagulation	PO	by mouth
DAT	direct antiglobulin test	RBC	red blood cell
Hb	hemoglobin	SC	subcutaneous
IV	intravenous	WBC	white blood cell

IMMUNOLOGICAL DISEASES AND IMMUNOPATHOLOGY

16

- ● **IMMUNOLOGICAL DISEASES**
- ● **SYSTEMIC AUTOIMMUNE DISEASES**
- ● **GAMMOPATHIES**

IMMUNOLOGICAL DISEASES

Immunological diseases include those conditions in which there is either an aberration in the immune response or the immune response to the disease agent leads to pathological changes. This category includes diseases with an immunological etiology or pathogenesis, immunodeficiency, hyperactivity of the immune response, or autoimmunity that leads to pathological sequelae.

Immunological diseases involving the blood

Autoimmune neutropenia (AIN) and immune thrombocytopenic purpura (ITP) are both cytopenias believed to be caused by an autoimmune mechanism. AIN is similar to ITP, which is a more common cytopenia. The blood component cells are destroyed by autoantibodies. This is followed by cytopenia and symptoms and signs associated with dysfunction of the blood cells.

Tumors of the haematopoietic and lymphoid tissues are also discussed under this category.

Table 16.1 lists autoimmune neutropenia, idiopathic thrombocytopenic purpura (ITP), B cell acute lymphoblastic leukemia (ALL), B cell chronic lymphocytic leukemia (B-CLL), mantle cell lymphoma (MCL), follicular lymphoma (FL), diffuse large B cell lymphoma (DLBCL), adult T cell leukemia/lymphoma (ATLL), acute myelogenous leukemia (AML), angioimmunoblastic lymphadenopathy (AILA), and Hodgkin's disease. Their etiology, genetics such as genetic predisposition associated human leukocyte histocompatibility antigens and chromosomal rearrangement, antibody/cellular involvement, clinical manifestations, laboratory manifestations and treatment are listed and compared.

Immunological diseases involving the skin

Human skin provides a unique opportunity to study organ-specific autoimmunity with a target tissue that is visually accessible. These diseases are defined by specific autoantibodies for identification of structural proteins of the dermis and epidermis. Bullous pemphigoid is an autoimmune disease affecting the dermal/epidermal junction, whereas pemphigus vulgaris is an autoimmune disease with disorders of cell-to-cell adhesion. Features of some autoimmune skin diseases are presented in **Table 16.2**.

Immunological diseases involving the vasculature

Autoimmune vasculitis is a broad and heterogeneous group of diseases characterized by inflammation and injury to the blood vessels, thought to be brought on by an autoimmune response. Any type, size, and location of blood vessel may be involved. Vasculitis may occur alone or in combination with other diseases, and may be confined to one organ or involve several organ systems. Injury to the vascular lumen leads to distal ischemia to the tissue perfused by the involved vessel and results in clinical symptoms and signs. **Table 16.3** summarizes some autoimmune vascular diseases including polyarteritis nodosa (necrotizing vasculitis of small- and medium-sized muscular arteries), leukocytoclastic vasculitis (small-vessel vasculitis), Henoch–Schoenlein purpura (systemic small vessel vasculitis), and Wegener's granulomatosis.

Immunological diseases of other organs and systems

Aberration in the immune response, hyperactivity of the immune response, or autoimmunity can target various

Table 16.1 Immunological diseases involving the blood

	Etiology	Antibodies/cellular reaction	Clinical manifestation	Laboratory findings	Treatment
Autoimmune neutropenia	Secondary to autoimmune diseases, e.g. systemic lupus erythematosus and Felty's syndrome (rheumatoid arthritis, splenomegaly, and severe neutropenia); myeloid cell growth suppressed by autoantibodies	Anti-granulocyte antibodies	Asymptomatic; recurrent infections	Normal bone marrow function with a shift to the left	Immunosuppressive drugs, corticosteroids, splenectomy
Idiopathic thrombocytopenic purpura (ITP)	Platelets destroyed by antiplatelet autoantibodies: circulating platelets coated with IgG autoantibodies removed by splenic macrophages at an accelerated rate	Accelerated platelet removal by splenic macrophages	Decreased blood platelets; hemorrhage and extensive thrombotic lesions (bleeding and purpura)	Platelet count < 20 000 to 30 000/μl; detectable antiplatelet antibodies in the serum and on platelets	Corticosteroids, splenectomy recommended in adults
B cell acute lymphoblastic leukemia (ALL)	Arrest of lymphoid precursor cells (i.e. lymphoblasts) in an early stage of development caused by an abnormal expression of genes, often as a result of chromosomal translocations	Lymphoblasts accumulation in the bone marrow and suppression of normal hemopoietic cells	Anemia, granulocytopenia, and thrombocytopenia; weakness, malaise, and pallor secondary to anemia; bleeding secondary to thrombocytopenia; bacterial infections secondary to neutropenia; bone pain; generalized lymphadenopathy especially affecting the cervical lymph nodes; hepatosplenomegaly and leukemic meningitis	Presence of lymphoblasts in the bone marrow; normal or decreased total leukocyte count with or without lymphoblasts in the peripheral blood; elevated leukocyte count accompanied by lymphoblasts in the peripheral blood; CD19+, cytoplasmic CD79a, CD10, CD24, variable CD22 and CD20, CD45 may be absent, CD10 negative, possible expression of myeloid antigen CD13 and CD33	Standard chemotherapy: a 4-drug regimen of vincristine, prednisone, anthracycline, and cyclophosphamide or L-asparaginase or a 5-drug regimen of vincristine, prednisone, anthracycline, cyclophosphamide, and L-asparaginase

Table 16.1 Immunological diseases involving the blood (*continued*)

	Etiology	Antibodies/cellular reaction	Clinical manifestation	Laboratory findings	Treatment
B cell chronic lymphocytic leukemia (B-CLL)	Clonal aberrations, the most common involving deletions at 13q	Exposure of a memory B cell to antigen in germinal centers of secondary follicles	Predisposed to repeated infections; abdominal discomfort and bleeding from mucosal surfaces; general localized or generalized lymphadenopathy, splenomegaly, hepatomegaly, petechiae and pallor, hypogamma-globulinemic in $\frac{1}{2}$ to $\frac{3}{4}$ of cases; autoimmune hemolytic anemia, neutropenia, or thrombocytopenia in 15% to 30% of cases	Lymphocytosis $> 4 \times 10^9/l$; hypogammaglobulinemia in 60% cases; mature small lymphocytes with condensed nuclear chromatin and sparse cytoplasm; coexpression of CD5 with CD19 and CD20 with very faint amounts of monoclonal surface immunoglobulin; CD23 and negative for CD10 and usually FMC7	None in early stage; chemotherapy afterwards
Mantle cell lymphoma (MCL)	Associated with chromosome translocation t(11;14)(q13;q32) involving the immunoglobulin heavy chain gene on chromosome 14 and the *BCL1* locus on chromosome 11; associated with viral infection EBV, HIV, human T-lymphotropic virus type I (HTLV-1), human herpesvirus 6 (HHV-6)	Overexpression of the protein cyclin D1 (coded by *PRAD1* gene located close to the breakpoint) which plays a key role in cell cycle regulation and progression of cells from G1 phase to S phase by activation of cyclin-dependent kinases	Fatigue, fever, night sweats, and weight loss; generalized lymphadenopathy; hepatosplenomegaly	Elevated lactate dehydrogenase (LDH), possible elevated beta2-microglobulin, lymphocytosis of more than 4000/μl; tumor cells are monoclonal B cells that express surface immunoglobulin, IgM, or IgD; CD5+ and pan B-cell antigen positive (e.g., CD19, CD20, CD22) and FMC7 positive but lack expression of CD10 and CD23, cyclin D1 is overexpressed.	Combination chemotherapy.

Table 16.1   Immunological diseases involving the blood (*continued*)

	Etiology	Antibodies/cellular reaction	Clinical manifestation	Laboratory findings	Treatment
Follicular lymphoma (FL)	Acquired nonrandom chromosomal translocations t(14;18)(q32;q21); bringing the *bcl2* protooncogene under the transcriptional influence of the immunoglobulin heavy chain gene and leading to the overexpression of a functionally normal bcl-2 protein	Overexpression of bcl-2 protein which confers a survival advantage to the cancer cells by inhibiting apoptosis	Painless slowly progressive adenopathy; fever, night sweats, weight loss in later stage	Abnormal lymphocytes identified in the blood smear; a follicular or nodular pattern of growth reminiscent of germinal centers; sIg+ (IgM+/-IgD, IgG or rarely IgA) Bcl2+ CD10+ CD5- and CD43- and expression of B cell associated antigens (CD19, CD20, CD22, CD79a), occasionally CD43+.	Radiation therapy; single-agent oral chemotherapy, such as with chlorambucil; immunotherapy such as rituximab (Rituxan), a monoclonal antibody directed against the CD20 antigen; bone marrow/ stem cell transplantation
Diffuse large B cell lymphoma (DLBCL)	Nonrandom chromosomal and molecular rearrangements t(3;22)(q27;q11)	Mutations or allelic losses of the *p53* tumor suppressor gene or 17p13.1	Lymphadenopathy; fevers, night sweats, weight loss, and fatigue; organ-specific symptoms, such as shortness of breath, chest pain, cough, abdominal pain and distension, or bone pain	Presence of abnormal lymphoid cells in peripheral blood smear; consistent expression of B cell restricted markers (CD19, CD20, CD22, CD79a), frequent expression of HLA-DR and uncommon expression of CD23, presence of CD10 or CD5	Radiation therapy and chemotherapy
Adult T-cell leukemia/ lymphoma (ATLL)	Human T-cell leukemia virus type I (HTLV-I)	Proliferative disorder of T cells	Elevated white blood cell count, skin lesions, lymphadenopathy, hepatosplenomegaly, lytic bone lesions, and hypercalcemia	HTLV-I and II infection detected with ELISA; peripheral blood lymphocytes found to have convoluted nuclei ('clover leaf' or 'flower' lymphocytes); expression of T cell associated antigens (CD2, CD3, CD5), may lack or have decreased CD7, CD4+/CD8[minus] in most cases, CD4-/ CD8+ or CD4+/CD8+ in rare cases, expression of CD25 in nearly all cases, CD30 positive but ALK negative	Chemotherapy; immunotherapy with interferon plus zidovudine or monoclonal antibodies

Table 16.1 Immunological diseases involving the blood (*continued*)

	Etiology	Antibodies/cellular reaction	Clinical manifestation	Laboratory findings	Treatment
Acute myelogenous leukemia (AML)	Neoplastic transformation in a multipotential hematopoietic stem cell or in one of restricted lineage potential	Arrest of differentiation of hematopoietic stem cells at the blast stage causing myeloblasts to accumulate in the bone marrow	Thrombocytopenia, neutropenia, and anemia; fatigue, weakness, and pallor due to anemia; bleeding and GI tract or CNS hemorrhage secondary to thrombocytopenia; increase secondary infections due to decreased neutrophil counts; hepatosplenomegaly	20% myeloblasts in the bone marrow with or without peripheral blood presence; expression of myeloid antigens CD13, CD33, CD15, MPO, and CD117 and CD34 variable	Chemotherapy and bone marrow transplantation
Angioimmunoblastic lymphadenopathy (AILA)	Proliferation of hyperimmune B cells	Formation of pleomorphic infiltrate by immunoblasts, both large and small, together with plasma cells in lymph nodes revealing architectural effacement	Fever, night sweats, hepatosplenomegaly, generalized lymphadenopathy, weight loss, hemolytic anemia, polyclonal gammopathy, and skin rashes	Arborization of newly formed vessels and proliferating vessels with hyperplasia of endothelial cells; amorphous eosinophilic PAS positive deposits in the interstitium	Chemotherapy with CHOP-like regimens (cyclophosphamide, doxorubicin/adriamycin, oncovin/vincristine, and prednisolone)
Hodgkin's disease	Transformation of germinal center B cells (rarely T cells) to malignant Hodgkin's cell or Reed Sternberg cell under certain transforming event(s), e.g. EBV infection	A defect in cell-mediated immunity; relative T lymphocytopenia, T cell dysfunction, and a serum factor that interferes with normal T cell reactions; B cell function is normal	Painless swelling in the neck, armpits or groin; night sweats or unexplained fever; weight loss and tiredness; cough or breathlessness; increased susceptibility to opportunistic infections	Presence of the malignant Reed Sternberg cells with an appropriate cellular background at lymph node biopsy	Radiotherapy and chemotherapy

Table 16.2 Immunological diseases involving the skin

	Etiology	Antibodies/cellular reaction	Clinical manifestation	Laboratory findings	Treatment
Allergic contact dermatitis	Covalent linkage of low mol wt chemicals to proteins in the skin	Delayed-type hypersensitivity mediated by specifically sensitized T cells	Erythema and swelling; blister formation; crust formation and weeping of the lesion	Perivascular cuffing with lymphocytes, vesiculation, and necrosis of epidermal cells	Systemic corticosteroids or the application of topical steroid cream to localized areas
Bullous pemphigoid	Reaction of autoantibody and complement with a 230 kD basic glycoprotein antigen produced by keratinocytes in the epidermis	Antigen-antibody-complement interaction and mast cell degranulation	Blistering skin lesion with fluid filled bullae developing at flexor surfaces of extremities, groin, axillae, and inferior abdomen	IgG and C3 deposition in a linear pattern at the lamina lucida of the dermal-epidermal junction	Immunosuppressive agents in combination with oral steroids; prevention of infection in combination with potent topical steroid creams for more rapid relief
Pemphigus vulgaris	May be associated with autoimmune diseases, thymoma, and myasthenia gravis; or induced by certain drugs	Autoantibodies to intercellular substance with activation of classic pathway-mediated immunologic injury	Blisters prominent on both the oral mucosa and anal/genital mucous membranes	IgG, C1q, and C3 in the intercellular substance between epidermal cells by immunofluorescence staining; circulating pemphigus antibodies in 80–90% cases	Corticosteroids, immunosuppressive therapy, and plasmapheresis
Psoriasis vulgaris	Associated with relatively high instance of HLA-B13 and -B17 antigen and decreased T suppressor cell function	Reaction of IgG/IgA to stratum corneum antigens; fixation of C3 and properdin and activation of the alternative complement pathway	Discrete, papulosquamous plaque on areas of trauma	Significantly reduced peripheral blood helper CD4+ T cell; focal granular or globular deposits of immunoglobulins (IgG, IgA), and C3 in the stratum corneum by immunofluorescence	Psoralens and long-wave ultraviolet radiation

organs and systems, such as muscle, neuromuscular junction, thyroid, lung, digestive system, liver, pancreas, kidney, nervous system, eye, and cartilage. Various immunological diseases involving these organs and systems are shown in **Table 16.4**, which summarizes the etiology, antibodies, and/or cellular reactions, clinical manifestations, laboratory findings, and treatment.

SYSTEMIC AUTOIMMUNE DISEASES

Major systemic autoimmune diseases such as systemic lupus erythematosus and rheumatoid arthritis and progressive sys-

temic sclerosis are listed in **Table 16.5**, comparing their etiology, antibodies, and/or cellular reactions produced, clinical manifestations, laboratory findings, relationship to other autoimmune diseases, genetics, and other features.

GAMMOPATHIES

A gammopathy is an abnormal increase in immunoglobulin synthesis. Gammopathies that are monoclonal usually signify malignancy such as multiple myeloma, Waldenström's disease, heavy chains disease, or chronic lymphocytic leukemia. Benign gammopathies occur in amyloidosis and

Table 16.3 Immunological diseases involving the vasculature

	Etiology	Antibodies/cellular reaction	Clinical manifestation	Laboratory findings	Treatment
Leukocytoclastic vasculitis (small vessel vasculitis)	Idiopathic; antibiotics, particularly beta-lactam drugs, nonsteroidal anti-inflammatory drugs, and diuretics; foods or food additives; various infections; hepatitis C; collagen vascular diseases; inflammatory bowel diseases	Circulating immune complexes; autoantibodies such as antineutrophil cytoplasmic antibody (ANCA), other inflammatory mediators, and local factors that involve the endothelial cells and adhesion molecules	May be localized to the skin and manifested as palpable purpura, urticarial lesions, nodular lesions; or may manifest in other organs	Deposition of fragments of neutrophil nuclei and immune complexes (IgM, IgG, C3, C4) and fibrin in vessel walls by direct immunofluorescence; presence of vascular and perivascular infiltration of polymorphonuclear leukocytes with formation of nuclear dust (leukocytoclasis), extravasation of erythrocytes, and fibrinoid necrosis of the vessel walls in skin biopsy	Treatment for identifiable cause; colchicine or dapsone for skin involvement; antihistamines for urticarial lesions; high doses of corticosteroids (1–2 mg/kg/d) with or without an immunosuppressive agent (e.g. cyclophosphamide, azathioprine, methotrexate) for visceral involvement
Polyarteritis nodosa (necrotizing vasculitis of small- and medium-sized muscular arteries)	Associated with hepatitis B antigenemia; a state of relative antigen excess	Formation of circulating antigen–antibody complexes; release of vasoactive amines from platelets and IgE-triggered basophils; and deposition of immune complexes in blood vessel walls	Weakness, abdominal pain, leg pain, fever, cough, and neurologic symptoms; kidney involvement, arthritis, arthralgia, or myalgia, and hypertension; skin involvement manifested as a maculopapular rash	Elevated erythrocyte sedimentation rate, leukocytosis, anemia, thrombocytosis, and cellular casts in the urinary sediment; presence of immune complexes, cryoglobulins, rheumatoid factor, and diminished complement component levels; presence of aneurysm and changes in vessel caliber on angiography	Cyclophosphamide, corticosteroids

Table 16.3 Immunological diseases involving the *vasculature* (*continued*)

	Etiology	Antibodies/ cellular reaction	Clinical manifestation	Laboratory findings	Treatment
Henoch–Schoenlein purpura (systemic small vessel vasculitis)	Upper respiratory infections; certain drugs, food, and immunizations	Immune complexes containing IgA activation of the alternative complement pathway	Arthralgias, nonthrombocytopenic purpuric skin lesions, abdominal pain with bleeding, and renal disease	Increased serum IgA (IgA1, not IgA2) concentrations, presence of IgA-containing circulating immune complexes, and IgA deposition in vessel walls and renal mesangium; vasculitis revealed by skin biopsy; IgA deposition in vessel walls by immunofluorescence staining	Cyclophosphamide, azathioprine (Imuran), and plasmapheresis; nephropathy treated supportively with fluid and electrolyte balance, monitoring of salt intake, and antihypertensives
Wegener's granulomatosis	Cause unknown; hypersensitivity postulated as the basis for the disease	Anti-neutrophil cytoplasmic antibodies (c-ANCA); immune complexes precipitated by C1q; an immune complex reaction on the epithelial side of the basement membrane	Systemic symptoms, e.g. fatigue, malaise, fever, anorexia, weight loss; hemorrhagic rhinorrhea, paranasal sinusitis, nasal mucosal ulcerations, and serous or purulent otitis mecia with hearing loss, cough, hemoptysis, and pleuritis; glomerulonephritis	Normal or elevated serum complement levels, elevated ESR, leukocytosis; high titers of ANCA; inflammatory perivascular exudate and fibrin deposition in small arteries, capillaries, and venules of pulmonary and skin biopsies; focal and segmental glomerulonephritis of varying severity, occasionally with necrotizing vasculitis at renal biopsy; scattered deposits of complement and IgG by immunofluorescence staining	Corticosteroids, cyclophosphamide, methotrexate, or azathioprine

Table 16.4 Immunological diseases

	Etiology	Antibodies/cellular reaction	Clinical manifestation	Laboratory findings	Treatment
Muscle					
Dermatomyositis	Idiopathic; possible existence of a link to certain human leukocyte antigen (HLA-DR3, DQA1*0501) types; triggering factors, e.g. infectious agents including viruses and drugs including hydroxyurea penicillamine, statin drugs, quinidine, and phenylbutazone	Circulating autoantibodies; abnormal T cell activity; complement-mediated damage to endomysial vessels and microvasculature of the dermis	A purple-tinged skin rash (heliotrope rash) that is prominent on the superior eyelids, extensor joints surfaces, and base of the neck; calcinosis often over bony prominences; weakness, muscle pain, increasing fatigue, and loss of proximal (thighs and shoulders) muscle strength	Autoantibodies against tRNA synthetases in the serum; abnormal muscle enzyme levels (creatine kinase, aldolase, aspartate, aminotransferase, and lactate dehydrogenase); positive myositis-specific antibodies (MSAs) (antinuclear antibody, anti-Mi-1, antisignal recognition protein, and anti-Ku)	Corticosteroids, immunosuppressants such as azathioprine and methotrexate
Polymyositis	Idiopathic	Production of a cytotoxin by lymphocytes against autologous muscle	Shoulder or pelvic girdle weakness; electromyographic evidence of myopathic abnormalities	Positive antinuclear antibody; anti-Jo-1, anti-PM-Scl, and anti-RNP; polyclonal hypergammaglobulinemia; increased levels of muscle enzymes (creatine kinase, aldolase, myoglobin, lactate dehydrogenase, aspartate aminotransferase, and alanine aminotransferase); elevated ESR	Corticosteroids, methotrexate or other cytotoxic agents
Neuromuscular					
Myasthenia gravis (MG)	Unknown; 75% associated with thymus abnormality	Formation of antibodies against acetylcholine (ACh) nicotinic postsynaptic receptors at the myoneural junction	Weakness of bulbar muscles; neck, and proximal limb weakness; respiratory weakness; generalized weakness	No specific lab findings	Cholinesterase-inhibiting medications (edrophonium and pyridostigmine); and corticosteroids

Table 16.4 Immunological diseases (*continued*)

	Etiology	Antibodies/ cellular reaction	Clinical manifestation	Laboratory findings	Treatment
Thyroid					
Hashimoto's disease (Hashimoto's thyroiditis)	A genetic predisposition associated with HLA-DR3, -DR4 and -DR5; susceptibility loci on chromosome 13 (HT1, 96cM), (HT2, 97cM), (AITD-1)	Recruitment of NK cells to the thyroid by antibodies against thyroid-specific antigens leading to tissue injury and inflammation: lymphocyte (B cell and CD4+ T cells), plasma cell, and macrophage infiltration and formation of lymphoid germinal centers	Initial hyperthyroidism and later on hypothyroidism	Circulating autoantibodies against thyroid peroxidase (microsomal antibodies), thyroglobulin, and colloid	Hormone replacement therapy
Graves' disease	Idiopathic, influenced by a combination of environmental and genetic factors, susceptibility loci on (AITD-1)	Circulating autoantibody to autothyroid antigens, i.e. TSH receptor (primary autoantigen), thyroglobulin, thyroperoxidase, and sodium-iodide symporter; expression of molecules that mediate T cell adhesion and complement regulation (Fas and cytokines) by thyroid cells	Thyrotoxicosis, hyperthyroidism	High titer of anti-TSH receptor (IgG1), antisodium-iodide, antithyroglobulin, and antithyroperoxidase antibodies; positive TSH-receptor antibodies (particularly TSIs)	Antithyroid medications (thiomides), radioactive iodine
Lung					
Usual interstitial pneumonitis (idiopathic pulmonary fibrosis)	Idiopathic	Activation of alveolar macrophages after phagocytizing immune complexes and release of cytokines that attract neutrophils	Progressive dyspnea upon exertion, interstitial infiltrates on chest radiographs, and a restrictive ventilatory defect found on pulmonary function tests	Positive anti-nuclear antibodies and rheumatoid factor; immune complexes in blood, alveolar walls and bronchoalveolar lavage fluid	Systemic corticosteroids and/or other immunosuppressants

Table 16.4 Immunological diseases (*continued*)

	Etiology	Antibodies/cellular reaction	Clinical manifestation	Laboratory findings	Treatment
Farmer's lung (extrinsic allergic alveolitis)	Intense or repeated exposure to organic dust and fungi such as the *Aspergillus* species	Type III hypersensitivity mechanism with the deposition of immune complexes in the lung; cell-mediated, delayed-type hypersensitivity (type IV hypersensitivity)	Breathlessness within hours after inhaling the dust and interstitial pneumonitis	Leukocytosis with neutrophilia (but not eosinophilia) and elevated ESR, C-reactive protein level, and quantitative immunoglobulin level; presence of antigen-specific immunoglobulin and complement activation and deposition in the lung; presence of lymphocytes, macrophages, and granulomas in alveolar spaces and interstitium	Corticosteroids
Digestive system					
Crohn's disease	Idiopathic; a complex set of interactions among susceptibility genes, environment, and the immune system; a genetic predisposition associated with HLA DR1-DQW5 complex and DRB3*0301, susceptibility locus on chromosome 16 centered around D16S409 and D16S419	Transmural granulomatous inflammation of the bowel wall characterized by lymphocyte, plasma cell, and eosinophil infiltration	Abdominal pain and diarrhea, which may be complicated by intestinal fistulization, obstruction, or both; involving the entire GI tract	Small IgG containing complexes which are merely aggregates of IgG in the blood; elevated serum concentrations of C3, factor B, C1 inhibitor, and C3b inactivator; high titer of antibodies to the yeast *Saccharomyces cerevisiae* (i.e. anti-*S cerevisiae* antibodies [sboASCAsbc])	Antidiarrheal agents; anti-inflammatory drugs or antibiotics; a short course of steroid therapy indicated in patients with severe systemic symptoms (e.g., fever, nausea, weight loss) and in those who do not respond to anti-inflammatory agents; immunotherapy using anti-tumor necrosis factor (TNF) monoclonal antibody

Table 16.4 Immunological diseases (*continued*)

	Etiology	Antibodies/cellular reaction	Clinical manifestation	Laboratory findings	Treatment
Ulcerative colitis (immunologic colitis)	A genetic predisposition associated with HLA-DR2	Exposure of mucosal immune system of the large intestine to a wide array of antigens; activation of cells of the mucosal immune system and release of cytokines that recruit inflammatory cells	Abdominal cramping, diarrhea, and bloody stools	High titer of perinuclear antineutrophil cytoplasmic antibody (p-ANCA)	Anti-inflammatory therapy with 5-aminosalicylate (5-ASA) preparations; corticosteroids; immunosuppressive agents, e.g. 6-mercaptopurine (Purinethol) and azathioprine (Imuran), in IBD patients who are steroid dependent or refractory to steroid treatment; therapeutic monoclonal antibody treatment
Liver					
Chronic active hepatitis (autoimmune)	Unknown etiology; genetic predisposition associated with HLA-DR3 and HLA-B8	Interaction of CD4$^+$ T cells and a self-antigenic peptide which is embraced by an HLA class II molecule and presented to uncommitted helper T cells (T$_H$0) by antigen-presenting cells; activation of T$_H$0, and functional differentiation of T$_H$0 into T$_H$1 and T$_H$2: T$_H$1 secretion of IL–2 and γ-IFN which activate macrophages and enhance expression of HLA classes I and II, thus perpetuating the immune recognition cycle; T$_H$2 secretion of IL-4, IL–5, and IL-10 which stimulate autoantibody production by B cells	Asymptomatic; nonspecific symptoms (eg, fatigue, anorexia, weight loss, behavioral changes, amenorrhea); mild jaundice to hepatomegaly, spleromegaly, ascites, cutaneous manifestations of chronic liver disease	Autoantibodies against liver-specific and non-liver-specific antigens: circulating anti-smooth muscle antibodies (ASMAs) (F-antibodies (actin target antigen) and/or antinuclear antibodies (ANAs) (heterogeneous target antigen) in type 1 AIH; presence of circulating liver-kidney microsomal type 1 (LKM-1) antibody (CYP2D6 target antigen) or anti-liver cytosol 1 (anti-LC1) antibody in type 2 AIH; presence of autoantibodies to soluble liver proteins or liver-pancreas antigen in type 3 AIH; increased IgG levels	Corticosteroid administration, either alone or in combination with azathioprine

Table 16.4 Immunological diseases (*continued*)

	Etiology	Antibodies/ cellular reaction	Clinical manifestation	Laboratory findings	Treatment
Primary biliary cirrhosis	Unknown etiology; increased expression of HLA class II antigens in the liver rendering hepatocytes and bile duct epithelial cells more susceptible to activated T cells and perhaps exacerbating immunologically mediated cytotoxicity; genetic predisposition associated with HLA-DR8 and, for some populations, HLA-DPB1	Continuous destruction of small and medium bile ducts mediated by activated CD4 and CD8 lymphocytes	Pruritis, fatigue, steatorrhea, renal tubular acidosis, hepatic osteodystrophy, chronic cholestasis, and increased incidence of hepatocellular carcinoma and breast carcinoma	Elevated serum levels of immunoglobulins, mainly IgM; multiple circulating autoantibodies (e.g. antinuclear antibodies); presence of antimitochondrial antibodies (AMAs) in the sera (hallmark)	Ursodeoxycholic acid (UDCA); immunosuppressive agents; liver transplantation
Pancreas					
Insulin-dependent (type I) diabetes mellitus	Environmental factors (infections and diet) interacting with a genetically susceptible person; genetic predisposition associated with HLA-DR3 and HLA-DR4	IgG autoantibodies against glucose transport proteins and anticytoplasmic and antimembrane antibodies directed to antigens in the pancreatic islets of Langerhans	Polyuria, nocturia, increased thirst, weight loss, nonspecific malaise; diabetic ketoacidosis; long-term complications (retinopathy, cataracts, hypertension, progressive renal failure, early coronary artery disease, peripheral vascular disease, peripheral and autonomic neuropathy, increased risk of infection)	A random whole-blood glucose concentration >200 mg/dl (11 mmol/l) a fasting whole-blood glucose concentration > 120 mg/dl (7 mmol/l); presence of islet cell antibodies	Insulin therapy; diet; activity
Kidney					
Immune complex disease	Antigens from microorganisms such as streptococci or endogenous antigens such as DNA or nuclear antigens in systemic lupus erythematosus leading to subepithelial deposits of immune complex in renal glomeruli	Type III hypersensitivity reaction mechanism: activation of the complement system by immune complexes lodged in the microvasculature such as the renal glomeruli; attraction of polymorphonuclear neutrophils, initiation of an inflammatory reaction	Fever, joint pain, lymphadenopathy, eosinophilia, hypocomplementemia, proteinuria, purpura, and urticaria	Renal biopsy: granular deposits of immune complexes (white) along the glomerular basement membrane (adjacent to and beneath the endothelium) and within the mesangium by fluorescent staining; electron dense deposits along the GBM mainly as subendothelial deposits by EM	Removal of external antibody or antigen source and symptomatic relief of the fever and kidney damage; immune suppression therapy for endogenous antigens

Table 16.4 Immunological diseases (*continued*)

	Etiology	Antibodies/cellular reaction	Clinical manifestation	Laboratory findings	Treatment
Poststreptococcal glomerulonephritis	Infection with group A beta hemolytic streptococci; staphylococcal and pneumococcal infections, coxsackievirus B, echovirus type 9, influenza virus, and mumps	Formation of antigen–antibody complexes by antibodies (IgG) against a cytoplasmic antigen termed endostreptosin together with some cationic streptococcal antigens; subepithelial deposition of immune complex in the mesangium or occasionally in a subendothelial or intramembranous position	Fever, nausea, oliguria, and hematuria; erythrocyte casts and mild proteinuria	Elevated antistreptolysin-O (ASO) titers; decreased serum complement levels; granular immune deposits that contain immunoglobulin and complement in the glomeruli by immunofluorescence of renal biopsies	Supportive treatment directed toward the potential complications
Membranous glomerulonephritis	Idiopathic; exposure to gold, mercury, penicillamine, or captopril; sequela of autoimmune diseases, infections (hepatitis B, E), metabolic disorders, or malignancy	Deposition of electron-dense, immune (Ag–Ab) deposits in the glomerular basement membrane in a subepithelial location; activation of complement pathway; and triggering of the biosynthesis of oxygen radical-producing enzymes within the glomerular epithelial cells by complement membrane attack complex (C5b–9)	Edema or generalized anasarca; proteinuria; renal insufficiency	Proteinuria with oval fat bodies and fatty casts; renal biopsy: immunofluorescence staining of granular capillary wall for IgG with C3 and both kappa and lambda light chains; electron microscopy showing electron-dense deposits	Low-salt diet; diuretics; NSAIDS; ACE inhibitors; immunosuppressive therapy
IgA nephropathy (Berger's disease)	Idiopathic	Accumulation of IgA, predominantly IgA1, in renal mesangial cells; release of cytokines; and activation of the complement system via the alternative pathway	Gross or microscopic hematuria; mild proteinuria	Mesangial IgA deposition, elevated serum IgA level, and IgA circulating immune complexes; of electron-dense deposits in mesangial areas	Reducing inflammatory-mediated renal injury (omega-3, polyunsaturated fatty acids, corticosteroids), controlling hypertension (ACE inhibitors), decreasing proteinuria, and managing sequelae of reduced renal function

Table 16.4 Immunological diseases (*continued*)

	Etiology	Antibodies/ cellular reaction	Clinical manifestation	Laboratory findings	Treatment
Nervous system					
Multiple sclerosis	Genetic predisposition associated with HLA-A3, B7, and Dw2 haplotypes; virus infection	Elevation of co-stimulatory factor B7-1; release of pro-inflammatory cytokines (e.g. IL-12) by lymphocytes; infiltration of lymphocytes and macrophages in the nervous system which facilitates demyelination	Paresthesias, muscle weakness, visual and gait disturbances, ataxia, and hyperactive tendon reflexes	Oligoclonal increase in CSF IgG; presence of anti-HTLV-I GAG (p24) protein antibody in CSF; characteristic lesions of high T2 signal intensity of variable location in the white matter of the brain, brain stem, optic nerves, or spinal cord by MRI	Cop-1, a polypeptide mixture that resembles myelin basic protein; immunomodulatory agents (i.e. interferon beta-1a and -1b, glatiramer acetate)
Eye					
Vogt–Koyanagi–Harada (VKH) syndrome (uveoencephalitis)	Genetic predisposition associated with HLA-DR4	Autoimmune reaction directed against an antigenic component shared by uveal, dermal, and meningeal melanocytes, possibly tyrosinase or a tyrosinase-related protein with involvement of T cell-mediated cytotoxicity	Headache, dysacusis, vertigo; patchy loss of scalp hair or whitening, vitiligo, poliosis	Pleocytosis with the presence of melanin-laden macrophages in CSF	Systemic corticosteroids; immunosuppression with cyclosporine or other antimetabolites (azathioprine, cyclophosphamide, methotrexate)
Cicatrical ocular pemphigoid	Unknown; genetic predisposition associated with HLA-DR2, HLA-DR4, HLA-DQw7; some triggered by systemic practolol therapy and topical antiglaucoma drugs	T cell dysregulation, the production of circulating autoantibodies directed against a variety of adhesion molecules ($\beta4$ subunit of $\alpha6\beta4$ integrin, $\alpha3$, $\beta3$, or $\gamma2$ subunits of laminin 5) in the hemidesmosome–epithelial membrane complex, and the production of proinflammatory cytokines and immune system activation markers	Blistering of conjunctiva; cicatrizing conjunctivitis	Decreased serum levels of IL-6 and increased serum levels of TNF-α; a diffuse, linear deposition of immunoglobulins and components, mainly IgG and C3 at the epithelial-subepithelial junction by immunoflourescence staining	Systemic corticosteroids

Table 16.4 Immunological diseases (*continued*)

	Etiology	Antibodies/ cellular reaction	Clinical manifestation	Laboratory findings	Treatment
Cartilage					
Relapsing polychondritis	Unknown; genetic predisposition associated with HLA-DR4	Infiltrating T cells, the presence of antigen–antibody complexes in affected cartilage, cellular and humoral responses against collagen type II and other collagen antigens	Inflammation of cartilaginous structures, predominantly those of the ear, nose, and laryngotracheobronchial tree, causing them to lose their structural integrity and collapse	Presence of circulating antibodies to cartilage-specific collagen types II, IX, and XI; early RP characterized by a mixed inflammatory infiltrate of lymphocytes, neutrophils, and plasma cells in the perichondrium; mononuclear cells and macrophage infiltration in later cartilage degeneration	Systemic corticosteroids; immunosuppressive agents, e.g. dapsone, azathioprine, methotrexate, cyclophosphamide, and cyclosporin A

Table 16.5   Systemic autoimmune diseases

	Etiology	Antibodies/ cellular reaction	Clinical manifestation	Laboratory findings	Treatment
Systemic lupus erythematosus (SLE)	Unknown; genetic predisposition associated with HLA-DR2 and HLA-DR3	Polyclonal B cell activation leading to formation of antibodies against self and nonself antigens; formation of immune complexes in the microvasculature, leading to complement activation and inflammation; 4 types of antinuclear antibodies: (1) antibodies against double stranded DNA, (2) antibodies against histones, (3) antibodies to nonhistone proteins bound to RNA, and (4) antibodies against nucleolar antigens	Fever, malaise, loss of weight, joint pain, butterfly rash over the bridge of the nose and lethargy; exacerbations and remissions of injuries to the skin, kidneys, joints, and serosal membranes; manifestations of multisystem involvement	Presence of serum antinuclear antibodies; antidouble-stranded DNA and anti-Sm antibodies; depressed serum complement (C3, C4, CH50) levels; immune deposits in glomerular basement membranes and at the dermal-epidermal junction, and the presence of multiple other autoantibodies	Corticosteroids; cytotoxic agents such as cyclophosphamide, chlorambucil, and azathioprine in more severe cases
Rheumatoid arthritis	Unknown; genetic predisposition associated with HLA-DR4/DR1	CD4+ T cells, mononuclear phagocytes, fibroblasts, osteoclasts, and neutrophils; production of autoantibodies (i.e. rheumatoid factors) by activated B cells; production of cytokines, chemokines, and other inflammatory mediators (e.g. TNF-α, IL-1, IL-6, TGF-β, IL-8, FGF, PDGF	Pain, stiffness, and swelling of joints; deformity and ankylosis in late stages; rheumatoid nodules; roentgenographic erosions of joints	Presence of rheumatoid factors in serum; presence of CD4$^+$ T cells, activated B cells, and plasma cells in the inflamed joint synovium and multiple proinflammatory cytokines such as IL-1 and TNF in synovial joint fluid	DMARD therapy; glucocorticoids; nonsteroidal anti-inflammatory drugs; immunotherapy (e.g. BRMs and monoclonal antibodies)
Ankylosing spondylitis	Unknown; genetic predisposition associated with HLA-B27	Cellular infiltration by lymphocytes, plasma cells, and polymorphonuclear leukocytes; release of cytokines such as TNF-α	Predilection for the axial skeleton, affecting particularly the sacroiliac and spinal facet joints and the paravertebral soft tissues; extraspinal manifestations of the disease including peripheral arthritis, iritis, pulmonary involvement, and systemic upset	Elevated ESR, negative rheumatoid factor, and antinuclear antibodies; radiographical signs: indistinctness of the joint, subchondral bony erosions, bony fusion, sacroiliitis, bony ankylosis	Regular lifelong exercises; diminishing inflammation and pain (nonsteroidal anti-inflammatory drugs, biological agents such as enbrel, a TNF-α antagonist); providing physical therapy

Table 16.5 Systemic autoimmune diseases (*continued*)

	Etiology	Antibodies/ cellular reaction	Clinical manifestation	Laboratory findings	Treatment
Sjögren's syndrome	Genetic predisposition associated with HLA-DR3/DR4	Lymphocytic infiltration (i.e. CD4$^+$ cells); release of proinflammatory cytokines such as IL-1, IL-6, and IL-8 and adhesion molecules, such as intercellular adhesion molecule-1 (ICAM-1); production of autoantibodies, antinuclear antibody (ANA), rheumatoid factor, or SS-specific antibodies (eg, anti-RO [SS-A], anti-LA [SS-B])	Dry eyes (keratoconjunctivitis sicca) and dry mouth (xerostomia) with associated visual or swallowing difficulties	Presence of circulating autoantibodies, including ANA or SS antibodies (i.e. SS-A, SS-B)	Artificial tears; immunomodulatory agents (e.g. topical cyclosporine A); topical androgens
Progressive systemic sclerosis	Unknown	Activation of T cells and/or complement; stimulation of the release of endothelial cytokines with subsequent endothelial damage, which facilitates adhesion and transvascular migration of CD4$^+$CD8$^-$ T cells and monocytes with the help of integrins and cell adhesion molecules	Fibrotic changes of the skin, subcutaneous tissue, and viscera; involving virtually any organ of the body, including the skin, gastrointestinal tract, lungs, heart, kidneys, and musculoskeletal system	Presence of speckled antinuclear antibodies, antitopoisomerase, anti-RNA polymerase, antiribonucleoprotein, anticentromere, anti-ku, anti-Th, anti-PM-Scl; positive rheumatoid factor; polyclonal hypergamma-globulinemia; presence of lymphokines such as IL-2, IL-4, and IL-6 in the sera	Immunomodulatory therapy including cyclosporin A, antilymphocyte globulin, intravenous immune serum globulin (IVIG), plasma exchange, methotrexate, and cyclophosphamide

monoclonal gammopathy of undetermined etiology. Inflammatory disorders are often accompanied by benign polyclonal gammopathies. These include rheumatoid arthritis, lupus erythematosus, tuberculosis, cirrhosis, and angioimmunoblastic lymphadenopathy. **Table 16.6** shows some common gammopathies, comparing their etiology, antibodies and/or cellular reactions produced, clinical manifestations, laboratory findings, type of excessive immunoglobulins, and other features.

Table 16.6 Gammopathies

	Etiology	Antibodies/ cellular reaction	Clinical manifestation	Laboratory findings	Treatment
Multiple myeloma	Unknown; chronic antigenic stimulation of a plasma cell, which results in transformation of plasma cells; may have genetic predisposition associated with HLA-Cw5 or HLA-Cw2	Presence of plasma cell activating factor IL-6 within bone marrow, resulting in plasma cell proliferation; production of innumerable clones, which spread hematogenously to other myelogenous areas; monoclonal proliferation of B cells resulting in increase of a single immunoglobulin and its fragments; myeloma cells {?omission?} production of osteoclast-stimulating factor, a cytokine that results in bone destruction	Bone destruction and pain; risk of compression fractures of the spine and pathologic fractures of the major weight-bearing bones; radiologically, multiple destructive lesions of the skeleton as well as severe demineralization demonstrated as multiple, discrete, small, lytic lesions	Increased levels of immunoglobulins in the blood and light chains (Bence–Jones protein) in the urine; monoclonal hypergamma-globulinemia; increased level of myeloma protein (i.e. M protein level)	Chemotherapy (melphalan and prednisone); radiation; peripheral blood or bone marrow stem cell transplantation
Waldenström's macroglobulinemia	Unknown	Abnormal proliferation of plasma cells; oversynthesis of monoclonal IgM		Presence of transitional or intermediate cells with characteristics of plasma cells and lymphocytes constituting so-called lymphoplasmacytoid cells on bone marrow examination	Plasmapheresis; chemotherapy; immunotherapy (IFN-α)

17

IMMUNODEFICIENCIES: CONGENITAL AND ACQUIRED

- **DEFECTS OF B CELL MATURATION**

- **DEFECTS OF T CELL MATURATION**

- **DEFECTS OF LYMPHOCYTE MATURATION – SEVERE COMBINED IMMUNODEFICIENCY**

- **DEFECTS IN LYMPHOCYTE ACTIVATION**

- **DEFECTS IN B CELL DIFFERENTIATION: COMMON VARIABLE IMMUNODEFICIENCY**

- **DEFECTIVE CLASS I MAJOR HISTOCOMPATIBILITY COMPLEX EXPRESSION**

- **IMMUNODEFICIENCY ASSOCIATED WITH OTHER INHERITED DISEASES**

- **DEFECT IN T CELL-DEPENDENT B CELL ACTIVATION: THE X-LINKED HYPER-IGM SYNDROME**

- **CONGENITAL DISORDERS OF INNATE IMMUNITY**

- **ACQUIRED IMMUNODEFICIENCIES**

Immunodeficiency is a failure in humoral antibody or cell-mediated limbs of the immune response. If attributable to intrinsic defects in T and/or B cells, the condition is termed a primary immunodeficiency. If the defect results from loss of antibody and/or lymphocytes, the condition is termed a secondary immunodeficiency.

Immunodeficiency disorders are conditions characterized by decreased immune function. They may be grouped into four principal categories based on recommendations from a committee of the World Health Organization. They include: antibody (B cell) deficiency, cellular (T cell) deficiency, combined T cell and B cell deficiencies, and phagocyte dysfunction. The deficiency can be congenital or acquired. It can be secondary to an embryologic abnormality, an enzymatic defect or may be attributable to an unknown cause. Types of infections produced in the physical findings are characteristic of the type of immunodeficiency disease. Screening tests identify a number of these conditions whereas others have an unknown etiology. Antimicrobial agents for the treatment of recurrent infections, immunotherapy, bone marrow transplantation, enzyme replacement, and gene therapy are all modes of treatment.

Immunodeficiencies are classified as either primary diseases with a genetic origin or those that are secondary to an underlying disorder.

DEFECTS OF B CELL MATURATION

X-linked (congenital) agammaglobulinemia results from a failure of pre-B cells to differentiate into mature B cells. The defect in **Bruton's disease** is in rearrangement of immunoglobulin heavy chain genes. It occurs almost entirely in males and is apparent after 6 months of age following disappearance of the passively transferred maternal immunoglobulins. Patients have recurrent sino-pulmonary infections caused by *Haemophilus influenzae*, *Streptococcus pyogenes*, *Staphylococcus aureus*, and *Streptococcus pneumoniae*. These patients have absent or decreased B cells and decreased serum levels of all immunoglobulin classes. The T cell system and cell-mediated immunity appear normal.

X-linked agammaglobulinemia (Bruton's X-linked agammaglobulinemia) affects males who develop recurrent sino-pulmonary or other pyogenic infections at 5–6 months of age after disappearance of maternal IgG. There is defective B cell gene (Chromosome Xq21.3-22). Whereas B cells and immunoglobulins are diminished, there is normal T cell function. Supportive therapy includes gammaglobulin injections and antibiotics. Repeated infections may lead to death in childhood. Their bone marrow contains pre-B cells with constant regions of immunoglobulin μ chains in the cytroplasm. There may be defective VH-D-JH gene rearrangement.

Btk is a protein tyrosine kinase encoded for by the defective gene in X-linked agammaglobulinemia (XLA). B cells and polymorphonuclear neutrophils express the btk protein. In XLA (Bruton's disease) patients, only the B cells manifest the defect, and the maturation of B cells stops at the pre-B cell stage. There is rearrangement of heavy chain genes but not of the light chain genes. The btk protein might have a role in linking the pre-B cell receptor to nuclear changes that result in growth and differentiation of pre-B cells.

Transient hypogammaglobulinemia of infancy is a temporary delay in the onset of antibody synthesis during the first 12 months or even 24 months of life. This leads only to a transient, physiologic immunodeficiency following catabolism of maternal antibodies passed to the infant across the placenta to the fetal circulation. Helper T cell function is impaired, yet B cell numbers are at physiologic levels.

DEFECTS OF T CELL MATURATION

Thymic hypoplasia (DiGeorge syndrome) occurs when the immune system in infants is deprived of thymic influence. T cells are absent or deficient in the blood and thymus-dependent areas of lymph nodes and spleen. Infants with this condition are highly susceptible to infection by viruses, fungi, protozoa, or intracellular bacteria due to

Table 17.1   T and B cell immunodeficiencies

	B cell deficiency	T cell deficiency
Features Serum Ig concentration DTH response to antigens	Diminished Normal	Normal or diminished Diminished
Lymphoid tissue histology	Few or no follicles and germinal centers	Normal follicles, possible decreased paracortical areas
Infection susceptibility	Pyogenic microorganisms (otitis, pneumonia, meningitis, osteomyelitis), enteric bacteria and viruses, some parasites	*Pneumonocystis carinii*, multiple viruses, atypical mycobacteria, fungi

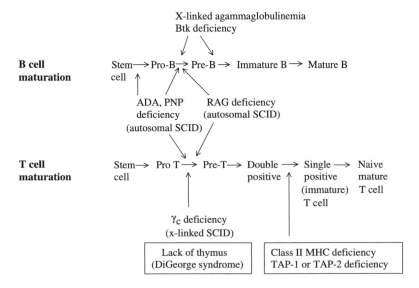

Figure 17.1   Defects in B and T cell maturation that lead to immunodeficiency

defective intracellular microbial killing by phagocytic cells with interferon. By contrast, B cells and immunoglobulins are not affected.

DiGeorge syndrome is a T cell immunodeficiency in which there is failure of T cell development, but normal maturation of stem cells and Bcells. This is attributable to failure in the development of the thymus, depriving the individual of the mechanism for T cell development. DiGeorge syndrome is a recessive genetic immunodeficiency characterized by failure of the thymic epithelium to develop. Maldevelopment of the thymus gland is associated with thymic hypoplasia. Anatomical structures derived from the third and fourth pharyngeal pouches during embryogenesis fail to develop. This leads to a defect in the function of both the thymus and parathyroid glands. DiGeorge syndrome is believed to be a consequence of intrauterine malfunction. It is not familial. Tetany and hypocalcemia, both characteristics of hypoparathyroidism,

are observed in DiGeorge syndrome in addition to the defects in T cell immunity. Peripheral lymphoid tissues exhibit a deficiency of lymphocytes in thymic-dependent areas. By contrast, the B or bursa equivalent-dependent areas, such as lymphoid follicles, show normal numbers of B cells and plasma cells. Serum immunoglobulin levels are within normal limits, and there is a normal immune response following immunization with commonly employed immunogens. A defect in delayed-type hypersensitivity is demonstrated by the failure of affected patients to develop positive skin tests to commonly employed antigens such as candidin or streptokinase, and the inability to develop an allograft response. Defective cell-mediated immunity may increase susceptibility to opportunistic infections and render the individual vulnerable to a graft-versus-host reaction in blood transfusion recipients. There is also minimal or absent *in vitro* responsiveness to T cell antigens or mitogens. The most significant advance has been the identification of

Table 17.2   Defects of lymphocyte maturation

Disease	Functional deficiencies	Presumed mechanism of defect
Severe combined immunodeficiency		
X-linked	Markedly decreased T cells, normal or increased B cells, reduced serum Ig	Cytokine receptor common γ chain gene mutations, defective T cell maturation from lack of IL-7 signals
ADA, PNP deficiency (autosomal recessive)	Progressive decrease in T and B cells (mostly T); reduced serum Ig in ADA deficiency, normal B cells and serum Ig in PNP deficiency	ADA or PNP deficiency leading to accumulation of toxic metabolites in lymphocytes
Other autosomal recessive	Decreased T and B cells, reduced serum Ig	Defective maturation of T and B cells; genetic basis unknown in most cases, may be mutations in *RAG* genes
B cell immunodeficiencies		
X-linked agammaglobulinemia	Decrease in all serum Ig isotypes, reduced B cell numbers	Block in maturation beyond pre-B cells because of mutation in B cell tyrosine kinase
Ig heavy chain deletions	IgG1, IgG2, or IgG4 absent; sometimes associated with absent IgA or IgE	Chromosomal deletion at 14q32 (Ig heavy chain locus)
T cell immunodeficiencies		
DiGeorge syndrome	Decreased T cells, normal B cells, normal or decreased serum Ig	Anomalous development of 3rd and 4th branchial pouches leading to thymic hypoplasia

Abbreviations:
ADA   adenosine deaminase
PNP   purine nucleoside phosphorylase
RAG   recombinase-activing gene

micro-deletions on human chromosome 22q in most DiGeorge syndrome patients. Considerable success in treatment has been achieved with fetal thymic transplants and by the passive administration of thymic humoral factors.

DEFECTS OF LYMPHOCYTE MATURATION – SEVERE COMBINED IMMUNODEFICIENCY

Severe combined immunodeficiency (Swiss type agammaglobulinemia): Comprises a group of conditions manifesting variable defects in both B and T cell immunity. In general, there is a lymphopenia with deficiency of T and B cell numbers and function. The thymus is hypoplastic or absent. Lymph nodes and other peripheral lymphoid tissues reveal depleted B and T cell regions. Infants with severe combined immunodeficiency show increased susceptibility to infections by viruses, fungi and bacteria, and often succumb during the first year.

Severe combined immunodeficiency syndrome (SCID): A profound immunodeficiency characterized by functional impairment of both B and T cell limbs of the immune response. It is inherited as an X-linked or autosomal recessive disease. The thymus has only sparse lymphocytes and Hassal's corpuscles or is bereft of them. Several congenital immunodeficiencies are characterized as SCID. There is T and B cell lymphopenia and decreased production of IL-2. There is an absence of delayed-type hypersensitivity, cellular immunity, and of normal antibody synthesis following immunogenic challenge. SCID is a disease of infancy with failure to thrive. Affected individuals frequently die during the first 2 years of life. Clinically, they may develop a measles-like rash, show hyperpigmentation, and develop severe recurrent (especially pulmonary) infections. These subjects have heightened susceptibility to infectious disease agents such as *Pneumocystis carinii, Candida albicans,* and others. Even attenuated microorganisms, such as those used for immunization, e.g., attenuated poliomyelitis viruses, may induce infection in SCID patients. Graft-versus-host disease is a problem in SCID patients receiving unirradiated blood transfusions. Maternal–fetal transfusions during gestation or at parturition or blood transfusions at a later date provide sufficient immunologically competent cells entering the SCID patient's circulation to induce graft-versus-host disease. SCID may be manifested in one of several forms. SCID is classified as a defect in adenosine deaminase (ADA) and purine nucleoside phosphorylase (PNP) enzymes and in a DNA-binding protein needed for HLA gene expression. Treatment is by bone marrow transplantation or by gene therapy and enzyme reconstitution in those cases caused by a missing gene, such as adenosine deaminase deficiency.

DEFECTS IN LYMPHOCYTE ACTIVATION

Selective IgA deficiency: The most frequent immunodeficiency disorder. It occurs in approximately 1 in 600 individuals in the population. It is characterized by nearly absent serum and secretory IgA. The IgA level is less than 5 mg/dl, whereas the remaining immunoglobulin class levels are normal or elevated. The disorder is either familial or it may be acquired in association with measles, other types of virus infection, or toxoplasmosis. The patients may appear normal and asymptomatic or they may have some form of an associated disease. IgA is the principal immunoglobulin in secretions and is an important part of the defense of mucosal surfaces. Thus, IgA-deficient individuals have an increased incidence of respiratory, gastrointestinal, and urogenital infections. They may manifest sino-pulmonary infections and diarrhea. Selective IgA deficiency is diagnosed by the demonstration of less than 5 mg/dl of IgA in serum. The etiology is unknown, but is believed to be arrested B cell development. The B cells are normal with surface IgA and IgM or surface IgA and IgD. Some patients also have an IgG2 and IgG4 subclass deficiency. They are especially likely to develop infections. IgA-deficient patients have an increased incidence of respiratory allergy and autoimmune disease such as systemic lupus erythematosus and rheumatoid arthritis. The principal defect is in IgA B cell differentiation. The 12-week-old fetus contains the first IgA B cells that bear IgM and IgD as well as IgA on their surface. At birth, the formation of mature IgA B cells begins. Most IgA B cells express IgA exclusively on their surface, with only 10% expressing surface IgM and IgD in the adult. Patients with selective IgA deficiency usually express the immature phenotype, only a few of which can transform into IgA-synthesizing plasma cells. Patients have an increased incidence of HLA-A1, -B8, and -Dw3. Their IgA cells form, but do not secrete IgA. There is an increased incidence of the disorder in certain atopic individuals. Some selective IgA deficiency patients form significant titers of antibody against IgA. They may develop anaphylactic reactions upon receiving IgA-containing blood transfusions. The patients have an increased incidence of celiac disease and several autoimmune diseases as indicated above. They synthesize normal levels of IgG and IgM antibodies. Autosomal recessive and autosomal dominant patterns of inheritance have been described. It has been associated with several cancers, including thymoma, reticulum cell sarcoma, and squamous cell carcinoma of the esophagus and lungs. Certain cases may be linked to drugs such as phenytoin or other anticonvulsants. Some individuals develop antibodies against IgG, IgM, and IgA. Gammaglobulin should not be administered to selective IgA-deficient patients.

Table 17.3 Defects in lymphocyte activation

Disease	Functional deficiencies	Mechanisms of defect
Selective Ig isotype deficiencies	Reduced or no production of selective isotypes or subtypes of Ig (IgA deficiency most common isotype deficiency, IgG3 deficiency most common subtype deficiency); susceptibility to bacterial infections or no clinical problems	Defect in B cell differentiation or T cell help; rare cases of homozygous deletions/mutations of Ig constant region genes
X-linked hyper-IgM syndrome	Defects in helper T cell-dependent B cell and macrophage activation	Mutation in CD40 ligand
Common variable immunodeficiency	Variable reductions in multiple Ig isotypes; normal or decreased B cells	Defect in B cell activation, usually caused by intrinsic B cell abnormality (nature unknown)
T cell receptor complex expression or signalling defects	Decreased T cells or abnormal ratios of CD4$^+$ and CD8$^+$ subsets; decreased cell-mediated immunity	Rare cases caused by mutations or deletions in genes encoding CD3 proteins, ZAP-70
X-linked lymphoproliferative syndrome	Uncontrolled B cell proliferation in the setting of Epstein–Barr virus infection leading to B cell lymphomas and hypogammaglobulinemia	Mutation in the gene encoding the SAP adapter protein, which is normally required for inhibiting signalling by the SLAM molecule
Defective class II MHC expression: the bare lymphocyte syndrome	Lack of class II expression and impaired CD4$^+$ T cell development and activation; defective cell-mediated immunity and T cell-dependent humoral immunity	Mutation in genes encoding transcription factors required for class II MHC gene expression
TAP deficiency	Lack of class I MHC molecule expression, decreased number of CD8$^+$ T cells; susceptibility to bacterial infections	Mutation in the *TAP* genes preventing peptide loading of class I MHC molecules

Abbreviations:
SAP, SLAM-associated protein
SLAM, signalling lymphocyte activation molecule
TAP, transporter associated with antigen processing
ZAP-70, zeta-associated protein of 70-kD

DEFECTS IN B CELL DIFFERENTIATION: COMMON VARIABLE IMMUNODEFICIENCY

Common variable immunodeficiency (CVID): A relatively common congenital or acquired immunodeficiency that may be either familial or sporadic. The familial form may have a variable mode of inheritance. Hypogammaglobulinemia is common to all of these patients and usually affects all classes of immunoglobulin, but in some cases only IgG is affected. The World Health Organization (WHO) classifies three forms of the disorder: (1) an intrinsic B cell defect; (2) a disorder of T cell regulation that includes deficient T helper cells or activated T suppressor cells; and (3) autoantibodies against T and B cells. The majority of patients have an intrinsic B cell defect with normal numbers of B cells in the circulation that can identify antigens and proliferate, but cannot differentiate into plasma cells. The ability of B cells to proliferate when stimulated by antigen is evidenced by hyperplasia of B cell regions of lymph nodes, spleen, and other lymphoid tissues. Yet, differentiation of B cells into plasma cells is blocked. The deficiency of antibody that results leads to recurrent bacterial infections, as well as intestinal infestation by *Giardia lamblia,* which produces a syndrome that resembles sprue. Noncaseating granulomas occur in many organs. There is an increased incidence of autoimmune diseases, such as pernicious anemia, rheumatoid arthritis, and hemolytic anemia. Lymphomas also occur in these immunologically deficient individuals.

DEFECTIVE CLASS I MAJOR HISTOCOMPATIBILITY COMPLEX EXPRESSION

Bare lymphocyte syndrome (BLS): Causes failure to express class I HLA-A, -B, or -C major histocompatibility antigens due to defective β_2-microglobulin expression on the cell surface. This immune deficiency is inherited as an autosomal recessive trait. In some individuals, the class II HLA-DR molecules are likewise not expressed. Patients may be asymptomatic or manifest respiratory tract infections, mucocutaneous candidiasis, opportunistic infections, chronic diarrhea and malabsorption, inadequate responsiveness to antigen, aplastic anemia, leukopenia, decreased T cells and normal or elevated B cells. The mechanism appears to be related to either defective gene activation or inaccessibility of promoter protein. DNA techniques are required for tissue typing.

IMMUNODEFICIENCY ASSOCIATED WITH OTHER INHERITED DISEASES

Wiskott–Aldrich syndrome: An X-linked recessive immunodeficiency disease of infants characterized by thrombocytopenia, eczema, and increased IgA and IgE levels. There is decreased cell-mediated immunity (and delayed hypersensitivity), and the antibody response to polysaccharide antigens is defective, with only minute quantities of IgM appearing in the serum. There may be an inability to recognize processed antigen. Male patients may have small platelets with absent surface glycoprotein Ib. Whereas IgA and IgE are increased, IgM is diminished, although IgG serum concentrations are usually normal. By electron microscopy, T cells appear to be bereft of the markedly fimbriated surface of normal T cells. T cells have abnormal sialophorin. Patients may have an increased incidence of malignant lymphomas. Bone marrow transplantation corrects the deficiency.

DEFECT IN T CELL-DEPENDENT B CELL ACTIVATION: THE X-LINKED HYPER-IGM SYNDROME

Hyperimmunoglobulin M syndrome: An immunodeficiency disorder in which the serum IgM level is normal or elevated. By contrast, the serum IgG and IgA levels are strikingly diminished or absent. These patients have repeated infections and may develop neoplasms in childhood This syndrome may be transmitted in an X-linked or autosomal dominant fashion. It may also be related to congenital rubella. The condition is produced by failure of the T cells to signal IgM-synthesizing B cells to switch to IgG- and IgA-producing cells. In this X-linked disease in boys who are unable to synthesize immunoglobulin isotypes other than IgM, there is a defect in the gene encoding the CD40 ligand. The T_H cells fail to express CD40L. These patients fail to develop germinal centers or displaced somatic hypermutation. They do not form memory B cells and are subject to pyogenic bacterial and protozoal infections.

Table 17.4 Congenital disorders of innate immunity

Disease	Functional deficiencies and clinical problems	Mechanisms of defect
Chronic granulomatous disease	Defective production of reactive oxygen intermediates by phagocytes; recurrent intracellular bacterial and fungal infections	Mutations in genes encoding components of the phagocyte oxidase enzyme, most often cytochrome b558
Leukocyte adhesion deficiency-1	Absent or deficient expression of β_2 integrins causing defective leukocyte adhesion-dependent functions; recurrent bacterial and fungal infections	Mutations in gene encoding the β chain (CD18) of β_2 integrins
Leukocyte adhesion deficiency-2	Absent or deficient expression of leukocyte ligands for endothelial E- and P-selectins causing failure of leukocyte migration into tissues; recurrent bacterial and fungal infections	Mutations in gene encoding a GDP-fucose transporter required for synthesis of the sialyl Lewis X component of E- and P-selectin ligands
Chédiak–Higashi syndrome	Defective lysosomal function in neutrophils, macrophages, and dendritic cells; defective granule function in natural killer cells; recurrent infections by pyogenic bacteria	Mutation in a gene of unknown function leading to increased fusion of cytoplasmic granules

CONGENITAL DISORDERS OF INNATE IMMUNITY

Phagocyte disorders: Conditions characterized by recurrent bacterial infections that can involve the skin, respiratory tract, and lymph nodes. Evaluation of phagocytosis should include tests of motility, chemotaxis, adhesion, intracellular killing (respiratory burst), enzyme testing, and examination of the peripheral blood smear. Phagocyte disorders include the following conditions: chronic granulomatous disease; myeloperoxidase deficiency; Job syndrome (hyperimmunoglobulin E syndrome); Chédiak–Higashi syndrome; leukocyte adhesion deficiency, together with less common disorders.

Chronic granulomatous disease (CGD): A disorder that is inherited as an X-linked trait in two-thirds of the cases and as an autosomal recessive trait in the remaining one-third. Clinical features are usually apparent before the end of the second year of life. There is an enzyme defect associated with NADPH oxidase. This enzyme deficiency causes neutrophils and monocytes to have decreased consumption of oxygen and diminished glucose utilization by the hexose monophosphate shunt. Although neutrophils phagocytize microorganisms, they do not form superoxide and other oxygen intermediates that usually constitute the respiratory burst. Neutrophils and monocytes also form a smaller amount of hydrogen peroxide, have decreased iodination of bacteria, and have diminished production of superoxide anions. All of this leads to decreased intracellular killing of bacteria and fungi. Thus, these individuals have an increased susceptibility to infection with microorganisms that normally are of relatively low virulence. These include *Aspergillus, Serratia marcescens,* and *Staphylococcus epidermidis.* Patients may have hepatosplenomegaly, pneumonia, osteomyelitis, abscesses, and draining lymph nodes. The quantitative nitroblue tetrazolium (NBT) test and the quantitative killing curve are both employed to confirm the diagnosis. Most microorganisms that cause difficulty in CGD individuals are catalase positive. Therapy includes interferon γ, antibiotics, and surgical drainage of abscesses.

Myeloperoxidase (MPO) deficiency: A lack of 116 kD myeloperoxidase in both neutrophils and monocytes. This enzyme is located in the primary granules of neutrophils. It possesses a heme ring, which imparts a dark-green tint to the molecule. MPO deficiency has an autosomal recessive mode of inheritance. Clinically, affected patients have a mild version of chronic granulomatous disease. *Candida albicans* infections are frequent in this condition.

Chédiak–Higashi syndrome: A childhood disorder with an autosomal recessive mode of inheritance that is identified by the presence of large lysosomal granules in leukocytes that are very stable and undergo slow degranulation. Multiple systems may be involved. Repeated bacterial infections with various microorganisms, partial albinism, central nervous system disorders, hepatosplenomegaly, and an inordinate incidence of malignancies of the lymphoreticular tissues may occur. The large cytoplasmic granular inclusions that appear in white blood cells may also be observed in blood platelets and can be seen by regular light microscopy in peripheral blood smears. There is defective neutrophil chemotaxis and an altered ability of the cells to kill ingested microorganisms. There is a delay in the killing time, even though hydrogen peroxide formation, oxygen consumption, and hexose monophosphate shunt are all within normal limits. There is also defective microtubule function, leading to defective phagolysosome formation. Cyclic AMP levels may increase. This causes decreased neutrophil degranulation and mobility. High doses of ascorbic acid have been shown to restore normal chemotaxis, bactericidal activity, and degranulation. Natural killer cell numbers and function are decreased. There is an increased incidence of lymphomas in Chédiak–Higashi patients. There is no effective therapy other than the administration of antibiotics for the infecting microorganisms. The disease carries a poor prognosis because of the infections and the neurological complications. The majority of affected individuals die during childhood, although occasional subjects may live longer.

Lazy leukocyte syndrome: A disease of unknown cause in which patients experience an increased incidence of pyogenic infections such as abscess formation, pneumonia, and gingivitis which is linked to defective neutrophil chemotaxis in combination with neutropenia. Random locomotion of neutrophils is also diminished and abnormal. This is demonstrated by the vertical migration of leukocytes in capillary tubes. There is also impaired exodus of neutrophils from the bone marrow.

Leukocyte adhesion deficiency (LAD): A recurrent bacteremia with staphylococci or *Pseudomonas* linked to defects in the leukocyte adhesion molecules known as integrins. These include the CD11/CD18 family of molecules. CD18 β chain gene mutations lead to a lack of complement receptors CR3 and CR4 to produce a congenital disease marked by recurring pyogenic infections. Deficiency of p150,95, LFA-1, and complement receptor 3 (CR3) membrane proteins leads to diminished adhesion properties and mobility of phagocytes and lymphocytes. There is a flaw in synthesis of the 95 kD β chain subunit that all three of these molecules share. The defect in mobility is manifested as altered chemotaxis, defective random migration, and faulty spreading. Particles coated with C3 are not phagocytized and therefore fail to activate a respiratory burst. The CR3 and p150,95 deficiency account for the defective phagocytic activity. LAD patients' T cells fail to respond normally to antigen or mitogen stimulation and are also unable to provide helper function for B cells producing immunoglobulin. They are ineffective in fatally injuring target cells, and they do not produce the lymphokine, γ interferon. LFA-1 deficiency accounts for the defective response of these T cells as well as all natural killer cells, which also have impaired ability to fatally injure target cells. Clinically, the principal manifestations are a consequence of defective phagocyte function rather than of defective T cell function. Patients may have recurrent severe infections, a defective inflammatory response, abscesses, gingivitis, and periodontitis. There are two forms of leukocyte adhesion deficiency. Those with the severe deficiency do not express the three α and four β chain complexes, whereas those with moderate deficiency express 2.5–6 percent of these complexes. There is an autosomal recessive mode of inheritance for leukocyte adhesion deficiency.

Job's syndrome: Refers to cold staphylococcal abscesses or infections by other agents that recur. There is associated eczema, elevated levels of IgE in the serum, and phagocytic dysfunction associated with glutathione reductase and glucose-6-phosphatase deficiencies. The syndrome has an autosomal recessive mode of inheritance.

Chemotactic disorders: Conditions attributable to abnormalities of the complex molecular and cellular inter actions involved in mobilizing an appropriate phagocytic cell response to injuries or inflammation. This can involve defects in either the humoral or cellular components of chemotaxis that usually lead to recurrent infections. The process begins with the generation of chemoattractants. Among these chemoattractants that act *in vivo* are the anaphylatoxins (C3a, C4a and C5a), leukotriene B$_4$ (LTB$_4$), IL-8, GM-CSF, and platelet activating factors. Once exposed to chemoattractant, circulating neutrophils embark upon a four-stage mechanism of emigration through the endothelial layer to a site of tissue injury where phagocytosis takes place. The four stages include (1) rolling or initial margination by the selectins (L-, P-, E-); (2) stopping on the endothelium by CD18 integrins and ICAM-1; (3) neutrophil–neutrophil adhesion by CD11b/CD18; and (4) transendothelial migration by CD11b/CD18, CD11a/CD18, ICAM-1. Chemotactic defects can be either acquired or inherited. Specific disorders are listed separately.

ACQUIRED IMMUNODEFICIENCIES

Acquired immunodeficiency describes a decrease in the immune response to immunogenic (antigenic) challenge as a consequence of numerous diseases or conditions that include acquired immunodeficiency syndrome (AIDS), chemotherapy, immunosuppressive drugs such as corticosteroids, psychological depression, burns, nonsteroidal antiinflammatory drugs, radiation, Alzheimer's disease, coeliac disease, sarcoidosis, lymphoproliferative disease, Waldenstrom's macroglobulinemia, multiple myeloma, aplastic anemia, sickle cell disease, malnutrition, aging, neoplasia, diabetes mellitus, and numerous other conditions.

Acquired immune deficiency syndrome (AIDS): A retroviral disease marked by profound immunosuppression that leads to opportunistic infections, secondary neoplasms, and neurologic manifestations. It is caused by the human immunodeficiency virus HIV-1, the causative agent for most cases worldwide with a few in western Africa attributable to HIV-2. Principal transmission routes include sexual contact, parenteral inoculation and passage of the virus from infected mothers to their newborns. Although originally recognized in homosexual or bisexual men in the United States, it is increasingly a heterosexual disease. It appears to have originated in Africa, where it is a heterosexual disease and has been reported from more than 193 countries. The CD4 molecule on T cells serves as a high-affinity receptor for HIV. HIVgp 120 must also bind to other cell surface molecules termed coreceptors for cell entry. They include CCR5 and CXCR4 receptors for β chemokines and α chemokines. Some HIV strains are macrophage-tropic whereas others are T cell-tropic. Early in the disease HIV colonizes the lymphoid organs. The striking decrease in CD4$^+$ T cells is a hallmark of AIDS that accounts for the immunodeficiency late in the course of HIV infection but qualitative defects in T cells can be discovered in HIV-infected persons who are asymptomatic. Infection of macrophages and monocytes is very important and the dendritic cells in lymphoid tissues are the principal sites of HIV infection and persistence. In addition to the lymphoid system, the nervous system is the major target of HIV infection. It is widely accepted that HIV is carried to the brain by infected monocytes. The microglia in the brain are the principal cell type infected in that tissue. The natural history of HIV infection is divided in three phases that include (1) an early acute phase, (2) a middle chronic phase, and (3) a final crisis phase. Viremia, measured as HIV-1 RNA, is the best marker of HIV disease progression and it is valuable clinically in the management of HIV-infected patients. Clinically, HIV infection can range from a mild acute illness to a severe disease. The adult AIDS patient may present with fever, weight loss, diarrhea, generalized lymphadenopathy, multiple infections, neurologic disease, and, in some cases, secondary neoplasms. Opportunistic infections account for 80 percent of deaths in AIDS patients. Prominent among these is pneumonia caused by *Pneumocystis carinii* as well as other common pathogens. AIDS patients also have a high incidence of certain tumors, especially Kaposi sarcoma, non-

Table 17.5 Clinical characteristics of human immunodeficiency virus infection

Disease phase	Clinical aspect
Acute HIV infection	Headaches
	Fever
	Sore throat with pharyngitis
	Generalized lymphadenopathy
	Rashes
Clinical latency period	Decreasing absolute CD4$^+$ lymphocyte count
AIDS	Opportunistic infections

	• Protozoal and Helminthic infections
	  ○ Cryptosporidium (enteritis)
	  ○ Pneumocystis carinii (pneumonia)
	  ○ Toxoplasma (pneumonia or CNS infection)

	• Fungal infections
	  ○ Candida (esophageal)
	  ○ Cryptococcus neoformans (CNS)
	  ○ Coccidioides immitis
	  ○ Histoplasma capsulatum

	• Bacterial infections
	  ○ Mycobacterium avium-intracellulare
	  ○ Mycobacterium tuberculosis
	  ○ Nocardia
	  ○ Salmonella

	• Viral infections
	  ○ Cytomegalovirus
	  ○ Herpes simplex
	  ○ Varicella zoster
	  ○ Progressive multifocal leukoencephalopathy

	Tumors • Kaposi's sarcoma • B cell non-Hodgkin lymphomas • Invasive cervical carcinoma
	Encephalopathy
	Wasting syndrome

Hodgkin lymphoma and cervical cancer in women. No effective vaccine has yet been developed.

Acute AIDS syndrome: Within the first to sixth week following HIV-1 infection, some subjects develop the flu-like symptoms of sore throat, anorexia, nausea, and vomiting, lymphadenopathy, maculopapular rash, wasting, and pain in the abdomen, among other symptoms. The total leukocyte count is slightly depressed with possible CD4 to CD8 ratio inversion. Detectable antibodies with specificity for HIV constituents gp120, gp160, p24, and p41 are not detectable until at least six months following infection. Approximately 33 percent of the infected subjects manifest the acute AIDS syndrome.

AIDS serology: Three to six weeks after infection with HIV-1 there are high levels of HIV p24 antigen in the plasma. One week to three months following infection there is an HIV-specific immune response resulting in the formation of antibodies against HIV envelope protein gp-120 and HIV core protein p24. HIV-specific cytotoxic T cells are also formed. The result of this adaptive immune response is a dramatic decline in viremia and a clinically asymptomatic phase lasting from 2 to 12 years. As CD4+ T cell numbers decrease the patient becomes clinically symptomatic. HIV-specific antibodies and cytotoxic T cells decline, and p24 antigen increases.

Human immunodeficiency virus (HIV): The retrovirus that induces acquired immune deficiency syndrome (AIDS) and associated disorders. It was previously designated as HTLV-III, LAV, or ARV. It infects CD4$^+$ T cells, mononuclear phagocytes carrying CD4 molecules on their surface, follicular dendritic cells, and Langerhans cells. It produces profound immunodeficiency affecting both humoral and cell-mediated immunity. There is a progressive decrease in CD4$^+$ helper/inducer T cells until they are finally depleted in many patients. There may be polyclonal

activation of B cells with elevated synthesis of immunoglobulins. The immune response to the virus is not protective and does not improve the patient's condition. The virus is comprised of an envelope glycoprotein (gp160) which is its principal antigen. It has a gp120 external segment and a gp41 transmembrane segment. CD4 molecules on CD4$^+$ lymphocytes and macrophages serve as receptors for gp120 of HIV. It has an inner core that contains RNA and is encircled by a lipid envelope. It contains structural genes designated *env, gag,* and *pol* that encode the envelope protein, core protein, and reverse transcriptase, respectively. HIV also possesses at least six additional genes, i.e., *tat,* that regulate HIV replication. It can increase production of viral protein several thousand-fold. *rev* encodes proteins that block transcription of regulatory genes. *vif (sor)* is the virus infectivity gene whose product increases viral infectivity and may promote cell to cell transmission. *nef* is a negative regulatory factor that encodes a product that blocks replication of the virus. *vpr* (viral protein R) and *vpu* (viral protein U) genes have also been described. No successful vaccine has yet been developed, although several types are under investigation.

Table 17.6   Immune dysfunction in AIDS

Lymphopenia

- Decreased CD4$^+$ helper-inducer T cells
- Inverted CD4:CD8 ratio

Diminished T cell function *in vivo*
- Loss of memory T cells
- Susceptibility to opportunistic infections
- Susceptibility to tumors
- Diminished delayed-type hypersensitivity

Altered T cell function *in vivo*
- Diminished proliferative response to mitogens, soluble antigens and alloantigens.
- Diminished specific cytotoxicity
- Diminished helper activity for pokeweed mitogen-induced B cell immunoglobulin synthesis
- Diminished IL-2 and IFN-γ synthesis

Polyclonal B cell activation
- Hypergammaglobulinemia; circulating immune complexes
- Lack of *de novo* antibody response to new antigen
- Lack of response to normal signals for B cell activation *in vivo*

Altered monocyte or macrophage functions
- Diminished chemotaxis and phagocytosis
- Diminished HLA class II antigen expression
- Decreased ability to present antigen to T cells
- Elevated secretion of IL-1, TNF-α, IL-6

TRANSPLANTATION IMMUNOLOGY   18

- **TISSUE COMPATIBILITY**

- **TRANSPLANTATION**

- **HOST-VERSUS-GRAFT DISEASE (HVGD)**

- **GRAFT-VERSUS-HOST REACTION (GVHR)**

- **IMMUNOSUPPRESSION**

Transplantation is the replacement of an organ or other tissue, such as bone marrow, with organs or tissues (grafts) derived ordinarily from a nonself source such as an allogeneic donor. Organs include kidney, liver, heart, lung, pancreas (including pancreatic islets), intestine, or skin. In addition, bone matrix and cardiac valves have been transplanted. Bone marrow transplants are given for nonmalignant conditions such as aplastic anemia, as well as to treat certain leukemias and other malignant diseases.

The transplantation of organs has been possible surgically since the early 1900s, when Alexis Carrel perfected the triangulation suture to sew blood vessels together. Yet, significant advances in immunology and immunosuppression required another 75 years.

This chapter defines key terms from the field of transplantation immunology, which is the study of immunologic reactivity of a recipient to transplanted organs or tissues from a histo-incompatible recipient. Effector mechanisms of transplantation rejection or transplantation immunity consist of cell-mediated immunity and/or humoral antibody immunity, depending upon the category of rejection. For example, hyperacute rejection of an organ such as a renal allograft is mediated by preformed antibodies and takes place soon after the vascular anastomosis is completed in transplantation. By contrast, acute allograft rejection is mediated principally by T cells and occurs during the first week after transplantation. There are instances of humoral vascular rejection mediated by antibodies as a part of the acute rejection response. Chronic rejection is mediated by a cellular response.

TISSUE COMPATIBILITY

Histocompatibility is tissue compatibility as in the transplantation of tissues or organs from one member to another of the same species, an *allograft*, or from one species to another, a *xenograft*. The genes that encode antigens that should match if a tissue or organ graft is to survive in the recipient are located in the major histocompatibility complex (MHC) region. This is located on the short arm of chromosome 6 in man and of chromosome 17 in the mouse. Class I and class II MHC antigens are important in tissue transplantation. The greater the match between donor and recipient, the more likely the transplant is to survive. For example, a six-antigen match implies sharing of two HLA-A antigens, two HLA-B antigens, and two HLA-DR antigens between donor and recipient. Even though antigenically dissimilar grafts may survive when a powerful immunosuppressive drug such as cyclosporine is used, the longevity of the graft is still improved by having as many antigens match as possible.

A **histocompatibility locus** is a specific site on a chromosome where the histocompatibility genes that encode histocompatibility antigens are located. There are major histocompatibility loci such as HLA in man and H-2 in the mouse across which incompatible grafts are rejected within 1–2 weeks. There are also several minor histocompatibility loci, with more subtle antigenic differences, across which only slow, low-level graft rejection reactions occur.

Histocompatibility antigen is one of a group of genetically encoded antigens present on tissue cells of an animal that provoke a rejection response if the tissue containing them is transplanted to a genetically dissimilar recipient. These antigens are detected by typing lymphocytes on which they are expressed. These antigens are encoded in man by genes at the HLA locus on the short arm of chromosome 6. In the mouse, they are encoded by genes at the H-2 locus on chromosome 17.

Histocompatibility testing is a determination of the MHC class I and class II tissue type of both donor and

Table 18.1   Grafts

Type of graft	Definition
Autograft	A graft of tissue taken from one part of the body and placed in a different site on the body of the same individual, such as grafts of skin from unaffected areas to burned areas in the same individual
Syngraft	A transplant from one individual to another within the same strain; also called isograft
Isograft	A tissue transplant from a donor to an isogenic recipient. Grafts exchanged between members of an inbred strain of laboratory animals, such as mice, are syngeneic rather than isogenic
Homograft	Allograft (i.e., an organ or tissue graft) from a donor to a recipient of the same species
Allograft	An organ, tissue, or cell transplant from one individual or strain to a genetically different individual or strain within the same species. Also called homograft
Xenograft	A tissue or organ graft from a member of one species (i.e., the donor) to a member of a different species (i.e., the recipient); also called a heterograft. Antibodies and cytotoxic T cells reject xenografts several days following transplantation
Heterograft	Refer to Xenograft
Orthotopic graft	An organ or tissue transplant that is placed in the location that is usually occupied by that particular organ or tissue
Heterotopic graft	A tissue or organ transplanted to an anatomic site other than the one where it is usually found under natural conditions – i.e., the anastomosis of the renal vasculature at an anatomical site that would situate the kidney in a place other than the renal fossa, where it is customarily found

recipient prior to organ or tissue transplantation. In man HLA-A, HLA-B, and HLA-DR types are determined, followed by cross-matching donor lymphocytes with recipient serum prior to transplantation. A mixed lymphocyte culture (MLC) was formerly used in bone marrow transplantation, but has now been replaced by molecular DNA typing. The

Table 18.2   Major histocompatibility loci of various species

Species	Major histocompatibility locus
Human	HLA
Mouse	H-2
Dog	DLA
Rhesus monkey	RhLA
Chicken	B
Guinea pig	GP-LA
Pig	SLA
Rat	RT1

MLC may also be requested in living related organ transplants. As in renal allotransplantation, organ recipients have their serum samples tested for percent reactive antibodies, which reveals whether or not they have been presensitized against HLA antigens of an organ for which they may be the recipient.

HLA is an abbreviation for human leukocyte antigen. The HLA histocompatibility system in humans represents a complex of MHC class I molecules distributed on essentially all nucleated cells of the body and MHC class II molecules that are distributed on B cells, macrophages, and a few other cell types. These are encoded by genes at the major histocompatibility complex. In humans the HLA locus is found on the short arm of chromosome 6. This has now been well defined, and in addition to encoding surface isoantigens, genes at the HLA locus also encode immune response (*Ir*) genes. The class I region consists of HLA-A, HLA-B, and HLA-C loci and the class II region consists of the D region which is subdivided into HLA-DP, HLA-DQ, and HLA-DR subregions. Class II molecules play an important role in the induction of an immune response, since antigen-presenting cells must complex an antigen with class II molecules to present it in the presence of interleukin-1 to CD4$^+$ T cells. Class I molecules are important in presentation of intracellular antigen to CD8$^+$ T cells as well as for effector functions of target cells. Class III molecules encoded by genes located between those that encode class I and class II molecules include C2, BF, C4a, and C4b. Class I and class II molecules play an important role in the transplantation of organs and tissues. The microlymphocytotoxicity assay is used for HLA-A, -B, -C, -DR, and -DQ typing. The primed lymphocyte test is used for DP typing. Uppercase letters designate individual HLA loci such as HLA-B and alleles are designated by numbers such as in HLA-B*0701.

HLA-A is a class I histocompatibility antigen in humans. It is expressed on nucleated cells of the body. Tissue typing to identify an individual's HLA-A antigens employs lymphocytes.

HLA-B is a class I histocompatibility antigen in humans which is expressed on nucleated cells of the body. Tissue typing to define an individual's HLA-B antigens employs lymphocytes.

HLA-C is a class I histocompatibility antigen in humans which is expressed on nucleated cells of the body. Lymphocytes are employed for tissue typing to determine HLA-C antigens. HLA-C antigens play little or no role in graft rejection.

The human MHC class II region is the **HLA-D region**, which is comprised of three subregions designated DR, DQ, and DP. Multiple genetic loci are present in each of these.

DN (previously DZ) and DO subregions are each comprised of one genetic locus. Each class II HLA molecule is comprised of one α and one β chain that constitute a heterodimer. Genes within each subregion encode a particular class II molecule's α and β chains. Class II genes that encode α chains are designated A, whereas class II genes that encode β chain are designated B. A number is used following A or B if a particular subregion contains two or more A or B genes.

HLA-DR antigenic specificities are epitopes on DR gene products. Selected specificities have been mapped to defined loci. HLA serologic typing requires the identification of a prescribed antigenic determinant on a particular HLA molecular product. One typing specificity can be present on many different molecules. Different alleles at the same locus may encode these various HLA molecules. Monoclonal antibodies are now used to recognize certain antigenic determinants shared by various molecules bearing the same HLA typing specificity. Monoclonal antibodies have been employed to recognize specific class II alleles with disease associations.

An **extended haplotype** consists of linked alleles in positive linkage disequilibrium situated between and including HLA-DR and HLA-B of the major histocompatibility complex of man.

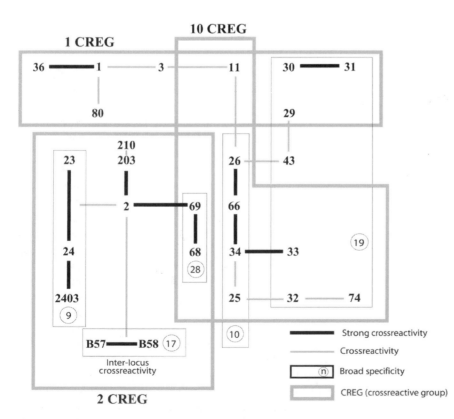

Figure 18.1 Serological cross-reactivity HLA-A locus

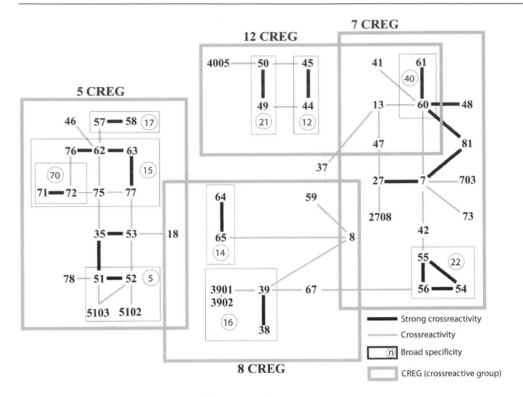

Figure 18.2   Serological cross-reactivity HLA-B locus

Linkage disequilibrium refers to the appearance of HLA genes on the same chromosome with greater frequency than would be expected by chance.

HLA disease association: certain HLA alleles occur in a higher frequency in individuals with particular diseases than in the general population. This type of data permits estimation of the 'relative risk' of developing a disease with every known HLA allele. For example, there is a strong association between ankylosing spondylitis, which is an autoimmune disorder involving the vertebral joints, and the class I MHC allele, HLA-B27.

HLA tissue typing refers to the identification of major histocompatibility complex class I and class II antigens on lymphocytes by serological and cellular techniques. *Class I* typing involves reactions between lymphocytes to be typed with HLA antisera of known specificity in the presence of complement. *Class II* typing detects HLA-DR antigens using purified B cell preparations. It is based on antibody-specific, complement-dependent disruption of the cell membrane of lymphocytes.

Antibody screening: candidates for organ transplants, especially renal allografts, are monitored with relative frequency for changes in their percent reactive antibody (PRA) levels. Obviously, those with relatively high PRA values are considered to be less favorable candidates for renal allotransplants than are those in whom the PRA values are low.

Microlymphocytotoxicity is a widely used technique for HLA tissue typing.

Molecular (DNA) typing: sequence specific priming (SSP) is a method that employs a primer with a single mismatch in the 3′-end that cannot be employed efficiently to extend a DNA strand because the enzyme Taq polymerase, during the PCR reaction, and especially in the first PCR cycles which are very critical, does not manifest 3′-5′ proofreading endonuclease activity to remove the mismatched nucleotide. If primer pairs are designed to have perfectly matched 3′-ends with only a single allele, or a single group of alleles and the PCR reaction is initiated under stringent conditions, a perfectly matched primer pair results in an amplification product, whereas a mismatch at the 3′-end primer pair will not provide any amplification product. A positive result, i.e., amplification, defines the specificity of the DNA sample. In this method, the PCR amplification step provides the basis for identifying polymorphism. The post-amplification processing of the sample consists only of a simple agarose gel electrophoresis to detect the presence or absence of amplified product. DNA amplified fragments are visualized by ethidium bromide staining and exposure to UV light. A separate technique detects amplified product by color fluorescence. The primer pairs are selected in such a manner that each allele should have a unique reactivity pattern with the panel of primer pairs employed. Appropriate controls must be maintained.

Table 18.3 HLA antigen specificities[a]

HLA-A	HLA-B	HLA-C	HLA-DR	HLA-DQ	HLA-DP
A1	B5	Cw1	DR1	DQ1	DPw1
A2	B7	Cw2	DR103	DQ2	DPw1
A203	B703	Cw3	DR2	DQ3	DPw3
A210	B8	Cw4	DR3	DQ4	DPw4
A3	B12	Cw5	DR4	DQ5 (1)	DPw5
A9	B13	Cw6	DR5	DQ6 (1)	DPw6
A10	B14	Cw7	DR6	DQ7 (3)	
A11	B15	Cw8	DR7	DQ8 (3)	
A19	B16	Cw9 (w3)	DR8	DQ9 (3)	
A23 (9)	B17	Cw10 (w3)	DR9		
A24 (9)	B18		DR10		
A2403	B21		DR11 (5)		
A25 (10)	B22		DR12 (5)		
A26 (10)	B27		DR13 (6)		
A28	B35		DR14 (6)		
A29 (19)	B37		DR1403		
A30 (19)	B38 (16)		DR1404		
A31 (19)	B39 (16)		DR15 (2)		
A32 (19)	B3901		DR16 (2)		
A33 (19)	B3902		DR17 (3)		
A34 (10)	B40		DR18 (3)		
A36	B4005				
A43	B41		DR51		
A66 (10)	B42				
A68 (28)	B44 (12)		DR52		
A69 (28)	B45 (12)				
A74 (19)	B46		DR53		
A80	B47				
	B48				
	B49 (12)				
	B50 (21)				

Table 18.3 HLA antigen specificities[a] (*continued*)

HLA-A	HLA-B	HLA-C	HLA-DR	HLA-DQ	HLA-DP
	B51 (5)				
	B5102				
	B5103				
	B52 (5)				
	B53				
	B54 (22)				
	B55 (22)				
	B56 (22)				
	B57 (17)				
	B58 (17)				
	B59				
	B60 (40)				
	B61 (40)				
	B62 (15)				
	B64 (14)				
	B65 (14)				
	B67				
	B71 (70)				
	B72 (70)				
	B73				
	B75 (15)				
	B76 (15)				
	B77 (15)				
	B7801				
	B8101				
	B8201				
	Bw4				
	Bw6				

Notes:
[a]Antigens as recognized by World Health Organization. Antigens in parentheses are the broad antigens. Antigens followed by broad antigens in parentheses are the antigen splits. Antigens of the Dw series are omitted.

Table 18.4 Comparison of HLA typing methods: DNA-based and serologic

Method	Serologic	DNA:SSP/SSOP	DNA:SBT
Number of identifiable types			
HLA-A	21	21–151	151
HLA-B	43	43–301	301
HLA-C	10	10–83	83
HLA-DR	18	18–282	282
HLA-DQ	9	9–43	43
HLA-DP	—	6–87	87
Sample material	2–3 million live lymphocytes	Minute amount of DNA	Minute amount of DNA
Reagents	Alloantisera (supply exhaustible), some monoclonals	Synthetic oligonucleotide primers/probes (supply unlimited)	Synthetic digonucleotide primers (supply unlimited)
Power to identify new alleles	Very limited: depends on availability and specificity of sera	Limited: based on knowledge of sequences and on novel reaction patterns	Unlimited: new alleles identified by their sequences
Level of resolving power for known alleles	Generic level	Generic to allele level	Allele level
Important factors	Expression of HLA on cell surface Viability of test cells	Quality and quantity of genomic DNA and amplification factors Stringency of test conditions	Quality and quantity of genomic DNA and amplification factors

Abbreviations:

SS/SSOP sequence-specific priming/sequence-specific oligonucleotide probing

SBT sequence-based typing

HLA human leukocyte antigen

PCR polymerase chain reaction

CREGs are cross-reactive groups. Public epitope-specific antibodies identify CREGs. Public refers to both similar (cross-reactive) and identical (public) epitopes shared by more than one HLA gene product.

Haplotype designates those phenotypic characteristics encoded by closely linked genes on one chromosome inherited from one parent.

Cross-match testing is an assay used in blood typing and histocompatibility testing to ascertain whether or not donor and recipient have antibodies against each other's cells that might lead to transfusion reaction or transplant rejection. Cross-matching reduces the chances of graft rejection by preformed antibodies against donor cell surface antigens which are usually MHC antigens. Donor lymphocytes are mixed with recipient serum, complement is added and the preparation observed for cell lysis.

Flow cytometry can also be used to perform the cross-matching procedure.

Splits are human leukocyte antigen (HLA) subtypes.

A private antigen is an antigen confined to one major histocompatibility complex (MHC) molecule.

A public antigen (supratypic antigen) is an epitope that several distinct or private antigens have in common.

TRANSPLANTATION

Immunologically privileged sites are certain anatomical sites within the animal body which provide an immunologically privileged environment that favors the prolonged survival of alien grafts. Immunologically privileged areas include: (1) the anterior chamber of the eye, (2) the substantia propria of the cornea, (3) the meninges of the brain,

Table 18.5   Molecular histocompatibility testing techniques

Name	Characteristic reagents	Characteristic processes	Polymorphisms detected
RFLP	Bacterial restriction endonucleases	Southern blotting	Restriction fragment length
SSP	Sequence-specific PCR primers	PCR/gel electrophoresis	Generic to allele-level
SSOP	Sequence-specific oligonucleotide probes	PCR/hybridization of probes to PCR product	Generic to allele-level
SBT	Labeled primers/labeled sequence terminators	PCR/nucleotide sequencing of PCR product	Allele-level
Heteroduplex analysis RSCA	Denatured, single-strand DNA/ Artificial universal heteroduplex generator (UHG)	Reannealing of strands of DNA/ electrophoresis of recombined DNA	DNA complexes characteristic of alleles

Abbreviations:

RFLP	restriction fragment-length polymorphism	SBT	sequence-based typing
SSP	sequence-specific priming	PCR	polymerase chain reaction
SSOP	sequence-specific oligonucleotide probing	HLA	human leukocyte antigen
		RSCA	Reference Strand Conformational Analysis

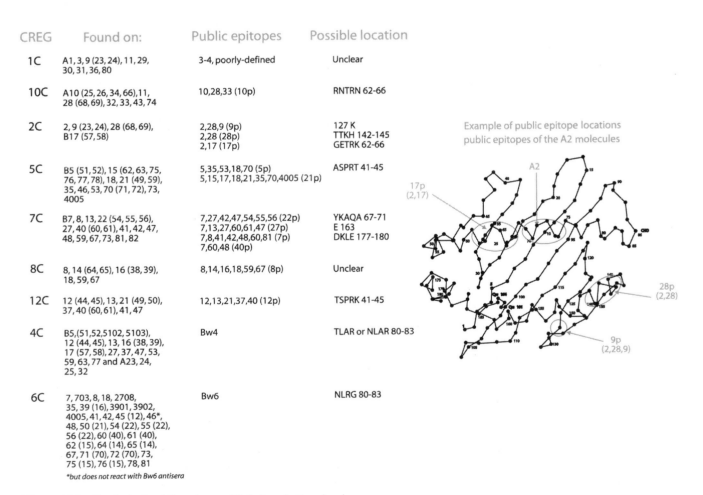

Figure 18.3   Serological public epitopes HLA-A and -B molecules

Table 18.6 HLA disease association

Disease	Antigen	Relative risk[a]
Ankylosing spondylitis	B27	87
Insulin-dependent diabetes mellitus[b]	DR3 + DR4	25
Goodpasture's syndrome	DR2	16
Pemphigus vulgaris	DR4	14
Acute anterior uveitis	B27	10
Systemic lupus erythematosus	DR3	6
Multiple sclerosis	DR2	5
Graves' disease	DR3	4
Rheumatoid arthritis	DR4	4
Myasthenia gravis	DR3	3

Notes:

[a]Relative risk (RR) is a measure of the strength of association and is defined as hK/Hk, where h is the frequency of patients with the antigen; k, is the frequency of patients without the antigen; H, is the frequency of healthy controls with the antigen; K is the frequency of controls without the antigen.

[b]This form of diabetes is associated independently with DR3 and DR4, However, the strongest association is with heterozygotes carrying both DR3 and DR4 as shown in the table.

(4) the testis, and (5) the cheek pouch of the Syrian hamster. Foreign grafts implanted in these sites show a diminished ability to induce transplantation immunity in the host.

Allogeneic bone marrow transplantation: hematopoietic cell transplants are performed in patients with hematologic malignancies, certain non-hematologic neoplasms, aplastic anemias and certain immunodeficiency states. In allogeneic bone marrow transplantation the recipient is irradiated with lethal doses either to destroy malignant cells or to create a graft bed. The problems that arise include graft-versus-host (GVH) disease and transplant rejection. GVH disease occurs when immunologically competent cells or their precursors are transplanted into immunologically crippled recipients. Acute GVH disease occurs within days to weeks after allogeneic bone marrow transplantation and primarily affects the immune system and epithelia of the skin, liver, and intestines. Rejection of allogeneic bone marrow transplants appears to be mediated by NK cells and T cells that survive in the irradiated host. NK cells react against allogeneic stem cells that are lacking self MHC Class I molecules and therefore fail to deliver the inhibitory signal to NK cells. Host T cells react against donor MHC antigens in a manner resembling their reaction against solid tissue grafts.

A **xenograft** is a tissue or organ graft from a member of one species, i.e., the donor, to a member of a different species, i.e., the recipient. It is also called a heterograft. Antibodies and cytotoxic T cells reject xenografts several days following transplantation.

Xenotransplantation is organ or tissue transplantation between members of different species.

An **isograft** is a tissue transplant from a donor to an isogenic recipient. Grafts exchanged between members of an inbred strain of laboratory animals such as mice are syngeneic rather than isogenic.

Adoptive transfer is a synonym for adoptive immunization – the passive transfer of lymphocytes from an immunized individual to a non-immune subject with immune system cells such as CD4[+] T cells. Tumor-reactive T cells have been adoptively transferred for experimental cancer therapy.

A **skin graft** uses skin from the same individual (autologous graft) or donor skin that is applied to areas of the body surface that have undergone third degree burns. A patient's keratinocytes may be cultured into confluent sheets that can be applied to the affected areas, although these may not 'take' because of the absence of type IV collagen 7 S basement membrane sites for binding and fibrils to anchor the graft.

Solid organ allotransplants include **kidney, heart, lung, liver, and pancreas. Pancreatic transplantation** is a treatment for diabetes. Either a whole pancreas or a large segment of it, obtained from cadavers, may be transplanted together with kidneys into the same diabetic patient. It is important for the patient to be clinically stable and for there to be as close a tissue (HLA antigen) match as possible. Graft survival is 50–80 percent at 1 year.

Islet cell transplantation is an experimental method aimed at treatment of type I diabetes mellitus. The technique has been successful in rats, but less so in man. It requires sufficient functioning islets from a minimum of two cadaveric donors that have been purified, cultured, and shown to produce insulin. The islet cells are administered into the portal vein. The liver serves as the host organ in the recipient who is treated with FK506 or other immunosuppressant drugs.

Autologous bone marrow transplantation (ABMT): leukemia patients in relapse may donate marrow which can be stored and readministered to them following a relapse. Leukemic cells are removed from the bone marrow which is cryopreserved until needed. Prior to reinfusion of the bone marrow, the patient receives supralethal chemoradiotherapy. This mode of therapy has improved considerably the survival rate of some leukemia patients.

Hematopoietic stem cell (HSC) transplants are used to reconstitute hematopoietic cell lineages and to treat neoplastic diseases. Twenty-five percent of allogeneic marrow transplants in 1995 were performed using hematopoietic stem cells obtained from unrelated donors. Since only 30 percent of patients requiring an allogeneic marrow transplant have a sibling that is HLA-genotypically identical, it became necessary to identify related or unrelated potential marrow donors. It became apparent that complete HLA compatibility between donor and recipient is not absolutely necessary to reconstitute patients immunologically. Transplantation of unrelated marrow is accompanied by an increased incidence of graft-versus-host disease (GVHD). Removal of mature T cells from marrow grafts decreases the severity of GVHD but often increases the incidence of graft failure and disease relapse. HLA-phenotypically identical marrow transplants among relatives are often successful. HSC transplantation provides a method to reconstitute hematopoietic cell lineages with normal cells capable of continuous self-renewal. The principal complications of HSC transplantation are graft-versus-host disease, graft rejection, graft failure, prolonged immunodeficiency, toxicity from radio-chemotherapy given pre- and post-transplantation, and GVHD prophylaxis. Methrotrexate and cyclosporine A are given to help prevent acute GVHD. Chronic GVHD may also be a serious complication involving the skin, gut, and liver and an associated sicca syndrome. Allogenic HSC transplantation often involves older individuals and unrelated donors. Thus, blood stem cell transplantation represents an effective method for the treatment of patients with hematologic and non-hematologic malignancies and various types of immunodeficiencies. The *in vitro* expansion of a small number of CD34$^+$ cells stimulated by various combinations of cytokines appears to give hematopoietic reconstitution when reinfused after a high-dose therapy. Recombinant human hematopoietic growth factors (HGF) (cytokines) may be given to counteract chemotherapy treatment-related myelotoxicity. HGF increase the number of circulating progenitor and stem cells, which is important for the support of high-dose therapy in autologous as well as allogeneic HSC transplantation.

Chimerism is the presence of two genetically different cell populations within an animal at the same time.

Corneal transplants are different from most other transplants in that the cornea is a 'privileged site'. These sites do not have a lymphatic drainage. The rejection rate in corneal transplants depends on vascularization; if vascularization occurs, the cornea becomes accessible to the immune system. HLA incompatibility increases the risk of rejection if the cornea becomes vascularized. The patient can be treated with topical steroids to cause local immunosuppression.

HOST-VERSUS-GRAFT DISEASE (HVGD)

Host-versus-graft disease is a consequences of humoral and cell-mediated immune response of a recipient host to donor graft antigens.

Graft rejection is an immunologic destruction of transplanted tissues or organs between two members or strains of a species differing at the major histocompatibility complex for that species (i.e., HLA in man and H-2 in the mouse). The rejection is based upon both cell-mediated and antibody-mediated immunity against cells of the graft by the histoincompatible recipient. First-set rejection usually occurs within 2 weeks after transplantation. The placement of a second graft with the same antigenic specificity as the first in the same host leads to rejection within one week and is termed second-set rejection. This demonstrates the presence of immunological memory learned from the first

Table 18.7 Renal allograft rejection

Type	Time after transplant	Mechanism	Histopathology
Hyperacute	Minutes	Preformed antibodies in recipient react with vascular endothelium	Attraction of polymorphonuclear neutrophils, denuding of vascular walls; platelets and fibrin plugs blocking blood flow
Acute	Days to weeks	Cellular (with humoral antibody episodes)	Cellular infiltration of interstitium. The cells are mostly mononuclear cells, plasma cells, lymphocytes, immunoblasts, some neutrophils. Endothelial cells swollen and vacuolated, vascular edema, renal tubular necrosis, sclerosed glomeruli
Chronic	>60 days	Cellular	Interstitial fibrosis, sclerosed glomeruli, mesangial proliferative glomerulonephritis, crescent formation

Table 18.8 Effect of HLA-A, -B and –DR mismatches on primary renal graft survival[a]

Number of mismatches	Estimated 10-year graft survival (%)		Half-life of graft (years)	
	Study 1	Study 2	Study 1	Study 2
0	53	65	12.3	20.3
1–2	—	47	—	10.4
3–4	42	38	9.4	8.4
5–6	32	32	7.5	7.7

Notes:
[a]Matching for split HLA-A and -B locus antigens.
Sources:
Study 1 data G Opelz, *Collaborative Transplant Study*, May 1992.
Study 2 data Zhou and Cecka, *Clinical Transplants*, 1993

experience with the histocompatibility antigens of the graft. When the donor and recipient differ only at minor histocompatibility loci, rejection of the transplanted tissue may be delayed, depending upon the relative strength of the minor loci in which they differ.

Rejection is an immune response to an organ allograft such as a kidney transplant. *Hyperacute rejection* is due to preformed antibodies and is apparent within minutes following transplantation. Antibodies reacting with endothelial cells cause complement to be fixed, which attracts polymorphonuclear neutrophils, resulting in denuding of the endothelial lining of the vascular walls. This causes platelets and fibrin plugs to block the blood flow to the transplanted organ, which becomes cyanotic and must be removed. Only a few drops of bloody urine are usually produced. Segmental thrombosis, necrosis, and fibrin thrombi form in the glomerular tufts. There is hemorrhage in the interstitium, mesangial cell swelling; IgG, IgM, and C3 may be deposited in arteriole walls. *Acute rejection* occurs within days to weeks following transplantation and is characterized by extensive cellular infiltration of the interstitium. These cells are largely mononuclear cells and include plasma cells, lymphocytes, immunoblasts, and macrophages, as well as some neutrophils. Tubules become separated, and the tubular epithelium undergoes necrosis. Endothelial cells are swollen and vacuolated. There is vascular edema, bleeding with inflammation, renal tubular necrosis, and sclerosed glomeruli. *Chronic rejection* occurs after more than 60 days following transplantation and may be characterized by structural changes such as interstitial fibrosis, sclerosed glomeruli, mesangial proliferative glomerulonephritis, crescent formation, and various other changes.

Orthoclone OKT3 is a commercial antibody against the T cell surface marker CD3. It may be used therapeutically to diminish T cell reactivity in organ allotransplant recipients experiencing a rejection episode.

Table 18.9 Effect of HLA matching on long-term renal allograft survival

Organ donor	Number of haplotypes matched[a]	% graft survival (10 year)	Transplant half-life (years)
HLA-identical sibling	2	74	24
Parent	1	54	12
Cadaver[b]	0	40	9

Notes:
[a]N = 40,765 transplants.
[b]Recipient treated with cyclosporine.
Source:
Data from PI Terasaki (ed.), *Clinical Transplants*, 1992; UCLA Tissue Typing Laboratory, 1993, p 501

GRAFT-VERSUS-HOST REACTION (GVHR)

The graft-versus-host reaction (GVHR) is the reaction of a graft containing immunocompetent cells against the genetically dissimilar tissues of an immunosuppressed recipient. Criteria requisite for a GVHR include: (1) histoincompatibility between the donor and recipient, (2) passively transferred immunologically reactive cells, and (3) a recipient host who has been either naturally immunosuppressed because of immaturity or genetic defect, or deliberately immunosuppresed by irradiation or drugs. The immunocompetent grafted cells are especially reactive against rapidly dividing cells. Target organs include the skin, gastrointestinal tract (including the gastric mucosa), and liver, as well as the lymphoid tissues. Patients often develop skin rashes and hepatosplenomegaly and may have aplasia of the bone marrow. GVHR usually develops within 7–30 days following the transplant or infusion of the lymphocytes. Prevention of the GVHR is an important procedural step in several forms of transplantation and may be accomplished by irradiating the transplant. The clinical course of GVHR may take a hyperacute, acute, or chronic form, as seen in graft rejection.

IMMUNOSUPPRESSION

Immunosuppression describes either the deliberate administration of drugs such as cyclosporine, azathioprine, corticosteroids, FK506 or rapamycin; the administration of specific antibody; the use of irradiation to depress immune reactivity in recipients of organ or bone marrow allotransplants; and the profound depression of the immune response that occurs in patients with certain diseases such as acquired immune deficiency syndrome in which the helper-inducer ($CD4^+$) T cells are destroyed by the HIV-1 virus. In addition to these examples of nonspecific immunosuppression, antigen-induced specific immunosuppression is associated with immunologic tolerance.

Clinical immunosuppression has been used to treat immunological diseases, including autoimmune reactions, as well as to condition recipients of solid organ allografts or of bone marrow transplants.

Table 18.10 Immunosuppressive agents used in organ and tissue transplantation

Agent	Mechanism of Action
Corticosteroids	Blocks cytokine gene expression
Azathioprine	Inhibits purine synthesis
Cyclosporine (CSA)	Suppresses interleukin-2 (IL-2) synthesis
	Blocks Ca^{2+}-dependent T cell activation pathway via binding to calcineurin
FK506 (pending FDAapproval)	Interferes with synthesis and binding of IL-2. It resembles cyclosporine, with which it may be used synergistically. Immunosuppressive properties 50 times greater than cyclosporine
Rapamycin	Inhibits the response of antigen-activated lymphocytes to growth factors; suppresses B and T cell proliferation, lymphokine synthesis, and T cell responsiveness to IL-2
OKT3 (Orthoclone)	Monoclonal antibody against T cell surface antigen CD3. Diminishes T cell reactivity. Used to treat post-rejection episodes in organ allotransplant recipients
Mycophenolate mofeteil	Induces reversible antiproliferative effects specifically on lymphocytes but does not induce renal, hepatic or neurologic toxicity. Inhibits a lymphocyte-specific guanosine synthesis pathway. Reversibly inhibits final steps in purine synthesis leading to depletion of guanosine and deoxyguanosine nucleotides

INDEX